T0141862

Lecture Notes in Electrical Engineering

Volume 427

About this Series

"Lecture Notes in Electrical Engineering (LNEE)" is a book series which reports the latest research and developments in Electrical Engineering, namely:

- Communication, Networks, and Information Theory
- Computer Engineering
- Signal, Image, Speech and Information Processing
- Circuits and Systems
- Bioengineering

LNEE publishes authored monographs and contributed volumes which present cutting edge research information as well as new perspectives on classical fields, while maintaining Springer's high standards of academic excellence. Also considered for publication are lecture materials, proceedings, and other related materials of exceptionally high quality and interest. The subject matter should be original and timely, reporting the latest research and developments in all areas of electrical engineering.

The audience for the books in LNEE consists of advanced level students, researchers, and industry professionals working at the forefront of their fields. Much like Springer's other Lecture Notes series, LNEE will be distributed through Springer's print and electronic publishing channels.

More information about this series at http://www.springer.com/series/7818

Kristiina Jokinen · Graham Wilcock
Editors

Dialogues with Social Robots

Enablements, Analyses, and Evaluation

Editors
Kristiina Jokinen
University of Helsinki
Helsinki
Finland

Graham Wilcock
University of Helsinki
Helsinki
Finland

ISSN 1876-1100 ISSN 1876-1119 (electronic)
Lecture Notes in Electrical Engineering
ISBN 978-981-10-9659-4 ISBN 978-981-10-2585-3 (eBook)
DOI 10.1007/978-981-10-2585-3

Softcover reprint of the hardcover 1st edition 2016

Printed on acid-free paper

This Springer imprint is published by Springer Nature
The registered company is Springer Nature Singapore Pte Ltd.
The registered company address is: 152 Beach Road, #22-06/08 Gateway East, Singapore 189721, Singapore

The original version of this book was revised: Book volume number has been changed from 999 to 427. The Erratum to this chapter is available at 10.1007/978-981-10-2585-3_40

Preface

Human–robot interaction (HRI) is a growing area in several research communities, and its research topics cut across various fields in interaction technology, speech technology, computer science, cognitive science, neuroscience, engineering, and many other related disciplines. It is expected that robots will play a bigger role in future environments and societies, and in particular their capabilities for natural language communication will increase: besides being able to perform various physical tasks, the robot will understand human social signals and support extensive, flexible, adaptable, and reliable spoken language interaction. Such communicating robot agents are generally called *social robots*, and they will provide interactive interfaces to digital data, adapting to the needs and requirements of different users.

In spoken dialogue research, interaction with robots is an increasingly active research and development area in both academia and industry. This is understandable, because intelligent communicative agents have been an application area for dialogue management since the late 1970s, when the first Belief–Desire–Intention (BDI) agents were developed with plan-based dialogue modelling as the mainstream technique. Nowadays spoken dialogue technology is ripe enough to be integrated in the robot platforms, while statistical and rule-based dialogue modelling techniques allow versatile experimentation with natural multimodal communication strategies between humans and intelligent agents. On the other hand, robot applications also provide useful testing grounds for applying dialogue models in practical contexts, as well as evaluating usability of spoken dialogue systems in real contexts. It should be noticed that "intelligent agent" here refers both to robots (situated agents able to move in the physical environment) and to animated and virtual agents (embodied conversational agents that appear on the screen).

Although the title of the book refers to robots, the research area also concerns chat systems. These are popular text-based dialogue systems whose development draws on dialogue modelling, and which can provide tools for data collection and further studies on dialogue management.

As an emergent field of research, human–robot interaction has many open questions. Important topics concern various aspects in multimodal communication, open-domain dialogue management, emotion detection, social acceptability, and natural interaction modelling. One of the major questions is how to evaluate and benchmark human–robot dialogues, and how to best model issues related to a social robot's dialogue capabilities. Studies need data from natural interaction situations, and the data is needed for two purposes: to build appropriate models for dialogue interactions, and to evaluate the models and modules, and the interaction as a whole with an intelligent agent system.

Communication via novel interfaces, especially with situated social robots, does not only involve interaction with an information-providing tool but also includes aspects which are typical of human–human communication: timing of responses, feedback-giving strategies, error management, use of multimodal behaviour, and understanding of the partner's contributions. The effect of a human-like appearance and the acceptability of the application in general are also crucial topics brought in by the development of novel autonomous interactive systems. It is important to study how dialogue strategies are carried over into new situations and new applications, i.e. to explore *affordance* of social robots. Moreover, advances in research and development of autonomous situated agents propel discussions on the ethics of social robots: their decision-making, application domains, conflicting goals between robots and their users, and the responsibilities of researchers. Such issues elevate research and development on social robotics to another level of social research: that of improving human well-being.

The book describes work on spoken dialogue systems and intelligent social agents. The chapters explore novel ways to model and use interaction strategies, and present improvements and new developments beyond previous work. The contributions are selected, adapted, and expanded from the papers presented at the Seventh International Workshop on Spoken Dialogue Systems (IWSDS 2016), and they offer different perspectives and solutions for the important questions, through papers by leading researchers covering key topics in the field. We hope the book will contribute to the design and study of interaction patterns that support successful human–robot communication, and will encourage ethical discussion and evaluation of the applications.

The IWSDS 2016 Conference

The series of IWSDS conferences aims to bring together researchers from both industry and academia working within the various disciplines that constitute the field of spoken dialogue systems. The Seventh IWSDS (IWSDS 2016) was held from January 13 to 16, 2016 at Hotel Riekonlinna in Saariselkä, Finland. This was the northernmost spoken dialogue workshop ever, and it allowed participants to discuss implementation and analytical work as well as to enjoy beautiful winter scenes and activities in Finnish Lapland. The *Kaamos*, the Arctic night when the sun stays below the horizon, had just ended, and at the time of the conference the daylight was less than three hours long; however, it contributed to wonderful sunrises and sunsets.

The special theme of IWSDS 2016 was *Dialogues with Social Robots: Enablements, Analyses, and Evaluation*. The full programme included presentations and demonstrations, a gala dinner and a trip to Inari to learn about the indigenous Sami culture at the Siida Museum for the Sami people and at Sajos, the Sami Parliament in Finland.

The conference also featured two invited speakers. Riitta Hari (Aalto University) spoke *About the Primacy of Social Interaction in Human Brain Function*, describing how current neuroscience, even "social neuroscience", examines brain functions of isolated persons who observe other persons' actions from a third-person view. Professor Hari advocated the idea that research on the brain basis of social cognition and interaction should move from this "spectator science" to studies of engaged participants. Recent advances in neuroimaging now allow study of the brain basis of *social interaction* by simultaneous recordings of brain activity of two participants engaged in natural communication. An important research question is whether social interaction emerges from lower-level perceptual, motor and cognitive functions, as is typically assumed, or whether social interaction is in fact the primary default mode governing perception and action, which would challenge many current ideas about human brain function.

David Traum (ICT, University of Southern California) spoke on *The Role of a Lifetime: Dialogue Models for Virtual Human Role-players*. Some dialogue systems are instruments to allow a user to efficiently solve a task, but other dialogue systems act as role-players in learning exercises, games, or other activities. For these cases, human-like dialogue is more important, especially if the purpose of the dialogue activity is to learn or practice interactive skills that should transfer to other humans. However, for role-play dialogue the most appropriate metric is not the familiar "Turing test" of indistinguishability from humans, but rather activation of the same sociocognitive skills that are appropriate for human interaction. Professor Traum presented examples of role-play dialogue systems from a wide variety of activities, genres, and roles, focussing on virtual humans created at the USC Institute for Creative Technologies.

Besides the normal sessions of papers and posters, the conference included three special sessions: *Dialogue State Tracking Challenge 4*, *Evaluation of Human–Robot Dialogue in Social Robotics*, and *Sociocognitive Language Processing*, which are briefly described below.

Dialogue State Tracking Challenge 4

Organisers: Luis F. D'Haro, Seokhwan Kim, Rafael E. Banchs, Matthew Henderson and Jason Williams.

Dialogue state tracking is one of the key sub-tasks of dialogue management, which defines the representation of dialogue states and updates them at each moment in a given ongoing conversation. In this challenge, participants used the TourSG corpus to develop the components. TourSG consists of dialogue sessions about touristic information for Singapore collected from Skype calls between 3 tour guides and 35 tourists. The challenge included a main task on dialogue state tracking at sub-dialogue level; four optional pilot tasks on spoken language understanding, speech act prediction, spoken language generation, and the implementation of end-to-end systems; and an optional open track for participants to explore any task of their interest over the provided dataset.

Evaluation of Human–Robot Dialogue in Social Robotics

Organisers: Laurence Devillers, Kristiina Jokinen, Joseph Mariani, Haizhu Li, Alex Waibel and Wolfgang Minker.

The aim of this special theme was to make a bridge between the social robotics and spoken dialogue research communities. In the spoken dialogue community the focus is on evaluation of verbal interaction including spontaneous speech recognition and understanding, whereas in the social robotics community the focus is on evaluation of engagement measures in interaction, with non-verbal features such as acoustics and gestures but without spontaneous speech information. Several robotics competitions already incorporate human-robot interaction (such as RoboCup@Home, which has benchmarks to evaluate robots in realistic home environments, or the AAAI Grand Challenge in which the robot attends a conference and delivers a talk) but these competitions do not focus on interactive spoken dialogues. Similarly, several spoken dialogue competitions have been proposed by Darpa (such as ATIS or Communicator), but as yet there are no established protocols and no ongoing evaluation campaigns for human-robot interaction.

Sociocognitive Language Processing

Organisers: Björn Schuller and Michael McTear.

Sociocognitive Language Processing (SCLP) is the idea of coping with everyday language, including slang and multilingual phrases and cultural aspects, and in particular with irony/sarcasm/humour and paralinguistic information such as the physical and mental state and traits of the dialogue partner (affect, age group, personality dimensions) and social aspects. Multimodal aspects such as facial expression, gestures, or bodily behaviour should also be included in the analysis. At the same time, SCLP can render future dialogue systems more "chatty" by not only feeling natural but also being emotionally and socially competent, ideally leading to a more symmetrical dialogue. For that, the computer should itself have a "need for humour" and an "increase of familiarity", enabling computers to experience or at least better understand emotions and personality.

We warmly thank the invited speakers, the steering committee, the reviewers in the scientific committee, the organisers of the special sessions, the members of the local arrangements committee, and all the participants for their contributions to the success of IWSDS 2016.

Helsinki, Finland
July 2016

Kristiina Jokinen
Conference Chair IWSDS 2016
Graham Wilcock
Local Arrangements Chair IWSDS 2016

Organization of IWSDS 2016

Steering Committee

Gary Geunbae Lee, POSTECH, Pohang, Korea
Ramón López-Cózar, University of Granada, Spain
Joseph Mariani, LIMSI and IMMI-CNRS, Orsay, France
Wolfgang Minker, Ulm University, Germany
Satoshi Nakamura, NAIST, Nara, Japan

Conference Chair

Kristiina Jokinen, University of Helsinki, Finland, and University of Tartu, Estonia

Local Arrangements Committee

Katri Hiovain, University of Helsinki, Finland
Mikko Kurimo, Aalto University, Finland
Niklas Laxström, University of Helsinki, Finland
Juho Leinonen, Aalto University, Finland
Ilona Rauhala, University of Helsinki, Finland
Markku Turunen, University of Tampere, Finland
Graham Wilcock, University of Helsinki, Finland

Scientific Committee

Samer Al Moubayed, Disney Research, USA
Masahiro Araki, Kyoto Institute of Technology, Japan
Rafael E. Banchs, Institute for Infocomm Research, Singapore
Frédéric Béchet, Aix-Marseille University, France
Felix Burkhard, Deutsche Telekom, Germany
Zoraida Callejas, University Granada, Spain
Nick Campbell, Trinity College Dublin, Ireland
Heriberto Cuayahuitl, Heriot-Watt University, UK
Ramón López-Cózar, University of Granada, Spain
Laurence Devillers, LIMSI-CNRS, France
Luis Fernando D'Haro, Institute for Infocomm Research, Singapore
Giuseppe Di Fabbrizio, Amazon Research, USA
Jens Edlund, KTH, Sweden
Julien Epps, The University of New South Wales, Australia
Benoit Favre, Aix-Marseille University, France
Milica Gašić, Cambridge University, UK
Kallirroi Georgila, ICT USC, USA
David Griol, Universidad Carlos III de Madrid, Spain
Dilek Hakkani-Tur, Microsoft Research, USA
Matthew Henderson, Google, USA
Ryuichiro Higashinaka, NTT, Japan
Anna Hjalmarsson, KTH, Sweden
Kristiina Jokinen, University of Helsinki, Finland
Pamela Jordan, University of Pittsburgh, USA
Tatsuya Kawahara, Kyoto University, Japan
Hong Kook Kim, Gwangju Institute of Science and Technology, Korea
Seokhwan Kim, Institute for Infocomm Research, Singapore
Kazunori Komatani, Osaka University, Japan
Stefan Kopp, Bielefeld University, Germany
Mikko Kurimo, Aalto University, Finland
Fabrice Lefèvre, University of Avignon, France
Rivka Levitan, Brooklyn College, USA
Haizhou Li, Institute for Infocomm Research, Singapore
Joseph Mariani, LIMSI & IMMI-CNRS, Orsay, France
Michael McTear, University of Ulster, UK
Wolfgang Minker, Ulm University, Germany
Teruhisa Misu, Honda Research Institute, USA
Satoshi Nakamura, Nara Institute of Science and Technology, Japan
Mikio Nakano, Honda Research Institute, Japan
Olivier Pietquin, University of Lille, France
Paul Piwek, The Open University, UK
Matthew Purver, Queen Mary University of London, UK

Sakriani Sakti, Nara Institute of Science and Technology, Japan
David Schlangen, Bielefeld University, Germany
Björn Schuller, University of Passau, Germany
Gabriel Skantze, KTH, Sweden
Amanda Stent, Yahoo! Labs, USA
Svetlana Stoyanchev, AT&T Labs Research, USA
David Suendermann-Oeft, ETS Research, USA
David Traum, ICT USC, USA
Markku Turunen, University of Tampere, Finland
Carl Vogel, Trinity College Dublin, Ireland
Nigel Ward, University of Texas at El Paso, USA
Graham Wilcock, University of Helsinki, Finland
Jason D. Williams, Microsoft Research, USA

Sponsors

We gratefully thank the Federation of Finnish Learned Societies and the DigiSami Project for their financial support for the conference. The workshop was endorsed by the Special Interest Group on Discourse and Dialogue (SIGDIAL), the Robotics Society of Finland, and the International Speech Communication Association (ISCA).

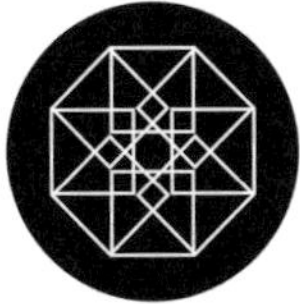

FEDERATION OF FINNISH LEARNED SOCIETIES

Delegation of the Finnish Academies of Science and Letters

SIGdial
Special Interest Group on Discourse and Dialogue

Suomen
Robotiikkayhdistys ry

Contents

Part I
The Northernmost Spoken Dialogue Workshop

DigiSami and Digital Natives: Interaction Technology for the North Sami Language

Kristiina Jokinen, Katri Hiovain, Niklas Laxström, Ilona Rauhala and Graham Wilcock

Abstract The DigiSami project operates in the general context of revitalisation of endangered languages, and focuses on the digital visibility and viability of the North Sami language in particular. The goal is to develop technological tools and resources that can be used for speech and language processing and for experimenting with interactive applications. Here we propose an interactive talking robot application as a means to reach these goals, and present preliminary analyses of a spoken North Sami conversational corpus as a starting point for supporting interaction studies. These are first steps in the development of SamiTalk, a Sami-speaking robot application which will allow North Sami speakers to interact with digital resources using their own language. The on-going work addresses the challenges of the digital revolution by showing that multilingual technology can be applied to small languages with promising results.

Keywords Language revitalisation · Digital resources · Interaction technology

K. Jokinen (✉) · K. Hiovain · N. Laxström · I. Rauhala · G. Wilcock
University of Helsinki, Helsinki, Finland
e-mail: kristiina.jokinen@helsinki.fi

K. Hiovain
e-mail: katri.hiovain@helsinki.fi

N. Laxström
e-mail: niklas.laxstrom@helsinki.fi

I. Rauhala
e-mail: ilona.rauhala@helsinki.fi

G. Wilcock
e-mail: graham.wilcock@helsinki.fi

K. Jokinen
University of Tartu, Tartu, Estonia

K. Jokinen and G. Wilcock (eds.), *Dialogues with Social Robots*,
Lecture Notes in Electrical Engineering 427,
DOI 10.1007/978-981-10-2585-3_1

1 Introduction

The digital revolution has made a dramatic impact on nearly all aspects of society. Global economics, technology and politics produce interdependence of countries world-wide, while everyday life is drastically changed by a media-rich environment where communication technology brings people speaking different languages together in new ways. New genres of discourse emerge through social media platforms and applications, collaboratively edited content, and user-generated online materials. The role of language in the novel communicative situations is prominent, since language is the vehicle that manifests these changes. On the other hand, the new technological paradigms affect language use as the language also adapts itself to these changes: by being frequently used in the new digital contexts, the language adopts new functions and forms, and is thus able to maintain its suppleness and flexibility. However, minority or lesser-used languages lack resources as well as speakers who could actively use the language in everyday and in novel communicative contexts, and they are thus most affected by these new paradigms in communication technology. The smaller language communities are the most sensitive to outside forces, and therefore also most endangered in the new digital world.

A language can survive only if it is in active use in a variety of interactive contexts, including social media networks, business and commerce, blogs, etc. In other words, a language should be viable in the digital world: it needs to have a function that is performed digitally. In order to enable digital presence and have such a function, it is important to have tools and applications which support use of the language in the new communication paradigms, and in order to develop these, it is necessary to have high quality corpora and resources. However, cutting-edge enabler technologies of language processing applications are typically available only for widely-spoken languages (so called "comfort zone" languages), while smaller communities are often left to their own resources, or they need to translate or localise information that is unavailable in their native language. The needs of less-resourced languages are to be specifically considered, in order to reduce the unbalanced situation among languages, see [1].

Nowadays much effort is directed to enable less-resourced languages to build necessary resources and to support their digital viability. Revitalisation of endangered languages is done through various projects and workshops, community efforts, influencing legislation, teaching, and cultural activities. In this article we describe our work in the DigiSami project and discuss especially its goals to apply language and speech technologies to support and revitalise the North Sami language community. We present a preliminary analysis of a spoken North Sami conversational corpus, and propose that an interactive talking robot application can make an important contribution towards the goals of the project.

The article is structured as follows. Section 2 presents the project goals to improve digital visibility and viability of the North Sami language. We also provide an overview of the Sami languages and of existing tools and resources for North Sami that can be used for speech and language processing. Section 3 describes the DigiSami

Corpus of spoken North Sami and gives preliminary analyses of the conversations in terms of laughter, speech properties, and change in the function of adjective forms. Section 4 describes first steps towards developing SamiTalk, a spoken dialogue system for Sami-speaking humanoid robots. Based on previous work on the WikiTalk system, SamiTalk will provide spoken information access from Sami Wikipedia. The DigiSami Corpus is being used to develop the speech components and to model dialogue. Section 5 presents conclusions and future work.

2 DigiSami and North Sami Language Resources

2.1 The DigiSami Project

The DigiSami project, within its larger framework of Academy of Finland and Hungarian Science Academy cooperation, sets out to investigate how modern language technology and corpus-based linguistic research can contribute to facing digitalisation challenges with lesser-resourced Fenno-Ugric languages. As described in [2], the DigiSami project focuses on North Sami, and aims to

- collect data from dialogue-related genres and annotate it on a range of levels from grammatical to discourse phenomena,
- experiment with language technology applications that can strengthen the user's interest in using the language in various interactive contexts, focussing in particular on the SamiTalk application,
- alleviate barriers in accessing information from user-generated content, supporting community-based generation of translated material on the web, based partially on existing language resources and technology.

This article concentrates on the first two goals of the project and describes how the SamiTalk robot application can be used to support revitalisation of the North Sami language. The robot application is chosen because it is an interface to collaboratively edited Wikipedia information (based on the WikiTalk application [3, 4]), and also, as a novel application, it is expected to increase the visibility of the languages by kindling interest in learning how the North Sami language can be used in the novel interactive technology (see discussion in Sect. 4). In particular, it is expected that young people may become more interested in using and studying the language, which is an important and effective strategy for language revitalisation in general.

2.2 The Sami Languages

We have chosen North Sami as our target language, since it is the largest of the Sami languages and generally used as a lingua franca among the Sami people. It is one of the Sami languages spoken in the area that extends from Middle Scandinavia to the Kola Peninsula, spanning four countries: Norway, Sweden, Finland, and Russia, as shown on the map in Fig. 1.

The Sami languages are divided into eastern and western groups, shown in Table 1. Both language groups are represented in Finland where North Sami, Inari Sami and Skolt Sami are spoken. All Sami languages are endangered, some even critically, although continuous efforts are made for their revitalisation and documentation.

The Sami languages are close relatives and the distinction between a dialect and a language is sometimes vague. The differences mainly concern morphophonetic variation whereas syntactic changes are fairly small [7]. The closer the languages are geographically, the more easily speakers understand each other, but also the majority language affects understandability since it is reflected in differences in both vocabulary and pronunciation (see more in Sect. 3.2).

The Sami languages use the Latin alphabet with various diacritics to represent phonological differences. An exception is Kildin Sami which is spoken only in Russia and written using the Cyrillic alphabet. Although written Sami texts have existed for more than 200 years, their orthographies were irregular and were stabilised only in late 1970s when the Sami Council (then: the Nordic Sami Council) decided to adopt a new uniform orthography for North Sami spoken in Finland, Norway and Sweden

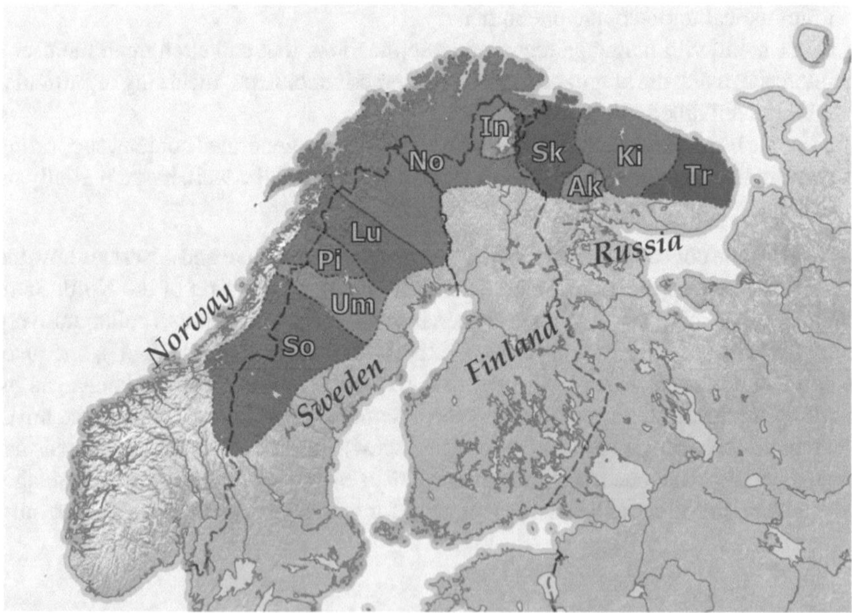

Fig. 1 The Sami language areas at the beginning of the 20th century. *Source* [5]. *So* South Sami, *Um* Ume Sami, *Pi* Pite Sami, *Lu* Lule Sami, *No* North Sami, *In* Inari Sami, *Sk* Skolt Sami, *Ak* Akkala Sami (now extinct), *Ki* Kildin Sami, *Tr* Ter Sami

[8]. This coincides with national educational reforms in the 1970s and 1980s which gave a legal basis for the use of Sami language in education.

The normative work continues today in collaboration with representatives from all the relevant countries. However, even today older Sami speakers who did not learn to read and write in Sami at school may feel uncertain of the written language conventions. The different majority languages (Finnish, Norwegian, and Swedish) also have influence on the written language due to their different phonetic rules and orthographic conventions: it is not uncommon to find "mistakes" in crowd-sourced North Sami texts such as Wikipedia articles, due to irregularities in the way the words are spelled. This makes automatic processing of digital texts a challenge.

Nowadays the Sami language (the term used to refer collectively to all the Sami language variations and to emphasise the Sami as a nation) is officially recognised in Norway, Finland, and Sweden. The respective Language Acts (1992 in Finland and Norway, 2000 in Sweden) guarantee the official status of the Sami language and the right of the Sami to use the Sami languages in all official encounters. Moreover, as a result of the national education reform in Finland, the Basic Education Act 1998 entitles Sami children who live in the Sami Homeland area and speak the Sami language to receive the main part of their basic education in the Sami language.

2.3 *The North Sami Language*

North Sami divides into three main dialects: Tornio, Finnmark and Sea Sami dialects, and the Finnmark dialect is further divided into western and eastern groups. The Finnmark dialect has the most speakers of North Sami, so the DigiSami corpus collection (see more in Sect. 3) was conducted especially in the locations which are representative of the main Finnmark dialect variations (see the map in Fig. 2).

Table 1 The Sami languages, their estimated number of speakers in 2012 [6] and their region (No: Norway, Sw: Sweden, Fi: Finland, Ru: Russia)

Western Sami languages	Region	Eastern Sami languages	Region
South Sami (500 speakers)	No, Sw	Inari Sami (350 speakers)	Fi
Ume Sami (<10 speakers)	Sw	Skolt Sami (300 speakers)	Fi, Ru
Pite Sami (40 speakers)	Sw	Kildin Sami (700 speakers)	Ru
Lule Sami (1000 speakers)	No, Sw	Ter Sami (<10 speakers)	Ru
North Sami (30,000 speakers)	No, Sw, Fi		

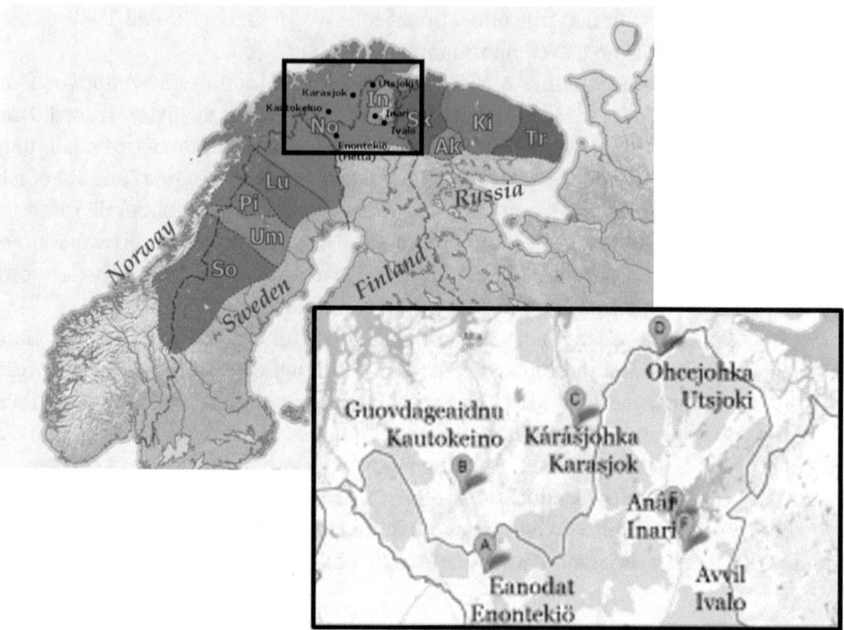

Fig. 2 The DigiSami data collection locations

Like all the Sami languages, North Sami belongs to the Fenno-Ugric branch of the Uralic language family, which also includes Finnish, Estonian, and Hungarian. Due to common origin and close contacts, North Sami shares many similarities especially with Finnish and Estonian, although the languages are rather distantly related. North Sami and Estonian resemble each other in morphophonological aspects, since both have a rich phoneme repertoire and complicated morpho-phonetic variations. While Finnish has preserved much of the original agglutinative marking in inflection, both Estonian and North Sami have changed over time towards fusional inflected morphology: instead of marking morphemes by suffixes only, the languages exploit phonetic modifications to the root, in particular consonant gradation which deals with consonant stem alternations between strong and weak grades.

The main linguistic similarities between Finnish and North Sami include the following [5]:

- Inflection with cases, persons, tense and mood,
- Similar morpheme and word order,
- Negation by using a negation verb as in Fenno-Ugric languages in general (except Estonian)
- A number of common words, both from a common origin and through loans from Finnish to North Sami.

There are also many differences between North Sami and Finnish. For instance, compared with Finnish, North Sami has:

- Separate dual verb forms and pronouns in addition to singular and plural forms,
- Six inflectional cases for nouns, as opposed to 15 in Finnish,
- No vowel harmony, whereas Finnish features back/front vowel harmony (typical for Fenno-Ugric languages in general excluding Estonian)
- Complex repertoire of phonemes and morphophonological variation:
  - 31 consonants including voiced and voiceless nasals as well as pre-aspirated consonants and affricates. The latter are not found in the Finnish system of only 17 consonants.
  - Consonant gradation concerns all consonant phonemes, as opposed to Finnish where alternations always involve a stop at least in either grade (weak or strong). According to [5], the number of stem gradation patterns in North Sami is counted to be at least 175.
  - The 5 vowels (/u, o, a, e, i/) do not contain front vowels /y/, /ä/, and /ö/ which are found in Finnish (however, these occur in recent loan words in North Sami, like *fysihkka*), and some dialects do not distinguish between /a/ and /ä/. The length of vowels is phonemic (meaning distinguishing) as in Finnish and Estonian.
  - Numerous diphthongs have phonetic length unlike Finnish and Estonian. North Sami also features triphthongs.

2.4 Existing North Sami Language Resources

Technical development has played a significant role in revitalising and modernising the Sami languages. Nowadays several language tools are available for speakers and language learners to check spelling, look for suitable words, and learn the language. Special keyboards adapted to the Sami orthographies are available for computers and cell-phones [9], so as to facilitate Sami language text input, which can be challenging due to various diacritics and their different encodings.

While the Sami languages have been fairly well documented and studied, North Sami enjoys a relatively favourable situation as it also has various tools for automatic language analysis. They have been developed at Giellatekno at University of Tromsø and are available from the website [10]. For instance, the spell-checker *Divvun* supports the writing of North Sami, and has an important role in taking the written language norms into use. Morphological and syntactic parsers for text corpora are also available, as well as a translation tool from North Sami to Norwegian Bokmål and experimental translation systems to Finnish and to other Sami languages.

A new approach to Sami morphology is taken in [11] which studies how to use an Active Learning approach to morphological segmentation. Since high-quality morphological analyzers require a significant amount of expert labour, data-driven approaches may provide sufficient quality for many applications. Grönroos et al. [11] describes how the semi-supervised Morfessor FlatCat method is used to create a statistical model for morphological segmentation for a large unannotated corpus, with a small amount of annotated word forms which are selected using an active learning approach.

Revitalisation of a language also means gaining new speakers via language learning. For this there are digital dictionaries and also a language learning website *Oahpa!* [12]. It offers different ways to learn and test language skills, with exercises for testing and practising morphology, vocabulary and syntax. The digital dictionaries [13] include more updated vocabulary than printed ones, and it is possible to search for a word, both from a majority language to Sami and vice versa.

Various spoken and text corpora are also available in North Sami. These are at the Sami culture archive [14] in the Giellagas Institute at University of Oulu. The collection of audio and video material as well as photographs and written documents supports research infrastructure and documentation of the Sami culture. The spoken corpora consist of interviews and official texts, such as the corpus of Yle Sápmi radio programs [15], but there are no natural conversations in the corpora. However, the DigiSami project has collected a small corpus of spoken conversational data in North Sami (see Sect. 3). The corpus has been transcribed and translated into Finnish, and is unique in that it contains spoken, non-scripted conversations between groups of people.

2.5 Existing North Sami Speech Technology

Current speech technology applications for the Sami languages still require development and are not yet commonly in use. However, advances have already been made in the development of a speech synthesizer and a speech recognizer for North Sami. A big challenge in the development is the limited speech corpora available. Some speech/voice data is available for North Sami under licence of the Sami Parliament of Norway, but speech corpora for other Sami languages remain in limited use at the moment.

A North Sami speech synthesizer *North Sami Infovox 4* (Windows) or *North Sami iVox* (OS X) was developed by Divvun and the Norwegian Sami Parliament, in cooperation with the voice and speech technology company Acapela [16]. It was released in May 2015. The system has both a female and a male voice, and they can be adapted to the user's needs. North Sami speech synthesis has also been studied in the Simple4All project [17] which focuses on creating methods that enable speech synthesis systems to be built by little or no supervised learning from the data.

A North Sami speech recognizer has been developed in the DigiSami project in collaboration with our partners at Aalto University. Leinonen [18] describes building an automatic speech recognizer for North Sami and discusses its further development. This is a notable work since to the best of our knowledge, this is the first and only speech recognizer for any of the Sami languages today.

3 The DigiSami Corpus

The DigiSami Corpus of spoken North Sami consists of both read speech (257 min of annotated data) and conversations of two or three persons (195 min of annotated data). The speakers are all native speakers of North Sami, and their ages vary between 16 and 65 years. The corpus was collected in five locations that represent the main communities of the Finnmark variation of North Sami: Enontekiö, Utsjoki, Inari and Ivalo in Finland, as well as Kautokeino and Karasjok in Norway. Figure 2 shows the locations for the data collection. More details about the collection and analysis of the corpus are given in [2, 19].

The DigiSami corpus is unique among the Sami language corpora because it contains natural spoken conversations between groups of participants. The interlocutors discuss freely about their own interests but also about the Wikipedia articles they were to write (such as Sami language, Sami costume, music, reindeer herding, and snowmobiles). Conversations between young students concern their everyday life, and the topics include the next vacation, driving school, and cars. Two adult men, who have known each other for a long time, converse about translation between Sami and other languages, and venture on with the technological tools that have been made to help writing North Sami more correctly.

The styles of the conversations differ depending on the age of the speakers and their hierarchical status. Participants who are familiar with each other have casual conversations and they often refer to things they had been talking about earlier. The conversations between a pupil and a teacher, however, are more formal and resemble interviews rather than conversations; the topics stick to the forthcoming task, i.e. things that one could write a Wikipedia article about.

The DigiSami corpus is multimodal, i.e. conversations are both recorded and videotaped. Thus it is possible to study non-verbal as well as verbal communication. Furthermore, the spoken non-scripted conversations allow us to study the language as it is used, not as it should be in formal grammars and dictionaries, and thus it forms a basis for studies on spoken colloquial North Sami. Moreover, it helps to create a model of the North Sami language for the use of speech technology applications.

Below we briefly discuss three aspects of the on-going DigiSami corpus analysis: the participants' engagement and the role of laughter in interactions, the influence of the majority language on spoken North Sami, and the apparent change in the use of adjective forms in the present-day spoken North Sami.

3.1 Preliminary Analysis: Engagement and Interaction

The annotation of the corpus was done with Praat and consists of 5 time-aligned tiers: a phonological/phonetic transcription, the words, the sentence in orthographic form, a Finnish translation, and remarks on things like dialectal variation. Challenges

Table 2 Laughter and overlapping speech

Conversation code	Informant code	Laughter	Laughter/min	Overlapping speech
01_S	S-1	9	0,54	15
01_S	S-2	25	1,52	3
01_S	S-3	9	0,54	4
02_V	V-1	75	5,53	1
02_V	V-2	34	2,51	–
02_V	V-3	63	4,64	1
03_V	V-2	7	0,87	1
03_V	V-3	6	0,75	–
04_S	S-1	0	0	–
04_S	S-2	6	1,72	–
04_S	S-3	1	0,29	–
05_TP	TP-2	21	2,45	–
05_TP	TP-3	34	3,97	–
06_PS	PS-1	5	0,43	–
06_PS	PS-3	4	0,34	–
07_SX	SX-1	3	0,44	–
07_SX	SX-X	1	0,15	–
08_VV	VV-Vih	15	1,2	–
08_VV	VV-Vio	19	1,53	–

in the annotation included unclear speech, unknown proper names, as well as some insider jokes, besides some Norwegian words in the corpus collected in Norway.

The participants' engagement in the conversation and mutual bonding can be measured using multimodal and non-verbal cues, such as the amount of laughing or chuckling, and overlapping speech. The analysis of different types of laughter does not only show humour and joking, but also indicates connections to how well participants know each other, if they are nervous or embarrassed, and what kind of relationship they have with each other. Overlapping speech, on the other hand, indicates how involved the speakers are in contributing to the shared goal of the conversation. For the purposes of measuring this kind of engagement and studying the roles of laughing and overlapping speech in conversations, we annotated the data with these features on the remarks tier in Praat. The basic statistics are shown in Table 2.

As can be seen, the number of laughter occurrences varies much depending on the dialogue, but interestingly, there is not much overlapping speech in the conversations, except in *01_S*. For example, in the conversation *02_V*, in which the participants laugh and chuckle the most, there are only two occurrences of overlapping speech. This leads us to conclude that laughing and overlapping speech indeed have different functions in the conversations, and although they can both be regarded as signs of the interlocutors' engagement, the preconditions for their occurrence are different. In *01_S*, both laughter and overlapping speech signal similar positive engagement in the conversation, whereas in *02_V*, the participants seem nervous and their conversation topics change very fast. Despite the many laughs, there is practically no overlapping speech, which seems to indicate that the constant chuckling might be related to a relief of embarrassment and nervousness in the situation. In fact, a closer analysis shows that although most of the instances in *02_V* are free or mirthful laughs, one fourth of them occurs when the speaker is embarrassed. This is in contrast with the conversation *01_S*, where laughter is more evenly distributed among free and mirthful types, and embarrassed laughs are a small minority.

In our other study [20], we noticed that laughter is also related to the participants' social roles: lack of laughter may also indicate rather formal conversations and an asymmetrical relationship between the speakers, such as in teacher-pupil conversation. In other words, laughter creates mutual bonds among the interlocutors and reinforces shared experience even if the situation is embarrassing, whereas lack of laughter is often a signal of distancing relation.

In fluent conversations, where people know each other and show no impression of nervousness, both laughter and high amount of overlapping speech seem to indicate enthusiasm and close relationship, as well as elaborated coordination of the conversation by smooth turn-taking: laughter occurs when the participants share jokes or funny stories, and turn-taking is timed so as not to have long silences.

These observations will be substantiated with deeper analysis and statistical modelling. Precise conditions for turn-taking, laughing and generally positive attitude will be explored further so as to enable appropriate interaction models be implemented in the SamiTalk application. A useful case is for instance to be able to recognise the user's embarrassment or uncertainty, and alleviate such situations accordingly.

3.2 Preliminary Analysis: Influence of Majority Language

North Sami is spoken in three countries: Finland, Norway and Sweden, and the language is thus in contact with a Baltic-Finnic language (Finnish) and with North Germanic (Scandinavian) languages (Norwegian and Swedish). As discussed in [21], the fact that North Sami is spoken in three countries makes describing the variation of the language multidimensional. Different parts of the speech community are influenced by different majority languages, and practically all speakers are bilingual or even trilingual. Language contacts with majority languages have made a strong impact on dialect diversification, resulting in changes in various linguistic features. Such features include phonetics and intonation, syntax and lexical expansion with words which denote new concepts but do not follow the traditional dialectological analysis. Consequently, [21] suggest a new way to classify the present-day dialects on the basis of both traditional regional features and the contact influence of the state language.

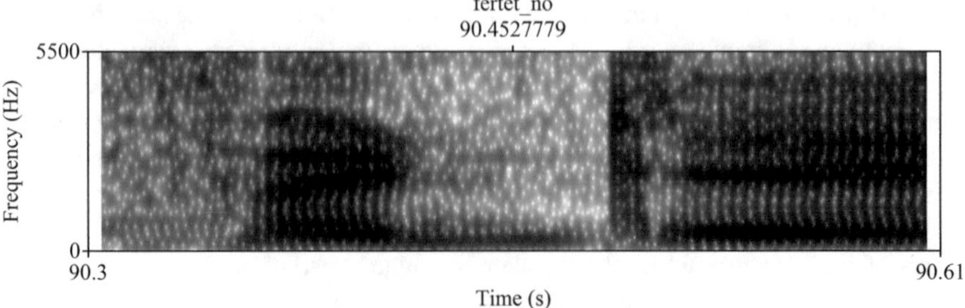

Fig. 3 Sami spoken in Norway: [fəɹdːeː]

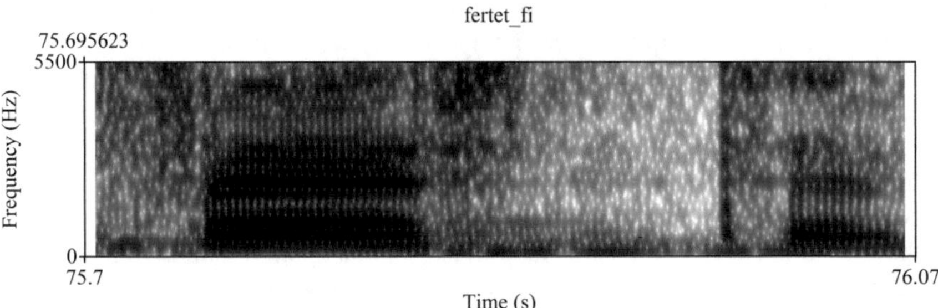

Fig. 4 Sami spoken in Finland: [feːrhtːe]

The DigiSami corpus indicates remarkable differences between Sami spoken in Norway and Finland, although dialectally the regions of Ivalo, Utsjoki and Karasjok, where the DigiSami conversation corpus was collected, represents the same eastern Finnmark dialect of North Sami. The differences can be heard in different pronunciation of various phonemes, most remarkably of /r/. In our data from Karasjok in Norway, /r/ has several allophones, occurring as [ø], [r], [ɹ] and [ɾ], while in our data from Finland it occurs only as [r] with no allophonic variation. This allophonic variation seems to follow the patterns of the majority languages. The IPA transcriptions and spectrogram pictures from the speech of two young girls in Figs. 3 and 4 demonstrate this difference (the example word is *fertet* 'have to').

The DigiSami conversations also exemplify differences in the influence of the majority languages on vocabulary. The use of Norwegian words in Sami spoken in Karasjok is much more common than the use of Finnish words in data collected in Ivalo and Utsjoki. In absolute counts the frequencies in the corpus are 60 versus 8 words, respectively.

An important reason to study the influence of the majority languages is that it affects the mutual comprehension more than the old regional dialect differences (see [5]). Especially different lexicon and idiomatic expressions have posed problems in cross-border communication between Finnish and Norwegian Sami speakers. Interestingly, these differences are often considered as interference rather than dialectal variation, although the features are already established in the language [21].

Using the acoustic language recognition technique of i-vector modelling, we studied if North Sami speech samples could be distinguished from each other automatically [22]. The average recognition error EER is 17.31 %, and the results support the view that the majority language has an impact on the dialect among bilingual speakers: mis-identification rates are significantly lower for Utsjoki and Ivalo than for Kautokeino and Karasjok. Moreover, since training of the recogniser was done with respect to a Finnish corpus, the better performance of the classifier for Ivalo and Utsjoki samples indicates their closer relation to Finnish. This research is important for the SamiTalk application, and will be further used in enhancing speech recognition and other speech tools for North Sami.

3.3 Preliminary Analysis: Adjectives in Spoken Language

In North Sami the adjective attributive form tends to be different from the nominal predicate form, e.g. *fiinnis* (pred.) and *fiinna* (attr.) 'fine, nice', as in *Beaivi lea fiinnis* 'the day is nice') and *Fiinna beaivi* 'a nice day'. The attributive marking system is complex [5], and in practice, a speaker has to know both forms by heart. However, the DigiSami corpus shows that use of adjectives in spoken language often differs from what is presented in grammars and dictionaries, and the speakers use attribute forms in nominal predicate position and vice versa. The preliminary findings seems to indicate that the adjective system is changing, and that the main factor affecting the preservation of the form is the frequency of an adjective in a certain position.

For example, the adjective *ođas* 'new' has the attributive form *ođđa*, and its base form *ođas* is also used as the nominal predicate form. The adjective has 8 occurrences in the corpus, of which 6 occur in attributive position. The adjective occurs only twice as nominal predicate, and in both cases, the speaker sounds seemingly hesitant of the correct form. In one of these cases, the speaker hesitates and is looking for the right form, while in the other case, the speaker actually produces the attributive form *ođđa*, which is wrong as the nominal predicate form. Another example is the adjective *váttis* 'difficult' which has the opposite distribution: the adjective occurs 5 times as a nominal predicate and once as an attribute, which occurrence has a wrong form.

The change in adjective attribute system has been recognized in other Sami languages [23], such as South Sami [24]. Our examples of non-normative use of adjectives forms may indicate a change in the adjective attribute system also in North Sami. The fact that the preserved form seems to depend on the frequency of the adjective form allows us to hypothesize that the change may be triggered by the frequency of the adjective in a certain position: it is plausible that adjectives such as 'new' *ođđa* are more frequently used as attributes and thus tend to lose their nominal predicate form, whereas adjectives such 'difficult' *váttis* are used mostly as nominal predicates and thus tend to lose their attribute form. On the other hand, the cause for the change may also be due to an influence of the majority language: neither Finnish nor Norwegian distinguishes adjective forms based on their syntactic function. Interestingly, it seems that the age of a speaker does not affect the use of the forms.

However, more data is essential in order to produce a reliable view of the adjective system and its change. Also further analysis of the changing mechanism is crucial for deeper understanding of language evolution in general. However, observations of the changing adjective system are crucial for producing good language models in order to develop speech applications such as SamiTalk.

4 Towards SamiTalk: A Sami-Speaking Robot Application

Robot applications are leaving the research laboratories and reaching the general population, both in homes and outside home. For example, WikiTalk [3, 4] is a multilingual spoken dialogue system that runs on a Nao robot. The user and the robot have a dialogue in which the robot talks fluently about an unlimited range of topics using information from Wikipedia.

The DigiSami project is working towards the creation of SamiTalk, an interactive robot application in the North Sami language, as part of its support for language revitalisation using speech and language technologies. The SamiTalk application will be based on existing WikiTalk technology and will provide access to Sami Wikipedia information via a dialogue in North Sami with a humanoid robot. This work is described in more detail by [25].

Localisation [26] of robot applications to an endangered language benefits the language in multiple ways. As mentioned in Sect. 2.1 this can provide motivation to use the language in novel communication frameworks. Localised robot applications can have a favourable effect on the prestige of the language, by showing that efforts have been made to support the language in the new technology. For instance, at home, applications such as WikiTalk can prevent bottom-to-top language death by encouraging use of the language in the family.

Of course, with a multilingual application there is a risk that the users will just switch it to some other language, or even more likely, will not switch it to the local language from the default language, because they do not know how, or do not even know that their language is available. For this reason, language selection is a crucial function in localisation [26] for robot applications, and is an important topic to study further.

New technology also offers opportunities for language revitalisation including existing or proposed Wikipedias for endangered languages. Wikipedias can bring together speakers of endangered languages even if they are not physically close to each other, can give motivation by Wikipedia's mission to provide free knowledge to everyone [27] and can foster collaboration between speakers and scholars studying the endangered language. There is a risk, however, that a Wikipedia started by scholars may fail to attract native speakers, for example if they are unable to access Wikipedia via desktop computers or mobile devices.

Up to now, the available Wikipedias in the languages supported by WikiTalk have been very large. English Wikipedia has almost 5 million articles, Japanese Wikipedia has almost 1 million articles and Finnish Wikipedia has over 350,000 articles. In these languages, WikiTalk can talk about almost any topic the user is interested in. By contrast Sami Wikipedia is much smaller, with about 7000 articles. This means that there are many topics that SamiTalk will not be able to talk about using existing methods. To address this problem, the DigiSami project is supporting initiatives to encourage the North Sami community to create new articles in Sami Wikipedia [19], and is investigating methods for on-line translation of Wikipedia articles [28] into under-resourced languages.

5 Conclusions and Future Work

Often endangered languages face a gradual language death by assimilation. The ability to use one's own language with new technology, in the modern world, is almost a necessity to prevent gradual language death. In these cases the motivation to continue using the endangered language is a most important factor.

In this article, we have discussed revitalisation of North Sami, and presented our work within the DigiSami project, the goal of which is to support North Sami digital natives and their communication in the North Sami language in new digital contexts. We are taking steps to address these issues by developing SamiTalk, an interactive robot application for an endangered language, North Sami. This application is based on the existing WikiTalk system. We have also collected a conversational spoken language corpus for North Sami. The DigiSami corpus has been transcribed and annotated, and preliminary analysis already shows interesting properties concerning laughing and joking, as well as turn-taking and overlapping speech. The corpus also provides examples of non-normative use of adjectives, which leads to a preliminary hypothesis that the alleged change in the North Sami adjective system is taking place, and that the frequency of an adjective in certain syntactic function affects the preservation of the form.

To develop a robust SamiTalk system, further work is required especially on the speech interface: comprehensive speech components for North Sami are needed for enabling natural conversations, and more spoken data is needed to cover language variation. We are developing speech technology for the language with our collaborators. Future work will deal with integration of the speech components on the robot, as well as deeper analysis of the conversational corpus.

Acknowledgements The work is funded by the Academy of Finland through the project Fenno-Ugric Digital Citizens (grant n°270082). The first author would also like to thank support of the Estonian Science Foundation project (grant n°IUT 20-56).

References

1. Soria, C., Mariani, J., Zoli, C.: Dwarfs sitting on the giants' shoulders—how LTs for regional and minority languages can benefit from piggybacking major languages. In: Norris, M., Anonby, E., Junker, M.O., Ostler, N., Patrick, D. (eds.) Proceedings of XVII FEL Conference, pp. 73–79. Ottawa (2013)
2. Jokinen, K.: Open-domain interaction and online content in the Sami language. In: Calzolari, N., Choukri, K., Declerck, T., Loftsson, H., Maegaard, B., Mariani, J., Moreno, A., Odijk, J., Piperidis, S. (eds.) Proceedings of Ninth International Conference on Language Resources and Evaluation (LREC 2014). European Language Resources Association (ELRA), Reykjavik (2014)
3. Wilcock, G.: WikiTalk: a spoken Wikipedia-based open-domain knowledge access system. In: Proceedings of the COLING 2012 Workshop on Question Answering for Complex Domains, pp. 57–69. Mumbai (2012)
4. Jokinen, K., Wilcock, G.: Multimodal open-domain conversations with the Nao robot. In: Mariani, J., Rosset, S., Garnier-Rizet, M., Devillers, L. (eds.) Natural Interaction with Robots, Knowbots and Smartphones: Putting Spoken Dialogue Systems into Practice, pp. 213–224. Springer (2014)
5. Sammallahti, P.: The Saami Languages: An Introduction. Davvi Girji, Kárášjohka (1998)
6. Grünthal, R., Siegl, F.: Uralilaisten kielten pensasmalli ja arvioidut puhujamäärät (The "bush model" of the Uralic languages and the estimated numbers of speakers). University of Helsinki, Department of Finno-Ugric Studies (2012)
7. Palismaa, M., Eira, I.M.G.: Gielas gillii, mielas millii 9 - Davvisámegiela suopmanat (From language to language, from mind to mind 9—The dialects of North Sami). Davvi Girji, Kárášjohka (2001)
8. Kulonen, U.M., Seurujärvi-Kari, I., Pulkkinen, R. (eds.): The Saami—A Cultural Encyclopedia. Suomalaisen Kirjallisuuden Seura, Helsinki (2005)
9. Divvun, University of Tromsø: Keyboards (2015). http://divvun.no/keyboards/index.html. Accessed 2 December 2015
10. Giellatekno, University of Tromsø: Programs for analysing North Saami (2015). http://giellatekno.uit.no/cgi/d-sme.eng.html. Accessed 2 December 2015
11. Grönroos, S.A., Jokinen, K., Hiovain, K., Kurimo, M., Virpioja, S.: Low-resource active learning of North Sámi morphological segmentation. In: Presentation at First International Workshop on Computational Linguistics for the Uralic Languages. Tromsø, Norway (2015)
12. Divvun, University of Tromsø: Welcome to the OAHPA! portal (2015). http://oahpa.no. Accessed 2 December 2015
13. Giellatekno, University of Tromsø: North Saami dictionaries (2015). http://dicts.uit.no/smedicts.eng.html. Accessed 2 December 2015
14. Giellagas Institute, University of Oulu: The Saami Culture Archive of University of Oulu (2015). http://www.oulu.fi/giellagasinstitute/the_saami_culture_archive. Accessed 2 December 2015
15. Giellagas Institute, University of Oulu: A. Äänitteet (Recordings) (2015). http://www.oulu.fi/giellagasinstituutti/aanitteet. Accessed 2 December 2015
16. Divvun, University of Tromsø: Text-to-speech (2015). http://divvun.no/en/tale/tale.html. Accessed 2 December 2015
17. Simple4All Consortium: Simple4All: developing automatic speech synthesis technology (2015). http://simple4all.org/. Accessed 2 December 2015
18. Leinonen, J.: Automatic speech recognition for human-robot interaction using an under-resourced language. Master's thesis, Aalto University, School of Electrical Engineering, Department of Signal Processing and Acoustics, Espoo (2015)
19. Jokinen, K., Wilcock, G.: Community-based resource building and data collection. In: Proceedings of 4th Workshop on Spoken Language Technologies for Under-resourced Languages (SLTU-2014), pp. 201–206. St. Petersburg (2014)

20. Hiovain, K., Jokinen, K.: Acoustic features of different types of laughter in North Sami conversational speech. In: Gilmartin, E., Campbell, N. (eds.) Proceedings of the LREC 2016 Workshop "Just Talking—Casual Talk among Humans and Machines", pp. 16–20. Portorož (2016)
21. Aikio, A., Arola, L., Kunnas, N.: Variation in North Saami. In: Smakman, D., Heinrich, P. (eds.) Globalising Sociolinguistics: Challenging and Expanding Theory, pp. 243–255. Routledge (2014)
22. Jokinen, K., Trung, N.T., Hautamäki, V.: Variation in spoken North Sami language. In: Proceedings of INTERSPEECH 2016. San Francisco (2016)
23. Rießler, M.: Om samiskans attributive adjektivform. In: Amft, A., Svonni, M. (eds.) Sápmi Y1K. Livet i samernas bosättningsområde för ett tusen år sedan, pp. 135–150. Sámi dutkan 3. Umeå: Umeå universitet (2006)
24. Bergsland, K.: Sydsamisk grammatikk, 2nd edn. Davvi Girji, Karasjok (1994)
25. Wilcock, G., Laxström, N., Leinonen, J., Smit, P., Kurimo, M., Jokinen, K.: Towards SamiTalk: a Sami-speaking Robot linked to Sami Wikipedia. In: Jokinen, K., Wilcock, G. (eds.) Dialogues with Social Robots, Springer (2016) pp. 343–351 (this volume)
26. Laxström, N., Wilcock, G., Jokinen, K.: Internationalisation and localisation of spoken dialogue systems. In: Jokinen, K., Wilcock, G. (eds.) Dialogues with Social Robots, Springer (2016) pp. 207–219 (this volume)
27. Wikipedia: Motivation (2015). http://wikipapers.referata.com/wiki/Motivation. Accessed 2 December 2015
28. Laxström, N., Giner, P., Thottingal, S.: Content translation: computer assisted translation tool for Wikipedia articles. In: Proceedings of 18th Annual Conference of the European Association for Machine Translation, pp. 194–197 (2015)

Part II
Methods and Techniques for Spoken Dialogue Systems

A Comparative Study of Text Preprocessing Techniques for Natural Language Call Routing

Roman Sergienko, Muhammad Shan and Alexander Schmitt

Abstract The article describes a comparative study of text preprocessing techniques for natural language call routing. Seven different unsupervised and supervised term weighting methods were considered. Four different dimensionality reduction methods were applied: stop-words filtering with stemming, feature selection based on term weights, feature transformation based on term clustering, and a novel feature transformation method based on terms belonging to classes. As classification algorithms we used k-NN and the SVM-based algorithm Fast Large Margin. The numerical experiments showed that the most effective term weighting method is Term Relevance Ratio (TRR). Feature transformation based on term clustering is able to significantly decrease dimensionality without significantly changing the classification effectiveness, unlike other dimensionality reduction methods. The novel feature transformation method reduces the dimensionality radically: number of features is equal to number of classes.

Keywords Call routing · Text classification · Term weighting · Dimensionality reduction

1 Introduction

Natural language call routing is an important problem in the design of modern automatic call services [1]. Generally natural language call routing can be considered as two different problems. The first one is speech recognition of calls and the second

R. Sergienko (✉) · M. Shan · A. Schmitt
Institute of Communications Engineering, Ulm University,
Albert-Einstein-Allee 43, 89081 Ulm, Germany
e-mail: roman.sergienko@uni-ulm.de

M. Shan
e-mail: muhammad.shan@uni-ulm.de

A. Schmitt
e-mail: alexander.schmitt@uni-ulm.de

K. Jokinen and G. Wilcock (eds.), *Dialogues with Social Robots*,
Lecture Notes in Electrical Engineering 427,
DOI 10.1007/978-981-10-2585-3_2

one is topic categorization of user utterances for further routing. Topic categorization of user utterances can be also useful for multi-domain spoken dialogue system design [2]. In this work we treat call routing as an example of a natural language understanding application based on text classification.

In the vector space model [3] text classification is considered as a machine learning problem. The complexity of text categorization with a vector space model is compounded by the need to extract the numerical data from text information before applying machine learning algorithms. Therefore, text classification consists of two parts: text preprocessing and classification algorithm application using the obtained numerical data.

Text preprocessing comprises three stages. The first one is the textual feature extraction based on raw preprocessing of the documents. This process includes deleting punctuation, transforming capital letters to lowercase, and additional procedures such as stop-words filtering [4] and stemming [5]. Stop-words list contains pronouns, prepositions, articles and other words that usually have no importance for the classification. Using stemming it is possible to join different forms of the same word into one textual feature.

The second stage is the numerical feature extraction based on term weighting. For term weighting we use "bag-of-words" model, in which the word order is ignored. There exist different unsupervised and supervised term weighting methods. The most well-known unsupervised term weighting method is TF-IDF [6]. The following supervised term weighting methods are also considered in the paper: Gain Ratio (GR) [7], Confident Weights (CW) [8], Term Second Moment (TM2) [9], Relevance Frequency (RF) [10], Term Relevance Ratio (TRR) [11], and Novel Term Weighting (NTW) [12]; these methods involve information about the classes of the documents.

As a rule, the dimensionality for text classification problems is high even after stop-words filtering and stemming. Due to the high dimensionality, the classification may be inappropriate time-consuming, especially for real-time systems such as natural language call routing. Therefore, the next stage of preprocessing is the dimensionality reduction based on numerical features; it is possible with feature selection or feature transformation. Term weighting methods provide a natural feature selection method; it is possible to ignore terms with the lowest weights. Feature transformation for text classification can be performed with clustering of terms. Also we propose a novel feature transformation method for text classification that reduces dimensionality radically; number of features will be equal to number of classes.

As classification algorithms we use the k-NN algorithm and the SVM-based algorithm Fast Large Margin [13].

The main goal of our work is to perform a comparative study of text preprocessing techniques (term weighting and dimensionality reduction methods) for natural language call routing. We use seven term weighting methods, four dimensionality reduction methods (stop-words filtering with stemming, feature selection based on term weights, feature transformation based on clustering of the terms, and the novel feature transformation method), and two classification algorithms.

This article is organized as follows: In Sect. 2, we describe the problem and the database. Section 3 describes the considered term weighting methods. The dimensionality reduction methods are explained in Sect. 4. Section 5 contains short description of classification algorithms. The results of numerical experiments are presented in Sect. 6. Finally, we provide concluding remarks in Sect. 7.

2 Corpus Description

The data for testing and evaluation consists of 292,156 user utterances recorded in English language from caller interactions with commercial automated agents. Utterances are short and contain only one phrase for further routing. The database contains calls in textual format after speech recognition. The database is provided by the company *Speech Cycle* (New York, USA). Utterances from this database are manually labelled by experts and divided into 20 classes (such as appointments, operator, bill, internet, phone and technical support). One of them is a special class TE-NOMATCH which includes utterances that cannot be put into another class or can be put into more than one class.

The database contains 45 unclassified calls and they were removed. The database contains also 23,561 empty calls without any words. These calls were placed in the class TE-NOMATCH automatically and they were also removed from the database. As a rule, the calls are short in the database; many of them contain only one or two words. The average length of an utterance is 4.66 words, the maximal length is 19 words. There are a lot of identical utterances in the database; the corpus contains only 24,458 unique non-empty classified calls. The corpus is unbalanced. The largest class contains 27.05 % and the smallest one contains 0.04 % of the unique calls.

Due to the very high frequency of a small number of utterances in the corpus, we formulate two different data configurations.

Data configuration 1. The whole database with 268,550 classified non-empty calls is used for training and test sets forming. Numbers of repetitions of the utterances in training and test sets are used as weights for classification. This problem definition is the closest to the real situation. However, frequently repeated utterances decrease difference between preprocessing and classification methods. Additionally, there are some identical utterances in training and test sets simultaneously. In this case the overfitting problem of classification may be hidden. Therefore, this data configuration is not the most appropriate for the comparative study.

Data configuration 2. Before training and test samples forming, all utterance duplicates were removed from the database. It means that there is no intersection between training and test sets and frequency of utterances is ignored.

Therefore, data configuration 1 is suitable for the quality estimation of the real natural language call routing system. Data configuration 2 is the most appropriate for the comparative study of different preprocessing and classification methods.

For statistical analysis we performed 20 different divisions of the database into training and test samples randomly. This procedure was performed for two data configurations separately. The train samples contain 90 % of the calls and the test samples contain 10 % of the calls. For each training sample we have designed a dictionary of unique words which appear in the training sample after deleting punctuation and transforming capital letters to lowercase. The size of the dictionary varies from 3,275 to 3,329 words for data configuration 1 and from 3,277 to 3,311 for data configuration 2.

3 Term Weighting Methods

As a rule, term weighting is a multiplication of two parts: the part based on the term frequency in a document (TF) and the part based on the term frequency in the whole training database. The TF-part is fixed for all considered term weighting methods and is calculated as following:

$$TF_{ij} = \log\left(tf_{ij} + 1\right); \tag{1}$$

$$tf_{ij} = \frac{n_{ij}}{N_j}, \tag{2}$$

where n_{ij} is the number of times the ith word occurs in the jth document, N_j is the document size (number of words in the document).

The second part of the term weighting is calculated once for each word from the dictionary and does not depend on an utterance for classification. We consider seven different methods for the calculation of the second part of term weighting.

Inverse Document Frequency (IDF)

IDF is a well-known unsupervised term weighting method which was proposed in [6]. There are some modifications of IDF and we use the most popular one:

$$idf_i = \log\frac{|D|}{n_i}, \tag{3}$$

where $|D|$ is the number of documents in the training set and n_i is the number of documents that have the ith word.

Gain Ratio (GR)

Gain Ratio (GR) is mainly used in term selection [14]. However, in [7] it was shown that it could also be used for weighting terms. The definition of GR is as follows:

$$GR\left(t_i, c_j\right) = \frac{\sum_{c\in\{c_j,\bar{c}_j\}}\sum_{t\in\{t_j,\bar{t}_j\}} M\left(t, c\right)}{-\sum_{c\in\{c_j,\bar{c}_j\}} P\left(c\right)\cdot\log P\left(c\right)}; \tag{4}$$

$$M\left(t, c\right) = P\left(t, c\right)\cdot\log\frac{P\left(t, c\right)}{P\left(t\right)\cdot P\left(c\right)}, \tag{5}$$

where $P(t, c)$ is the relative frequency that a document contains the term t and belongs to the category c; $P(t)$ is the relative frequency that a document contains the term t and $P(c)$ is the relative frequency that a document belongs to category c. Then, the weight of the term t_i is the max value between all categories as follows:

$$GR\left(t_i\right) = \max_{c_j \in C} GR\left(t_i, c_j\right), \tag{6}$$

where C is a set of all classes.

Confident Weights (CW)

This supervised term weighting approach has been proposed in [8]. Firstly, the proportion of documents containing term t is defined as the Wilson proportion estimate $p(x, n)$ by the following equation:

$$p\left(x, n\right) = \frac{x + 0.5z^2_{\alpha/2}}{n + z^2_{\alpha/2}}, \tag{7}$$

where x is the number of documents containing the term t in the given corpus, n is the number of documents in the corpus and $\Phi\left(z_{\alpha/2}\right) = \alpha/2$, where Φ is the t-distribution (Students law) when $n < 30$ and the normal distribution when $n \geq 30$.

In this work $\alpha = 0.95$ and $0.5z^2_{\alpha/2} = 1.96$ (as recommended by the authors of the method). For each term t and each class c two functions $p_{pos}(x, n)$ and $p_{neg}(x, n)$ are calculated. For $p_{pos}(x, n)$ x is the number of documents which belong to the class c and have term t; n is the number of documents which belong to the class c. For $p_{neg}(x, n)$ x is the number of documents which have the term t but do not belong to the class c; n is the number of documents which do not belong to the class c.

The confidence interval (p^-, p^+) at 0.95 is calculated using the following equation:

$$M = 0, 5z^2_{\alpha/2}\sqrt{\frac{p(1-p)}{n + z^2_{\alpha/2}}}; \tag{8}$$

$$p^- = p - M;\ p^+ = p + M. \tag{9}$$

The strength of the term t in the category c is defined as the follows:

$$str(t, c) = \begin{cases} \log_2 \frac{2p^-_{pos}}{p^-_{pos} + p^+_{neg}}, & \text{if } p^-_{pos} > p^+_{neg}, \\ 0, & \text{otherwise.} \end{cases} \tag{10}$$

The maximum strength (Maxstr) of the term t_i is calculated as follows:

$$Maxstr(t_i) = \max_{c_j \in C} str\left(t_i, c_j\right)^2. \tag{11}$$

Term Second Moment (TM2)

This supervised term weighting method was proposed in [9]. Let $P(c_j|t)$ be the empirical estimation of the probability that a document belongs to the category c_j

with the condition that the document contains the term t; $P(c_j)$ is the empirical estimation of the probability that a document belongs to the category c_j without any conditions. The idea is the following: the more $P(c_j|t)$ is different from $P(c_j)$, the more important the term t_i is. Therefore, we can calculate the term weight as the following:

$$TM2(t_i) = \sum_{j=1}^{|C|} \left(P(c_j|t) - P(c_j)\right)^2, \tag{12}$$

where C is a set of all classes.

Relevance Frequency (RF)

The RF term weighting method was proposed in [10] and is calculated as the following:

$$rf(t_i, c_j) = \log_2\left(2 + \frac{a_j}{\max\{1, \bar{a_j}\}}\right); \tag{13}$$

$$rf(t_i) = \max_{c_j \in C} rf\left(t_i, c_j\right), \tag{14}$$

where a_j is the number of documents of the category c_j which contain the term t_i and $\bar{a_j}$ is the number of documents of all the other categories which also contain this term.

Term Relevance Ratio (TRR)

The TRR method [11] uses *tf* weights and it is calculated as the following:

$$TRR(t_i, c_j) = \log_2\left(2 + \frac{P(t_i|c_j)}{P(t_i|\bar{c_j})}\right); \tag{15}$$

$$P(t_i|c) = \frac{\sum_{k=1}^{|T_c|} tf_{ik}}{\sum_{l=1}^{|V|} \sum_{k=1}^{|T_c|} tf_{lk}}; \tag{16}$$

$$TRR(t_i) = \max_{c_j \in C} TRR\left(t_i, c_j\right), \tag{17}$$

where c_j is a class of the document, $\bar{c_j}$ is all of the other classes of c_j, V is the vocabulary of the training data and T_c is the document set of the class c.

Novel Term Weighting (NTW)

This method was proposed in [12, 15]. The details of the procedure are the following. Let L be the number of classes; n_i is the number of documents which belong to the i_{th} class; N_{ij} is the number of occurrences of the j_{th} word in all documents from the i_{th} class. $T_{ij} = N_{ij}/n_i$ is the relative frequency of occurrences of the j_{th} word in the i_{th} class; $R_j = \max_i T_{ij}$; $S_j = \arg\max_i T_{ij}$ is the class which we assign to the j_{th} word. The term relevance C_j is calculated by the following:

$$C_j = \frac{1}{\sum_{i=1}^{L} T_{ij}} \cdot \left(R_j - \frac{1}{L-1} \cdot \sum_{i=1, i \neq S_j}^{L} T_{ij} \right). \tag{18}$$

4 Dimensionality Reduction Methods

Stop-word filtering and stemming

We consider stop-word filtering with stemming as a language-based dimensionality reduction method which is performed before numerical feature extraction. We used special libraries ("tm", "SnowballC") in the programming language *R* for stop-word filtering and stemming for English.

Feature selection based on term weights

Term weighting methods provide a natural feature selection method; it is possible to ignore terms with the lowest weights.

For RF, TM2, and TRR methods we decreased the dictionary size from 100 to 10 % with the interval equals 10. It means deleting the corresponding number of the terms with the lowest weights. It is better to perform class-based feature selection, i.e. deleting the corresponding number of terms with the lowest weights for each class independently because the distribution of terms and weights can vary significantly for different classes. Therefore, it is necessary to assign each term from the dictionary to one corresponding class. During supervised term weighting methods CW, GR, RF, NTW, and TRR such an assignment is performed automatically. With IDF and TM2 we can also assign one class for each term using the relative frequency of the word in classes:

$$S_j = \arg\max_{c \in C} \frac{n_{jc}}{N_c}, \tag{19}$$

where S_j is the most appropriate class for the jth term, c is an index of a class, C is a set of all classes, n_{jc} is number of documents of the cth class which contain the jth term, N_c is the number of all documents of the cth class.

IDF and NTW provide getting a lot of terms with the equal highest value. For IDF the highest weight means that the term occurs only in one document from the training sample, for NTW it means that the term occurs only in documents of one class. Therefore, for these two methods we used different constraints for the value of weights; the predefined percentage of the dictionary size is not appropriate for NTW and IDF.

CW and GR provide getting a lot of terms with zero weights; it means that these two methods provide feature selection automatically. For our problem we have 43.5 % of the dictionary as terms with non-zero weights for GR and 20.4 % for CW on the average. We also decreased the size of the dictionary for CW and GR with the class-based approach.

Feature transformation based on term clustering

The idea of using class-based language model by applying term clustering was proposed in [16]. It is possible to use the term clustering in the dictionary for dimensionality reduction. In this case we suggest preprocessing our dictionary such that words of equal or similar weights are placed in the same cluster and one common weight (a new feature) will be assigned to all words in this cluster.

Term clustering is performed for each class separately. We use the assignment of terms to classes as it was described in paragraph 4.2. In order to reduce the dictionary size we take hierarchical agglomerative clustering [17] with Euclidean metric. As a common weight of the cluster we calculate the arithmetic mean of all term weights from this cluster. We set the maximal number of clusters for each class 10, 20, 50 and 100.

Novel feature transformation based on terms belonging to classes

We propose a novel feature transformation method based on terms belonging to classes. After the assigning of each term to one class, we can calculate the sums of term weights in a document for each class. We can consider these sums as new features of the text classification problem. Therefore, such a method reduces the dimensionality radically: the dimensionality equals the number of classes.

5 Classification Algorithms

As classification algorithms we use the k-NN algorithm with weight distance (k from 1 to 15) and the SVM-based algorithm Fast Large Margin (SVM-FLM) [13]. A lot of investigations [18–21] have shown effectiveness of the k-NN algorithm and SVM-based algorithms for text classification. *RapidMiner* with standard setting [22] was used as software for classification algorithm application. The classification criterion is the macro F-score [23] which is appropriate for classification problems with unbalanced classes. For k-NN we performed averaging by 20 different test samples for each value of k and after that we chose the best F-score by k.

6 Numerical Experiments

Tables 1 and 2 show the results of the numerical experiments for data configurations 1 and 2 with three situations: without dimensionality reduction (all terms are used), with stop words filtering + stemming, and with the novel feature transformation method (novel FT). The procedure with stop words filtering and stemming was not combined with other dimensionality reduction methods. For all situations the ranking of term weighting methods was performed with t-test (the confidence probability equals 0.95). The ranks are illustrated in brackets. Other comparisons were also performed with t-test. The best results in tables are bold. The results of feature

Table 1 Average F-score for data configuration 1

Term weighting method	All terms with *k*-NN	Stop-word + stemming with *k*-NN	Novel FT with *k*-NN	All terms with SVM-FLM	Stop-word + stemming with FLM	Novel FT with SVM-FLM
IDF	0.859 (5–6)	0.783 (7)	0.820 (7)	**0.873 (1)**	0.836 (1)	0.544 (7)
GR	0.858 (7)	0.792 (6)	0.841 (6)	0.670 (7)	0.680 (7)	0.621 (4–6)
CW	0.873 (4)	0.799 (2–4)	0.855 (2–3)	0.835 (4)	0.801 (4)	0.747 (2–3)
RF	0.862 (5–6)	0.795 (5)	0.849 (4–5)	0.864 (2–3)	0.819 (3)	0.744 (2–3)
TM2	**0.876 (1–3)**	0.801 (2–4)	0.853 (2–3)	0.734 (6)	0.720 (6)	0.618 (4–6)
TRR	**0.878 (1–3)**	0.805 (1)	0.865 (1)	0.865 (2–3)	0.823 (2)	0.792 (1)
NTW	**0.876 (1–3)**	0.800 (2–4)	0.844 (4–5)	0.825 (5)	0.797 (5)	0.621 (4–6)

Table 2 Average F-score for data configuration 2

Term weighting method	All terms with *k*-NN	Stop-word + stemming with *k*-NN	Novel FT with *k*-NN	All terms with SVM-FLM	Stop-word + stemming with FLM	Novel FT with SVM-FLM
IDF	0.639 (7)	0.645 (7)	0.554 (7)	**0.721 (1–2)**	0.704 (1–3)	0.370 (7)
GR	0.655 (6)	0.675 (6)	0.662 (3)	0.478 (7)	0.521 (7)	0.512 (5–6)
CW	0.710 (4)	0.699 (4–5)	0.653 (4)	0.674 (4)	0.675 (4)	0.605 (1–2)
RF	0.692 (5)	0.703 (4–5)	0.672 (1–2)	0.715 (3)	0.700 (1–3)	0.561 (3)
TM2	**0.718 (1–2)**	0.708 (2–3)	0.641 (5)	0.563 (6)	0.586 (6)	0.510 (5–6)
TRR	**0.721 (1–2)**	0.712 (1)	0.675 (1–2)	**0.721 (1–2)**	0.701 (1–3)	0.611 (1–2)
NTW	0.714 (3)	0.707 (2–3)	0.631 (6)	0.650 (5)	0.660 (5)	0.528 (4)

selection are presented in Figs. 1 and 2, the results for feature transformation based on term clustering are illustrated in Figs. 3 and 4.

Using the obtained results of the numerical experiments we may formulate the following:

TRR is the best term weighting method with *k*-NN for both problem definitions and with SVM-FLM for data configuration 2. IDF is the best term weighting with SVM-FLM for both data configurations. However, the best result for data configuration 1 with *k*-NN statistically significantly overcomes the best result with SVM-FLM. Therefore, we may conclude that TRR is the best term weighting method for the considered problem.

Stop-words filtering with stemming reduces the average dimensionality from 3,304 to 2,482 (for GR from 1,436 to 1,076; for CW from 673 to 507). In the same time significant decrease of the classification effectiveness is observed; it means that useful information is lost after this procedure. The reason lies in the fact that only very short utterances for classification (not more than 20 words) are used. Every word and every form of a word in the utterance may be useful for classification.

The similar situation is observed with feature selection. We do not observe a statistically significant decrease of F-score only for GR (one of the worst methods) and for TRR with 90 % features. Feature selection can be useful in the case of large documents with a redundant dictionary [24].

The most effective dimensionality reduction method is feature transformation based on term clustering. We obtain the best result for data configuration 1 only

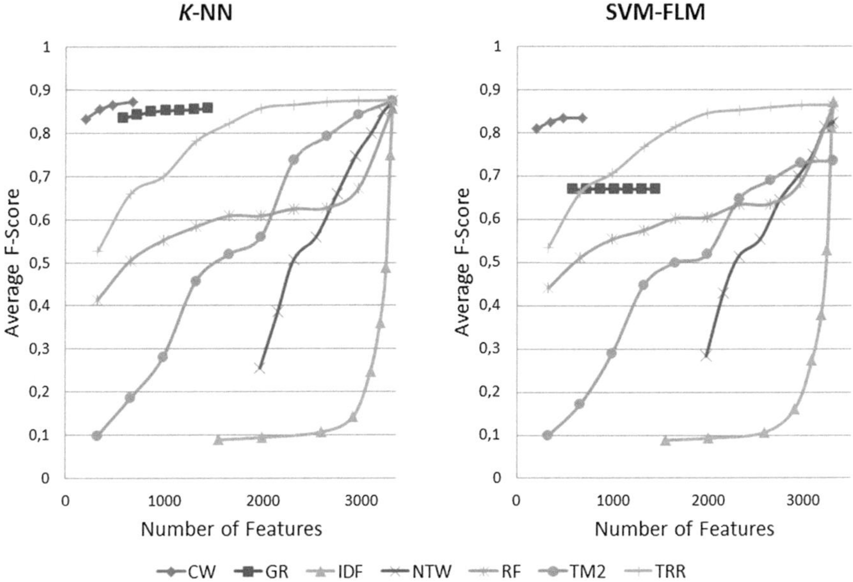

Fig. 1 Feature selection for data configuration 1

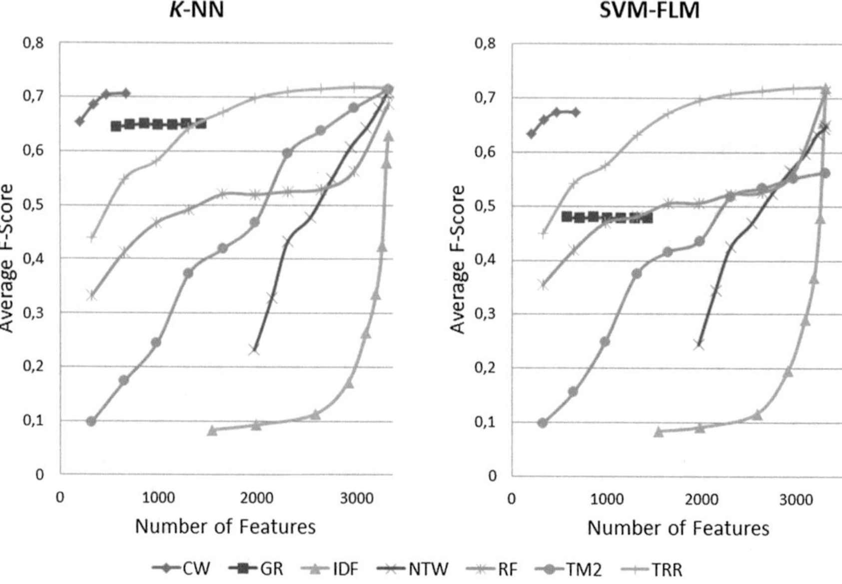

Fig. 2 Feature selection for data configuration 2

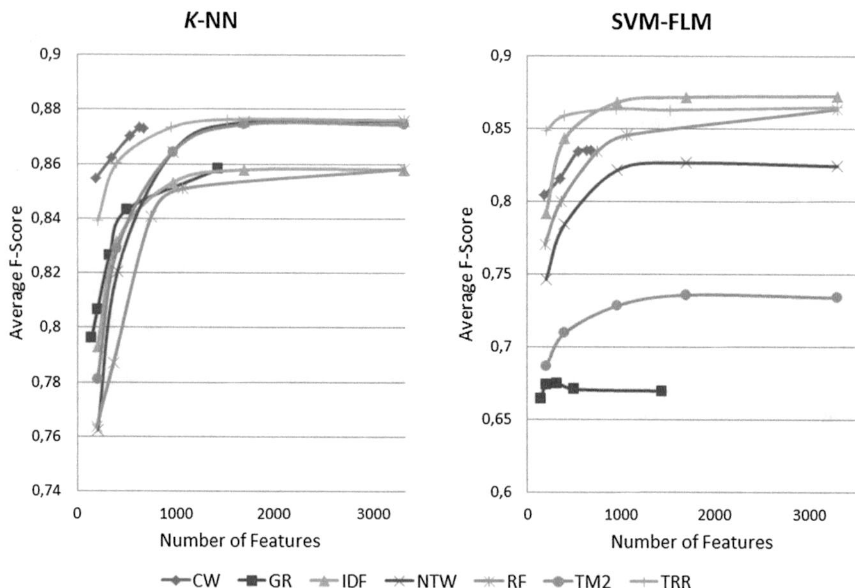

Fig. 3 Feature transformation based on term clustering for data configuration 1

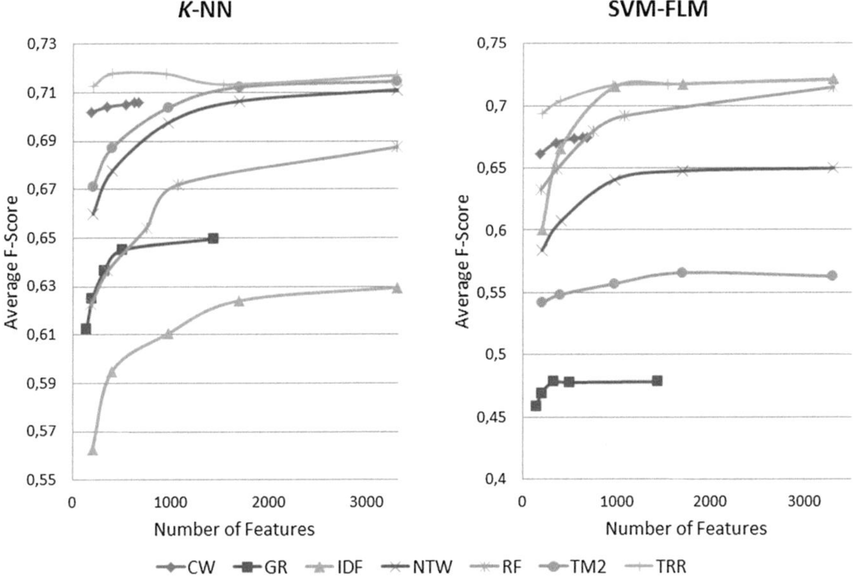

Fig. 4 Feature transformation based on term clustering for data configuration 2

with 1532 features (TRR, *k*-NN) and for data configuration 2 only with 399 features (TRR, *k*-NN).

The novel feature transformation method provides significant decrease of the classification effectiveness. However, for data configuration 1 with *k*-NN the novel feature transformation provides a very small decrease of F-score in comparison with the best result obtained with all terms: 0.013. It means that the novel FT is appropriate for real natural language call routing system because the dimensionality is reduces radically (20 features), which increases the performance in real-time systems.

7 Conclusions

The investigation has shown that the most effective term weighting method for the considered natural language call routing problem is Term Relevance Ratio (TRR). The best dimensionality reduction method is feature transformation based on term clustering which is able to decrease dimensionality significantly without decrease of the classification effectiveness. The novel feature transformation method reduces the dimensionality radically with slight decrease of the classification effectiveness and can be useful for real-time classification systems.

References

1. Suhm, B., Bers, J., McCarthy, D., Freeman, B., Getty, D., Godfrey, K., Peterson, P.: A comparative study of speech in the call center: natural language call routing vs. touch-tone menus. In: Proceedings of the SIGCHI Conference on Human Factors in Computing Systems, pp. 283–290. ACM (2002)
2. Lee, C., Jung, S., Kim, S., Lee, G.G.: Example-based dialog modeling for practical multi-domain dialog system. Speech Commun. **51**(5), 466–484 (2009)
3. Sebastiani, F.: Machine learning in automated text categorization. ACM Comput. Surv. (CSUR) **34**(1), 1–47 (2002)
4. Fox, C.: A stop list for general text. In: ACM SIGIR Forum, vol. 24, pp. 19–21. ACM (1989)
5. Porter, M.F.: Snowball: a language for stemming algorithms (2001)
6. Salton, G., Buckley, C.: Term-weighting approaches in automatic text retrieval. Inf. Proc. Manag. **24**(5), 513–523 (1988)
7. Debole, F., Sebastiani, F.: Supervised term weighting for automated text categorization. In: Text Mining and its Applications, pp. 81–97. Springer (2004)
8. Soucy, P., Mineau, G.W.: Beyond TFIDF weighting for text categorization in the vector space model. IJCAI **5**, 1130–1135 (2005)
9. Xu, H., Li, C.: A novel term weighting scheme for automated text categorization. In: Seventh International Conference on Intelligent Systems Design and Applications, ISDA 2007, pp. 759–764. IEEE (2007)
10. Lan, M., Tan, C.L., Su, J., Lu, Y.: Supervised and traditional term weighting methods for automatic text categorization. IEEE Trans. Pattern Anal. Mach. Intell. **31**(4), 721–735 (2009)
11. Ko, Y.: A study of term weighting schemes using class information for text classification. In: Proceedings of the 35th International ACM SIGIR Conference on Research and Development in Information Retrieval, pp. 1029–1030. ACM (2012)
12. Gasanova, T., Sergienko, R., Akhmedova, S., Semenkin, E., Minker, W.: Opinion mining and topic categorization with novel term weighting. In: Proceedings of the 5th Workshop on Computational Approaches to Subjectivity, Sentiment and Social Media Analysis, pp. 84–89. ACL (2014)
13. Fan, R.E., Chang, K.W., Hsieh, C.J., Wang, X.R., Lin, C.J.: Liblinear: a library for large linear classification. J. Mach. Learn. Res. **9**, 1871–1874 (2008)
14. Yang, Y., Pedersen, J.O.: A comparative study on feature selection in text categorization. ICML **97**, 412–420 (1997)
15. Sergienko, R., Gasanova, T., Semenkin, E., Minker, W.: Text categorization methods application for natural language call routing. In: 11th International Conference on Informatics in Control, Automation and Robotics (ICINCO), vol. 2, pp. 827–831. IEEE (2014)
16. Momtazi, S., Klakow, D.: A word clustering approach for language model-based sentence retrieval in question answering systems. In: Proceedings of the 18th ACM Conference on Information and Knowledge Management, pp. 1911–1914. ACM (2009)
17. Ward Jr., J.H.: Hierarchical grouping to optimize an objective function. J. Am. Stat. Assoc. **58**(301), 236–244 (1963)
18. Han, E.H.S., Karypis, G., Kumar, V.: Text Categorization Using Weight Adjusted k-Nearest Neighbor Classification. Springer (2001)
19. Baharudin, B., Lee, L.H., Khan, K.: A review of machine learning algorithms for text-documents classification. J. Adv. Inf. Tech. **1**(1), 4–20 (2010)
20. Joachims, T.: Learning to Classify Text Using Support Vector Machines: Methods. Kluwer Academic Publishers, Theory and Algorithms (2002)
21. Morariu, D.I., Vintan, L.N., Tresp, V.: Meta-classification using SVM classifiers for text documents. Int. J. Appl. Math. Comput. Sci. **1**(1) (2005)
22. Shafait, F., Reif, M., Kofler, C., Breuel, T.M.: Pattern recognition engineering. In: RapidMiner Community Meeting and Conference, vol. 9. Citeseer (2010)
23. Goutte, C., Gaussier, E.: A probabilistic interpretation of precision, recall and f-score, with implication for evaluation. In: Advances in Information Retrieval, pp. 345–359. Springer (2005)

24. Gabrilovich, E., Markovitch, S.: Text categorization with many redundant features: using aggressive feature selection to make svms competitive with c4. 5. In: Proceedings of the Twenty-First International Conference on Machine learning, p. 41. ACM (2004)

Compact and Interpretable Dialogue State Representation with Genetic Sparse Distributed Memory

Layla El Asri, Romain Laroche and Olivier Pietquin

Abstract User satisfaction is often considered as the objective that should be achieved by spoken dialogue systems. This is why the reward function of Spoken Dialogue Systems (SDS) trained by Reinforcement Learning (RL) is often designed to reflect user satisfaction. To do so, the state space representation should be based on features capturing user satisfaction characteristics such as the mean speech recognition confidence score for instance. On the other hand, for deployment in industrial systems there is a need for state representations that are understandable by system engineers. In this article, we propose to represent the state space using a Genetic Sparse Distributed Memory. This is a state aggregation method computing state prototypes which are selected so as to lead to the best linear representation of the value function in RL. To do so, previous work on Genetic Sparse Distributed Memory for classification is adapted to the Reinforcement Learning task and a new way of building the prototypes is proposed. The approach is tested on a corpus of dialogues collected with an appointment scheduling system. The results are compared to a grid-based linear parametrisation. It is shown that learning is accelerated and made more memory efficient. It is also shown that the framework is scalable in that it is possible to include many dialogue features in the representation, interpret the resulting policy and identify the most important dialogue features.

L.El. Asri (✉) · R. Laroche
Orange Labs, Chatillon, France
e-mail: layla.elasri@gmail.com

R. Laroche
e-mail: romain.laroche@orange.com

L.El. Asri
UMI 2958 (CNRS-GeorgiaTech), Metz, France

O. Pietquin
University of Lille, CNRS, Lille, France
e-mail: olivier.pietquin@univ-lille1.fr

O. Pietquin
UMR 9189—CRIStAL, 59000 Lille, France

O. Pietquin
Institut Universitaire de France (IUF), Paris, France

K. Jokinen and G. Wilcock (eds.), *Dialogues with Social Robots*,
Lecture Notes in Electrical Engineering 427,
DOI 10.1007/978-981-10-2585-3_3

Keywords Spoken dialogue systems · Reinforcement learning · State space representation · Genetic algorithms · Sparse distributed memory

1 Introduction

Reinforcement Learning (RL) [1] is now a state of the art method to learn optimal policies for dialogue systems [2–6]. To do so, a reward function, describing how good a decision made by the system is, has to be designed. It encodes the goal of the system. On the other hand, a commonly used metric to assess dialogue management quality is user satisfaction [7, 8]. Since in RL the reward function defines the task of the system, it is natural to have the rewards reflect user satisfaction. There has been extensive research on automatically estimating user satisfaction for a given dialogue [9–11]. These studies have shown that many dialogue features (duration, mean speech recognition scores, number of help requests, ...) could play an important role in user satisfaction [12, 13]. Because RL relies on a representation of the dialogue state, if the rewards depend on dialogue features then the state space representation should also include these features [14, 15]. Some of these features are continuous and in this case it is not possible to learn by estimating the Q-function for each possible value, a parametrisation is needed.

In addition, if one wants to have his learning algorithm deployed in big industrial systems, there is a need for having dialogue state representations that are interpretable by a human designer. Therefore, if the representation is learnt, it should not transform the original features (no projection, mixing etc.). This might lead to intractable dimensions in the representation.

In the RL community, the parametric approximation (and especially linear ones) of the Q-function according to a set of basis functions is the most used framework because theoretical convergence properties can be proved [16]. Generally, the basis functions are assumed to be known in advance. Yet there is a parallel trend that learns basis functions from data. This trend is state aggregation which learns state prototypes by aggregating states into homogeneous clusters. Once the prototypes are learned, the Q-function is estimated as a linear function of their features. Linear representation are interpretable and come with theoretical guarantees [17–19]. However, linear representations are subject to the problem known as the *curse of dimensionality*: learning might become inconveniently slow as the number of dimensions of the state space increases [1]. This problem has already been addressed in dialogue [20, 21] but proposed approaches lead to features that can not be interpreted. Here, we proposed to perform state aggregation based on a Sparse Distributed Memory (SDM, [22]) so as to learn a compact and interpretable state representation.

Contrary to [23–27] where states are aggregated according to their similarity or to [28–31] where states are aggregated according to the similarity of the optimal policy in those states, the proposed approach aggregates states according to the similarity of the Q-function in those states. To do so, a new parametrisation using an SDM is proposed. An SDM manipulates binary data vectors called *prototypes* and assesses the similarity between two prototypes with the Hamming distance. The contributions of this paper are an adaptation of the work on combining SDM and genetic programming for classification [32] to the RL problem and a novel method to incrementally build the set of prototypes in the SDM, according to the states observed by the learning agent. This representation is called Genetic Sparse Distributed Memory for Reinforcement Learning (GSDMRL). Advantages of GSDMRL over previous representations is that it can combine continuous, discrete and symbolic features, that it can be used in both an online and a batch learning setting and that it performs feature selection according to the value function returned by the chosen RL algorithm. GSDMRL is applied to the DINASTI (Dialogues with a Negotiating Appointment Setting Interface) corpus [33]. It is shown that learning is more efficient with GSDMRL and scales well with the number of dialogue features. Then, an analysis of the policy learnt with GSDMRL highlights some important features to take into account for decision making.

2 Background

2.1 The Reinforcement Learning Framework

Dialogue management is modelled as a Markov Decision Process (MDP) which is a five-tuple (S, A, T, R, γ). Elements of this tuple are respectively the state space, the action space, state transition probabilities, the reward function and a discount factor $\gamma \in [0, 1[$. The discounted cumulative reward at time t is the return: $r_t = \sum_{k \geq 0} \gamma^k R_{t+k}$ A deterministic policy π maps each state to one action. The Q-function for π is $Q^\pi(s, a) = E[r_t \mid s_t = s, a_t = a, \pi]$ The dialogue manager seeks an optimal policy π^*, which maximises the expected return starting from any state-action pair. The associated Q-function is: $Q^{\pi^*}(s, a) \in \operatorname{argmax}_\pi Q^\pi(s, a)$

2.2 The Sparse Distributed Memory Model

An SDM [22, 34] is defined on a n-dimensional binary space. It is initialised with a set of addresses. At each address, a vector of counters is stored. Each counter corresponds to a bit in the stored data and all counters are first set to 0. The writing process is illustrated in steps 1 and 2 in Fig. 1. Let r be the address register corresponding to the data vector d to be written. The Hamming distance between r and the addresses in the memory is computed. The set of addresses for which the distance is below a given threshold Δ form the *selection set* (step 1).

The data vector d is then written in the vectors of counters linked to the addresses in the selection set (step 2): each counter of each vector is incremented if the corresponding bit in d is 1 and it is decremented otherwise. Reading from the memory

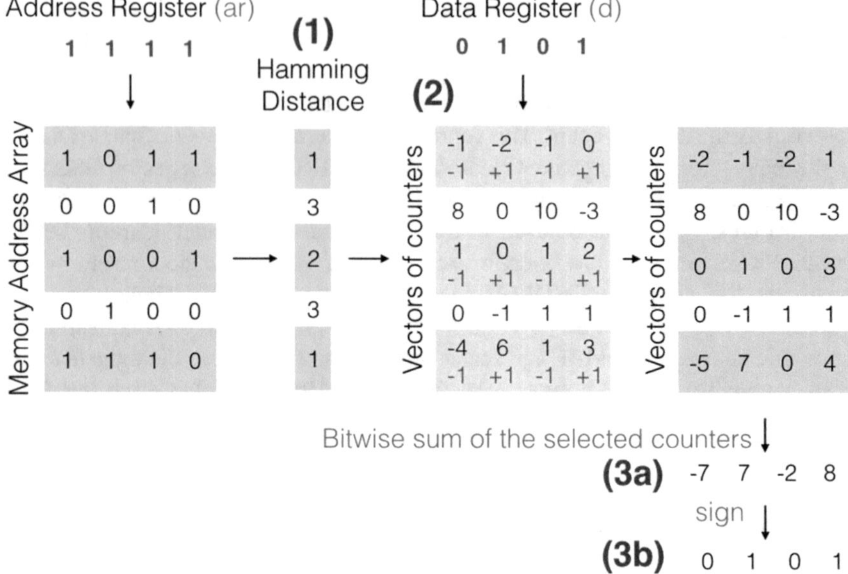

Fig. 1 A schematic display of the reading process in a sparse distributed memory

with r as input is illustrated in step 3: the selection set is computed as before and the vectors of counters linked to these addresses are summed bitwise. If a counter of the resulting sum is positive, the returned bit is 1, otherwise it is 0.

The SDM was designed to store large binary vectors. It takes advantage of the fact that in large vector spaces, vectors tend to be orthogonal. When a data vector d is written with an address r, all the addresses in r's selection set receive a copy of d. Then, if an address r' close to r is presented to the memory for reading, virtually the same selection set will be computed and copies of d will be numerous in this set. Therefore, d will be output with high probability.

Initialisation of the addresses is an issue. If the patterns stored in the memory are large binary vectors, a random initialisation implies to sample a great number of addresses. There exists a method to infer a set of addresses from data [35–38]. Yet they are not relevant for RL applications where only a small region of the state space is visited by most of the acceptable policies. A novel method for initialisation, based on the observations of the learning agent, is proposed in Sect. 3.

In the RL context, the addresses in the memory represent state prototypes and the stored data are no longer counters representing binary vectors but continuous values representing the Q-function. The memory is randomly initialised [29, 39]. Then, prototypes are re-engineered, for instance, according to visit frequencies. However, this can be problematic as some states might not be visited often but still be crucial like catastrophic states or even goal states [40].

2.3 Genetic Sparse Distributed Memory for Classification

Rogers and then Das and Whitley [32, 41, 42] used the SDM model for binary classification, combining SDM with genetic search. In this context, each address in the memory is linked to a unique counter where the class (0 or 1) is written as explained in Fig. 1. Once enough data has been written in the memory, the value of the counter of an address indicates the probability that the class is 1 knowing that the address was selected. It is also possible to know if an address is relevant for class prediction or not. Each address in the memory has an *activation set* which is the set of addresses which would be selected if the address was fed to the memory (the activation set of an address in the memory is equivalent to the selection set of an address register presented to the memory). An address is relevant for classification if the probability that the class is 1 knowing the activation set is significantly different than the probability that the class is 1 given all the addresses in the memory. Based on this, Rogers defined a measure of the *fitness* of a given address. This fitness was then used to rank the addresses in the memory. The address with the lowest ranking was suppressed and replaced by the crossover of the two fittest addresses. The fitness scores were also used to weight the counters: when a new address was presented to the memory for reading, the counters of the addresses in the selection set were weighted according to their fitness so that more importance was given to the fittest addresses.

3 Genetic Sparse Distributed Memory for Reinforcement Learning (GSDMRL)

Let's start with some notations. A prototype p^j is composed of n bits $p^j = (p_1^j, \ldots, p_n^j)$. The Hamming distance between two prototypes p^1 and p^2 is $d(p^1, p^2) = \sum_{k=1}^{n}(1 - \delta_{p_k^1, p_k^2})$ where δ is the Kronecker symbol. A prototype p^2 belongs to the activation set (resp. selection set) of a prototype p^1 (resp. a state s) if $d(p^1, p^2) \leq \Delta$ (resp. if $d(p^1, s) \leq \Delta$). The notation $|S|$ will be used for the cardinal of a set S.

3.1 Building the Set of Prototypes

GSDMRL builds an SDM M with three layers: each prototype p^j is linked to a vector of counters $c(p^j) = c^j$ and an estimation of the Q-function for each action. The set of prototypes is built incrementally. GSDMRL starts with an empty memory. Each time a new state s_t is observed during learning, if the selection set X_t for this state is empty, the state is added to X_t and to the address set of M and linked to a vector of counters and Q-values initialised to 0:

$$\forall i \in 1, \dots, |A| \; M(s_t)_i = \{\forall k \in 1, \dots, n \; c(s_t)_k = 0, \; Q^{\pi}(s_t, a^i) = 0\}. \tag{1}$$

The state s_t is then written in the counters corresponding to the prototypes in X_t, like in step 2 in Fig. 1:

$$\forall p^j \in X_t, k \in 1, \dots, n \; c(p^j)_k = c_k^j = c_k^j + (-1)^{(s_t)_k + 1}. \tag{2}$$

3.2 Q-Function Parametrisation

We use a linear representation for the Q-function: $Q_\theta(s_t, a^i) = \sum_{j=1}^{|X_t|} \theta_{j,i} \phi_j(s_t, a^i)$. The ϕ_j are basis functions and the $\theta_{j,i}$ will be updated by the reinforcement learning algorithm. The basis functions are defined as follows:

$$\phi_j(s_t, a^i) = \frac{w^{j,i}}{\sum_{k=1}^{|X_t|} w^{k,i}} \qquad Q_\theta(s_t, a^i) = \sum_{j=1}^{|X_t|} \theta_{j,i} \frac{w^{j,i}}{\sum_{k=1}^{|X_t|} w^{k,i}}. \tag{3}$$

The Q-function is thus estimated as the weighted average of the values of the weights at $(p^j) \in X_t$. This weight is denoted by $w^{j,i}$. Following the same idea as the genetic sparse distributed memory proposed by Rogers, the prototypes in X_t are weighted according to their relevance to estimate the Q-values. The weight $w^{j,i}$ is a function of the fitness $f^{j,i}$ of (p^j, a^i). Rogers [43] showed that the weight which should be given to (p^j, a^i) is $w^{j,i} = \frac{f^{j,i}}{(1 - f^{j,i})^2}$.

The fitness score measures the relevance of p^j with respect to the regression of the Q-function. It is computed by looking at the prototypes in the activation set of p^j and checking whether the confidence intervals for the Q-function estimated at action a^i overlap with the confidence interval for $Q^{\pi}(p^j, a^i)$, noted $CI(Q^{\pi}(p^j, a^i))$. Confidence intervals are defined as [44]:

$$\begin{aligned} CI(Q^{\pi}(p^j, a^i)) = CI^{j,i} &= [Q^{\pi}(p^j, a^i) - \epsilon^{j,i}, Q^{\pi}(p^j, a^i) + \epsilon^{j,i}] \\ \epsilon^{j,i} &= t_{\alpha/2}^{n^{j,i}-1} \frac{\sigma^{j,i}}{\sqrt{n^{j,i}}}. \end{aligned} \tag{4}$$

In this equation, $n^{j,i}$ is the number of visits to the pair (p^j, a^i), $\sigma^{j,i}$ is the standard deviation of the returns observed after visiting (p^j, a^i) and $t_{\alpha/2}^{n^{j,i}-1}$ is Student's t function with $n^{j,i} - 1$ degrees of freedom. Parameter α sets the confidence with which observed returns will be comprised in the interval. Since the prototypes in M are added only if their distance to all the other prototypes is greater than Δ, the activation set of p^j, $AS(p^j)$ only contains p^j. Nevertheless, two distant prototypes might be activated by the same states and they should predict the same Q-values. The vectors of counters which constitute the second layer of M are used as averages of the states which activate the prototypes. Indeed, every state which activates a prototype is written in the prototype's vector of counters according to Eq. 2. So, if two prototypes are linked to similar vectors of counters, it means these prototypes tend to be jointly activated and they should be linked to similar Q-values. The fitness $f^{j,i}$ of (p^j, a^i) is computed by first finding the activation set of p^j's vector of counters c^j. Then, c^j is compared to the other vectors of counters in Hamming distance and the vectors closer than Δ form the activation set $AS(p^j)$. The fitness of (p^j, a^i) is the ratio of the number of confidence intervals overlapping on the size of $AS(p^j)$:

$$f^{j,i} = \frac{|\{p^k \in AS(p^j)\ ,\ (CI^{j,i} \cap CI^{k,i}) \neq \emptyset\}|}{|AS(p^j)|} \tag{5}$$

Besides weighting the prototypes, the fitness scores are also used to re-engineer the prototypes in M.

3.3 Re-engineering the Prototypes

We define the set-policy π_s for a prototype p^j as:

$$\pi_s(p^j) \in \underset{a}{\operatorname{argmax}}\ Q^\pi(AS(p^j), a^i) = \frac{\sum_{k=1}^{|AS(p^j)|} w^{k,i}\theta_{k,i}}{\sum_{k=1}^{|AS(p^j)|} w^k}$$

The re-engineering rule is only based on the confidence interval for the set-policy of each prototype. The fitness score of (p^j, π_s) measures the relevance of p^j for predicting $Q^\pi(s, \pi_s(p^j))$ where s is a state activating p^j. A prototype with a low fitness adds noise to learning. Therefore, the prototype p^- with the lowest fitness score for its set-policy is suppressed and replaced by the result of a random crossover between $c^{+,1}$ and $c^{+,2}$ which are respectively the vectors of counters for the two fittest prototypes $p^{+,1}$ and $p^{+,2}$. The crossover process is detailed in Algorithm 1.

Algorithm 1 Crossover process

Choose random crossover size s in 1, ..., n
Choose random crossover point c in 1, ..., $n - s$
$p^{\text{cross}} = \text{concat}(c^{+,1}_{1,\ldots,d}, c^{+,2}_{d+1,\ldots,d+1+s}, c^{+,1}_{d+s+2,\ldots,n})$
Add p^{cross} to the set of addresses in M
Remove p^- from M

Once the radii of the confidence intervals (ϵ in Eq. 4) have reached a low value, the confidence intervals in an activation set will no longer overlap. Therefore, a parameter β is added to the algorithm as a lower bound for the confidence intervals. When the radius of a confidence interval goes under β, it is set to β.

4 Experiments with the NASTIA Dialogue System

NASTIA (Negotiating Appointment Setting Interface) is a French-speaking SDS. Its task is an appointment scheduling for landline maintenance. It was tested on 1734 scenario-based dialogues with 385 volunteers who interacted at most 5 times with the system [45]. After each dialogue, the user was asked to fill an evaluation questionnaire and rate the dialogue on a scale of 1 to 10. This corpus is named DINASTI (Dialogues with a Negotiating Appointment Setting Interface) [33].

NASTIA has five dialogue phases where decisions should be taken. We want to use the ratings to infer a state space for each of NASTIA's dialogue phases with GSDMRL. First the system chooses a negotiation strategy. If User Initiative (UI) is chosen, the user is asked: "When would you like to book an appointment?". System Initiative (SI) asks a sequence of questions in three different dialogue turns, filling successively the day, week and half-day slots. The third option consists of proposing directly a List of Availabilities (LA) to the user, waiting for her/him to interrupt the list as soon as an appropriate appointment has been proposed. A second dialogue phase is the help phase, it is encountered either because the user has requested help or because the system decided to play a help message. Three lengths of messages are available. The third phase is visited after the user has proposed a time slot. NASTIA chooses between three confirmation strategies. The fourth dialogue phase is visited after a rejection from speech recognition or a user time out. In this case, NASTIA may prompt a help message or inform the user that s/he was not understood or heard. The last phase is visited after an appointment setting failure or after the user has expressed some constraints. The system can decide to provide information about its availabilities based on what it has understood of the user's constraints.

Two experiments were conducted using Fitted-Q Iteration [17, 46]. First we compared GSDMRL to a grid-based state representation. For GSDMRL, a dialogue state is the concatenation of 32-bit vectors, each vector being the binary representation of one of the 120 features in the corpus (see the list in [33]). We set $\beta = 0.01$ and $\gamma = 0.99$. The grid was built by performing entropy-based discretisation on the dialogue feature [47, 48]. With this kind of representation, it is not possible to use all the 120 features because the size of the state space, defined as a Cartesian product of the intervals, would become inconveniently high. Therefore, to compare these two approaches, the first experiment only used two features, which are well-known to be important for user satisfaction: the number of dialogue turns and the mean speech recognition confidence score. GSDMRL was thus composed of 64-bit vectors and Δ was set to 15. The second experiment highlights the scalable and interpretable powers of GSDMRL. To do this, the original 120 features of the DINASTI corpus were included in the state space and the threshold Δ was set to 150.

4.1 Comparing GSDMRL to a Grid-Based Representation

The comparison of GSDMRL and the grid is based on the expected cumulative reward per dialogue. Each dialogue starts with deciding the negotiation strategy so they start with the same state s_0. This state s_0 is the first prototype to be written in GSDMRL and it is added to the grid. For both representations, it is ensured that this state would only be visited (or selected in the case of GSDMRL) at the beginning of each dialogue. The reward function gives $\frac{P_i}{\gamma^{t_f-1}}$ at the end of each dialogue D_i where P_i is the performance rating for D_i and t_f is the last dialogue turn (Dialogue turns start at time $t_0 = 0$). For the other intermediate dialogue turns, the reward is equal to 0. These rewards were designed so that the value of the Q-function at the initial state is an estimate of the expected performance for the dialogue:

$$Q^{\pi}(s_0, a) = E[r_t = \gamma^{tf-1} \times \frac{P_i}{\gamma^{t_f-1}} \mid \mathcal{D}, s_0, a, \pi]$$

$$Q^{\pi}(s_0, a) = E[P_i \mid \mathcal{D}, s_0, a, \pi] \tag{6}$$

Both representations are compared on $\hat{r}_0$, the expected cumulative reward starting from state s_0 and following the distribution of actions in the corpus, that is to say:

$$\hat{r}_0^{\pi} = \sum_a \frac{n(s_0, a)}{\sum_a n(s_0, a)} Q^{\pi}(s_0, a), \tag{7}$$

where $n(s_0, a)$ is the number of visits to (s_0, a). In order to organize a fair comparison, the policies learnt with both representations were evaluated on the same state space representation. Two policies π_{GSDMRL} and π_{Grid} were learnt respectively with the GSDMRL and the grid representation. Then the expected cumulative reward for both policies was computed by projecting the policies on the simple state space representation composed of NASTIA's dialogue phases. Thus, $\hat{r}_0^{\pi}$ was computed as the expected cumulative reward given the policy π_{GSDMRL} or π_{Grid} and given the probability transitions between the dialogue phases. This evaluation tests which mapping from dialogue phases to a higher-dimensional space is the most efficient.

4.2 Scalability and Interpretability of GSDMRL

In the second experiment, a policy is learnt with 120 features. It is possible to easily analyse the policy, based on the prototypes. To achieve this, for each action a, all the prototypes whose optimal action is a are grouped together. Then, the mean feature-wise Hamming distance is computed between the prototypes of one group and the prototypes of the other groups. These distances are normalised by the variances of the values in the corpus. The highest distances are representative of the features which play an important role in choosing action a. For instance, for the strategy negotiation, if the mean speech recognition score of the prototypes choosing the LA action is, in normalised distance, far from the mean recognition scores of the prototypes choosing another action, then it means that the mean speech recognition score is important when it comes to choosing LA. Prototypes within a range of recognition scores will tend to have LA as optimal action. From this, the policy can be analysed by identifying how features influence the system's behaviour.

4.3 Results

The results of experiment 1 are presented in Table 1. First, the number of states varies considerably less with GSDMRL than with the grid. The number of states in GSDMRL is also significantly lower with this value of Δ. Even with a smaller number of states, the average expected cumulative reward with GSDMRL is at least as good as the one with the grid and even a little higher. This shows that GSDMRL enables to build a small set of states upon which a good policy will be learnt in a more efficient way than it is possible to do with a grid-based representation.

In experiment 2, a total number of 899 prototypes were added to the memory. An analysis of these prototypes enables to highlight the most important features in the policy NASTIA learnt with GSDMRL. These features and their values explain the circumstances behind the system's decision.

Concerning the negotiation phase, the LA action is linked to short dialogues (35.23 s *versus* 67.45), with few rejections from speech recognition (1.26 *versus* 4.68), and where, in average, less than one confirmation occurred. This strategy should thus rather be chosen at the beginning of a dialogue. As expected, the SI strategy should be chosen if the dialogue is problematic. In average, SI should be chosen after 5.43 speech recognition rejections *versus*. 3.92 for UI and 1.26 for LA. UI seems more fit after a list of availabilities was proposed with LA but none of the propositions suited the user. As for the phase recovering from ASR rejections and user inactivities, the two most important features are the number of user dialogue turns and dialogue duration. It is better to ask the user to repeat after dialogue length

Table 1 Comparison of GSDMRL and a grid-based representation

Training corpus	Test corpus	#states grid	#states GSDMRL	$\hat{r}_0^{\pi Grid}$	$\hat{r}_0^{\pi GSDMRL}$
531	1200	427.07	**63.9**	7.867	7.878
731	1000	282.23	**66.67**	7.868	**7.885**
831	900	271.99	**66.85**	7.869	**7.891**
931	800	266.84	**62.58**	7.872	**7.900**

All values are averages on 500 runs. The results in bold are statistically significant (p-value under 0.01 for Student's t-test)

has gone above a certain threshold (183.42 s *versus* 52.97). The implicit confirmation strategy is chosen after dialogue duration has gone above 200 s and when, in average, there has been less than 1 speech recognition rejection. The explicit strategy should be chosen if more than 2.5 help messages have been played and the number of user turns is high. This means that the dialogue might be problematic and it is better to choose a more conservative strategy. Information about the system's calendar should not be provided to the user if the system has already tried the LA strategy at least once. This reflects the fact that the user is already partially aware of system availabilities and might be annoyed by a reminder. On the other hand, information is given after, in average, 0.47 dialogue acts notifying the user that a slot is not available have been observed. This means that after UI or SI, if the user proposes unrealisable constraints, the system should always provide information about its calendar. Finally, indications of a long dialogue with several unsuccessful negotiation rounds should push to choose the shortest help message.

5 Conclusion

This article proposed a new model for state aggregation in reinforcement learning-based dialogue management. The model, Genetic Sparse Distributed for Reinforcement Learning (GSDMRL) incrementally builds a set of prototypes. GSDMRL can include features from speech recognition, natural language understanding and dialogue management. Compared to a grid-based representation, GSDMRL was shown to be more efficient both in terms of memory and learning. In addition, GSDMRL is scalable and can handle many dialogue features. Finally, the set of prototypes provides a valuable insight on the system's learning and is easily interpretable.

References

1. Sutton, R.S., Barto, A.G.: Reinforcement Learning. MIT Press, An introduction (1998)
2. Levin, E., Pieraccini, R., Eckert, W.: Learning dialogue strategies within the Markov decision process framework. In: Proceedings of the IEEE ASRU (1997)
3. Singh, S., Kearns, M., Litman, D., Walker, M.: Reinforcement learning for spoken dialogue systems. In: Proceedings of the NIPS (1999)
4. Williams, J.D., Young, S.: Partially observable Markov decision processes for spoken dialog systems. Comput. Speech Lang. **21**, 231–422 (2007)
5. Laroche, R., Putois, G., Bretier, P.: Optimising a handcrafted dialogue system design. In: Proceedings of the Interspeech (2010)
6. Daubigney, L., Geist, M., Chandramohan, S., Pietquin, O.: A comprehensive reinforcement learning framework for dialogue management optimisation. IEEE J. Sel. Top. Sign. Proc. **6**(8), 891–902 (2012)
7. Dybkjaer, L., Bernsen, N.O., Minker, W.: Evaluation and usability of multimodal spoken language dialogue systems. Speech Commun. **43**, 33–54 (2004)
8. Lemon, O., Pietquin, O.: Data-Driven Methods for Adaptive Spoken Dialogue Systems. Springer (2012)

9. Walker, M., Hindle, D., Fromer, J., Fabbrizio, G., Mestel, C.: Evaluating competing agent strategies for a voice e-mail agent. In: Proceedings of the EuroSpeech (1997)
10. Schmitt, A., Schatz, B., Minker, W.: Modeling and predicting quality in spoken human-computer interaction. In: Proceedings of the SIGDIAL (2011)
11. El Asri, L., Khouzaimi, H., Laroche, R., Pietquin, O.: Ordinal regression for interaction quality prediction. In: Proceedings of the ICASSP (to be published) (2014)
12. Larsen, L.B.: Issues in the evaluation of spoken dialogue systems using objective and subjective measures. In: Proceedings of the IEEE ASRU, pp. 209–214 (2003)
13. Walker, M.A., Langkilde-Geary, I., Hastie, H.W., Wright, J., Gorin, A.: Automatically training a problematic dialogue predictor for a spoken dialogue system. J. Artif. Intell. Res. **16**, 293–319 (2002)
14. Paek, T., Pieraccini, R.: Automating spoken dialogue management design using machine learning: An industry perspective. Speech Commun. **50** (2008)
15. Paek, T., Chickering, D.M.: The markov assumption in spoken dialogue management. In: Proceedings of the SIGdial Workshop on Discourse and Dialogue, pp. 35–44 (2005)
16. Geist, M., Pietquin, O.: Algorithmic survey of parametric value function approximation. IEEE Trans. Neural Netw.ne Learn. Syst. (2013)
17. Gordon, G.J.: Stable function approximation in dynamic programming. In: Proceedings of the ICML (1995)
18. Tsitsiklis, J., Van Roy, B.: An analysis of temporal-difference learning with function approximation. IEEE Trans. Autom. Control (1997)
19. Gordon, G.J.: Reinforcement learning with function approximation converges to a region. In: Proceedings of the NIPS (2001)
20. Li, L., Williams, J.D., Balakrishnan, S.: Reinforcement learning for dialog management using least-squares policy iteration and fast feature selection. In: Proceedings of the Interspeech (2009)
21. Chandramohan, S., Geist, M., Pietquin, O.: Sparse approximate dynamic programming for dialog management. In: Proceedings of the SIGDIAL (2010)
22. Kanerva, P.: Associative Neural Memories: Theory and Implementation. Oxford University Press (1993)
23. Broomhead, D., Lowe, D.: Multivariable functional interpolation and adaptive networks. Complex Syst. (1988)
24. Albus, J.S.: A theory of cerebellar function. Math. Biosci. (1971)
25. Singh, S., Sutton, R.S.: Reinforcement learning with replacing eligibility traces. In: Mach. Learn. (1996)
26. Forbes, J.R.: Reinforcement learning for autonomous vehicles. Ph.D. thesis, University of California at Berkeley (2002)
27. Mahadevan, S., Maggioni, M., Guestrin, C.: Proto-value functions: a Laplacian framework for learning representation and control in markov decision processes. J. Mach. Learn. Res. (2006)
28. Bernstein, A., Shimkin, N.: Adaptive aggregation for reinforcement learning with efficient exploration: Deterministic domains. In: Proceedings of the COLT (2008)
29. Wu, C., Meleis, W.: Adaptive fuzzy function approximation for multi-agent reinforcement learning. In: Proceedings of the IEEE/WIC/ACM IAT (2009)
30. Baumann, M., Buning, H.K.: State aggregation by growing neural gas for reinforcement learning in continuous state spaces. In: Proceedings of the ICMLA (2011)
31. Baumann, M., Klerx, T., Büning, H.K.: Improved state aggregation with growing neural gas in multidimensional state spaces. In: Proceedings of the ERLARS (2012)
32. Rogers, D.: Weather prediction using a genetic memory. Tech. Rep., NASA (1990)
33. El Asri, L., Laroche, R., Pietquin, O.: DINASTI: dialogues with a negotiating appointment setting interface. In: Proceedings of the LREC (to be published) (2014)
34. Kanerva, P.: Hyperdimensional computing: An introduction to computing in distributed representation with high-dimensional random vectors. Cogn. Comput. (2009)
35. Hely, T.A., Willshaw, D.J., Hayes, G.M.: A new approach to Kanerva's sparse distributed memory. IEEE Trans. Neural Netw. (1997)

36. Rao, R.P.N., Fuentes, O.: Hierarchical learning of navigational behaviors in an autonomous robot using a predictive sparse distributed memory. Mach. Learn. (1998)
37. Anwar, A., Dasgupta, D., Franklin, S.: Using genetic algorithms for sparse distributed memory initializations. In: Proceedings of the GECCO (1999)
38. Hart, E., Ross, P.: Exploiting the analogy between immunology and sparse distributed memories: A system for clustering non-stationary data. In: Proceedings of the ICAIS (2002)
39. Kostiadis, K., Hu, H.: KaBaGe-RL: Kanerva based generalisation and reinforcement learning for possession football. In: Proceedings of the IEEE IROS (2001)
40. Ratitch, B., Precup, D.: Sparse distributed memories for on-line value-based reinforcement learning. In: Proceedings of the ECML (2004)
41. Rogers, D.: Statistical prediction with Kanerva's sparse distributed memory. Tech. Rep., NASA (1989)
42. Das, R., Whitley, D.: Genetic sparse distributed memory. In: Proceedings of the COGANN (1992)
43. Rogers, D.: Using data-tagging to improve the performance of the sparse distributed memory. Tech. Rep., NASA (1988)
44. Kaelbling, L.P.: Learning in embedded systems. Ph.D. thesis (1990)
45. El Asri, L., Lemonnier, R., Laroche, R., Pietquin, O., Khouzaimi, H.: NASTIA: negotiating appointment setting interface. In: Proceedings of the LREC (to be published) (2014)
46. Chandramohan, S., Geist, M., Pietquin, O.: Optimizing spoken dialogue management with fitted value iteration. In: Proceedings of the Interspeech (2010)
47. Fayyad, U.M., Irani, K.B.: Multi-interval discretization of continuous-valued attributes for classification learning. In: Proceedings of UAI, pp. 1022–1027 (1993)
48. Rieser, V., Lemon, O.: Learning and evaluation of dialogue strategies for new applications: empirical methods for optimization from small data sets. Comput. Linguist. **37** (2011)

Incremental Human-Machine Dialogue Simulation

Hatim Khouzaimi, Romain Laroche and Fabrice Lefèvre

Abstract This chapter introduces a simulator for incremental human-machine dialogue in order to generate artificial dialogue datasets that can be used to train and test data-driven methods. We review the various simulator components in detail, including an unstable speech recognizer, and their differences with non-incremental approaches. Then, as an illustration of its capacities, an incremental strategy based on hand-crafted rules is implemented and compared to several non-incremental baselines. Their performances in terms of dialogue efficiency are presented under different noise conditions and prove that the simulator is able to handle several configurations which are representative of real usages.

Keywords Incremental dialogue · Dialogue simulation

1 Introduction

Spoken Dialogue Systems (SDSs) still offer poor turn-taking capabilities as, in general, the listener must wait for the speaker to release the floor before reacting. Building on the recent ASR processing delay reductions [1], a new research field has emerged where SDSs offer the ability to be interrupted and to interrupt the user at any point, allowing a more natural floor management between speakers. For this purpose, the system should be able to incrementally process the user's request without waiting for the end of the user's turn [2–4], like in human-human dialogue [5]. Such systems are called *incremental dialogue systems* [6].

H. Khouzaimi (✉) · R. Laroche
Orange Labs, Châtillon, France
e-mail: hatim.enst@gmail.com

R. Laroche
e-mail: romain.laroche@orange.com

H. Khouzaimi · F. Lefèvre
CERI-LIA, University of Avignon, Avignon, France
e-mail: fabrice.lefevre@univ-avignon.fr

K. Jokinen and G. Wilcock (eds.), *Dialogues with Social Robots*,
Lecture Notes in Electrical Engineering 427,
DOI 10.1007/978-981-10-2585-3_4

Designing optimal dialogue management strategies based on hand-crafted rules only is a difficult task. Therefore data-driven methods, mostly using reinforcement learning [7], are now largely applied [8, 9]. However, these techniques require a big amount of dialogue data which are very costly to collect. A common way to deal with this problem is to build a user simulator that emulates real users' behaviours [10, 11]. The motivation behind this approach is to validate that learning algorithms scale well and that they can learn in a context that is similar to reality, while avoiding the costs caused by real corpora gathering. Moreover, even though the learnt strategies are not guaranteed to perform well with real users, they can be used as a source policy for online learning, in a Transfer Learning approach [12].

To the best of our knowledge, the only former proposition of an incremental simulated environment is in [13]. However it did not simulate the ASR instability phenomenon: as the ASR output is updated while the user is speaking, the best hypothesis is likely to be partially or totally revised. This is a critical problem in incremental dialogue processing as shown in [14, 15]. One must wait for last partial hypothesis to get stabilised before making a decision that is based on it. We show how to use information from the understanding module to emulate this mechanism.

The aim of this article is to introduce an incremental user simulator in detail, so that it can be easily replicated by the reader. This simulator integrates the ASR instability phenomenon. A simple personal agenda management task is used here for illustration, however, the ideas developed here still hold on a wider scope. Moreover, some parameters are empirically set here, nevertheless, other values can be specified in order to test different configurations.

Section 2 presents the simulated environment and Sect. 3 describes a functioning illustration experiment. Finally, Sect. 4 concludes and provides an overview of future work.

2 Simulated Environment

Our user simulator interacts with a dialogue system that is composed of a service and a Scheduler [16]. The first component handles high-level conceptual decisions whereas the second is in charge of floor management. An overview of the simulated environment architecture is given in Fig. 1. Without loss of generality we will directly describe our simulator in light of the currently implemented service (personal agenda handling), allowing to give more practical cues on the implementations of the various components including parameter's true order of magnitude. Dialogues are in French, but they are translated here into English for the sake of clarity.

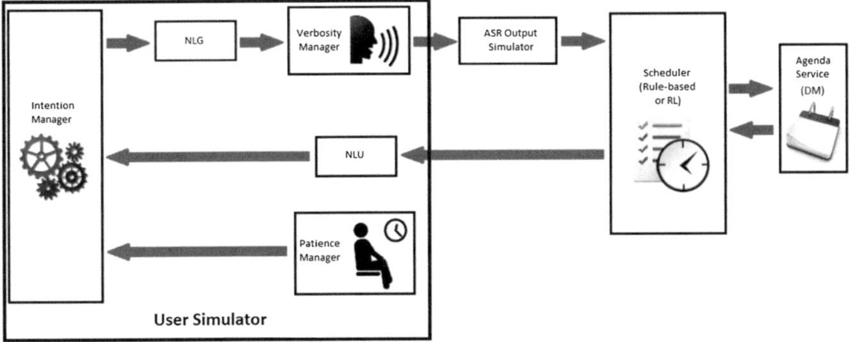

Fig. 1 Simulated environment architecture

2.1 Service

2.1.1 Dialogue Manager (DM)

The domain used for the simulation is an assistant that allows the user to add, modify or delete events in her personal agenda, as long as there is no overlap between them. In order to perform a task, the user must specify 4 slots: the action type (add, modify or delete), the event title (for example: *football game*), its date and its time window.

In its initial setting a mixed initiative strategy is used in order to gather theses pieces of information [17]. At first, the user is asked to formulate a request providing all the slot values and in the case there is still missing information (because of ASR noise or incomplete request), the system asks the user for the missing slots in turn.

2.1.2 Natural Language Understanding (NLU)

The service parses the user simulator's utterances by iteratively applying a set of rules (in the order they appear in a rules file). As new concepts appear, some of them are regrouped in further iterations giving birth to other concepts. We show the details of parsing the sentence *I want to add the event birthday party on January 6th from 9pm to 11pm* (at each step, the parsing output is provided followed by the applied rules):

1. **I want to ADD the TAG_EVENT birthday party on MONTH(January) NUMBER(6) from TIME(9,0) to TIME(11,0)**
    - add : [ADD]
    - event : [TAG_EVENT]
    - Regex(January|...|December) : MONTH($word)

- Regex([0-9]+) : NUMBER($word)
- Regex((([0-1]?[0-9])|(2[0-3]))h([0-5][0-9])?) : TIME($word)[1]

2. **I want to ADD the TAG_EVENT birthday party on DATE(6,1) WINDOW(TIME(21,0),TIME(23,0))**
   - Combine(NUMBER,MONTH) : DATE(NUMBER,MONTH)
   - Combine(from,TIME_1,to,TIME_2) : WINDOW(TIME_1,TIME_2)
3. **I want to ADD EVENT(birthday party, DATE(6,1), WINDOW(TIME(21,0),TIME(23,0)))**
   - Combine(TAG_EVENT,$x,on,DATE,WINDOW) : EVENT($x,DATE, WINDOW)
4. **I want ACTION(ADD, EVENT(birthday party, DATE(6,1), WINDOW(TIME(21,0),TIME(23,0))))**
   - Combine(ADD,EVENT) : ACTION(ADD,EVENT)

2.2 *User Simulator*

2.2.1 Intent Manager

The Intent Manager module computes the user simulator's next dialogue move at each turn. Its decision is based on an initial *dialogue scenario* defined by two lists of events: the events that are supposed to initially be present in the agenda before the beginning of the dialogue (*InitList*) and those the user simulator is supposed to add to the agenda during the dialogue (*ToAddList*). Moreover, each event is given a priority and potentially a list of alternative dates and time windows in case the main time window is not free. A dialogue scenario example is provided in Table 1 (we reduce each list to a single event to make it simpler to read, but it is not always the case).

The Intent Manager runs a recursive algorithm on a list of actions to perform: *actionList* (an action corresponds to either ADD, MODIFY, or DELETE). Initially, *actionList* is the list of the ADD actions corresponding to *ToAddList*. It starts by trying to perform the first action of the list (*actionList(0)*) and if it is successful, then this action is removed from *actionList* and the algorithm continues, until *actionList* is empty (stopping criterion). If the action could not be performed (because of a conflict with another event in the agenda for instance), the algorithm will try all the alternatives of this action (by successively adding the corresponding ADD actions in *actionList*) until one of them is successful. If it is still not the case, that means that each alternative alt_i is in conflict with a set S_i of events (most of the time, S_i

[1]This rule is kept as an indication, it is the French way of telling time and it does not apply to English.

Table 1 A dialogue scenario example

List	Title	Alternative	Date	Time window	Priority
InitList	House cleaning	0	January 6th	18:00 to 20:00	3
		1	January 7th	18:00 to 20:00	
		2	January 9th	10:00 to 12:00	
ToAddList	Birthday party	0	January 6th	18:00 to 23:00	2

contains only one element). Let m_i be the maximum priority of the events in S_i, then the algorithm focuses on the alternative indexed by $i_{min} = \arg\min m_i$.[2] If $alt_{i_{min}}$ has a lower priority than $m_{i_{min}}$ then it is discarded. Otherwise, the algorithm tries to recursively move the events in $S_{i_{min}}$ given their alternatives (the corresponding MODIFY actions are placed on top of *actionList*). The events in $S_{i_{min}}$ that do not have alternatives and the ones that could not be moved are deleted.[3]

For instance, in the previous example, the Intent Manager will reschedule *house cleaning* to *January 7th from 18:00 to 20:00* before adding the *birthday party* on *January 6th from 18:00 to 23:00.*

Finally, during a dialogue, as long as the system declares misunderstandings, asks for specific slots that it did not catch or asks for a confirmation, the intent is constant. It only changes when the system fully understands that intent and confirms that fact. Therefore, the Intent Manager is able to provide an answer to an open question (where the user is supposed to provide all the slots at once) as well as intermediate dialogue acts (specific slot questions, misunderstandings and confirmations).

2.2.2 Natural Language Generation (NLG)

At each dialogue turn, the user simulator knows which action has to be performed given the Intent Manager's output. The latter is then transformed into a simple straightforward request in natural language (NLG module), for instance: *add the event birthday party on January 6th from 18:00 until 23:00* in the case of a response to an open question, *January 6th* in the case where only the date is to be specified (same for the other slots) or *yes* after a yes/no question.

Then, the user's request is fed to an internal Verbosity Manager module. In [18], a corpus study with task-oriented dialogues showed that more than 10 % of the users' requests were off-domain and that users often repeat the same information several times in a sentence especially after a misunderstanding. The Verbosity Manager

[2]Here, the value of the priority variable designate its importance. Therefore, the higher the priority, the more important the task is.

[3]We do not claim that the Intent Manager algorithm solves the task in an optimal way. Moreover, there are some pathological examples that are not handled at all. Yet it complies with our objective to have a simple algorithm able to run realistic dialogues to study turn-taking mechanisms.

outputs an off-domain sentence with a 0.1 probability, and when the system does not understand a request, it is repeated once with a 0.7 probability and twice in the same utterance with the remaining 0.3 probability (For instance, *January 6th, I said, January 6th*). Also, sentence prefixes such as *I would like to* and suffixes like *please* may be added.

2.2.3 Duration Computation and Patience Manager

To evaluate a dialogue strategy in the simulated environment, we estimate the mean dialogue duration as well as the task completion ratio. At the end of a dialogue, the total number of words is calculated and a speech rate of 200 words per minute is assumed to compute the dialogue duration [19]. A silence of 1 s is assumed at each regular system/user transition and a 2 s silence is assumed the other way round. In the case of interruptions and accurate end-point detection, no silence is taken into account.

The task completion rate is the ratio between the number of dialogues where the user did not hang up before a system's bye and the total number of dialogues. The user hangs up if the dialogue lasts for too long. For each new dialogue, the patience duration threshold is computed as $T_{max} = 2\mu_{pat} sigmoid(X)$ where X is sampled from a standard Gaussian distribution and μ_{pat} is the mean threshold ($\mu_{pat} = 180\,\text{s}$). The user hangs up as soon as the dialogue duration reaches that threshold. Real users' patience is complex and very difficult to model, hence, this approach is only a rough empirical approximation among other possibilities.

2.3 ASR Output Simulator

The output of the Verbosity Manager is fed to the ASR Output Simulator word by word. Each step is called a *micro-turn*. A scrambling function is used to simulate errors by replacing incoming words. The frequency of these replacements is following the Word Error Rate (WER) parameter which determines the noise level. Once a word w^* is elected for scrambling, the replacement acts as follows: with a 0.7 probability, a random word is taken from a dictionary (different from w^*), with a 0.15 probability, a new word is added to w^* and with the remaining 0.15 probability, the word is deleted (like the other parameters specified in this paper, the values are empirical and they can be changed to model different configurations). At time t, the ASR output is an N-Best (best recognition hypotheses with their confidence scores) corresponding to all the words that have been pronounced so far $\{(s_1^{(t)}, hyp_1^{(t)}), (s_2^{(t)}, hyp_2^{(t)}), \ldots, (s_N^{(t)}, hyp_N^{(t)})\}$. At time t+1, a new word w_{t+1}^* pops

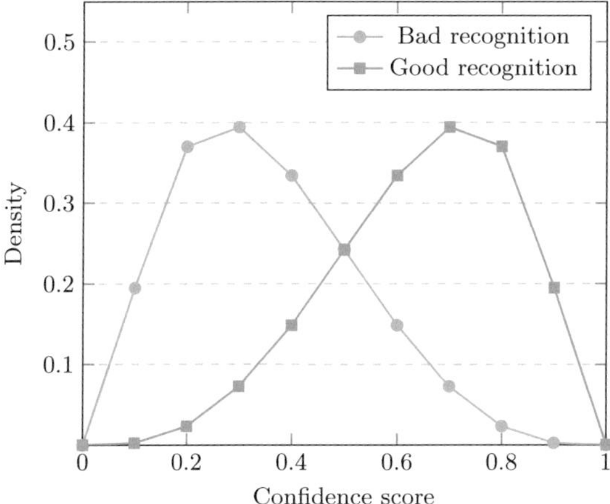

Fig. 2 ASR score sampling distribution

up as an input and its N-Best[4] is calculated as follows (the best ASR hypothesis for the word w_t^* is called w_t):

1. Randomly determine whether w_{t+1}^* is among the N-Best with a specific probability. A parameter called INBF in [0, 1] is used (In N-Best Factor, here set to 0.7): this probability is set to $(1 - \text{WER}) + \text{INBF.WER}$. If w_{t+1}^* is not among the N-Best, set the latter to a list of Scrambler samples and jump to step 4.
2. $w_{t+1} = w_{t+1}^*$ with a probability $(1 - \text{WER})$. Otherwise, w_{t+1} is a Scrambler sample.
3. If the first hypothesis is the correct one, then the other $N - 1$ hypotheses are populated with Scrambler samples. Otherwise, the correct hypothesis position is taken randomly between 2 and N.
4. If the first hypothesis is the correct one, X is sampled from a standard Gaussian centered in 1. Otherwise, it is sampled from a standard Gaussian centered in -1. s_1 is then set to $sigmoid(X)$ (in order for it to be in $]0, 1[$). Following this method, correct scores are sampled from a distribution with a mean of 0.7, and 0.3 for incorrect scores. The standard deviation is 0.18 for both (it can be changed to simulate different confidence score model qualities, the closest to 0, the most discriminating the score confidence). The two distributions are drawn in Fig. 2.
5. For each i between 2 and N, the scores are calculated in an iterative way: s_i is sampled uniformly between 0 and s_{i-1}.

In [13], the probability for outputting the correct word is taken as a basis for the confidence score. Here, a different point of view is adopted. It is an adaptation of the simple ASR introduced in [20] to the incremental case. Confidence estimation is a

[4]This N-Best corresponds to the last input word only. It is important to make the distinction between this N-Best and the one corresponding to the last partial utterance as a whole. In Fig. 3, the block *New word N-Best* is a word N-Best whereas the other three blocks are partial utterances N-Best.

complex problem and it is still a research field [21, 22]. In addition, the confidence score is not always viewed as a probability, that is why the only assumption made here is that misrecognised words have a higher chance of having a score below 0.5 whereas well recognised words have scores above that threshold. This justifies the model presented on step 4 above. The score of the whole partial utterance is the product of the scores associated with all the $\tilde{w}_t$ that constitutes it (no language model is used for the sake of simplicity). Hence, at each micro-turn, the combinations that have the best scores form the N-Best corresponding to the current partial utterance.

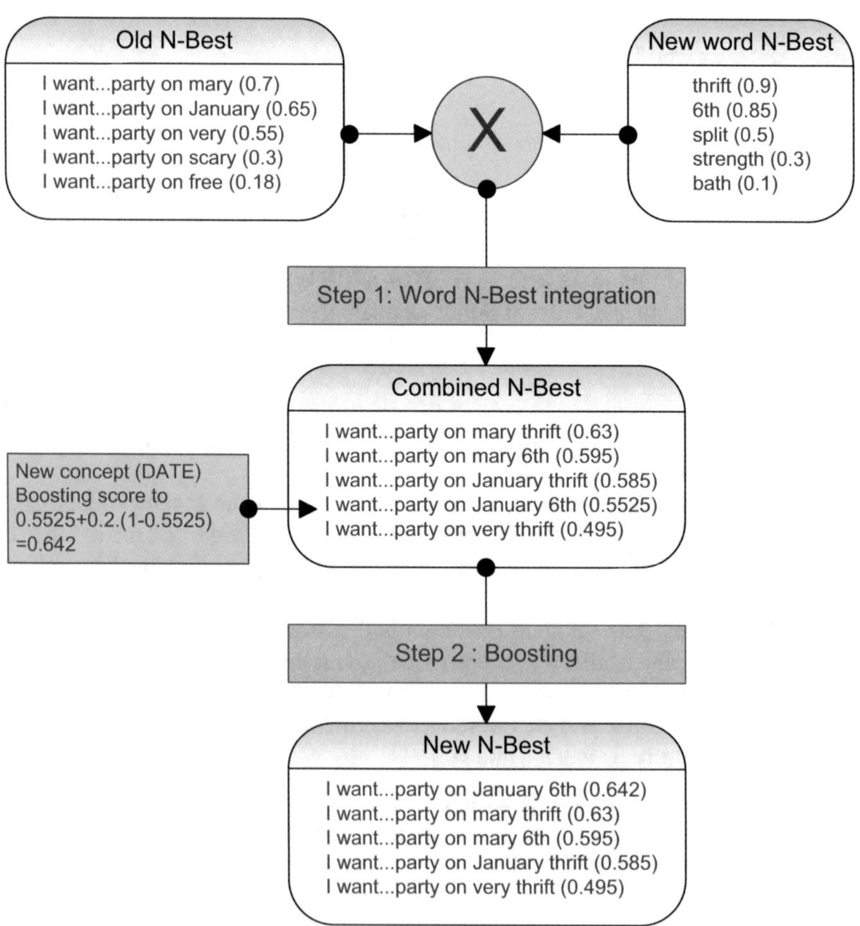

Fig. 3 Illustration of the N-Best incremental update

In real human-machine dialogue, a new increment in the ASR input can completely or partially change the best hypothesis (ASR instability [14, 15]). In order to simulate this instability, the scores of the hypotheses containing concepts that are correctly recognised by the Natural Language Understanding (NLU) module—thanks to the last ASR input—are boosted (a similar idea is used in the shift-reduce parser introduced in [23]). A parameter, called the *boost factor* (BF, here empirically set to 0.2) controls this mechanism: if a new NLU concept has just been recognised during the last micro-turn in the ith hypothesis of the N-Best, then $s_i \leftarrow s_i + BF.(1 - s_i)$. An illustration of this mechanism is provided in Fig. 3: at time t, the best hypothesis is *I want to add the event birthday party on mary* and at time $t + 1$ it becomes *I want to add the event birthday party on January 6th.*

2.4 Scheduler

2.4.1 Functioning

The Scheduler module principles have been introduced in [16] in order to transform a traditional dialogue system into an incremental one and at the same time, to offer an architecture where dialogue management in the traditional sense (dialogue act, conceptual decisions) is separated from turn-taking decisions. Here, the Scheduler is the interface between the user simulator (and the ASR Output Simulator) and the service. As mentioned previously the ASR Output Simulator sends growing partial results to the Scheduler. The latter requests the service immediately for a response and then picks an action among three options: to wait (WAIT), to convey the last service's response to the client (SPEAK) or to repeat the last word of the last stable partial request (REPEAT). It can be extended to repeating k words (REPEAT(k)). Because of the ASR instability, the Scheduler first removes some words at the end of the current hypothesis as a stability margin (SM) so that, if x_i is the ith word of the current partial utterance, the decision is based only on $(x_1, ..., x_{n_t - SM})$ (called *the last stable partial request*) where n_t is the current number of words. So, the last SM words $(x_{n_t - SM+1}, ..., x_{n_t})$ will have time to get stable before being acted on. In this work, we suppose that the speech rate is 200 words per minute [19]. So a word is pronounced in 0.3 s. In [14], a partial request that lasted for more than 0.6 s has more than 90 % chance of staying unchanged. Hence, we set $SM = 2$.

3 Illustration

3.1 Turn-Taking Strategy Example

Humans use a large set of turn-taking behaviours when talking to each other [24, 25]. In [26], we introduced a new taxonomy of these phenomena. The five turn-taking phenomena that are the most likely to improve the dialogue efficiency have been implemented in the Scheduler module of the simulated environment. They are described (the labels from our taxonomy are used) in the following.[5]

FAIL_RAW: The first phenomenon corresponds to the cases where the system interrupts the user because it could not understand the meaning of the partial utterance. Depending on the last system's dialogue act, a threshold relative to the number of words without detecting a key concept in the utterance has been set. In the case of an open question (where the systems waits for all the information needed in one request), if no action type (ADD, MODIFY or DELETE) has been detected after 6 words, a SPEAK is performed. The Scheduler waits for 3 words in the case of a yes/no question, for 4 words in the case of a date and for 6 words in the case of time windows (some concepts need more words to be detected and the user may use additional off-domain words).

INCOHERENCE_INTERP: Secondly, the system promptly reacts on partial requests that would lead to an error, not because they were not understood, but because they are in conflict with the current dialogue state. The Scheduler decides to speak as soon as the last stable partial request generates an overlap with another event in the agenda, or when the intent is to delete or modify a non existing event

FEEDBACK_RAW: When the system is not sure that it got the right message, it can repeat it. Here, when a new word is added to the partial utterance and the ratio between s_t and s_{t-1} is lower that 0.5, the Scheduler waits for SM words, and if the word is still in the partial utterance, a REPEAT action is performed.

BARGE_IN_RESP (System): Moreover, as soon as the user's sentence contains all the information needed by the system [27], the latter can immediately take the floor. That depends on the last system dialogue act as it determines which kind of NLU concept the system is waiting for. Once it is detected in the last stable partial request, the Scheduler performs a SPEAK.

BARGE_IN_RESP (User): The last phenomenon is the symmetric of the previous one [28, 29]. It is triggered directly by the user (no Scheduler decision is involved). The more familiar she is with the system, the earlier she will understand the system's intention when it has the floor, and the earlier she could interrupt. The moment where each dialogue act is interrupted is manually set.

[5]Here, we only use the best hypothesis of the N-Best. However, the others are indirectly used through the boost mechanism.

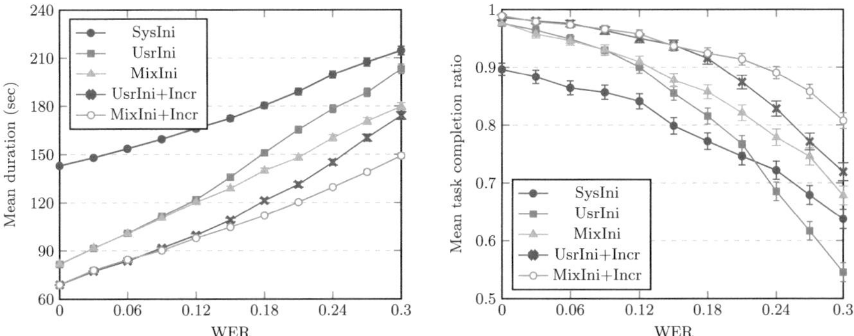

Fig. 4 Mean dialogue duration and task completion for generic strategies

3.2 Evaluation

In the first part of the experiment, the user simulator communicates directly with the service, with no Scheduler involved. Several non-incremental dialogue strategies have been simulated. In the System Initiative strategy (SysIni), the user is asked to fill the different slots (action type, description, date and time window) one by one. On the contrary, in the User Initiative strategy (UsrIni), she formulates one request that contains all the information needed to fill all the slots. In the Mixed Initiative strategy (MixIni), the user tries to fill all the slots at once, and if she fails, the system switches to SysIni to gather the missing information.

Three dialogue scenarios and different error rate levels were tested. For each strategy and each word error rate (WER), 1000 dialogues have been simulated for each scenario. Figure 4 represents the mean duration and the mean task completion, with the corresponding 95 % confidence intervals, for the different strategies and for WER varying between 0 and 0.3.

SysIni is more tedious in low noise conditions. For WER above 0.24, its task completion becomes significantly higher than UsrIni even though its mean duration is still slightly higher (the duration distribution in the case of UsrIni is centered on short dialogues with a thick tail, whereas SysIni is more centered on average length dialogues). MixIni brings the best of each world as it behaves similarly to SysIni for low WERs and performs better than both SysIni and UsrIni in noisy environments.

In the second part of the experiment, the Scheduler is added between the user simulator and the service in order to simulate incremental dialogue. UsrIni+Incr (resp. MixIni+Incr) denotes the User Initiative (resp. Mixed Initiative) strategy with incremental behaviour (ability to interrupt the user through the four first phenomena described earlier and letting him interrupt the system through the last one). Incrementality is another way of making UsrIni more robust to noise, even better than MixIni. But, in noisy environment, the difference is slighter. Finally, combining the mixed initiative strategy and incrementality is the best strategy as MixIni+Incr achieves the same performance as UsrIni+Incr for low WERs but outperforms it for noisier dialogues.

4 Conclusion and Future Work

This article describes an incremental dialogue simulator. The overview architecture as well as a detailed description of the different modules is provided, so that it can be easily replicated. Its functioning is illustrated through the implementation of an incremental strategy that replicates some human turn-taking behaviours in personal agenda management domain. Nevertheless, the framework introduced here is aimed to provide the reader with a general tool that can be adapted to different domains and different situations. It is particularly useful for testing data-driven approaches as they require an important amount of training data which is costly when it comes to spoken dialogue systems. From the experiment results it can be observed that the proposed implementation truly offers a convenient way to simulate an improved floor management capacity and with parameters which will allow to adjust its global behavior to several targeted situations with different configurations.

The simulator has already been used to build a reinforcement learning model for optimising turn-taking decisions in order to improve dialogue efficiency [17]. In an on-going work, the turn-taking phenomena are tested with human users so as to validate that the simulator can effectively reflect real conditions. In future work, the data gathered from that experiment will be used to adjust the simulator's parameters in order to better replicate human behaviours.

Acknowledgements This work is part of the FUI project VoiceHome.

References

1. Plátek, O., Jurčíček, F.: Free on-line speech recogniser based on Kaldi ASR toolkit producing word posterior lattices. In: Proceedings of the 15th Annual Meeting of the Special Interest Group on Discourse and Dialogue (SIGDIAL) (2014)
2. Allen, J., Ferguson, G., Stent, A.: An architecture for more realistic conversational systems. In: Proceedings of the 6th International Conference on Intelligent User Interfaces (2001)
3. Dohsaka, K., Shimazu, A.: A system architecture for spoken utterance production in collaborative dialogue. In: Working Notes of IJCAI 1997 Workshop on Collaboration, Cooperation and Conflict in Dialogue Systems (1997)
4. Skantze, G., Schlangen, D.: Incremental dialogue processing in a micro-domain. In: Proceedings of the 12th Conference of the European Chapter of the ACL (EACL) (2009)
5. Tanenhaus, M.K., Spivey-Knowlton, M.J., Eberhard, K.M., Sedivy, J.C.: Integration of visual and linguistic information in spoken language comprehension. Science **268**, 1632–1634 (1995)
6. Schlangen, D., Skantze, G.: A general, abstract model of incremental dialogue processing. Dialogue and Discourse **2**, 83–111 (2011)

7. Sutton, R.S., Barto, A.G.: Reinforcement Learning, An Introduction. The MIT Press, Cambridge (1998)
8. Levin, E., Pieraccini, R.: A stochastic model of computer-human interaction for learning dialogue strategies. In: Proceedings of the 5th Biennial European Conference on Speech Communication and Technology (Eurospeech) (1997)
9. Lemon, O., Pietquin, O.: Data-Driven Methods for Adaptive Spoken Dialogue Systems. Springer Publishing Company, Incorporated (2012)
10. Eckert, W., Levin, E., Pieraccini, R.: User modeling for spoken dialogue system evaluation. In: Proceedings of the IEEE Workshop on Automatic Speech Recognition and Understanding (1997)
11. Pietquin, O., Hastie, H.: A survey on metrics for the evaluation of user simulations. Knowl. Eng. Rev. (2013)
12. Taylor, M.E., Stone, P.: Transfer learning for reinforcement learning domains: a survey. J. Mach. Learn. Res. **10**, 1633–1685 (2009)
13. Selfridge, E.O., Heeman, P.A.: A temporal simulator for developing turn-taking methods for spoken dialogue systems. In: Proceedings of the 13th Annual Meeting of the Special Interest Group on Discourse and Dialogue (2012)
14. McGraw, I., Gruenstein, A.: Estimating word-stability during incremental speech recognition. In: Proceedings of the 12th Annual Conference of the International Speech Communication Association (Interspeech) (2012)
15. Selfridge, E.O., Arizmendi, I., Heeman, P.A., Williams, J.D.: Stability and accuracy in incremental speech recognition. In: Proceedings of the 12th Annual Meeting of the Special Interest Group on Discourse and Dialogue (SIGDIAL) (2011)
16. Khouzaimi, H., Laroche, R., Lefèvre, F.: An easy method to make dialogue systems incremental. In: Proceedings of the 15th Annual Meeting of the Special Interest Group on Discourse and Dialogue (SIGDIAL) (2014)
17. Khouzaimi, H., Laroche, R., Lefèvre, F.: Optimising turn-taking strategies with reinforcement learning. In: Proceedings of the 16th Annual Meeting of the Special Interest Group on Discourse and Dialogue (SIGDIAL) (2015)
18. Ghigi, F., Eskenazi, M., Torres, M.I., Lee, S.: Incremental dialog processing in a task-oriented dialog. In: Proceedings of the 15th Annual Conference of the International Speech Communication Association (Interspeech) (2014)
19. Yuan, J., Liberman, M., Cieri, C.: Towards an integrated understanding of speaking rate in conversation. In: Proceedings of the 9th International Conference on Spoken Language Processing (Interspeech-ICLSP) (2006)
20. Pietquin, O., Beaufort, R.: Comparing asr modeling methods for spoken dialogue simulation and optimal strategy learning. In: Proceedings of the 9th European Conference on Speech Communication and Technology (Eurospeech/Interspeech) (2005)
21. Jiang, H.: Confidence measures for speech recognition: a survey. Speech Commun. **45**, 455–470 (2005)
22. Seigel, M.S., Woodland, P.C.: Combining information sources for confidence estimation with crf models. In: Proceedings of the 11th Annual Conference of the International Speech Communication Association (Interspeech) (2011)
23. Nakano, M., Miyazaki, N., Hirasawa, J.I., Dohsaka, K., Kawabata, T.: Understanding unsegmented user utterances in real-time spoken dialogue systems. In: Proceedings of the 37th Annual Meeting of the Association for Computational Linguistics (ACL) (1999)
24. Clark, H.H.: Using Language. Cambridge University Press (1996)
25. Sacks, H., Schegloff, E.A., Jefferson, G.: A simplest systematics for the organization of turn-taking for conversation. Language **50**, 696–735 (1974)
26. Khouzaimi, H., Laroche, R., Lefèvre, F.: Turn-taking phenomena in incremental dialogue systems. In: Proceedings of the 2015 Conference on Empirical Methods in Natural Language Processing (2015)
27. DeVault, D., Sagae, K., Traum, D.: Incremental interpretation and prediction of utterance meaning for interactive dialogue. Dialogue and Discourse **2**, 143–170 (2011)

28. El Asri, L., Lemonnier, R., Laroche, R., Pietquin, O., Khouzaimi, H.: NASTIA: Negotiating Appointment Setting Interface. In: Proceedings of the 9th International Conference on Language Resources and Evaluation (LREC) (2014)
29. Selfridge, E.O., Arizmendi, I., Heeman, P.A., Williams, J.D.: Continuously predicting and processing barge-in during a live spoken dialogue task. In: Proceedings of the 14th Annual Meeting of the Special Interest Group on Discourse and Dialogue (SIGDIAL) (2013)

Active Learning for Example-Based Dialog Systems

Takuya Hiraoka, Graham Neubig, Koichiro Yoshino, Tomoki Toda and Satoshi Nakamura

Abstract While example-based dialog is a popular option for the construction of dialog systems, creating example bases for a specific task or domain requires significant human effort. To reduce this human effort, in this paper, we propose an active learning framework to construct example-based dialog systems efficiently. Specifically, we propose two uncertainty sampling strategies for selecting inputs to present to human annotators who create system responses for the selected inputs. We compare performance of these proposed strategies with a random selection strategy in simulation-based evaluation on 6 different domains. Evaluation results show that the proposed strategies are good alternatives to random selection in domains where the complexity of system utterances is low.

Keywords Active learning · Example-based dialog · Dialog management

1 Introduction

Example-based dialog is one of the popular methods for constructing dialog systems [1]. Example-based dialog managers store dialog examples, which consist of pairs of an example input and a corresponding system response, in a database, then generate

T. Hiraoka (✉) · G. Neubig · K. Yoshino · S. Nakamura
Nara Institute of Science and Technology, Ikoma, Japan
e-mail: hiraoka.et.al@gmail.com

G. Neubig
e-mail: neubig@is.naist.jp

K. Yoshino
e-mail: koichiro@is.naist.jp

S. Nakamura
e-mail: s-nakamura@is.naist.jp

T. Toda
Nagoya University, Nagoya, Japan
e-mail: tomoki@icts.nagoya-u.ac.jp

K. Jokinen and G. Wilcock (eds.), *Dialogues with Social Robots*,
Lecture Notes in Electrical Engineering 427,
DOI 10.1007/978-981-10-2585-3_5

system responses for input based on these dialog examples. Example-based dialog managers can be easily and flexibly modified by updating dialog examples in the database, and thus are effective in situations where (1) the domain or task of the dialog system is frequently expanded, or (2) constructing a sophisticated dialog manager a priori is difficult. In previous research, this variety of dialog managers has been used for information retrieval dialog systems [1–3], multi-domain dialog systems [4], question answering dialog systems [5], and chatter-oriented dialog systems [6–8].

Generally, in the construction of example-based dialog managers, a large number of dialog examples are required to cover a variety of inputs in the dialog. To deal with this problem, Banchs et al. [6] and Nio et al. [8] utilize dialog corpora acquired from the Web (e.g. Twitter posts or movie scripts) as dialog examples. However, generally, corpora on the Web include examples which might have a bad influence on the dialog system's performance (e.g. ungrammatical or impolite sentences), and manual screening by a human is needed. In addition, we cannot always find dialog corpora that match the dialog system's domain and style. Therefore, manual creation of dialog examples is still required in the development of practical example-based dialog managers.

In this article, we propose a method that reduces the human effort in creating dialog examples by using active learning [9] to construct an example-based dialog manager. Given (1) a prototype example-based dialog system with a small example base and (2) input logs of the prototype system, we focus on improving the example base in the prototype dialog manager by adding new dialog examples (pairs of an input and the corresponding system response) efficiently. At first, in Sect. 2, we propose an active learning framework for construction of example-based dialog managers that employs some strategy to determine which inputs should be labeled with system responses. In Sect. 3, a couple of strategies for selecting effective examples are proposed. In Sect. 4, we evaluate the proposed strategies with simulated active learning experiments.

The main contribution of this article is that, to our knowledge, this is the first work that applies active learning to construction of a database (such as dialog examples in this research) for a dialog system. In the context of dialog research, active learning has mainly been applied to construction of language understanding, [10–15] and speech recognition modules [16, 17]. Mairesse et al. [18] use active learning in construction of natural language generation module. Further, Gašic' and Young [19] use active learning to speed reinforcement learning of the dialog system policy. Unlike these related works, we apply active learning to the construction of a database (i.e. example base) for a dialog system.

2 An Active Learning Framework for Example-Based Dialog Managers

In this section, we describe example-based dialog managers, and the proposed active-learning framework.

2.1 Example-Based Dialog Managers and Their Evaluation

Example-based dialog managers utilize dialog examples to respond to input. Dialog examples $D := \{\langle u_i, s_i \rangle\}_{i=1}^{|D|}$ consist of pairs of an example input u_i (e.g., a user utterance or a system dialog state) and a corresponding system response s_i (left side of Fig. 1). Given the example base D, the dialog manager determines the system response s^* to input u by the following steps:

1. Calculate the similarity $\text{sim}(u_i, u)$ between all example inputs u_i in D, and input u. This is often defined as tf-idf weighed cosine similarity [20]:

$$\text{sim}(u_i, u) := \frac{\text{w}(u_i) \cdot \text{w}(u)}{|\ \text{w}(u_i)\ | \cdot |\ \text{w}(u)\ |} \tag{1}$$

   where the function w returns the vector representation of input (for example the frequency vector of the content words) weighted according to tf-idf.
2. Return system response s^* whose corresponding example input u^* has the highest similarity with u:

$$u^* = \underset{u_i \in D}{\arg\max}\ \text{sim}(u_i, u) \tag{2}$$

$$s^* = \{s_i | \langle u_i, s_i \rangle \in D \wedge u_i = u^*\} \tag{3}$$

The left side of Fig. 1 demonstrates how the system determines a response for the user input “That’s fun!”, calculating the similarity between this input and example user inputs in D based on Eq. (1). The similarity between “Football is fun!” (u_{54}) and the user input is 0.6, which is the highest of the example inputs in D. Therefore, based on Eqs. (2) and (3), “Seems to be fun.” (s_{54}), which is the system utterance

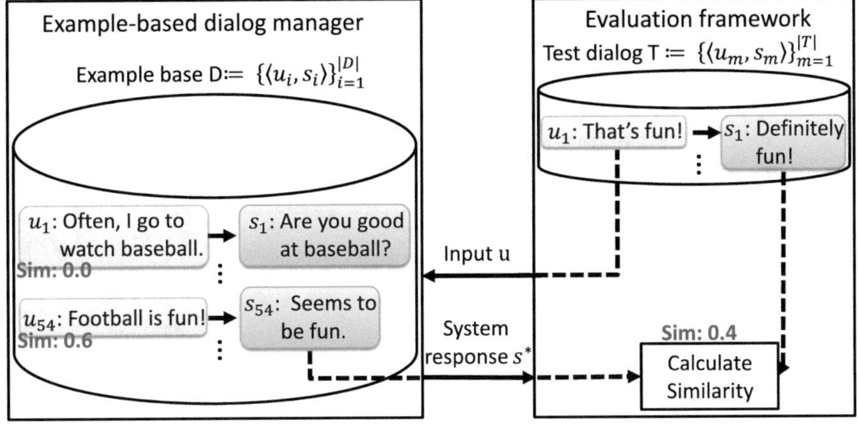

Fig. 1 Example-based dialog managers and their evaluation

corresponding to example user input u_{54}, is selected as system response s^*. This method is commonly used in the core of example-based dialog managers [1–4, 6, 8].

Given an example-based dialog manager, it is necessary to evaluate the quality of its responses. To maintain generality of our framework, we avoid using a domain specific evaluation framework (such as task-completion), and use reference-based evaluation [7, 8, 21–23] instead. In particular, we follow the evaluation framework of Nio et al. [8] and evaluate the dialog system with test examples (right side of Fig. 1). The test examples $T = \{\langle u_m, s_m\rangle\}_{m=1}^{|T|}$ consist of pairs of a test input u_m and the oracle system response s_m. Using these test examples, we calculate average similarities between the dialog system's responses and the oracle system responses for the evaluation. More concretely, given test examples T and the dialog system S, the performance p of S is calculated as follows:

$$p = \frac{1}{|T|} \sum_{m=0}^{|T|} \frac{\mathrm{w}(s_m^*) \cdot \mathrm{w}(s_m)}{|\ \mathrm{w}(s_m^*)\ | \cdot |\ \mathrm{w}(s_m)\ |}. \tag{4}$$

This is the average of cosine similarities between S's response s_m^*, calculated according to Eq. (1), and the oracle system response s_m over the test examples.[1] $|T|$ represents the total number of pairs in test set T. In the example in the right side of Fig. 1, the evaluation framework evaluates the system response to the test user input "That's fun!" (u_1). In this example, the system outputs "Seems to be fun." as its response to u_1. The similarity between "Seems to be fun." and oracle system response "Definitely fun!" is calculated according to Eq. (4).

2.2 Active Learning Framework

In this section, we propose an active learning framework for the example-based dialog managers described in Sect. 2.1. Starting from a prototype example-based dialog manager with a small number of dialog examples, the active learning problem is to improve the system as much as possible with minimal human effort. We focus on the situation where there are input logs collected by the prototype dialog system, and a human creator is required to create system responses for these inputs (Fig. 2). Therefore, given the example dialog $D := \{\langle u_i, s_i\rangle\}_{i=1}^{|D|}$ and input log $U := \{\langle u_j\rangle\}_{j=1}^{|U|}$, the goal is to select the *subset of input that yields the greatest improvement in system performance* from U to present to the human creator.

Algorithm 1 describes our active learning framework in detail. At first, we construct our initial system S with example base D, and evaluate its performance based on

[1]The experimental results of Nio et al. [8] indicate that the human subjective evaluation for naturalness and relevance of system response is correlated with the score calculated in Eq. (4).

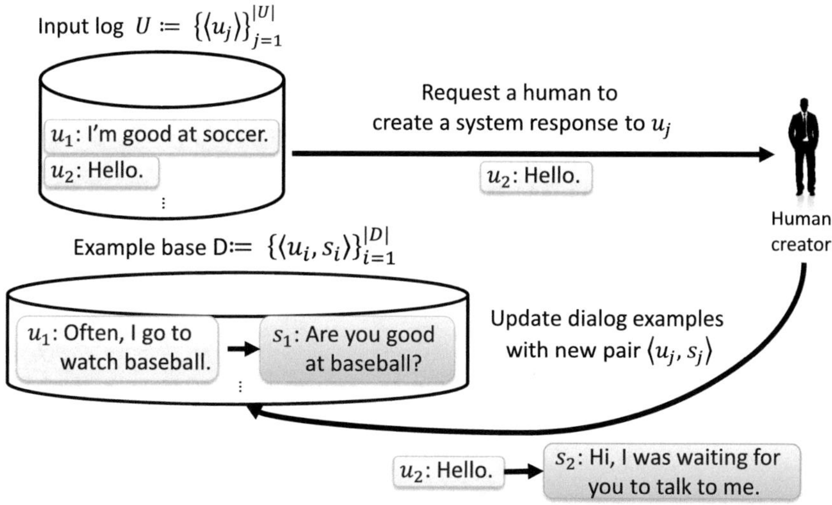

Fig. 2 Active learning for updating the example base given input logs

---

Algorithm 1 AL-EBDM

```
1: Input example base D, input log U
2: S ← constructSystem(D)
3: Evaluate the performance of S on the test data T
4: for e=1,2,... do
5:   Select k inputs {u_1, ..., u_k} from U, and request a human creator to create system response
     {s_1, ..., s_k}.
6:   Remove {u_1, ..., u_k} from U.
7:   D ← D ∪ {⟨u_1, s_1⟩, ..., ⟨u_k, s_k⟩}
8:   S ← constructSystem(D)
9:   Evaluate the performance of S on the test dialog T
10: end for
```

test data T using Eq. (4) (line 2 and line 3). Then, we continue to incrementally update dialog examples for S until training epoch e reaches a particular threshold (from line 4 to line 10). Lines 5 and 6 select and remove k inputs from U (i.e. $\{u_1, ..., u_k\}$), and request a human creator to create system responses for these inputs (i.e. $\{s_1, ..., s_k\}$). The strategies for selecting $\{u_1, ..., u_k\}$ are proposed in Sect. 3. Then, Lines 7 and 8 update example base D by adding created example pairs $\{\langle u_1, s_1 \rangle, ..., \langle u_k, s_k \rangle\}$, and reconstruct S with the updated dialog examples. Finally, Line 9 evaluates the performance of the updated S on the test data T.

3 Input Selection Strategies

The performance of Algorithm 1 heavily depends on how we select input from U to present to human creators (i.e. Line 5). In this section, we propose 2 strategies (**DUnc** and **PUnc**) for selecting effective input from U to present to a human creator. These methods are based on the intuition that we can expect that covering examples that are not yet well covered by the database will compensate for the current database's weaknesses. This strategy will improve the dialog system's ability to respond to a variety of inputs. This intuition is well known as uncertainty sampling in general active learning research [24].

Our proposed strategies select *the user input that is different from the dialog examples in D*. In this strategy, the similarity between (1) input u_j in U and (2) example input u_i in D are calculated as the score of u_j:

$$s(u_j) = 1 - \max_{u_i \in D} \frac{\mathrm{w}(u_j) \cdot \mathrm{w}(u_i)}{\mid \mathrm{w}(u_j) \mid \cdot \mid \mathrm{w}(u_i) \mid}. \tag{5}$$

After all of u_j in U are scored according to Eq. (5), k inputs in U are presented to the human creator according to two different sampling methods:

DUnc: samples the k inputs with the highest score.
PUnc: samples k inputs according their probabilities calculated by:

$$\frac{s(u_j)}{\sum_{u_k \in U} s(u_k)}. \tag{6}$$

Queries selected by DUnc are strongly biased by Eq. (5) because of deterministic sampling. To examine the effect of this bias, we additionally introduce the strategy based on probabilistic sampling (i.e., PUnc), which falls halfway between random sampling and the deterministic strategy.

4 Experiments

4.1 Experimental Setup

To evaluate the input selection strategies proposed in Sect. 3, we performed an experimental evaluation using a simulated active learning setup.[2] For simulation, we divided the dialog corpora into initial dialog examples D_0, simulated input logs U and test

[2]Source files to replicate these experiments are available: https://github.com/TakuyaHiraoka/Active-Learning-for-Example-based-Dialog-Systems.

dialogs T. Each set has oracle pairs of inputs and system responses. Each selection strategy selects the most appropriate input u from input log U to update the dialog manager, and the human creator simulator returns the oracle system response corresponding to the given input. For evaluation, in addition to 2 strategies proposed in Sect. 3, we use a **Random** baseline, which randomly selects input to present to the human creator simulator. We compared these 3 strategies based on system performance calculated with similarity between the system response and oracle system response (i.e. Eq. (4) in Sect. 2). Starting with different initial dialog examples D_0, we repeated this evaluation 50 times for each strategy, and used the average of system performance for each number of annotated examples. Note that, in simulations, if input in U already overlaps an example input in example database D, the overlapping input is deleted from U.

To ensure the portability of the proposed strategies, we prepared 6 simulation domains, which are based on different dialog corpora open to the public:

BusInfo: Human-system dialog for bus information retrieval [25].
Restaurant: Human-system dialog in restaurant information retrieval [26].
Tourist: Human-system dialog in tourist information retrieval [27].
ChatBot: Dialogs between humans and a Japanese chatbot [28].
CleverBot: Dialogs between humans and an English chatbot.[3]
Movie: Human-human dialog in movies and television [8, 29].

We calculated several properties of each domain as shown in Table 1. First we calculate complexity of the domain based on the entropy of the input $Hu_d = E[-log_2 P(u_d)]$ and the system response $Hs_d = E[-log_2 P(s_d)]$, where u_d represents all inputs in domain d, s_d represents all system responses in d, and P represents uni-gram probability. These entropies quantify how much variety exists in the input or system response appearing in each domain. In this research, we use Kylm[4] to calculate each entropy. Furthermore, we describe input types (2nd column of Table 1) in each domain. In the domain annotated with SDF, semantic and discourse features[5] are used as the input to the system. In addition, in the domain annotated with BOW, a bag of words in user utterance is used as the input to the system.

4.2 Experimental Results and Additional Analysis

Figure 3 shows the learning curves of each strategy, where the system performances of each strategy is plotted at each training epoch. In addition, to summarize the performance of each strategy, we define the improvement of the uncertainty-based

[3]We use dialog logs collected from http://www.cleverbot.com/j2convbydate-page1.

[4]http://www.phontron.com/kylm/.

[5]In our research, we use previous dialog act and slot filling status [4] as semantic and discourse features.

Table 1 Input type, number of examples, and entropy of each simulation domain

Domain	Input type	$\|D_0\|$	$\|T\|$	$\|U\|$	Hu_d	Hs_d
BusInfo	SDF	661	7000	6175	8.661	7.691
Restaurant	SDF	505	6500	6488	6.397	7.371
Tourist	SDF	418	8000	7534	6.518	7.832
ChatBot	BOW	500	6500	6363	6.297	6.863
CleverBot	BOW	248	9500	9593	8.677	8.415
Movie	BOW	753	8500	8540	8.39	8.402

strategies over Random in domain d as average ratio of performance of uncertainty-based strategies to that of Random:

$$AR_{d,sys} = \frac{1}{E_d} \sum_{e=1}^{E_d} \frac{p_{d,sys}}{p_{d,ran}}, \tag{7}$$

where E_d represents the maximum training epoch at domain d, $p_{d,sys}$ represents the system performance of the selected strategy at d according to Eq. (4), and $p_{d,ran}$ represents the score of Random calculated in the same manner as $p_{d,sys}$.

The experimental result in Fig. 3 indicate that *proposed strategies based on uncertainty (DUnc, PUnc) can be a good alternative to Random in some domains.* Performances of DUnc were better than those of Random in some domains (BusInfo and ChatBot), and especially its performance in BusInfo was much better than others. In addition, performance of PUnc was equal to or better than Random in all domains except for CleverBot. One of the reasons why these strategies outperformed Random in some domains is that these strategies tend not to select redundant inputs as queries to the dialog creator simulator. For example, in ChatBot, Random selected "(By the way, what are you doing now?)" and (What are you doing now?)". These inputs are not perfectly overlapped, but not very different, and thus we can not expect system performance to increase efficiently by creating system responses for these inputs. Note that the performance of each strategy is dependent on the domain, and these are not necessarily better than that of Random.

Additional analysis indicated that *the DUnc and PUnc strategies can be expected to achieve better performance than Random in domains where the complexity of the system utterance is low.* This was made clear by a correlation analysis between (1) the improvement of uncertainty based strategies according to Eqs. (7) and (2) properties of each domain (described in Table 1). The result of this analysis (Fig. 4) indicated that there is a strong correlation between the improvement of DUnc and PUnc from Random and entropy of system response Hs_d. If entropy of the system response is high, the system response may be different even if inputs are similar. For example, in CleverBot where the entropy of the system response is high, the system

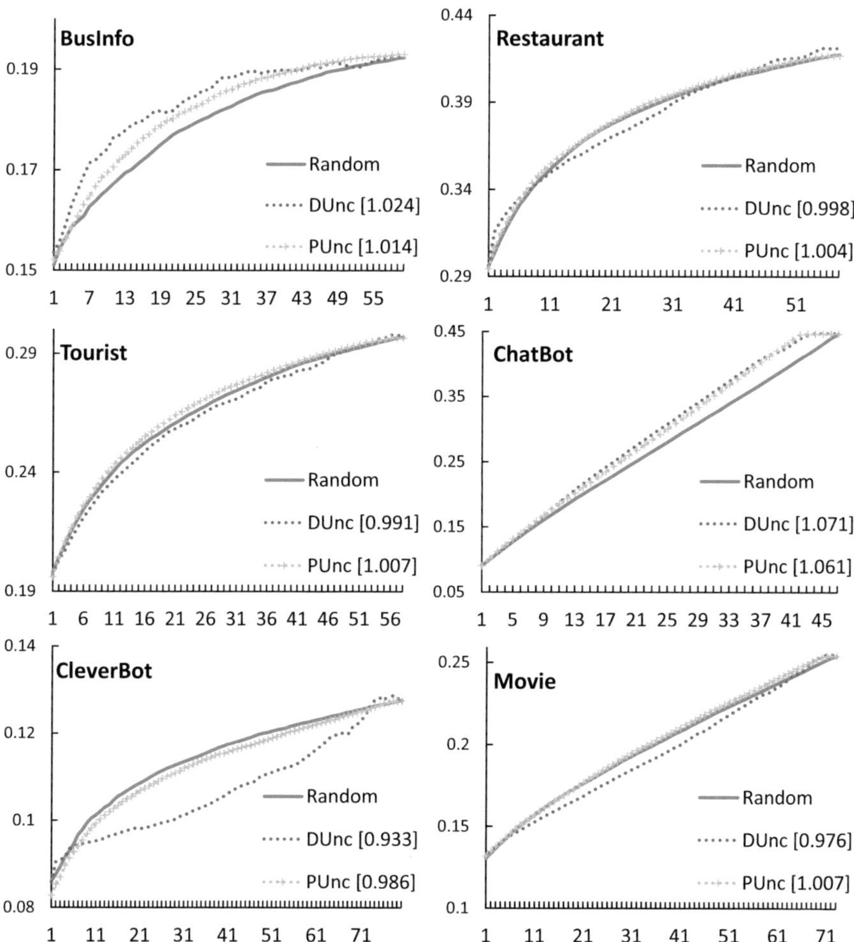

Fig. 3 Learning curves for each input selection strategy in six domains. In each figure, the vertical axes indicate similarity between system responses and oracle responses according to Eq. (4), and the horizontal axes indicate number of training epochs (i.e. e in Algorithm 1). In each training epoch 100 inputs are selected (i.e., $k = 100$). Values at the right side of each label in the legend represent the improvement of uncertainty based strategies from Random according to Eq. (7)

response to input "You're a stupid" is "No you are." whereas the response to the input "You are stupid bot" is "We are the robots". In such a case, considering only information of the input is not enough, and information about the system response is also required to make the proposed strategies be a good alternative to Random.

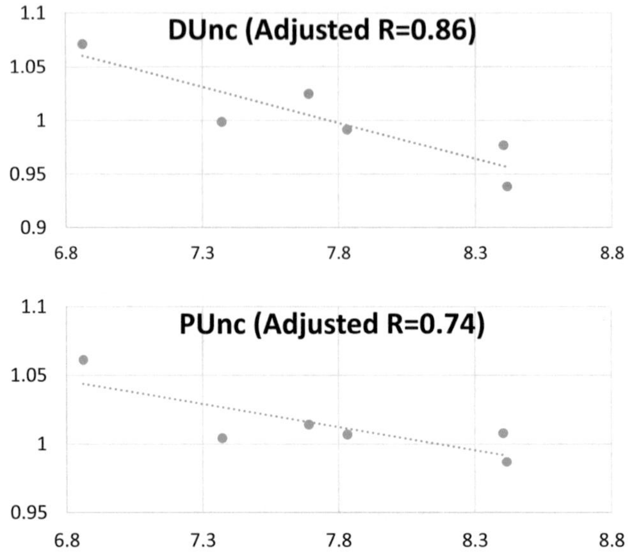

Fig. 4 Correlation between performance $AR_{d,sys}$ and entropy of the system response Hs_d. The top figure shows the case of $sys = DUnc$, and the bottom figure represents the case of $sys = PUnc$. In each figure, the vertical axis represents $AR_{d,unc}$, the horizontal axis represents Hs_d, and each dot represent the tuple $(Hs_d, AR_{d,unc})$ for d

5 Conclusion

In this article, we applied active learning to construct example-based dialog managers efficiently. To reduce human effort in creating example bases, we proposed an active learning-based framework, and proposed strategies for selecting input to present to a human creator to create dialog examples. Then, we performed evaluation based on simulation in 6 different domains. Experimental results and analysis indicated that (1) proposed strategies based on uncertainty can be a good alternative to Random in some domains and (2) these strategies (DUnc, PUnc) can be expected to achieve better performance than Random in domains where the uncertainty of the system utterance is low.

As future work, we plan to propose query selection strategies for domains where system responses are complex, and evaluate with a real human creator. Furthermore, we plan to expand our active learning framework to be more general and cover other response generation frameworks [21–23].

Acknowledgements Part of this research was supported by JSPS KAKENHI Grant Number 24240032, and the Commissioned Research of National Institute of Information and Communications Technology (NICT), Japan.

References

1. Murao, H., Kawaguchi, N., Matsubara, S., Inagaki, Y.: Example-based query generation for spontaneous speech. In: Proceedings of ASRU (2001)
2. Murao, H., Kawaguchi, N., Matsubara, S., Yamaguchi, Y., Inagaki, Y.: Example-based spoken dialogue system using WOZ system log. In: Proceeding of SIGDIAL (2003)
3. Nisimura, R., Nishihara, Y., Tsurumi, R., Lee, A., Saruwatari, H., Shikano, K.: Takemaru-kun: Speech-oriented information system for real world research platform. In: Proceedings of LUAR (2003)
4. Lee, C., Jung, S., Kim, S., Lee, G.G.: Example-based dialog modeling for practical multi-domain dialog system. Speech Commun. (2009)
5. Xue, X., Jeon, J., Croft, W.B.: Retrieval models for question and answer archives. In: Proceeding of SIGIR (2008)
6. Banchs, R.E., Li, H.: Iris: a chat-oriented dialogue system based on the vector space model. In: Proceedings of ACL (2012)
7. Nio, L., Sakti, S., Neubig, G., Toda, T., Nakamura., S.: Improving the robustness of example-based dialog retrieval using recursive neural network paraphrase identification. In: Proceedings of SLT (2014)
8. Nio, L., Sakti, S., Neubig, G., Toda, T., Nakamura, S.: Utilizing human-to-human conversation examples for a multi domain chat-oriented dialog system. Trans. IEICE (2014)
9. Settles, B.: Active learning literature survey. Computer Sciences Technical Report, vol. 1648, University of Wisconsin-Madison (2011)
10. Ghigi, F., Tamarit, V., Martinez-Hinarejos, C.D., Benedi, J.M.: Active learning for dialogue act labelling. Lect. Notes Comput. Sci. (2011)
11. Gotab, P., Bechet, F., Damnati, G.: Active learning for rule-based and corpus-based spoken language understanding models. In: Proceedings of ASRU (2009)
12. Hakkani-Tür, D., Riccardi, G., Tür, G.: An active approach to spoken language processing. ACM Trans. Speech Lang. Process. (2015)
13. Tür, G., Hakkani-Tür, D., Schapire, R.E.: Combining active and semi-supervised learning for spoken language understanding. Speech Commun. (2005)
14. Tür, G., Schapire, R.E., Hakkani-Tür, D.: Active learning for spoken language understanding. In: Proceedings of ICASSP (2003)
15. Williams, J.D., Niraula, N.B., Dasigi, P., Lakshmiratan, A., Suarez, C.G.J., Reddy, M., Zweig, G.: Rapidly scaling dialog systems with interactive learning. In: Proceedings of IWSDS (2015)
16. Riccardi, G., Hakkani-Tür, D.: Active learning: theory and applications to automatic speech recognition. Speech Audio Process. (2005)
17. Yu, D., Varadarajan, B., Deng, L., Acero, A.: Active learning and semi-supervised learning for speech recognition: a unified framework using the global entropy reduction maximization criterion. Comput. Speech Lang. (2010)
18. Mairesse, F., Gašić, M., Jurčíček, F., Keizer, S., Thomson, B., Yu, K., Young, S.: Phrase-based statistical language generation using graphical models and active learning. In: Proceedings of ACL (2010)
19. Gašić, M., Young, S.: Gaussian processes for POMDP-based dialogue manager optimisation. Audio Speech Lang. Process. (2014)
20. Leskovec, J., Rajaraman, A., Ullman, J.D.: Mining of Massive Datasets. Cambridge University Press (2014)
21. Ritter, A., Cherry, C., Dolan, W.B.: Data-driven response generation in social media. In: Proceedings of EMNLP (2011)
22. Shin, A., Sasano, R., Takamura, H., Okumura, M.: Context-dependent automatic response generation using statistical machine translation techniques. In: Proceedings of NAACL (2015)
23. Sordoni, A., Galle, M., Auli, M., Brockett, C., Mitchell, Y.M., Nie, J.Y., Gao, J., Dolan, B.: A neural network approach to context-sensitive generation of conversational responses. In: Proceedings of NAACL (2015)

24. Lewis, D.D., Gale, W.A.: A sequential algorithm for training text classifiers. In: Proceedings of SIGIR (1994)
25. Williams, J., Raux, A., Ramachandran, D., Black, A.: The dialog state tracking challenge. In: Proceedings of SIGDIAL (2013)
26. Henderson, M., Thomson, B., Williams, J.: The second dialog state tracking challenge. In: Proceedings of SIGDIAL (2014)
27. Henderson, M., Thomson, B., Williams, J.: The third dialog state tracking challenge. In: Proceedings of SLT (2014)
28. Higashinaka, R., Funakoshi, K., Araki, M., Tsukahara, H., Kobayashi, Y., Mizukami, M.: Towards taxonomy of errors in chat-oriented dialogue systems. In: Proceedings of SIGDIAL (2015)
29. Nio, L., Sakti, S., Neubig, G., Toda, T., Nakamura., S.: Conversation dialog corpora from television and movie scripts. In: Proceedings of O-COCOSDA (2014)

Question Selection Based on Expected Utility to Acquire Information Through Dialogue

Kazunori Komatani, Tsugumi Otsuka, Satoshi Sato and Mikio Nakano

Abstract We have been developing a method for a dialogue system to acquire information (e.g., cuisines of unknown restaurants) through dialogue by asking questions of the users. It is important that the questions are concise and concrete to prevent users from being annoyed. Our method selects the most appropriate question on the basis of expected utility calculated for four types of question: Yes/No, alternative, 3-choice, and Wh- questions. We define utility values for the four types and also derive the probability representing how likely each question is to contain a correct cuisine. The expected utility is then calculated as the sum totals of their products. We empirically compare several ways to integrate two previously proposed basic confidence measures (CMs) when deriving the probability for each question. We also examine the appropriateness of the utility values through questionnaires administered to 15 participants.

Keywords Dialogue system · Knowledge acquisition through dialogue · Expected utility

---

K. Komatani (✉)
Osaka University, Ibaraki, Osaka 567-0047, Japan
e-mail: komatani@sanken.osaka-u.ac.jp

T. Otsuka · S. Sato
Nagoya University, Nagoya, Aichi 464-8603, Japan
e-mail: t_ootuka@nuee.nagoya-u.ac.jp

S. Sato
e-mail: ssato@nuee.nagoya-u.ac.jp

M. Nakano
Honda Research Institute Japan Co. Ltd., Wako, Saitama 351-0188, Japan
e-mail: nakano@jp.honda-ri.com

K. Jokinen and G. Wilcock (eds.), *Dialogues with Social Robots*, Lecture Notes in Electrical Engineering 427,
DOI 10.1007/978-981-10-2585-3_6

1 Introduction

One of the most pressing issues of dialogue systems is how to handle words that are not in the system's knowledge. This is known as the out-of-vocabulary (OOV) word problem. Although the vocabulary size of an automatic speech recognizer (ASR) has increased and even the named entities (NEs) are correctly extracted thanks to machine learning techniques, a system cannot respond appropriately to user utterances containing words that are not in its knowledge, e.g., its backend relational database. For example, in the restaurant search task shown in the left side of Fig. 1, the system can only respond with "Sorry, I do not know that" to a user utterance that includes "Botan-tei", which is not in the system's database. A large-scale knowledge base has been constructed collaboratively [1] and there have been some research enhancing system knowledge by using public data such as Linked Open Data (LOD) [2], but not all data are included there. In particular, it is difficult to include words used in local communities in such a large database. These may include newly opened local store names and nicknames for stores or persons. The current solution is for the system developer to manually update the target database. However, this is problematic because it increases the system's maintenance costs and the developer needs to check the system log frequently.

In our earlier work, we proposed a method to acquire such information by having the system ask users questions during dialogues [3]. Our target has been the cuisines of unknown restaurants that are not in the system's target database (DB) in a restaurant search task. By estimating and obtaining the cuisines of an unknown restaurant, the system will also be able to recommend similar restaurants as well acquire the new information, as shown in the right side of Fig. 1.

In this work, we propose a method to select appropriate questions from four types of question on the basis of expected utility [4]. Our conjecture is that a question is appropriate if it is concise and cuisines in it seem correct. The latter is requisite because cuisine candidates are automatically estimated. We first prepare four question templates and set utility values for each type by considering user impressions of them. We also calculate the probability representing how reliable cuisine estimation results in each question are. We calculate the expected utility as the sum of the products

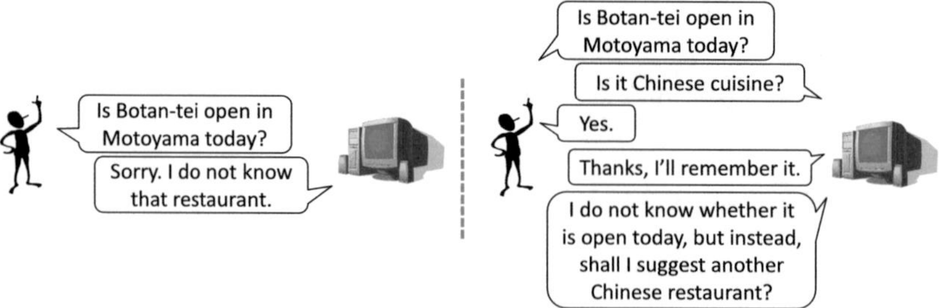

Fig. 1 Dialogue examples in current systems and enabled by our proposed method

between the utility value and the probability and then select the question type having the maximum expected utility.

Our current focus is how to select appropriate questions because annoying questions make users stop using the system. A related study focused on which kind of knowledge can be obtained through dialogue [5]. Some other studies tried to estimate the named entity class of an unknown word [6, 7]. This can be used before generating our questions because we assume the system already knows that the unknown word is a restaurant name.

2 Appropriate Questions for Acquiring Information

A question to obtain information should satisfy two requirements: (1) be as concise and concrete as possible and (2) not contain incorrect information.

A concise question will make the user less annoyed when it is displayed as a spoken utterance, and furthermore can narrow down the subsequent user responses. For example, the Yes/No question "Is it Chinese cuisine?" with only one specific cuisine is shorter than an alternative or 3-choice question, as shown in the right side of Fig. 2. It can narrow down the user's response to a positive or negative expression such as "Yes" or "No". This is a favorable feature to correctly understand the user's response. In contrast, a Wh-question such as "What cuisine is it?" does not narrow down the following user responses, thus, increasing the risk that the user may use unknown words like "Sichuan", as shown in the left side of Fig. 2. Our current database contains "Chinese", but does not contain "Sichuan".

On the other hand, there is another risk that no correct cuisine is contained in the system's question when the question is too concise by limiting the number of cuisines in it. This is because the candidate cuisines in the system's question may be erroneous.

We propose a question selection method by considering the balance between conciseness and the risk of no correct cuisine in it. Here, a question is selected from the four types listed in Table 1. We do not prepare questions with four or more choices

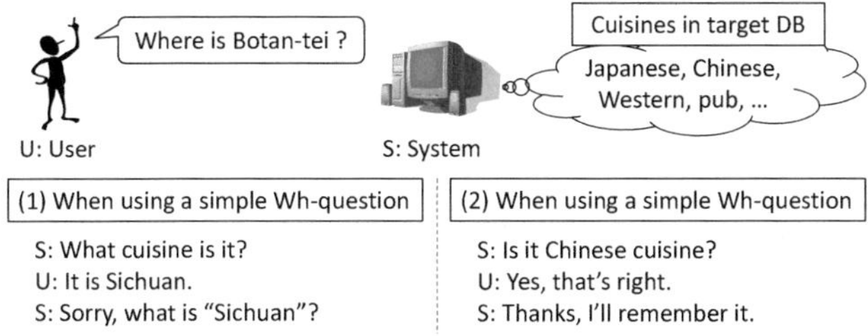

Fig. 2 Dialogue example (restaurant name is "Botan-tei" and its cuisine is Chinese)

Table 1 Four question types

Question types	No. of displayed cuisines	Example question
Yes/no	1	Is it Chinese cuisine?
Alternative	2	Which cuisine is it, Chinese or Japanese?
3-choice	3	Which cuisine is it, Chinese, Japanese, or pub?
Wh	–	What cuisine is it?

because this would be too annoying for the user, especially when they are provided via speech output. A Yes/No, alternative, and 3-choice question contains one, two, and three cuisine candidates, respectively. Wh-questions have no candidates.

3 Question Selection Based on Expected Utility

Here, we introduce our method to select a question from the four question types listed in Table 1 on the basis of expected utility. The notions, utility and expected utility, are those used in Game Theory [4]. First, in Sect. 3.1, we describe utility, which represents to which degree the user feels a question is intelligent or annoying. Then, we explain the probability representing how much each question seems to contain a correct cuisine in Sect. 3.2. In Sect. 3.3, we calculate expected utility and then select the question type having the maximum expected utility.

3.1 Utility for Each Question Type

Here, we represent the degree to which a question makes the user feel intelligent as utility. The utility is set for each question type and whether or not it contains correct information. In our restaurant task, a question is regarded as containing correct information if it contains the correct cuisine type of a target restaurant. For example, an alternative question is correct if either of the two candidate cuisines is correct.

We set positive and negative utility values for correct and wrong questions, respectively. $U^{\{+,-\}}_{x\in\{1,2,3,\mathrm{Wh}\}}$ in Table 2 denotes utility values. Here, U denotes the initial character of utility, $+$ and $-$ denote whether the question is correct or not, and x denotes the number of cuisines in the question. That is, $x = 1$, 2, and 3 correspond to a Yes/No, alternative, and 3-choice question, respectively. For example, U^-_2 represents the utility value for a wrong alternative question.

The values are set to $U^+_1 \geq U^+_2 \geq U^+_3 \geq U^+_{\mathrm{Wh}} = 0$ for correct questions and $U^-_{\mathrm{Wh}} = 0 \geq U^-_1 \geq U^-_2 \geq U^-_3$ for wrong questions. This is based on our assumption that a more concise question is better when it is correct, and a longer question

Table 2 Expected utility for each question type

Question type	Utility of correct question	Utility of wrong question	Probability	Expected utility
Yes/no	U_1^+	U_1^-	P_1	$U_1^+ P_1 + U_1^- (1 - P_1)$
Alternative	U_2^+	U_2^-	P_2	$U_2^+ P_2 + U_2^- (1 - P_2)$
3-choice	U_3^+	U_3^-	P_3	$U_3^+ P_3 + U_3^- (1 - P_3)$
Wh	U_{Wh}^+	U_{Wh}^-	–	0

$P_x = \sum_{i=1}^{x} CM(c_i)$, where c_i denotes a cuisine having the i-th largest $CM(c_i)$

$U_{\text{Wh}}^+ = U_{\text{Wh}}^- = 0$

is worse when it is wrong. Since a Wh-question contains no specific cuisine, we set $U_{\text{Wh}}^+ = U_{\text{Wh}}^- = 0$ as a standard.

3.2 *Probability Representing Content of Questions*

Here, we explain how to determine the probability that a question contains the correct cuisine. The probability is derived by confidence measures (CMs) [3] that represent how reliable a cuisine estimation result is for an input unknown restaurant name r. CMs are calculated for all cuisines c in the target database for retrieval. Their values $CM(r, c)$ are normalized so that their sum becomes 1. They can thus be regarded as a posterior probability when an unknown restaurant name r is given. Hereafter, we denote $CM(r, c)$ simply as $CM(c)$ for short.

We define this probability as the sum of all CM values for cuisines contained in the question. Cuisines used in a question are selected in descending order of $CM(c)$. More specifically, when we denote c_i as a cuisine having the i-th largest $CM(c_i)$, the probability that question x contains the correct cuisine is calculated by Eq. (1).

$$P_x = \sum_{i=1}^{x} CM(c_i) \tag{1}$$

For example, the probability for an alternative question containing c_1 and c_2 can be derived as $CM(c_1) + CM(c_2)$.

In this paper, we integrate two basic CMs [3] and use it to calculate the probability, as discussed in Sect. 4.2. The two basic CMs are $CM_D(c)$ and $CM_W(c)$, which respectively use character distribution in the target database and occurrence frequency of each cuisine on the Web [3] to estimate a cuisine from a given restaurant name.

3.3 Calculating Expected Utility

Next, we select the question type having the maximum expected utility. The expected utility is calculated as the sum of products of utility and its probability for each question type. It can thus be used for predicting how much utility will be obtained when each question type is selected.

More specifically, the expected utility is calculated as follows. An alternative question is used here as an example. First, we obtain the probability that an alternative question containing two cuisines c_1 and c_2 is correct as $P_2 = \sum_{i=1}^{2} CM(c_i)$. Thus, the probability that this question is wrong is $1 - P_2$. By using these two probabilities, its expected value $U_2^+ P_2 + U_2^-(1 - P_2)$ can be calculated as its expected utility.

Using the expected utility, we select the question type having the maximum expected utility by

$$\text{(Selected question type)} = \operatorname*{argmax}_{x \in \{1,2,3,\text{Wh}\}} \{U_x^+ P_x + U_x^-(1 - P_x)\}. \tag{2}$$

4 Empirical Setting of Parameters

Here, we empirically discuss appropriate utility values and how to derive the probability. In this section, we virtually assume that one restaurant in the target database is unknown and generate a question to ask its cuisine. Then we can determine the accuracy of whether the question contains its correct cuisine by consulting the database. We also calculate the total sum of obtained utility for the whole database.

The flow of generating a question is depicted in Fig. 3. First, a restaurant name is selected as an input and its two basic confidence measures CM_D and CM_W are calculated [3]. Here, we calculated CM_D by tenfold cross validation in the whole database since it needs training data for its classifier, while no training data is needed to calculate CM_W. Then, we obtain the probability that each question contains the correct cuisine by integrating the two basic confidence measures and adding them in descending order according to its question type. The integration method is discussed in Sect. 4.2. We can then obtain the question having the maximum expected utility. The utility values are discussed in Sect. 4.1.

As the target database, we used our restaurant database in Aichi Prefecture, which contains 1,656 pairs of restaurant names and their cuisines [3].

4.1 Setting Utility Values

We discuss appropriate utility values on the basis of the following two criteria. First is the number of wrong questions. It is better that such questions are fewer. Second is the number of Yes/No and Wh-questions. A Wh-question is safe, i.e., does not

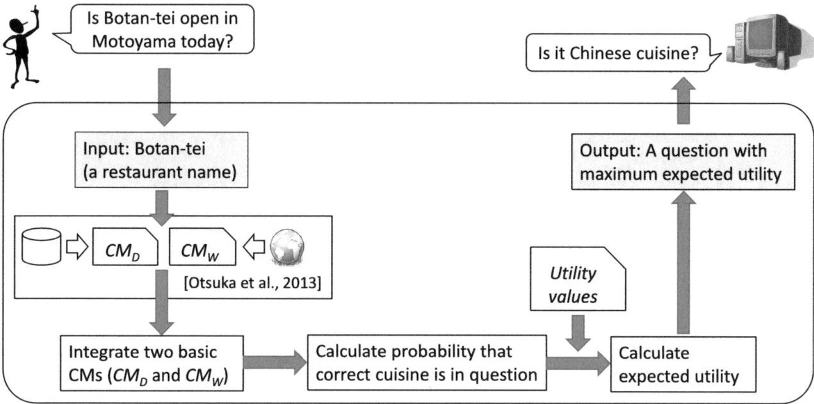

Fig. 3 Process flow to generate a question asking a restaurant's cuisine

contain erroneous cuisine, but it cannot narrow down the user's following responses. It also does not use cuisine estimation results even when the results seem reliable. Thus, we regard too many Wh-questions as inappropriate.

Here, we use five utility value patterns from (i)–(v), as shown in Table 3. These were manually determined from several value patterns. Investigating whether there are better patterns is among the future work. Pattern (i) represents that the user will feel more annoyed when more wrong cuisines are displayed. Specifically, values are equal when the correct cuisine is contained and the ones are different for wrong questions. Pattern (ii) represents that the user will feel the system is more intelligent when more concise questions are displayed, having equal values for wrong questions and different values for correct questions. Pattern (iii) considers the both cases above, i.e., the utility values are the sum of patterns (i) and (ii). In patterns (iv) and (v), utility values for wrong questions are decreased and increased from pattern (iii), respectively.

Table 4 shows the number of selected question types for each case of correct and wrong ones. Their totals and the numbers of Wh-questions are also shown at the bottom. Here we integrated the two confidence measures CM_D and CM_W by

Table 3 Utility values

	(i)	(ii)	(iii)	(iv)	(v)
U_1^+	1	1	2	2	2
U_2^+	1	1/2	3/2	3/2	3/2
U_3^+	1	1/3	4/3	4/3	4/3
U_1^-	−1	−1	−2	−1	−3
U_2^-	−2	−1	−3	−2	−4
U_3^-	−3	−1	−4	−3	−5

Table 4 Number of selected questions and their accuracies

		(i)	(ii)	(iii)	(iv)	(v)
Correct	**Yes/No**	**381**	**597**	**520**	**756**	**393**
	Alternative	221	229	224	251	166
	3-choice	416	160	258	198	263
Wrong	Yes/No	42	76	61	181	39
	Alternative	24	40	28	43	17
	3-choice	55	32	46	43	33
Total	Correct	1018	986	1002	1205	822
	Wrong	**121**	**148**	**135**	**267**	**89**
	Wh	**517**	**522**	**519**	**184**	**745**

avaraging them for simplicity. We can see that more Yes/No questions were selected in patterns (ii), (iii), and (iv) than in pattern (i). This is because pattern (i) has flat values for correct questions while utility values for Yes/No questions in the other three patterns are larger than alternative or 3-choice questions. Similarly, in pattern (v), fewer Yes/No questions were selected than in patterns (ii)–(iv). Second, in patterns (ii)–(vi), the number of wrong questions was the lowest in pattern (iii). Thus, we use the utility values in pattern (iii), hereafter.

4.2 Integration of Two Confidence Measures to Obtain Probability

We compared several integration methods of two previously developed basic confidence measures (CMs) [3]. More specifically, the total sum of utility values and the numbers of selected question types are compared for each integration method. The total sum of utility values represents how intelligent questions were actually selected in total because they considered the utility and whether or not the selected questions actually contained the correct cuisines. The same as in the previous section, we used our database having 1,656 entries.

Here, we compare five CMs: each of the two basic CMs (CM_D and CM_W) and three integration methods. The integration methods obtain a new CM(c) concerning cuisine c from $CM_D(c)$ and $CM_W(c)$. The three integration methods are shown in Eqs. (3)–(5). Note that Z_1, Z_2, and Z_3 are coefficients for normalizing the sum of probabilities to 1.

$$CM1(c) = \frac{1}{Z_1}\{\sum_{k\in\{D,W\}} CM_k(c)\} \tag{3}$$

$$CM2(c) = \frac{1}{Z_2}\{\sum_{k\in\{D,W\}} (1+\frac{1}{r_k(c)})CM_k(c)\} \tag{4}$$

$$CM3(c) = \frac{1}{Z_3}\{\sum_{k\in\{D,W\}} (1+\frac{Acc_k(c)}{r_k(c)})CM_k(c)\} \tag{5}$$

$CM1(c)$ is defined as a simple sum of $CM_D(c)$ and $CM_W(c)$. $CM2(c)$ reflects rank information $r_k(c)$ in the two $CM_k(c)$ as a bonus. $r_k(c)$ was set to give a larger bonus to cuisines in a higher rank when $CM_k(c)$ is estimated. This assumes that an estimation result in a higher rank tends to be correct. This tendency can be seen in Fig. 4, which shows accuracies per rank in each CM ($CM_D(c)$ and $CM_W(c)$) list calculated by using the whole 1,656 entries in our target database.

$CM3(c)$ reflects estimation accuracies per cuisine $Acc_k(c)$ together as a bonus in addition to $r_k(c)$. $Acc_k(c)$ is used for giving a larger bonus to cuisines that are easy to estimate. It considers that the estimation accuracies differ depending on the cuisine. This can be seen in Fig. 5, which shows the estimation accuracies when each cuisine

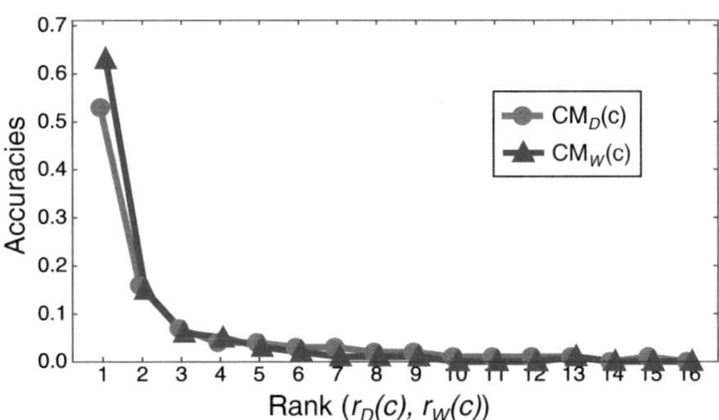

Fig. 4 Estimation accuracies per rank in each CM

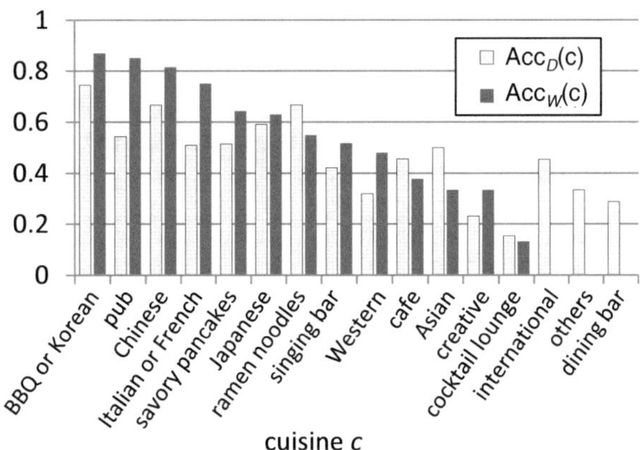

Fig. 5 Estimation accuracies when cuisine *c* had the highest CM

Table 5 Numbers of selected questions and sums of utility values

		CM_D	CM_W	$CM1$	$CM2$	$CM3$
Correct	Yes/No	556	659	522	660	**650**
	Alternative	155	149	226	335	**276**
	3-choice	186	238	254	277	**267**
Wrong	Yes/No	281	156	61	107	**90**
	Alternative	50	23	28	76	**45**
	3-choice	61	51	46	66	**55**
Total	Correct	897	1046	1002	1292	**1193**
	Wrong	392	230	135	249	**190**
	Wh	367	380	519	146	**273**
Sum of utility values		636	1273	1331	1515	**1534**

c had the highest CM. We can see here that the estimation accuracies were also not uniform; for example, the estimation accuracies for "BBQ or Korean" cuisine were high in both $CM_D(c)$ and $CM_W(c)$. Moreover, the estimation accuracy of "ramen noodles" was higher in $CM_D(c)$ than in $CM_W(c)$, and conversely that of "pub" was higher in $CM_W(c)$ than in $CM_D(c)$.

The results for the five CMs are listed in Table 5. We can see the maximum sum of utility values were obtained when $CM3(c)$ was used. This is because both the rank information $r_k(c)$ and estimation accuracies per cuisine $Acc_k(c)$ were taken into consideration. This corresponds to our giving prior distribution of cuisine estimation as weights to each cuisine.[1]

We also analyzed the output when $CM3(c)$ was used by examining the numbers of selected question types. When $CM3(c)$ was used, the number of correct questions increased and that of wrong questions decreased, compared to when either $CM_D(c)$ or $CM_W(c)$ was used. When we compared this with the case where $CM1(c)$ was used, the sum of utility values was larger with $CM3(c)$, because although the number of wrong questions increased, the number of correct questions increased more; in fact, the number of correct Yes/No questions by $CM3(c)$ was higher than $CM1(c)$ by 128. We can also see that the sum of utility values was higher for $CM3(c)$ than for $CM2(c)$. This was because more Wh-questions tend to be selected by $CM3(c)$ when questions seemed to be wrong by diminishing the bonus by $Acc_k(c)$. This was effective when an incorrect cuisine that was difficult to estimate got the highest CM, e.g., "cocktail lounge" cuisine.

[1]The prior information $r_k(c)$ and $Acc_k(c)$ should have been obtained by a held-out data. The information here was obtained in a closed manner, i.e., using the whole 1,656 entries.

5 Subjective Evaluation for Utility Values

We administered a questionnaire to participants in order to verify our conjecture on the utility values. Specifically, we compared the order of user impressions for the question types with that of utility values. Participants consisted of 15 individuals, nine of whom were students in our laboratory and six from among the general public. The latter was recruited because the students in our laboratory might have had some knowledge about the intention of this experiment. We used a system with the parameters discussed in the previous section for this experiment.

In the experiments, we showed several pairs of a restaurant name and its cuisine to participants, and asked for impressions of the four questions types by score, under situations where the question contains the correct cuisine and when it does not. Figure 6 shows an example of questions asking about "Azuma-zushi", which is "Japanese" cuisine. The order of questions was randomized when displayed to participants. In situations where the question contains the correct cuisine, we showed the four types of question to a participant: Yes/No, alternative, 3-choice questions containing the correct cuisine, and a Wh-question. Similarly, in situations where the question does not contain the correct cuisine, the four types of question were shown to participants. We asked participants their relative impressions for the questions by using an 11-point Likert scale from −5 to 5, to represent how intelligent they felt each question was.

The results of the questionnaire are summarized in Table 6, which shows the scores given by participants averaged by the cases of whether or not the question contained the correct cuisine and by each participant group. Table 7 shows the numbers of participants whose scores satisfied our conjecture, that is, how many participants felt the intelligence of the questions was on the same order as our setting of utility values. More specifically, in the notation used in Table 2, we counted the number of

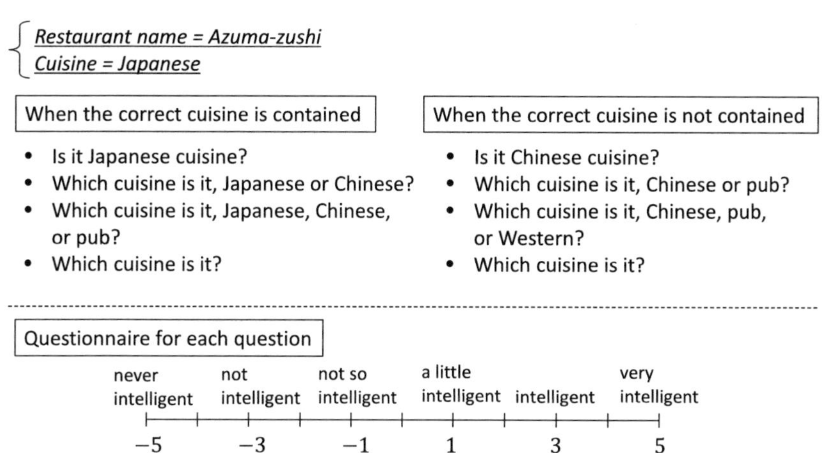

Fig. 6 Example of questions to ask user impressions and Likert scale for it

Table 6 Questionnaire results for four question types

	For correct questions				For wrong questions			
	Yes/no	Alternative	3-choice	Wh	Yes/no	Alternative	3-choice	Wh
In-lab.	3.8	1.7	0.6	−2.6	−2.4	−3.0	−3.7	−1.6
General	4.3	2.5	1.0	−0.3	−2.7	−3.3	−4.2	−2.5
Total	4.0	2.0	0.7	−1.7	−2.5	−3.1	−3.9	−1.9

Table 7 Numbers of participants whose scores satisfied our conjecture

	For correct questions			For wrong questions		
	In-lab.	General	Total	In-lab.	General	Total
w/o Wh-questions	8/9	6/6	14/15	8/9	5/6	13/15
w/ Wh-questions	8/9	4/6	12/15	5/9	3/6	8/15

participants whose scores satisfied $U_1^+ \geq U_2^+ \geq U_3^+ \geq U_{\text{Wh}}^+$ for correct questions, and $U_{\text{Wh}}^- \geq U_1^- \geq U_2^- \geq U_3^-$ for wrong questions. These are based on our conjecture that a more concrete question is better than a Wh-question when the question contains the correct cuisine and that a Wh-question is better than questions containing no correct cuisine.

We can see that utility values were almost always appropriate for the three types of questions (i.e., w/o Wh-questions) from the upper line in Table 7. More specifically, the order of impression scores was satisfied 93 % (14/15) for correct questions and 87 % (13/15) for wrong questions. That is, we empirically confirmed the tendency that a more concrete question is better when the question contains the correct cuisine and that a question containing more cuisines is worse when no correct cuisine is contained in it.

We can also see from the lower line of the table that the percentage decreased when we take relations between the three types of questions and Wh-questions into consideration. This was because two participants gave a higher score for Wh-questions than 3-choice questions for correct questions. Although we assume that Wh-questions are worse because the following user utterances may cause new OOV words, this should be verified through actual following dialogues in which participants will judge which questions are intelligent. On the other hand, for wrong questions, three participants gave the lowest score to Wh-questions. These participants said they felt that Wh-questions were not intelligent because the system was not thinking anything. This may have been caused by the wording on the questionnaire; that is, we should have phrased it "not annoying or annoying" instead of "intelligent or not".

6 Conclusion

Dialogue systems need to be capable of learning through dialogue. We have explored a scenario in which a system asks intelligent questions to obtain new information from the user. We developed a method to select the most appropriate question from four question types on the basis of expected utility and discussed the utility values and integration techniques of basic confidence measures (CMs) used in our method. Results of subjective analysis showed that our conjecture on utility values conforms to user impressions.

As future work, we will implement a dialogue system that incorporates the proposed method and conduct a user study to investigate how the user actually feels about the generated questions in actual sequences of dialogues. We will also clarify which types of question can actually obtain new information and whether the obtained information is usable in later dialogues. Some users may provide irrelevant information for our task especially when they answer a Wh-question.

Acknowledgments This study was partly supported by SCOPE of MIC.

References

1. Bollacker, K., Evans, C., Paritosh, P., Sturge, T., Taylor, J.: Freebase: a collaboratively created graph database for structuring human knowledge. In: Proceedings of the ACM SIGMOD international conference on Management of data, pp. 1247–1250 (2008)
2. Hakkani-Tur, D., Celikyilmaz, A., Heck, L., Tur, G., Zweig, G.: Probabilistic enrichment of knowledge graph entities for relation detection in conversational understanding. In: Proceedings of Annual Conference of the International Speech Communication Association (INTERSPEECH), pp. 2113–2117 (2014)
3. Otsuka, T., Komatani, K., Sato, S., Nakano, M.: Generating more specific questions for acquiring attributes of unknown concepts from users. In: Proceedings of Annual Meeting of the Special Interest Group on Discourse and Dialogue (SIGDIAL), pp. 70–77. Metz, France (2013)
4. von Neumann, J., Morgenstern, O.: Theory of Games and Economic Behavior. Princeton University Press (1944)
5. Pappu, A., Rudnicky, A.I.: Learning situated knowledge bases through dialog. In: Proceedings of Annual Conference of the International Speech Communication Association (INTERSPEECH), pp. 120–124 (2014)
6. Takahashi, Y., Dohsaka, K., Aikawa, K.: An efficient dialogue control method using decision tree-based estimation of out-of-vocabulary word attributes. In: Proceedings of International Conference on Spoken Language Processing (ICSLP), pp. 813–816 (2002)
7. Meng, H., Ching, P.C., Chan, S.F., Wong, Y.F., Chan, C.C.: ISIS: An adaptive, trilingual conversational system with interleaving interaction and delegation dialogs. ACM Trans. Comput. Hum. Interact. **11**(3), 268–299 (2004)

Separating Representation, Reasoning, and Implementation for Interaction Management: Lessons from Automated Planning

Mary Ellen Foster and Ronald P.A. Petrick

Abstract Numerous toolkits are available for developing speech-based dialogue systems. We survey a range of currently available toolkits, highlighting the different facilities provided by each. Most of these toolkits include not only a method for representing states and actions, but also a mechanism for reasoning about and selecting the actions, often combined with a technical framework designed to simplify the task of creating end-to-end systems. This near-universal tight coupling of representation, reasoning, and implementation in a single toolkit makes it difficult both to compare different approaches to dialogue system design, as well as to analyse the properties of individual techniques. We contrast this situation with the state of the art in a related research area—automated planning—where a set of common representations have been defined and are widely used to enable direct comparison of different reasoning approaches. We argue that adopting a similar separation would greatly benefit the dialogue research community.

Keywords Interaction management · Automated planning · Representation and reasoning · Systems integration

1 Introduction

A fundamental component of any dialogue system is the *interaction manager* [1], whose primary task is to carry out *action selection*: that is, based on the current state of the interaction and of the world, the interaction manager makes a high-level decision as to which spoken, non-verbal, and task-based actions should be taken next by the system as a whole. In contrast to more formal, descriptive accounts of dialogue

M.E. Foster (✉)
School of Computing Science, University of Glasgow, Glasgow, UK
e-mail: MaryEllen.Foster@glasgow.ac.uk

R.P.A. Petrick
Department of Computer Science, Heriot-Watt University, Edinburgh, UK
e-mail: R.Petrick@hw.ac.uk

K. Jokinen and G. Wilcock (eds.), *Dialogues with Social Robots*,
Lecture Notes in Electrical Engineering 427,
DOI 10.1007/978-981-10-2585-3_7

(e.g., [2]), which aim to model the full generality of language use, work on interaction management has concentrated primarily on developing end-to-end systems and on evaluating them through interaction with human users [3, 4].

A number of toolkits are available to support the construction of such end-to-end dialogue systems. Such a toolkit generally incorporates three main features. First, it provides a *representational formalism* for specifying states and actions. Second, the state/action representation is usually tightly linked to a *reasoning strategy* that is used to carry out action selection. Finally, most toolkits also include a set of *infrastructure building tools* designed to support modular system development. While these three features can clearly simplify the task of implementing an individual end-to-end system, the fact that the features are so tightly connected does complicate the task of comparing representational formalisms or reasoning strategies: in general, to carry out such a comparison, there is no alternative but to re-implement the entire system across multiple frameworks [5, 6].

In this chapter, we argue that the dialogue community could benefit from the wider use of system development techniques that break these tight connections among action selection, representation, and technical middleware. As motivation for this view, we use a related research field as as an example: automated planning. At a basic level, the core problem studied in automated planning is also one of context-dependent action selection. However, in the planning community, the focus has been on defining domains in common representation languages and on comparing different action-selection (i.e., planning) strategies within this common context, especially through a series of regularly organised planning competitions [7]. This has had important benefits for the planning community, such as allowing planning engines and domains to be directly compared and shared; also, the study of the representation languages themselves has led to a better understanding of the inherent trade-offs in choosing different representations. We believe that a similar approach could also benefit the dialogue community.

This chapter is structured as follows. We begin with a survey of available interaction management toolkits, summarising the representation, reasoning, and technical facilities provided by each. We then outline certain research directions in the automated planning community, concentrating on how common representations are used and exploited. We also present a recent example of relevant work where an off-the-shelf planner has been used to support interaction management for a socially intelligent robot bartender. Finally, we discuss the potential benefits of developing interaction management strategies that separate representation, reasoning, and integration, and outline our plans for future research in this area.

2 A Survey of Interaction Management Toolkits

In the traditional three-level architecture (Fig. 1) that is typical of a multimodal interactive system [8, 9], the interaction manager sits at the highest level and reasons about the most abstract structures, such as knowledge and action, usually

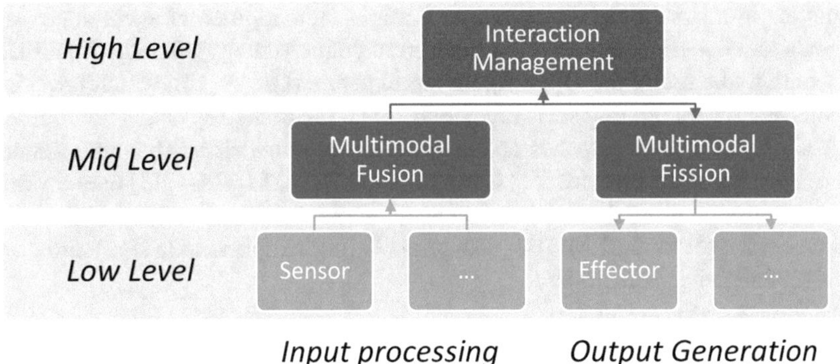

Fig. 1 Typical multimodal system architecture

represented in a logical form, and chooses high-level action specifications for the system to perform. The mid-level components deal with abstract, cross-modality representations of states and events: for input, multimodal fusion [10] combines continuous information from the low-level input sensors into a unified representation, while for output, multimodal fission [11] translates high-level communicative actions into concrete specifications for the individual output components. Finally, the components at the low level deal with modality-specific, highly detailed information: on the input side, this might include coordinates from a vision system or hypotheses from a speech recogniser, while the low-level output components would deal with instructions such as motion specifications for an embodied agent or content for a speech synthesiser.

In this section, we survey a representative set of dialogue systems toolkits, including several well-established, widely-used ones as well as a number of more recently developed toolkits. In particular, we concentrate on the representations used to support high-level interaction management, but also discuss the reasoning mechanisms and relevant details of any accompanying technical architecture.

2.1 TrindiKit/DIPPER

One of the widely used approaches to dialogue management is the Information State Update (ISU) approach, which is exemplified by TrindiKit [12] and its lighter-weight Java reimplementation DIPPER [13]. The core of this approach is the use of an *information state* which represents the state of the dialogue and which is updated by applying *update rules* following a given *update strategy*. The details of an information state are determined by the needs of a particular application. For example, the information state might include external aspects such as variables and their assignments (as in a slot-filling dialogue), or it might include internal agent states such as goals and beliefs (for a more plan-based dialogue strategy). TrindiKit and DIPPER both make use of the Open Agent Architecture (OAA) [14], which provides a middleware for integrating software agents into a distributed system.

A similar ISU approach has also been taken in more recent dialogue systems, but using other infrastructure. For example, the MATCH system [15] uses a similar approach to modelling the information state, while the Flipper toolkit [16] and the dialogue manager for the EMOTE robot tutor [17] both implement ISU-style dialogue management using XML rules.

2.2 *Ravenclaw*

Another widely-used toolkit is Ravenclaw [18], which is based around a *dialogue task specification* representing the domain-specific aspects of the control logic. This representation forms a hierarchical plan for the interaction and is executed by a domain-independent engine at run time. The specification consists of a tree of *dialogue agents*, each of which handles a sub-task of the dialogue (e.g., greeting the user). The dialogue engine traverses the tree in a depth-first order, putting agents from the tree onto an execution stack and removing them when they are completed. The agents are defined through C++ macros that communicate by exchanging user-defined data structures through a message-passing system.

2.3 *COLLAGEN/DISCO*

COLLAGEN [19] is a toolkit based on the *collaborative interface paradigm*, which assumes that a software agent is collaborating with a user to operate an application programme, with both agents communicating with each other as well as interacting with the application. COLLAGEN has been used to implement a range of interface agents, including ones for travel booking and for controlling a programmable thermostat. More recently, COLLAGEN has been extended into an open-source tool called DISCO [20], which combines hierarchical task networks (HTNs) with traditional dialogue trees to permit semi-automated dialogue authoring and dialogue structure reuse. The target scenario is specified as a collection of *recipes*—rules for decomposing a goal into subgoals and for accomplishing those subgoals. In contrast to Ravenclaw, where the dialogue flow must be specified, COLLAGEN and DISCO only need a specification of the tasks; the dialogue is then generated automatically via a generic rule framework.

2.4 OpenDial

OpenDial [21] is a domain-independent toolkit for developing spoken dialogue systems. Its primary goal is to support robust dialogue management, using a hybrid framework that combines logical and statistical approaches through probabilistic rules to represent the internal models of the framework. OpenDial also includes a Java-based blackboard architecture where all modules are connected to a central information hub which represents the dialogue state, along with a plugin framework allowing new modules to be integrated.

2.5 IrisTK

IrisTK [22] is a toolkit for the rapid development of real-time systems for face-to-face multi-party interaction which accompanies the Furhat robot head [23]. IrisTK provides an XML-based scripting language for defining statecharts [24] that map input events to output events depending on the system state, along with an event-based distributed architecture that allows a system to be built by integrating modules such as speech recognition and synthesis; it also incorporates pre-built modules for such common tasks.

2.6 InproTK

InproTK [25] is designed for the development of systems which are able to support incremental processing through a network of modules which continuously read input and output the processed result as Incremental Units (IUs). IUs are passed between modules, with the connections between modules specified through a configuration file. InproTK comes with a selection of pre-built modules, and also allows developers to write their own. While the original InproTK used its own internal middleware, a more recent system update [26] adds support for three message-passing protocols (XML-RPC, the Robotics Service Bus [27], and InstantReality [28]), along with a meta-communication layer which mediates among the three protocols.

2.7 Pamini

The Pamini approach [29] was specificially designed to support human-robot interaction. It includes a task-state protocol which abstracts away from the details of processing in the robot system (e.g., perceptual analysis, motor control, or output generation); it also provides a set of generic interaction patterns (such as action requests and clarification processes) which allow it to support rapid prototyping and combination. Pamini also includes support for mixed-initiative interaction and for online learning during an interaction. Action selection is carried out through interleaving of a set of patterns on a stack: whenever a new piece of input is detected, it is sent to all of the active patterns in turn until one is able to deal with it. Pamini makes use of the XCF middleware [30], which is written in C++ and based on the Internet Communications Engine (Ice) middleware protocol [31].

2.8 Summary

As highlighted throughout this section, and as summarised in Table 1, each of the described toolkits provides a different representation of the information needed for action selection, including declarative update rules, statecharts, interaction patterns, or the more procedural representations used by toolkits such as Ravenclaw and COLLAGEN. Each toolkit also incorporates its own reasoning mechanism to make use of the defined representation—in fact, the representation and reasoning components are often so tightly related that they cannot be fully distinguished. Finally, the majority of the toolkits described either provide or make use of a specific technical middleware framework. As a result, the task of choosing a specific toolkit generally also means adopting both its reasoning strategy and its associated technical infrastructure.

This diversity of toolkit approaches has had a clear impact on the research carried out in the dialogue systems field. In particular, while it is common to compare interaction management strategies within a single framework—for example, by comparing action-selection policies that are learnt from data against hand-coded policies

Table 1 Summary of toolkits considered

Toolkit	Representation	Reasoning	Technical
Trindikit/DIPPER	Information state	Update/selection rules	Open Agent Architecture (C++)
COLLAGEN/DISCO	Recipes	Generic rule framework	Java API
Ravenclaw	Task tree, agenda	Tree traversal	C++ macros, message passing
OpenDial	Probabilistic rules	Event-based state update	Java-based blackboard architecture
IrisTK	XML state charts	Event-based state update	Java event-based distributed architecture
InproTK	Modules, incremental units	Incremental processing	Various middleware options
Pamini	Interaction patterns	Stack of active patterns	XCF middleware (Ice/C++)

[32]—it is relatively uncommon to compare the representational ability and reasoning performance across different frameworks. One of the few studies that did carry out this sort of cross-toolkit comparison is [5], in which the same interactive system was separately implemented using Ravenclaw, DIPPER, Collagen/DISCO, and Pamini. The target domain was the "Curious Robot": a humanoid robot able to engage in an interactive object learning and manipulation scenario. Note that carrying out this comparison required the entire dialogue system to be implemented separately in each formalism; there was no possibility of transferring any representations or reasoning components across the implementations. The overall conclusion of this experiment was that, while all of the toolkits were able to support basic one-on-one spoken interactions, there was more diversity in their approach to and support for more advanced interactive tasks such as multimodal fusion/fission and multi-party interaction. In addition, the different systems were found to offer a range of methods for linking the dialogue state with the task state, which was a feature that was particularly relevant for the Curious Robot scenario. In a more recent study comparing DISCO and Ravenclaw [6], the comparison again required the entire dialogue system to be implemented end-to-end in each individual formalism; in a small-scale user study, few differences were found in the resulting implementations.

3 The Separation of Representation and Reasoning in Automated Planning

The overall problem of selecting high-level actions for an intelligent agent is not unique to dialogue systems, but is a problem addressed in a variety of research communities including automated planning. In planning, the emphasis is on applying problem-solving techniques to find an ordered sequence of actions (called a *plan*) that, when chained together, transform an initial state into a state where a set of specified goal objectives are achieved.[1] The general planning problem is usually divided into two parts: a description of the planning domain and the definition of a planning problem instance to be achieved within that domain [33]. A *planning domain* provides a definition of the symbols and actions used by the planner. Symbols are used to specify the objects, properties, states, and knowledge that make up the planning agent's operating environment, normally defined in a logic-like language. Actions are typically specified in terms of the state properties that must be true to execute that action (the action's *preconditions*) and the changes that the action makes to the state when it is executed (the action's *effects*). A *planning problem* provides a definition of the initial state the planner begins its operation in, and a description of the goals to be achieved. A central goal of planning research is to build *domain-*

[1]This differs somewhat from the task of interaction management, where the goal is (usually) to find the next system action, rather than a complete action sequence. However, we note that a system that is able to achieve the latter can also be used in the former context.

independent planning systems that are able to solve a range of planning problems in a variety of domains, rather than just a single problem in a single domain.

One important feature of research in automated planning is that the tools developed by this community usually support one of a number of common representation languages, such as PDDL [34], PPDDL [35], or RDDL [36], among others. Many of these languages have been developed or extended as part of the International Planning Competitions (IPC) [7, 37]—a series of competitions in which different planning systems compete against each other on a common set of planning problems—which have run approximately every other year since 1998 within the context of the International Conference on Automated Planning and Scheduling (ICAPS).[2] Even when some planners implement their own representation languages, which may differ (usually syntactically) from the standard planning languages, additional work is often performed to establish the relationship between such languages and the more common representations.

These activities have led to some important benefits for the planning community. First, by adopting common representations, the task of modelling a planning problem can be separated from the task of implementing an efficient engine for solving those problems. This allows different planning engines to be developed and directly compared, either quantitatively or qualitatively, on a common set of inputs (i.e., planning problems). Second, planning domains and planning engines can be shared, leading to the development of common benchmarks for future planning systems, as well as an improvement in the baseline systems that can solve problems in these domains. In particular, the IPC has contributed greatly to these activities by creating and requesting new domains, which has in turn helped spur the development of more powerful planning tools. The activities of the IPC have also resulted in a repository of planning domains which can be studied, analysed, and reused as necessary [37], with additional efforts from the community aimed at making planning tools and domains more accessible.[3] Finally, the representation languages themselves—and the planning problems they support—can be studied and compared, leading to a better understanding of the complexity of particular classes of domains and problems [38], and the tradeoffs of using one language over another. This work has close connections to related communities such as knowledge representation and reasoning (KR&R) formal logic. It has also produced some interesting research directions, such as a range of compilation approaches which seek to transform more complex planning problems into simpler forms that are solvable in an efficient manner with existing tools [39, 40].

We believe that similar approaches could be applied within the dialogue systems research community, leading to similar positive results. In the following section, we give a concrete example of this approach, where a domain-independent automated planner is used for the task of interaction management in a scenario involving a socially intelligent humanoid robot.

---

[2] http://www.icaps-conference.org/.

[3] http://planning.domains/.

4 Plan-Based Interaction Management in a Robot Bartender Domain

The JAMES robot bartender (Fig. 2)[4] has the goal of supporting socially appropriate multi-party interaction in a bartending scenario. Based on (uncertain) observations about the users in the scene provided by the vision and speech recognition components, the system maintains a model of the social context and the task state, and decides on the socially-appropriate responses that are required to respond to human users in that setting.

In this context, high-level action selection is performed by a domain-independent planner which manages the interactions with customers, tracks multiple drink orders, and gathers additional information as needed with follow-up questions [41]. In particular, the task of interacting with human customers is mixed with the physical task of ensuring that the correct drinks are delivered to the correct customers. Plans are generated using PKS (Planning with Knowledge and Sensing) [42, 43], a planner that works with incomplete information and sensing actions. Figure 3 shows an example of two actions from the robot domain, defined in the PKS representation. Here, `ask-drink` models an information-gathering dialogue action that asks a customer for their drink order, while `serve` is a physical robot action for serving a drink to a customer. The complete planning domain description also includes actions such

Fig. 2 The JAMES robot bartender

[4]http://jamesproject.eu/.

```
action ask-drink(?a : agent)
    preconds: K(inTrans = ?a) &
              !K(ordered(?a))
              !K(otherAttnReq) &
              !K(badASR(?a))
    effects:  add(Kf,ordered(?a)),
              add(Kv,request(?a))
```

```
action serve(?a : agent, ?d : drink)
    preconds: K(inTrans = ?a) &
              K(ordered(?a)) &
              Kv(request(?a)) &
              K(request(?a) = ?d) &
              !K(otherAttnReq) &
              !K(badASR(?a))
    effects:  add(Kf,served(?a))
```

Fig. 3 Example PKS actions in the JAMES bartender domain

Table 2 A partial list of actions in the robot bartender planning domain

Action	Description
`greet(?a)`	Greet agent `?a`
`ask-drink(?a)`	Ask agent `?a` for a drink order
`serve(?a,?d)`	Serve drink `?d` to agent `?a`
`bye(?a)`	End an interaction with agent `?a`
`wait(?a)`	Tell agent `?a` to wait
`ack-order(?a)`	Acknowledge agent `?a`'s order
`ack-wait(?a)`	Thank agent `?a` for waiting
`ack-thanks(?a)`	Acknowledge agent `?a`'s thanks
`inform-drinklist(?a,?t)`	Inform agent `?a` of the available drinks of type `?t`

Table 3 A plan for interacting with a single customer in the bartender domain

Plan steps	Description
`greet(a1),`	Greet agent `a1`
`ask-drink(a1),`	Ask agent `a1` for a drink order
`ack-order(a1),`	Acknowledge agent `a1`'s order
`serve(a1,request(a1)),`	Serve `a1` the drink that was ordered
`bye(a1).`	End the interaction with agent `a1`

as `greet(?a)` (a purely social action to greet customer `?a`), `bye(?a)` (end an interaction with `?a`), and `wait(?a)` (tell `?a` to wait, e.g., by nodding), among others. A partial list of the actions in the robot bartender domain is given in Table 2. The planner uses these actions to form plans by chaining together particular action instances to achieve the goals of a planning problem. For instance, Table 3 shows a five step plan for interacting with a single customer in the bartender domain.

An important design decision for the robot bartender was to define the state and action representations separately from the tools used to reason about them, and also from the infrastructure needed for the planner to communicate with the rest of the system (which employs the Ice object middleware [31]). In addition to supporting the modular, distributed development of the system, this also permitted the PKS planner to be exchanged with a completely separate interaction manager based on Markov Decision Processes [32], with no other changes needed to the system. In terms of our particular planning approach, PKS's representation can be compiled into a variant of standard PDDL, allowing the bartender domain to be tested with other planning systems. Using this approach, we performed a series of offline experiments with other planners to study particular aspects of the planner's performance in the bartender domain, for instance how it scales when the number of agents or the number of subdialogues is increased [44]. The modular design of our approach also means that, if necessary, the planner could be replaced in the robot system with an alternative domain-independent planners, with few changes needed to the underlying domain representation or the high-level software infrastructure. Similarly, our planning approach could be easily integrated into other interactive systems using its existing application programming interface [45].

5 Summary and Future Work

We have examined a number of toolkits designed to support the development of speech-based dialogue systems. While the features provided by these toolkits simplify the prototyping and deployment of an individual end-to-end system, each individual toolkit also tends to be very tightly linked to the representations, reasoning techniques, and even technical infrastructure used to connect the interaction manager to the rest of the system. This makes it difficult either to compare different approaches or to analyse the properties of individual techniques without re-implementing the entire system from the ground up in multiple frameworks.

We argue that the approach taken in the automated planning community—where domains have long been defined in common representation languages, and action-selection strategies compared within this common context—could also benefit the dialogue systems community, by permitting diverse approaches to be benchmarked and compared more easily. Although some early work was carried out in this area [46–50], the approach has for the most part been largely overlooked (with the exception of approaches like [51–55]). Note that common tasks such as the Dialogue State Tracking challenge [56] do exist in the dialogue community; however, to our knowledge, there has never been a successful effort to develop standard, high-level representations for use in interaction management.

There are also opportunities for the automated planning community to benefit from closer collaboration with the dialogue systems community. For instance, problems in dialogue systems can also serve as the basis for new challenge domains for planning, showcasing planning tools and techniques, and possibly extending the standard planning representations with the features needed to model new types of problems. Beyond the opportunity for novel test domains, there are also some important general lessons that the planning community can learn from dialogue systems.

For instance, dialogue systems are inherently application driven and, as such, any adoption of planning techniques must be situated in the context of larger, more complex systems of which planning is a single component. This often requires a degree of maturity in tool development that goes beyond offline lab-tested code, with a focus on robustness and the development of standard application programming interfaces (APIs). While there have been recent attempts to build such systems within the planning community [57], more work is still needed. Finally, the issue of user evaluation is at the heart of dialogue systems work, with a focus on (non-expert) users actually using the developed tools. As a result, dialogue systems domains are often driven by the needs of the real-world application, rather than lab-based assumptions, which could help facilitate the wider adoption of planning approaches in such settings.

More generally, the work described here is situated in the context of a larger research programme aimed at revisiting the use of techniques from automated planning in the context of natural language interaction. We believe that the time is right for the natural language community—and, in particular, the dialogue systems community—to benefit from recent advances in the area of automated planning. We have already demonstrated that components from the two communities can be successfully combined in the JAMES bartender [41]; we plan to continue our work in this area by exploring the challenges and opportunities that arise from the intersection of these two research fields. In particular, we believe that the adoption of common, formally understood representation languages for states and actions that are separated from reasoning mechanisms and technical infrastructures can facilitate closer links between the two research communities.

Acknowledgements This research has been partially funded by the European Union's Seventh Framework Programme for research, technological development and demonstration under grant no. 270435 (JAMES, http://james-project.eu/) and grant no. 610917 (STAMINA, http://stamina-robot.eu/), and by the European Union's Horizon 2020 research and innovation programme under grant no. 688147 (MuMMER, http://mummer-project.eu/).

References

1. Bui, T.H.: Multimodal dialogue management—state of the art. Technical Report 06–01, University of Twente (UT), Enschede, The Netherlands (2006)
2. Asher, N., Lascarides, A.: Logics of Conversation. Cambridge University Press (2003)
3. Jokinen, K., McTear, M.: Spoken Dialogue Systems. Morgan & Claypool (2009)
4. McTear, M., Callejas, Z., Griol, D.: The Conversational Interface. Springer International Publishing (2016)
5. Peltason, J., Wrede, B.: The curious robot as a case-study for comparing dialog systems. AI Mag. **32**(4), 85–99 (2011)
6. Olaso, J.M., Milhorat, P., Himmelsbach, J., Boudy, J., Chollet, G., Schlögl, S., Torres, M.I.: A multi-lingual evaluation of the vAssist spoken dialog system: comparing Disco and RavenClaw. In: Jokinen, K. & Wilcock, G. (Eds.) Dialogues with Social Robots, Springer pp.221–238 (this volume) (2016)
7. Coles, A., Coles, A., García Olaya, A., Jiménez, S., Linares López, C., Sanner, S., Yoon, S.: A survey of the seventh international planning competition. AI Mag. **33**(1), 83–88 (2012)

8. Gat, E.: Three-layered architectures. In: AI-Based Mobile Robots: Case Studies of Successful Robot Systems. MIT Press (1998)
9. Dumas, B., Lalanne, D., Oviatt, S.: Multimodal interfaces: a survey of principles, models and frameworks. In: Human Machine Interaction, Lecture Notes in Computer Science, vol. 5440, pp. 3–26 (2009)
10. Atrey, P.K., Hossain, M.A., El Saddik, A., Kankanhalli, M.S.: Multimodal fusion for multimedia analysis: a survey. Multimedia Syst. **16**(6), 345–379 (2010)
11. Foster, M.E.: State of the art review: multimodal fission. Deliverable 6.1, COMIC project (2002)
12. Larsson, S., Traum, D.R.: Information state and dialogue management in the TRINDI dialogue move engine toolkit. Nat. Lang. Eng. **6**(3&4), 323–340 (2000)
13. Bos, J., Klein, E., Lemon, O., Oka, T.: DIPPER: Description and formalisation of an information-state update dialogue system architecture. In: Proceedings of SIGdial, pp. 115–124 (2003)
14. Martin, D.L., Cheyer, A.J., Moran, D.B.: The open agent architecture: a framework for building distributed software systems. Appl. Artif. Intell. **13**(1–2), 91–128 (1999)
15. Johnston, M., Bangalore, S., Vasireddy, G., Stent, A., Ehlen, P., Walker, M., Whittaker, S., Maloor, P.: MATCH: an architecture for multimodal dialogue systems. In: Proceedings of ACL, pp. 376–383, Philadelphia, Pennsylvania, USA (2002)
16. ter Maat, M., Heylen, D.: Flipper: an information state component for spoken dialogue systems. In: Intelligent Virtual Agents. Lecture Notes in Computer Science, vol. 6895, pp. 470–472. Springer, Berlin (2011)
17. Janarthanam, S., Hastie, H., Deshmukh, A., Aylett, R., Foster, M.E.: A reusable interaction management module: use case for empathic robotic tutoring. In: Proceedings of goDIAL, Gothenburg, Sweden (2015)
18. Bohus, D., Rudnicky, A.I.: The RavenClaw dialog management framework: architecture and systems. Comput. Speech Lang. **23**(3), 332–361 (2009)
19. Rich, C., Sidner, C.L.: COLLAGEN: a collaboration manager for software interface agents. User Model. User-Adap. Inter. **8**(3–4), 315–350 (1998)
20. Rich, C., Sidner, C.L.: Using collaborative discourse theory to partially automate dialogue tree authoring. In: Intelligent Virtual Agents, Lecture Notes in Computer Science, vol. 7502, pp. 327–340 (2012)
21. Lison, P.: A hybrid approach to dialogue management based on probabilistic rules. Comput. Speech Lang. (2015)
22. Skantze, G., Al Moubayed, S.: IrisTK: a statechart-based toolkit for multi-party face-to-face interaction. In: Proceedings of ICMI, pp. 69–76 (2012)
23. Al Moubayed, S., Beskow, J., Skantze, G., Granström, B.: Furhat: a back-projected human-like robot head for multiparty human-machine interaction. In: Cognitive Behavioural Systems, Lecture Notes in Computer Science, vol. 7403, pp. 114–130 (2012)
24. Harel, D.:Statecharts: a visual formalism for complex systems. Sci. Comput. Program. **8**(3), 231–274 (1987). ISSN 0167-6423. http://dx.doi.org/10.1016/0167-6423(87)90035-9
25. Baumann, T., Schlangen, D.: The InproTK 2012 release. In: Proceedings of the NAACL-HLT Workshop on Future directions and needs in the Spoken Dialog Community: Tools and Data, pp. 29–32 (2012). http://projects.ict.usc.edu/nld/SDCTD2012/
26. Kennington, C., Kousidis, S., Schlangen, D.: InproTKs: a toolkit for incremental situated processing. In: Proceedings of SIGdial, pp. 84–88 (2014)
27. Wienke, J., Wrede, S.: A middleware for collaborative research in experimental robotics. In: Proceedings of the 2011 IEEE/SICE International Symposium on System Integration, pp. 1183–1190 (2011)
28. Kousidis, S., Kennington, C., Schlangen, D.: Investigating speaker gaze and pointing behaviour in human-computer interaction with the mint.tools collection. In: Proceedings of SIGDIAL, pp. 319–323, Metz, France (2013)
29. Peltason, J., Wrede, B.: Pamini: a framework for assembling mixed-initiative human-robot interaction from generic interaction patterns. In: Proceedings of SIGdial, pp. 229–232 (2010)

30. Wrede, S., Hanheide, M., Bauckhage, C., Sagerer, G.: An active memory as a model for information fusion. In: Proceedings of the 7th International Conference on Information Fusion, pp. 198–205 (2004)
31. Henning, M.: A new approach to object-oriented middleware. IEEE Internet Comput. **8**(1), 66–75 (2004)
32. Keizer, S., Foster, M.E., Lemon, O., Gaschler, A., Giuliani, M.: Training and evaluation of an MDP model for social multi-user human-robot interaction. In: Proceedings of SIGdial (2013)
33. Ghallab, M., Nau, D., Traverso, P.: Automated Planning: Theory and Practice. Morgan Kaufmann (2004)
34. McDermott, D., Ghallab, M., Howe, A., Knoblock, C., Ram, A., Veloso, M., Weld, D., Wilkins, D.: PDDL—The Planning Domain Definition Language (Version 1.2). Technical Report CVC TR-98-003/DCS TR-1165, Yale Center for Computational Vision and Control (1998)
35. Younes, H.L.S., Littman, M.L.: PPDDL1.0: an extension to PDDL for expressing planning domains with probabilistic effects. Technical Report CMU-CS-04-162, Carnegie Mellon University (2004)
36. Sanner, S.: Relational dynamic influence diagram language (RDDL): language description. http://users.cecs.anu.edu.au/~ssanner/IPPC_2011/RDDL.pdf (2010)
37. ICAPS: ICAPS competitions. http://www.icaps-conference.org/index.php/Main/Competitions (2015)
38. Rintanen, J.: Complexity of planning with partial observability. In: Proceedings of ICAPS, pp. 345–354 (2004)
39. Palacios, H., Geffner, H.: Compiling uncertainty away in conformant planning problems with bounded width. J. Artif. Intell. Res. **35**, 623–675 (2009)
40. Albore, A., Palacios, H., Geffner, H.: A translation-based approach to contingent planning. In: Proceedings of IJCAI, pp. 1623–1628 (2009)
41. Petrick, R.P.A., Foster, M.E.: Planning for social interaction in a robot bartender domain. In: Proceedings of ICAPS 2013 (2013)
42. Petrick, R.P.A., Bacchus, F.: A knowledge-based approach to planning with incomplete information and sensing. In: Proceedings of AIPS, pp. 212–221 (2002)
43. Petrick, R.P.A., Bacchus, F.: Extending the knowledge-based approach to planning with incomplete information and sensing. In: Proceedings of ICAPS, pp. 2–11 (2004)
44. Sharma, V.: Automated Planning for Natural Language Robot Dialogue. M.Sc. Project, University of Edinburgh, Edinburgh (2012)
45. Petrick, R.P.A., Gaschler, A.: Extending knowledge-level planning with sensing for robot task planning. In: Proceedings of PlanSIG (2014)
46. Perrault, C.R., Allen, J.F.: A plan-based analysis of indirect speech acts. Am. J. Comput. Linguist. **6**(3–4), 167–182 (1980)
47. Appelt, D.: Planning English Sentences. Cambridge University Press, Cambridge (1985)
48. Hovy, E.: Generating Natural Language Under Pragmatic Constraints. Lawrence Erlbaum Associates, Hillsdale (1988)
49. Cohen, P., Levesque, H.: Rational interaction as the basis for communication. In: Cohen, P., Morgan, J., Pollack, M. (eds.) Intentions in Communication, pp. 221–255. MIT Press, Cambridge (1990)
50. Young, R.M., Moore, J.D.: DPOCL: a principled approach to discourse planning. In: Proceedings of INLG, pp. 13–20, Kennebunkport, Maine, USA (1994)
51. Koller, A., Stone, M.: Sentence generation as planning. In: Proceedings of ACL, pp. 336–343, Prague, Czech Republic (2007)
52. Benotti, L.: Accommodation through tacit sensing. In: Proceedings of LONDIAL, pp. 75–82, London, UK (2008)
53. Brenner, M., Kruijff-Korbayová, I.: A continual multiagent planning approach to situated dialogue. In: Proceedings of LONDIAL, pp. 67–74 (2008)
54. Koller, A., Petrick, R.P.A.: Experiences with planning for natural language generation. Comput. Intell. **27**(1), 23–40 (2011)

55. Mackaness, W., Boye, J., Clark, S., Fredriksson, M., Geffner, H., Lemon, O., Minnock, M., Webber, B.: The SpaceBook project: pedestrian exploration of the city using dialogue based interaction over smartphones. In: Proceedings of the 8th Symposium on Location-Based Services, Vienna, Austria (2011)
56. Henderson, M., Thomson, B., Williams, J.D.: The second dialog state tracking challenge. In: Proceedings of SIGdial, pp. 263–272, Philadelphia, PA, USA (2014)
57. Cashmore, M., Fox, M., Long, D., Magazzeni, D., Ridder, B., Carrera, A., Palomeras, N., Hurtos, N., Carreras, M.: ROSPlan: planning in the robot operating system. In: Proceedings of ICAPS (2015)

SimpleDS: A Simple Deep Reinforcement Learning Dialogue System

Heriberto Cuayáhuitl

Abstract This article presents *SimpleDS*, a simple and publicly available dialogue system trained with deep reinforcement learning. In contrast to previous reinforcement learning dialogue systems, this system avoids manual feature engineering by performing action selection directly from raw text of the last system and (noisy) user responses. Our initial results, in the restaurant domain, report that it is indeed possible to induce reasonable behaviours with such an approach that aims for higher levels of automation in dialogue control for intelligent interactive systems and robots.

Keywords Dialogue systems · Reinforcement learning · Deep learning

1 Introduction

Almost two decades ago, the (spoken) dialogue systems community adopted the Reinforcement Learning (RL) paradigm since it offered the possibility to treat dialogue design as an optimisation problem, and because RL-based systems can improve their performance over time with experience. Although a large number of methods have been proposed for training (spoken) dialogue systems using RL, the question of "How to train dialogue policies in an efficient, scalable and effective way across domains?" still remains as an open problem. One limitation of current approaches is the fact that RL-based dialogue systems still require high-levels of human intervention (from system developers), as opposed to automating the dialogue design. Training a system of this kind requires a system developer to provide a set of features to describe the dialogue state, a set of actions to control the interaction, and a performance function to reward or penalise the action-selection process. All of these elements have to be carefully engineered in order to learn a good dialogue policy (or policies). This suggests that one way of advancing the state-of-the-art in this field is

H. Cuayáhuitl (✉)
School of Computer Science, University of Lincoln, Lincoln, UK
e-mail: HCuayahuitl@lincoln.ac.uk

K. Jokinen and G. Wilcock (eds.), *Dialogues with Social Robots*,
Lecture Notes in Electrical Engineering 427,
DOI 10.1007/978-981-10-2585-3_8

by reducing the amount of human intervention in the dialogue design process through higher degrees of automation, i.e. by moving towards truly autonomous learning.

Recent advances in artificial intelligence have proposed machine learning methods as a way to reduce human intervention in the creation of intelligent agents. In particular, the field of Deep Reinforcement Learning (DRL) targets feature learning and policy learning simultaneously—which reduces the effort in feature engineering [1]. This is relevant because the vast majority of previous RL-based dialogue systems make use of carefully engineered features to represent the dialogue state [2].

Motivated by the advantages of DRL methods over traditional RL methods, in this article we present a core domain-independent dialogue management framework, recently applied to strategic board games [3]. This article makes use of raw noisy text—without any engineered features to represent the dialogue state. By using this representation, the dialogue system does not require a Spoken Language Understanding (SLU) component. We bypass SLU by learning dialogue policies directly from (simulated) speech recognition outputs. The rest of the article describes a proof of concept system which is trained based on this idea.

2 Deep Reinforcement Learning for Dialogue Control

A Reinforcement Learning (RL) agent learns its behaviour from interaction with an environment and the physical or virtual agents within it, where situations are mapped to actions by maximising a long-term reward signal [4]. An RL agent is typically characterised by: (i) a finite or infinite set of states $S = \{s_i\}$; (ii) a finite or infinite set of actions $A = \{a_j\}$; (iii) a state transition function $T(s, a, s')$ that specifies the next state s' given the current state s and action a; (iv) a reward function $R(s, a, s')$ that specifies the reward given to the agent for choosing action a in state s and transitioning to state s'; and (v) a policy $\pi : S \rightarrow A$ that defines a mapping from states to actions. The goal of an RL agent is to select actions by maximising its cumulative discounted reward defined as $Q^*(s, a) = \max_\pi \mathbb{E}[r_t + \gamma r_{t+1} + \gamma^2 r_{t+1} + \cdots | s_t = s, a_t = a, \pi]$, where function Q^* represents the maximum sum of rewards r_t discounted by factor γ at each time step. While the RL agent takes actions with probability $Pr(a|s)$ during training, it takes the best actions $\max_a Pr(a|s)$ at test time.

To induce the Q function above we use Deep Reinforcement Learning as in [1], which approximates Q^* using a multilayer neural network. The Q function of a DRL agent is parameterised as $Q(s, a; \theta_i)$, where θ_i are the parameters (weights) of the neural net at iteration i. More specifically, training a DRL agent requires a dataset of experiences $D_t = \{e_1, ...e_t\}$ (also referred to as 'experience replay memory'), where every experience is described as a tuple $e_t = (s_t, a_t, r_t, s_{t+1})$. The Q function can be induced by applying Q-learning updates over minibatches of experience $MB = \{(s, a, r, s') \sim U(D)\}$ drawn uniformly at random from dataset D. A Q-learning update at iteration i is thus defined as the loss function

$$L_i(\theta_i) = \mathbb{E}_{MB}\left[(r + \gamma \max_{a'} Q(s', a'; \overline{\theta}_i) - Q(s, a; \theta_i))^2\right],$$

where θ_i are the parameters of the neural net at iteration i, and $\overline{\theta}_i$ are the target parameters of the neural net at iteration i. The latter are only updated every C steps. This process is implemented in the learning algorithm *Deep Q-Learning with Experience Replay* described in [1].

3 The *SimpleDS* Dialogue System

Figure 1 shows a high-level diagram of the *SimpleDS* dialogue system. At the bottom, the learning environment receives an action (dialogue act) and outputs the next environment state and numerical reward. To do that, the environment first generates the word sequence of the last system action, the user simulator generates a word sequence as a response to that action, and the user response is distorted given some noise level and word-level confidence scores. Based on the system's verbalisation and noisy user response, the next dialogue state and reward are calculated and given as a result of having executed the given action. At the top of the diagram, a Deep Reinforcement Learning (DRL) agent receives the state and reward, updates its policy during learning, and outputs an action according to its learnt policy.

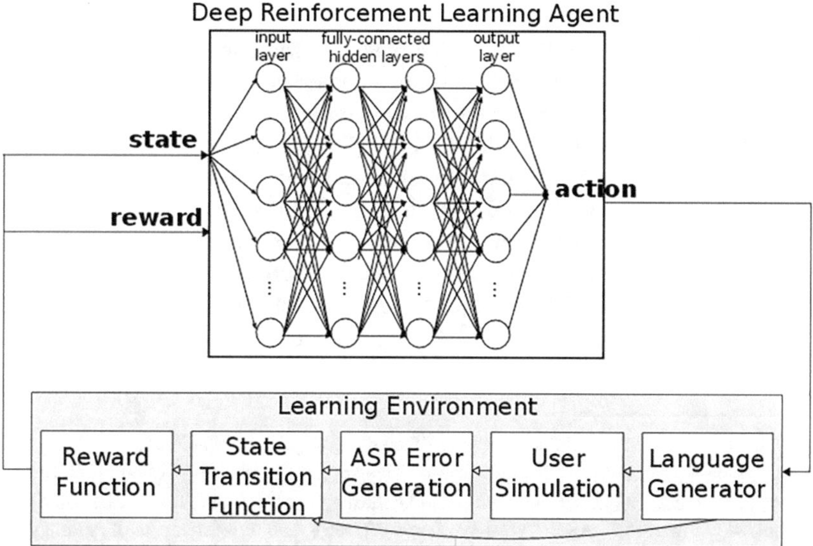

Fig. 1 High-level architecture of the *SimpleDS* dialogue system—see text for details

This system runs under a client-server architecture, where the environment acts as the *server* and the learning agent acts as the *client*. They communicate by exchanging messages, where the client tells the server the action to execute, and the server tells the client the dialogue state and reward(s) observed. The *SimpleDS* learning agent is based on the ConvNetJS tool [5], which implements the algorithm 'Deep Q-Learning with experience replay' proposed by [1]. We extended this tool to support multi-threaded and client-server processing with constrained search spaces.[1]

The *state space* includes up to 100 word-based features depending on the vocabulary of the *SimpleDS* agent in the restaurant domain. The initial release of *SimpleDS* provides support for English, German and Spanish. While words derived from system responses are treated as binary variables (i.e. word present or absent), the words derived from noisy user responses can be seen as continuous variables by taking confidence scores into account. Since we use a single variable per word, user features override system ones in case of overlaps.

The *action space* includes 35 dialogue acts in the Restaurant domain.[2] They include 2 salutations, 9 requests, 7 apologies, 7 explicit confirmations, 7 implicit confirmations, 1 retrieve information, and 2 provide information. Rather than learning with whole action sets, *SimpleDS* supports learning from constrained actions by applying Q-learning learning updates only on the set of valid actions. The constrained actions come from the most likely actions (e.g. $Pr(a|s) > 0.01$) with probabilities derived from a Naive Bayes classifier trained from example dialogues. The latter are motivated by the fact that a new system does not have training data apart from a small number of demonstration dialogues. In addition to the most probable data-like actions, the constrained action set is restricted to legitimate requests, apologies and confirmations in state s. The fact that constrained actions are data-driven and driven by application independent heuristics facilitates its usage across domains.

The *state transition function* is based on a numerical vector representing the last system and user responses. The former are straightforward, 0 if absent and 1 if present. The latter correspond to the confidence level [0..1] of noisy user responses. Given that *SimpleDS* targets a simple and extensible dialogue system, it uses templates for language generation, a rule-based user simulator, and confidence scores generated uniformly at random (words with scores under a threshold were distorted).

The *reward function* is motivated by the fact that human-machine dialogues should confirm the information required and that interactions should be human-like. It is defined as $R(s, a, s') = (CR \times w) + (DR \times (1 - w)) - DL$, where CR is the number of positively confirmed slots divided by the slots to confirm; w is a

[1]The code of *SimpleDS* is available at https://github.com/cuayahuitl/SimpleDS.

[2]Actions: Salutation(greeting), Request(hmihy), Request(food), Request(price), Request(area), Request(food, price), Request(food, area), Request(price, area), Request(food, price, area), AskFor(more), Apology(food), Apology(price), Apology(area), Apology(food, price), Apology(food, area), Apology(price, area), Apology(food, price, area), ExpConfirm(food), ExpConfirm(price), ExpConfirm(area), ExpConfirm(food, price), ExpConfirm(food, area), ExpConfirm(price, area), ExpConfirm(food, price, area), ImpConfirm(food), ImpConfirm(price), ImpConfirm(area), ImpConfirm(food,price), ImpConfirm(food, area), ImpConfirm(price, area), ImpConfirm(food, price, area), Retrieve(info), Provide(unknown), Provide(known), Salutation(closing).

weight over the confirmation reward (CR), we used $w = 0.5$; DR is a data-like probability of having observed action a in state s, and DL is used to encourage efficient interactions, we used $DL = 0.1$. The DR scores are derived from the same statistical classifier above, which allows us to do statistical inference over actions given states $(Pr(a|s))$.

The *model architecture* consists of a fully-connected multilayer neural net with up to 100 nodes in the input layer (depending on the vocabulary), 40 nodes in the first hidden layer, 40 nodes in the second hidden layer, and 35 nodes (action set) in the output layer. The hidden layers use Rectified Linear Units to normalise their weights [6]. Finally, the learning parameters are as follows: experience replay size = 10,000, discount factor = 0.7, minimum epsilon = 0.01, batch size = 32, learning step s = 30,000. A comprehensive analysis comparing multiple state representations, action sets, reward functions and learning parameters is left as future work.

Figure 2 shows the learning curve of a *SimpleDS* agent using 3000 simulated dialogues. This agent uses a smaller set of actions per state (between 4 and 5 actions on average) rather than the whole action set per state—according to the application-independent heuristics mentioned in the previous Section. This reduction has the advantage that policies can be learnt quicker, that more sensible dialogues can potentially be learnt, and that it is inherent that some domains make use of legitimate actions during the interaction. In the case of more complex systems, with higher amounts of features and actions, learning with valid actions (rather than all actions) can make a huge difference in terms of computational resources and learning time. The quality of the learnt policies will depend on the learning environment and given constraints.

Table 1 shows example dialogues of the learnt policies with user inputs derived from simulated speech recognition results. Our initial tests suggest that reasonable interactions can be generated using the proposed learning approach, using spoke interaction, and its application to human-robot dialogues is ongoing work.

4 Summary

We describe a publicly available dialogue system motivated by the idea that future dialogue systems should be trained with almost no intervention from system developers. In contrast to previous reinforcement learning dialogue systems, *SimpleDS* selects dialogue actions directly from raw (noisy) text of the last system and user responses. Although it remains to be demonstrated how far one can go with such an approach, *SimpleDS* is a domain independent framework that can be applied to other domains such as social robots. Future work includes to (a) compare different model architectures, training parameters and reward functions; (b) extend or improve the abilities of the proposed dialogue system; (c) train deep learning agents in other (larger scale) domains [7–9]; (d) evaluate end-to-end systems with real users; (e) compare or combine different types of neural nets [10]; and (e) perform fast learning based on parallel computing.

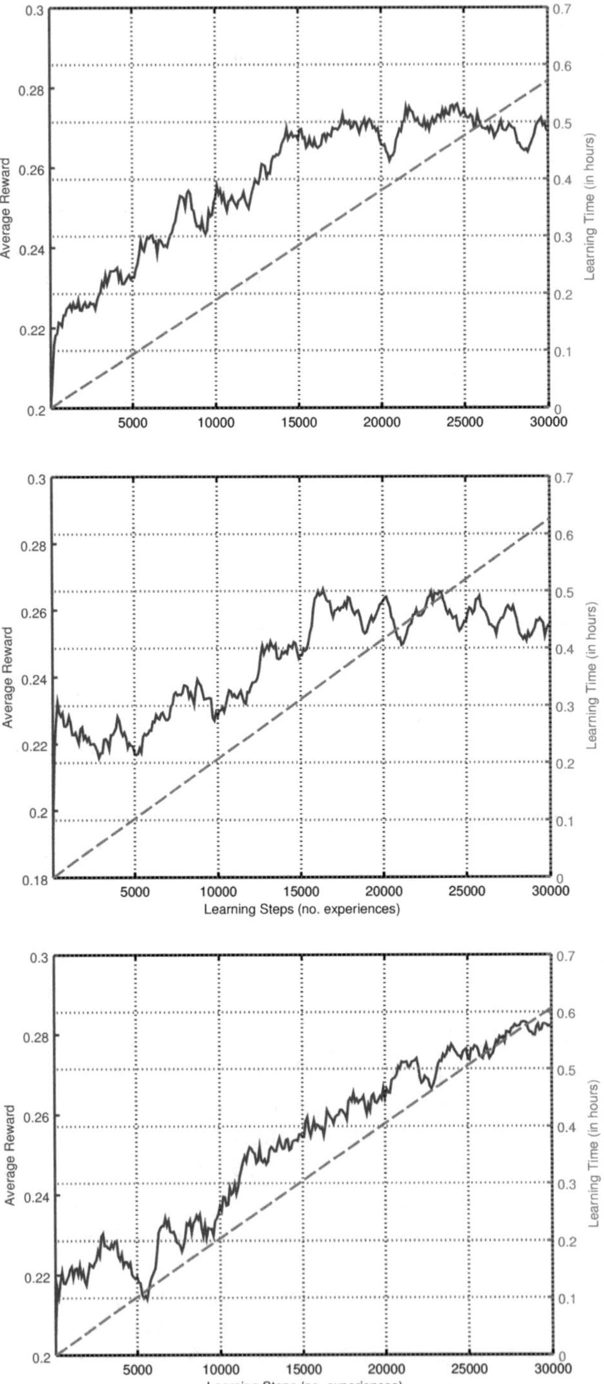

Fig. 2 Learning curves of *SimpleDS* Deep Reinforcement Learning agents in the restaurant domain for English (*top plot*), German (*middle plot*) and Spanish (*bottom plot*)—see text for details

Table 1 Example dialogues using the policies from Fig. 2, where states are numerical representations of the last system and user inputs, actions are dialogue acts, and user responses are in brackets

Language	Environment state	Action	Verbalisation
English	0,0,0,0,0,0,0,0,0,0,0,0,0,0,0,0,0,0,0,0,0,0,0,0,0,0,0,0,0,0,0,0,0,0,0,0,0,0,0,0,0, 0,0,0,0,0,0,0,0,0,0,0,0,0,0,0,0,0,0,0,0,0,0,0,0,0,0,0,0,0,0,0,0,0,0,0,0,0,0,0,0,0	Salutation (greeting)	Hello!
	0,0,0,0,0,0,0,0,0,0,0,0,0,0,0,0,0,0,0,0,0,0,0,0,0,0,1,0,0,0,0,0,0,0,0,0,0,0,0,0,0, 0,0,0,0,0,0,0,0,0,0,0,0,0,0,0,0,0,0,0,0,0,0,0,0,0,0,0,0,0,0,0,0,0,0,0,0,0,0,0,0,0	Request (food, price, area)	What type of food, price range, and area are you looking for? [reasonably priced mexican ...food in the east of town]
	0,0,0,1,0,1,1,0,0,0,0,0,0,0,0,0,0,0,0.53,0,0,0,0,0.63,1,0,0,0,0,0,0,0.83,0,0,0, 0,0,0,0,0,1,0,0,0,0.78,0,0,0,0,0.31,0,0,0,1,0.36,1,0.25,0,0,0,0,0,0,0,0,0,0.57, 0,0,0.82,1,0,0,0,0,1,0,0,0,0,0,1	ImpConfirm (food, price, area)	Okay, reasonably priced mexican food in the east.
	0,0,0,0,0,0,0,0,0,0,0,0,0,0,0,0,0,0,1,0,0,0,0,1,0,0,0,0,0,0,0,1,0,0,0,0,0,0,0,0,0, 0,0,0,1,0,0,0,0,0,1,0,0,0,1,0,1,0,0,0,0,0,0,0,0,0,1,0,0,0,0,0,0,0,0,0,0,0,0,0,0,0	Retrieve (info)	Let me see.
	0,0,0,0,0,0,0,0,0,0,0,0,0,0,0,0,0,0,0,0,0,0,0,0,0,0,0,0,0,0,0,0,0,0,0,0,0,1,0,0,0, 0,0,1,0,0,0,0,0,0,0,0,0,0,0,0,0,0,0,1,0,0,0,0,0,0,0,0,0,0,0,0,0,0,0,0,0,0,0,0,0,0	Provide (known)	Restaurant X is an excellent choice. It is located in Y.
	1,0,0,0,0,0,0,0,0,0,0,0,0,0,0,0,0,1,0,0,0,0,1,0,0,0,0,0,0,1,1,0,0,0,0,0,0,0,0,0,0, 0,1,0,0,0,0,0,0,0,0,0,1,0,0,0,0,0,0,0,0,0,1,0,0,0,0,1,0,0,0,1,0,0,0,0,1,0,0,0,0,0	AskFor (more)	Anything else? [no]
	0,0,0,0,1,0,0,0,0,0,0,0,0,0,0,0,0,0,0,1,0,0,0,0,0,0,0,0,0,0,0,0,0,0,0,0,0,0,0,0,0,0, 0,0,0,0,0.72,0,0,0,0,0,0,0,0,0,0,0,0,0,0,0,0,0,0,0,0,0,0,0,0,0,0,0,0,0,0,0,0,0,0	Salutation (closing)	Okay, talk to you soon. Bye!
	0,0,0,0,0,0,0,0,1,0,0,0,0,0,0,0,0,0,0,0,0,0,0,0,0,0,0,0,0,0,0,0,0,0,0,0,0,0,0,0,0, 0,0,0,0,0,0,0,0,0,1,0,0,0,0,0,0,0,0,0,0,1,0,0,1,0,0,0,1,0,0,0,0,0,0,0,0,0,0,0,0,1		

(continued)

Table 1 (continued)

Language	Environment state	Action	Verbalisation
German	0,0,0,0,0,0,0,0,0,0,0,0,0,0,0,0,0,0,0,0,0,0,0,0,0,0,0,0,0,0,0,0,0,0,0,0,0,0,0,0,0, 0,0,0,0,0,0,0,0,0,0,0,0,0,0,0,0,0,0,0,0,0,0,0,0,0,0,0,0,0,0,0,0,0,0,0,0,0,0,0,0,0, 0,0,0,0,0,0,0,0,0,0,0,0,0,0,0,0,0	Salutation (greeting)	Hallo!
	0,0,0,0,0,0,0,0,0,0,0,0,0,0,0,0,0,0,0,0,0,0,0,0,0,0,0,0,0,0,0,0,0,1,0,0,0,0,0,0,0, 0,0,0,0,0,0,0,0,0,0,0,0,0,0,0,0,0,0,0,0,0,0,0,0,0,0,0,0,0,0,0,0,0,0,0,0,0,0,0,0,0, 0,0,0,0,0,0,0,0,0,0,0,0,0,0,0,0,0	Request (hmihy)	Wie kann ich ihnen helfen? Ich suche nach teuer chinesischem ...Essen in der suedlichen Gegend
	0.14,0,0,0,0,0,0,0.15,0,0,0.66,0,0,0,0,0,0.83,0,0,0,0,0.05,0,0,0,0,0,0,0, 0,0,1,0,0,1,1,0,0,0.27,0,0,0,0,0,1,0,0,0,0,0,0,0.04,0,0,0,0,0,0,0,0,0,0,0,0, 0.93,0,0,0,0,0,0,0,0,0.08,0,0,0,0,0,0,0,0,0,1,0,0,0,0,0,0,0	ExpConfirm (food, price, area)	Haben Sie gesagt, sie moechten teuer chinesischem essen im sueden? Ja bitte
	1,0,0,0,0,0.42,0,1,0,0,0,0,0,0,0,0,1,0,0,0,0,0,0,0,0,1,0,1,0,0,0,0,0,0,0,0, 0,1,0,0,0,0,0.32,0,0,0,0,0,0,0,0,0,1,0,1,0,0,0,0,0,0,0,0,0,0,0,1,0,0,0,0,1,0,0,0, 0,1,0,0,0,0,0,0,0,0,0,0,0,0,0,0,0,0,0,0,0	Retrieve (info)	Lassen sie mich einmal nachsehen.
		Provide (known)	Restaurant X ist eine ausgezeichnete Wahl. Es ist in Y.
	0,0,0,0,0,0,0,0,0,0,0,0,0,0,0,1,1,0,0,0,1,0,1,0,0,0,0,0,1,0,0,0,0,0,0,0,1,0,1, 0,0,0,0,0,0,0,0,1,0,1,0,0,0,1,0,0,0,0,0,0,0,0,0,0,0,0,0,0,0,0,0,0,0,0,0,0,0,0, 0,0,1,0,0,0,0,0,0,0,0,0,0,0,0,1,0,0,0	AskFor (more)	Sonst noch etwas? [Nein]
	0,0,0,0,0,0,0,0,0,0,0,0,0,0,0,0,0,0,0,1,0,0,0,0,0,0,0,0,0,0,0,0,0,0,0,0,0,0,0,0, 0,0,0,0,0,0,0,0,0,0,0,0,0,0,0,0,0,0,0,0.8036069,0,1,0,0,0,0,0,0,0,0,0,0,1,0, 0,0,0,0,0,0,0,0,0,0,0,0,0,0,0,0,0,0,0,0,0,0	Salutation (closing)	Okay, auf Wiedersehen. Tschüss!
	0,0,0,1,0,0,0,0,0,0,0,0,0,0,0,0,0,0,0,0,0,0,0,0,0,0,0,0,0,0,0,0,0,0,0,0,0,0,0,0, 0,0,0,0,0,0,0,0,0,0,0,0,0,0,0,0,0,0,0,0,0,0,0,0,0,0,1,0,0,0,0,0,0,0,0,0,0,0,0,0, 0,1,0,0,0,0,0,0,0,0,0,1,0,0,0,0,0		

(continued)

Table 1 (continued)

Language	Environment state	Action	Verbalisation
Spanish	0,0,0,0,0,0,0,0,0,0,0,0,0,0,0,0,0,0,0,0,0,0,0,0,0,0,0,0,0,0,0,0,0,0,0,0,0, 0,0,0,0,0,0,0,0,0,0,0,0,0,0,0,0,0,0,0,0,0,0,0,0,0,0,0,0,0,0,0,0,0,0,0,0	Salutation (greeting)	Hola!
	0,0,0,0,0,0,0,0,0,0,0,0,0,0,0,0,0,0,0,0,0,0,0,0,0,0,0,0,0,0,0,0,0,0,0,0,0, 0,0,0,1,0,0,0,0,0,0,0,0,0,0,0,0,0,0,0,0,0,0,0,0,0,0,0,0,0,0,0,0,0,0,0,0	Request (food, price, area)	Que tipo de comida, rango de precio y area estas buscando? [Estoy buscando comida hindu ...de precio razonable en el... ...oeste de la ciudad]
	0,0,1,0,0,0,0,0.96,0,0,0,0.03,0,0.77,0,0,0.75,0,0,0,0,0,0.63,0.43,0,0,0,0,0, 0,1,0,0.13,0,0,0,0,0,0.01,0,0,0,0.85,0,0,0,0,0,0,0,0,0.48,0,0,0.68,0,0,0,1, 0,1,0.52,0,0,0,0,1,0,0,0,1	ImpConfirm (food, price)	Bien, comida hindu de precio razonable.
	0,0,0,0,0,0,1,0,0,0,0,0,1,0,0,1,0,0,0,0,0,0,0,0,0,0,0,0,0,0,0,0,0,0,0,0, 0,0,1,0,0,0,0,0,0,0,0,0,0,0,0,0,0,0,0,1,0,0,0,0,0,0,1,0,0,0,0,0,0,0,0,0	ExpConfirm (area)	Dijiste en el area oeste? [Si por favor]
	0,0,1,0,0,0,0,0,0,0,0,0,0,0,0,0,0,0,0,0,1,0,0,1,1,0,0,0,0,0,0,0,0,0,0,0,0.82,0, 0,0,0,0,0,0,0,0,0,0,0,0,0,0,0,0,1,0,0.67,0,0,0,0,0,0,0,0,0,0.13,0,0,0,0,0,0,0	Retrieve (info)	Dejame ver.
	0,0,0,0,0,0,0,0,0,0,0,0,0,0,0,0,0,1,0,0,0,0,0,0,0,0,0,0,0,0,0,0,0,0,0,0, 0,0,0,0,0,0,0,0,0,0,0,0,0,0,0,0,0,0,0,0,0,0,0,0,0,0,0,0,0,0,0,0,0,0,1,0,0	Provide (known)	El restaurante X es una excelente opción. Esta localizado en Y.
	0,0,0,0,0,0,0,0,0,0,0,0,0,0,1,0,0,0,0,0,0,0,1,0,0,0,1,0,1,0,0,0,0,0,0,0,0, 0,0,0,0,0,0,0,0,0,0,0,0,1,1,0,0,0,0,0,0,1,0,0,0,0,0,0,1,0,0,0,0,0,0,0,0	AskFor (more)	Algo más? [no]
	1,0,0,0,0,0,0,0,0,0,0,0,0,0,0,0,0,0,0,0,0,0,0,0,0,0,0,0,0,0,0,0,0,0,0,0,0,0, 0,0,0,0,0,0,0,0,0,0,1,0,0.11,0,0,0,0,0,0,0,0,0,0,0,0,0,0,0,0,0,0,0,0,0,0	Salutation (closing)	Muy bien. Hasta la próxima!
	0,0,0,0,0,0,1,0,0,0,0,0,0,0,0,0,0,0,0,0,0,0,0,0,0,0,0,0,0,0,0,0,0,0,0,0,0, 0,1,0,0,0,0,1,0,0,0,1,0,0,0,0,0,0,0,0,0,0,1,0,0,0,0,0,0,0,0,0,0,0,0,0,0		

Acknowledgements Funding from the European Research Council (ERC) project "STAC: Strategic Conversation" no. 269427 is gratefully acknowledged.

References

1. Mnih, V., Kavukcuoglu, K., Silver, D., Graves, A., Antonoglou, I., Wierstra, D., Riedmiller, M.: Playing atari with deep reinforcement learning. In: Proceedings of the NIPS Deep Learning Workshop (2013)
2. Paek, T., Pieraccini, R.: Automating spoken dialogue management design using machine learning: an industry perspective. Speech Commun. **50**(8–9) (2008)
3. Cuayáhuitl, H., Keizer, S., Lemon, O.: Strategic dialogue management via deep reinforcement learning. In: Proceedings of the NIPS Deep Reinforcement Learning Workshop (2015)
4. Szepesvári, C.: Algorithms for Reinforcement Learning. Morgan and Claypool Publishers (2010)
5. Karpathy, A.: ConvNetJS: Javascript library for deep learning (2015). http://cs.stanford.edu/people/karpathy/convnetjs/
6. Nair, V., Hinton, G.E.: Rectified linear units improve restricted boltzmann machines. In: Proceedings of ICML (2010)
7. Cuayáhuitl, H., Renals, S., Lemon, O., Shimodaira, H.: Evaluation of a hierarchical reinforcement learning spoken dialogue system. Comput. Speech Lang. **24**(2) (2010)
8. Cuayáhuitl, H., Dethlefs, N.: Spatially-aware dialogue control using hierarchical reinforcement learning. TSLP **7**(3) (2011)
9. Cuayáhuitl, H., Kruijff-Korbayová, I., Dethlefs, N.: Nonstrict hierarchical reinforcement learning for interactive systems and robots. TiiS **4**(3) (2014)
10. Sainath, T.N., Vinyals, O., Senior, A.W., Sak, H.: Convolutional, long short-term memory, fully connected deep neural networks. In: Proceedings of the ICASSP (2015)

Breakdown Detector for Chat-Oriented Dialogue

Tomo Horii, Hideaki Mori and Masahiro Araki

Abstract It is difficult to determine the cause of a breakdown in a dialogue when using a chat-oriented dialogue system due to a wide variety of possible causes. To address this problem, we analyzed a chat dialogue corpus and formulated a taxonomy of the errors that could lead to dialogue breakdowns. The experimental results demonstrated the effectiveness of the taxonomy to some degree. We also developed a breakdown detector that comprises combinations of classifiers for different causes of errors based on the taxonomy.

Keywords Chat-oriented dialogue system · Breakdown detection · Taxonomy of errors

1 Introduction

The chat functionality in a dialogue system is helpful in various ways. It is expected to encourage the use of the dialogue system on a daily basis and to establish a sense of closeness to the system [1]. With continued daily usage, novice and elderly users can become accustomed with the use of a speech interface, which makes it easier for them to enjoy the benefits of different services provided by the dialogue system.

Recently proposed statistical utterance generation methods in chat-oriented dialogue systems (e.g., [2]) have robust response generation functionalities. However, there is a tradeoff because these systems occasionally generate an inappropriate utter-

T. Horii (✉) · H. Mori · M. Araki
Kyoto Institute of Technology, Matsugasaki, Sakyo-ku, Kyoto 6068585, Japan
e-mail: horii@ii.is.kit.ac.jp
URL: http://www.ii.is.kit.ac.jp/araki/FSwiki/wiki.cgi?page=English+page

H. Mori
e-mail: mori@ii.is.kit.ac.jp

M. Araki
e-mail: araki@kit.ac.jp

K. Jokinen and G. Wilcock (eds.), *Dialogues with Social Robots*,
Lecture Notes in Electrical Engineering 427,
DOI 10.1007/978-981-10-2585-3_9

ance, which leads to a breakdown of the dialogue. Thus, it is important to detect a breakdown before an inappropriate system utterance is generated.

In the present study, we analyzed a Japanese text chat dialogue corpus and formulated a taxonomy of the errors that can cause dialogue breakdowns [3, 4]. Based on this analysis, we developed a breakdown detection method that comprises combinations of classifiers for different causes of errors based on the taxonomy. Our experimental results demonstrated the effectiveness of the taxonomy to some degree.

In this study, we describe a breakdown detector we implemented based on the proposed method.

2 Proposed Method

2.1 Dialogue Data Analysis and Proposed Taxonomy

We analyzed a Japanese text chat dialogue corpus collected by a web-based text dialogue data collection system using a chat API provided by NTT DOCOMO [5]. The system utterance is generated from (1) pre-stored knowledge (predicate-noun sets created by analyzing text on the Internet) to generate responses or (2) being selected from large-scale utterance data. The generated utterance is converted in pre-output conversion module in order to realize the gender-specific inflections at the end of sentences in the Japanese language and a more consistent personality.

The corpus contained 1,146 dialogues, each of which included 21 utterances (one system prompt followed by 10 user–system utterance pairs). Figure 1 shows an excerpt of a dialogue collected with the system.

This corpus also contained annotation data, which indicated three levels for a breakdown in a dialogue: "Not a breakdown," "Possible breakdown," and "Breakdown." About 60 % of the system utterances were annotated as "Not a breakdown,"

```
S: Hello! I'm bored.
U: Don't be lazy. Get down to work.
S: Don't you like boredom?
U: It's not good for you to take such frequent rests.
S: They don't like boredom.
U: You seem to have poor powers of concentration.
   I know it's after a long vacation.
S: They put up with boredom.
U: Did you go back to your hometown in the vacation?
S: I spend my time in my hometown.
...
```

Fig. 1 Excerpt of collected dialogue. S and U stand for system and user utterances, respectively. Dialogue was originally in Japanese and translated by authors

25 % as "Possible breakdown," and 15 % as "Breakdown." We also analyzed the "Possible breakdown" and "Breakdown" utterances to investigate the causes of the errors. Based on this analysis, we formulated an error taxonomy [3, 4].

This error taxonomy is hierarchical where the main categories represent the scope of error detection, i.e., utterance, response, context, and environment. Within each main category, we created the subcategories shown in Table 1.

The errors in utterance-level can be recognized as errors by looking at single utterances. They typically come from language-generation deficiencies of the system. (1) Syntax error: The utterance is ungrammatical or lacks important elements to be a valid sentence. (2) Semantic error: The utterance is semantically invalid such as when the combination of a predicate and its arguments cannot constitute any meaning. (3) Uninterpretable: The utterance is not understandable. There are no recognizable words or just a fragment of an utterance.

An example of the semantic error in the system utterance is as follows,

```
S: I am good at raining.
   (one cannot be good at raining)
```

The errors in response-level can be recognized as errors by looking at adjacency pairs. In this level, we identified five sub-categories in which the first four relate to following Grice's maxims.

Table 1 Taxonomy of errors that could cause a dialogue breakdown

Main category	Sub category	Type of errors
Utterance	Syntactic error	Ungrammatical sentence, missing necessary arguments
	Semantic error	Semantically invalid sentence
	Uninterpretable	No recognizable words, fragment of an utterance
Response	Excess/Lack of information	Error of maxim of quantity
	Not understood	Error of maxim of quality
	No relevance	Error of maxim of relation
	Unclear intention	Error of maxim of manner
	Misunderstanding	Failure to grasp the meaning of previous user utterance
Context	Excess/lack of proposition	Error of maxim of quantity
	Contradiction	Error of maxim of quality
	Non-relevant topic	Error of maxim of relation
	Unclear relation	Error of maxim of manner
	Topic switch error	Miss the change in topic by the user
Environment	Lack of common ground	Out of the blue
	Lack of common sense	Believed to be false in the real world
	Lack of sociality	Violate social common sense

- maxim of quantity
- maxim of quality
- maxim of relation
- maxim of manner

The last sub-category of this level is Misunderstanding. It means the utterance displays the misunderstanding of the previous user utterance such as when the system mishandles the sense of a polysemic word.

An example of the no-relevance error (i.e. violation of the maxim of relation) is as follows,

```
U: What is your hobby, then?
S: When was your last trip?
```

The errors in context-level can be recognized as errors by looking at the local context, not just by adjacency pairs. In this level, we identified five sub-categories in which the first four relate to Grice's maxims. The last sub-category of this level is Topic switch error. It means the utterance displays the fact that the system missed the switch in topic by the user, continuing with the previous topic.

An example of the contradiction error (i.e. violation of the maxim of quality) is as follows,

```
S: It's nice to go shopping alone.
U: I agree. That's nice.
S: Shopping takes time.
U: Window shopping is also fun.
S: It's fun to go shopping with somebody.
```

The errors in environment-level are those that cannot be recognized as errors by the local context but by other issues outside the dialogue. In this level, we identified three sub-categories, (1) Lack of common ground, (2) Lack of common sense and (3) Lack of sociality.

An example of the Lack of sociality error is as follows,

```
U: I will look for a good pool for you.
S: Why don't you know it!
```

The last system utterance satisfies the utterance-level, response-level and context-level constraint. However, such a rude system utterance must be filtered out especially in commercial service.

2.2 Breakdown Detection by Combining Different Error Detectors

Based on the above analysis of the types of errors, we concluded that a different approach was required to detect each type of error. For example, the utterance-

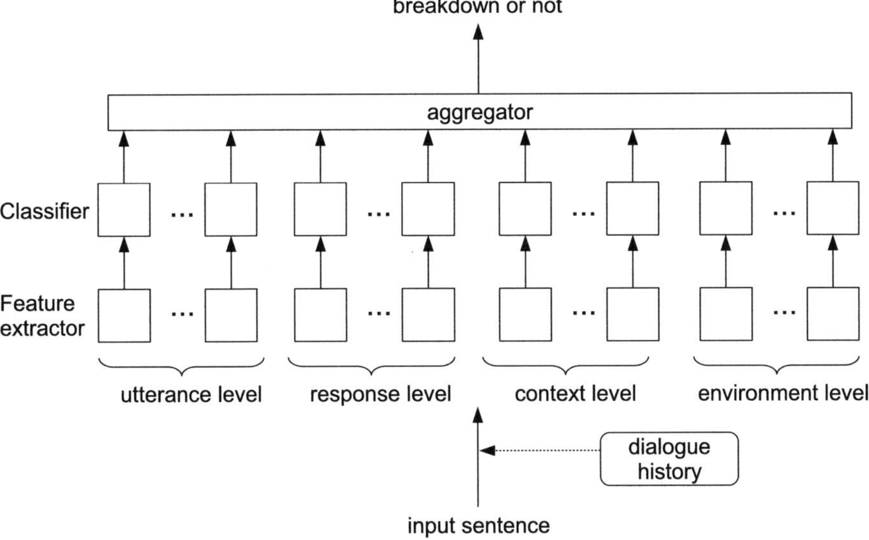

Fig. 2 Architecture of the proposed breakdown detection method

level errors can be identified based on the parsing score obtained from the syntactic parser, or matching score with the semantic case frame. The statistics of adjacent-pairs could be effective in identifying the response-level errors. On the other hand, dialogue history information is necessary for identifying the context-level errors.

Also, the features used for each error category could be different. The classifier learning algorithm for each category could also be different. Therefore, we decided to use an aggregation architecture for the classifiers shown in Fig. 2.

We selected this architecture to develop a good classification procedure but also because it allowed a loss coefficient to be attached to each classifier. Some errors, e.g., insulting the user ("Lack of sociality" subcategory in the "Environment" category), cannot be permitted in commercial services. Thus, we could incorporate the loss coefficient for each error into the classifier using this architecture.

3 Experiments

3.1 Baseline Classifier

In the first step, we defined 16 classifiers for each of the 16 sub-categories shown in Table 1. We used a word vector as a feature, which comprised the system utterance and the preceding user utterance. We used a degree-1 polynomial kernel SVM (support vector machine) as a classifier. The training data comprised 1,000 utterances taken

Table 2 Classification results

	Accuracy	Precision	Recall	F-measure
16 classifiers	0.56	0.76	0.29	0.42
4 classifiers	0.49	0.76	0.70	0.73
1 classifier	0.42	0.73	0.73	0.73

from the annotated corpus described in Sect. 2. The test data comprised 200 utterances taken from another chat corpus. The final outputs from this baseline classifiers were the breakdown results obtained from the sub-category classifier outputs. The results of the first experiment are shown in the first row (16 classifiers) in Table 2.

3.2 Aggregated Classifiers by Level

The recall rate was extremely low in the first experiment, so we conducted a follow-up experiment, where we implemented four classifiers for each main category, as shown in Table 1. The other experimental settings were the same as those used in the first experiment. The results of the second experiment are shown in the second row (four classifiers) of Table 2.

The recall rate was improved considerably in the second experiment and the same precision rate was maintained. This suggests that the poor performance in the first experiment was due to the low number of training data.

To confirm this result, we conducted a third experiment with one classifier. The results of the third experiment are shown in the third row (one classifier) of Table 2. The F-score was almost the same as that in the second experiment, but the accuracy was lower. Therefore, we demonstrated the advantage of using the aggregation classifier architecture. However, the results of the third experiment show that the number of training data may have been insufficient.

We also examined the following different features for each level classifier (four classifiers). However, also due to insufficient training data, they do not outperform the features consist of word vector of the system utterance and the preceding user utterance.

- Utterance-level: score of the syntactic parser KNP[1]
- Response-level: degree-2 polynomial kernel
- Context-level: cosine similarity between accumulated word vector of context and the word vector of the system utterance
- Environment-level: word vector of the system utterance

[1] http://nlp.ist.i.kyoto-u.ac.jp/?KNP.

4 Breakdown Detector

Based on the result of this experiment, we employed an architecture with the aggregated classifiers by level in our breakdown detection system. Figure 3 shows the architecture of the breakdown detection system with four classifiers, which corresponded to each of the main error categories in Table 1.

The frontend system was run on a web browser. We used the Google speech API as a speech recognition system. For demonstration purposes, we used the Microsoft translator API to translate Japanese sentences into English. The breakdown detection system only worked with Japanese sentences.

Figure 4 shows an example of a successful chat interaction.

Figure 5 shows an example of error detection at the response level.

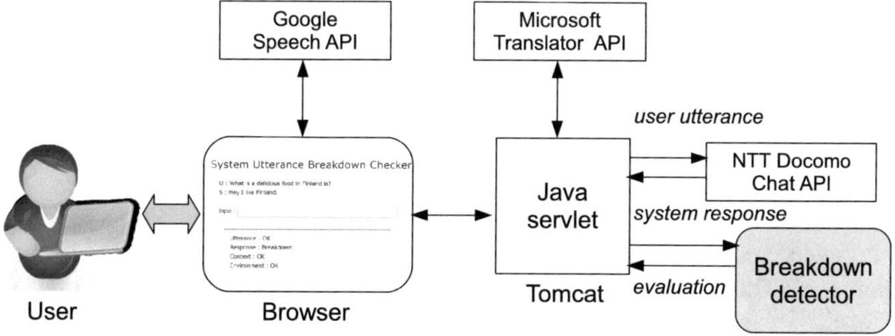

Fig. 3 Architecture of the demonstration system

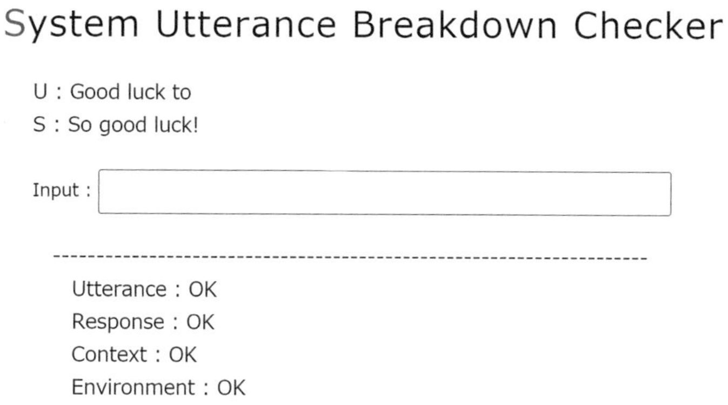

System Utterance Breakdown Checker

U : Good luck to

S : So good luck!

Input :

Utterance : OK

Response : OK

Context : OK

Environment : OK

Fig. 4 Screen shot of breakdown detection (1)

System Utterance Breakdown Checker

U : What is a delicious food in Finland is?
S : Hey I like Finland.

Input :

Utterance : OK
Response : Breakdown
Context : OK
Environment : OK

Fig. 5 Screen shot of breakdown detection (2)

5 Related Work

Previous studies have detected problematic situations in chat-oriented dialogue systems [2, 6].

Thus, Xiang et al. [6] proposed a method for identifying a user's unsatisfied utterance in a conversation via a chat-oriented dialogue system, where they used a sequence model of dialogue and included the features of the user's intent and sentiment. The difference between their approach and ours is that they used the utterance of the current user to detect a problematic situation. By contrast, in our approach, we tried to avoid producing an unacceptable system utterance before it was generated by the system. Therefore, we did not use the user utterance after the breakdown.

Higashinaka et al. [2] proposed a method for evaluating the coherence of a system utterance during a conversation via a chat-oriented dialogue system. They used various pieces of information, such as utterance pairs, dialogue acts, and predicate-argument structures, to evaluate the coherence of a system utterance. However, this rich feature extraction process requires strong support from linguistic tools, and thus this method depends partly on the language used. Our approach employs a combination of simple detectors for each type of error, so the method itself does not depend on a specific language.

6 Conclusion

In this study, we proposed a breakdown detector that uses combinations of classifiers to represent different causes of errors based on a taxonomy that we formulated. The experimental results demonstrated the effectiveness of the taxonomy to some extent.

In future research, we will consider appropriate features for each error category and create a new breakdown detector to improve the detection performance.

Acknowledgements We thank all members of the dialogue task in "Project Next NLP" for data collection, annotation, and productive discussions.

References

1. Bickmore, T.W., Picard, R.W.: Establishing and maintaining long-term human-computer relationships. ACM Trans. Comput.-Hum. Interact. **12**(2), 293–327 (Jun 2005). http://doi.acm.org/10.1145/1067860.1067867
2. Higashinaka, R., Meguro, T., Imamura, K., Sugiyama, H., Makino, T., Matsuo, Y.: Evaluating coherence in open domain conversational systems. In: Proceedings of of Interspeech, pp. 130–134 (2014)
3. Higashinaka, R., Funakoshi, K., Araki, M., Tsukahara, H., Kobayashi, Y., Mizukami, M.: Towards taxonomy of errors in chat-oriented dialogue systems. In: Proceedings of the 16th Annual Meeting of the Special Interest Group on Discourse and Dialogue. pp. 87–95. Association for Computational Linguistics, Prague, Czech Republic (September 2015). http://aclweb.org/anthology/W15-4611
4. Higashinaka, R., Mizukami, M., Funakoshi, K., Araki, M., Tsukahara, H., Kobayashi, Y.: Fatal or not? finding errors that lead to dialogue breakdowns in chat-oriented dialogue systems. In: Proceedings of the 2015 Conference on Empirical Methods in Natural Language Processing. pp. 2243–2248. Association for Computational Linguistics, Lisbon, Portugal (September 2015). http://aclweb.org/anthology/D15-1268
5. Onishi, K., Yoshimura, T.: Casual conversation technology achieving natural dialog with computers. NTT DOCOMO Tech. J **15**(4), 16–21 (2014)
6. Xiang, Y., Zhang, Y., Zhou, X., Wang, X., Qin, Y.: Problematic situation analysis and automatic recognition for chinese online conversational system. In: Proceedings of The Third CIPS-SIGHAN Joint Conference on Chinese Language Processing. pp. 43–51. Association for Computational Linguistics, Wuhan, China (October 2014). http://www.aclweb.org/anthology/W14-6808

User Involvement in Collaborative Decision-Making Dialog Systems

Florian Nothdurft, Pascal Bercher, Gregor Behnke and Wolfgang Minker

Abstract Mixed-initiative assistants are systems that support humans in their decision-making and problem-solving capabilities in a collaborative manner. Such systems have to integrate various artificial intelligence capabilities, such as knowledge representation, problem solving and planning, learning, discourse and dialog, and human-computer interaction. These systems aim at solving a given problem autonomously for the user, yet involve the user into the planning process for a collaborative decision-making, to respect e.g. user preferences. However, how the user is involved into the planning can be framed in various ways, using different involvement strategies, varying e.g. in their degree of user freedom. Hence, here we present results of a study examining the effects of different user involvement strategies on the user experience in a mixed-initiative system.

Keywords Human-computer interaction · Cooperative decision-making · User experience · Dialogue systems

---

F. Nothdurft (✉) · W. Minker
Institute of Communications Engineering, Ulm University, Ulm, Germany
e-mail: florian.nothdurft@uni-ulm.de

W. Minker
e-mail: wolfgang.minker@uni-ulm.de

P. Bercher · G. Behnke
Institute of Artificial Intelligence, Ulm University, Ulm, Germany
e-mail: pascal.bercher@uni-ulm.de

G. Behnke
e-mail: gregor.behnke@uni-ulm.de

K. Jokinen and G. Wilcock (eds.), *Dialogues with Social Robots*,
Lecture Notes in Electrical Engineering 427,
DOI 10.1007/978-981-10-2585-3_10

1 Introduction

Most contemporary dialog systems (DS) manage the interaction between human and machine in a uni-directional dependency. Most common, users interact with a DS to solve domain-dependent tasks. However, this is usually limited to information retrieval or exchange tasks, such as searching for bus connections or restaurants, where usually the user is exclusively in charge. Contrary to that, future DS will evolve towards being *companions* [1] for the user (i.e. intelligent personal assistants) that help the user not only in simple but also in complex tasks. These companions solve complex problems collaboratively with the user, if either the user or the DS is not able to solve the problem on its own, to the liking of the other, or simply if the user's load should be reduced. Companions may, for example, provide assistance in form of artificial intelligence problem solving skills (i.e. planning), that are intrinsically different from human problem solving skills. AI planning may solve combinatory problems that are too complex for humans, due to the need of mathematical computations or fast information storage and retrieval of large data. In addition, such systems may have knowledge and problem solving skills for a domain, the user is not an expert in.

However, the automatic and autonomous generation of a solution (i.e., a plan) to a user-provided problem by such artificial skills potentially does not respect the user's individual needs and preferences, and is per se not always the best solution to the problem. The generated plan is usually only *a* solution, and not one which is best-suited for the user. Integrating preferences into planning [2] is a solution, yet requires the user to specify his preferences a priori in an expressible (e.g. action or method costs) way, which is likely to result in user frustration or even interaction stop. In addition, the planning process is done autonomously and exclusively by the planning system. However, this exclusion of the user from the decision-making process will lead to a couple of problems, we described in previous work [3]. For example, if humans are not involved into a decision-making process, they are less likely to follow or execute a proposed plan or solution. In addition, in decisions that involve grave risks, e.g. in military settings [4] or spaceflight [5], humans must have the final decision on which actions are to be contained in the plan.

Therefore, we proposed a collaborative decision-making assistant, in Nothdurft et al. [3], that combines AI planning and human problem-solving skills into a collaborative decision-making process. This results in a mixed-initiative planning system (MIP) [4–6], or more general a mixed-initiative assistant (MIA) [7] that supports users in problem solving and finding appropriate solutions. A collaborative decision-making process has the intent of solving a problem the user is not able to solve at all or only with great effort. It aims at relieving the user's cognitive load and simplifying the problem at hand. In general, the intertwining of human and AI decision-making skills should lead to an increased user-experience by more preferable and individual solutions for the user. In addition, MIP also facilitates the adaptation of a companion to its owner. The companion may learn from past interaction episodes and direct the future decision-making processes to the user's liking. Since not only the companion may adapt to the user over time, but the user to the decision-making capabilities of the system as well, this process may be described as a *co-adaptation* of two parties.

However, intertwining user and AI planning systems into a MIP system, does not only facilitate more intelligent and competent systems, but does also raise new questions. In previous work we described the potentials, challenges, and risks involved in such MIP systems, along a prototypical MIP system architecture. Some of these challenges, for example, how to maintain coherent models for the participating components [8], or how to deal with occuring phenomena, such as *backtracking* in a collaborative MIP process [3], were already tackled.

2 Mixed-Initiative Planning

In general, the interaction between AI planning and user has to begin with a dialog to define the statement of an objective. This first dialog has the goal of defining the task in a way understandable for the planner. Once the problem is passed to the planner the interactive planning itself may start. Using a selected search strategy (here: *depth-first search*) the plan is refined by selecting appropriate modifications for open decisions. In order to decide whether to involve the user or not during this process, an elaborate decision model, integrating various information sources, is used. Relevant information sources are, e.g. the *dialog history* (e.g. was the user's decision the same for all past similar episodes?), the kind of *open plan decision* (e.g. is this decision relevant for the user?), the *user profile* (e.g. does the user have the competencies for this decision?), or the current *situation* (e.g. is the current cognitive load of the user low enough for interaction?). These information sources are used in a superordinate component, the *decision model*, to decide whether to involve the user. The decision model can either initiate a user interaction or determine by itself that the planner should make the decision. This is equivalent with the user signaling no explicit preference in the decision-making. Furthermore, it is important whether the additional interaction is critical and required, to successfully continue the dialog. Additional dialogs may contribute to achieving short-term goals, but risk the user's cooperativeness, in the long run, e. g. by overstraining his cognitive capabilities or boring him.

In case of user involvement the information on the current *plan decision* has to be communicated to the user. This means that the open decision and the corresponding choice between available *modifications* have to be represented in a dialog suitable for the user. Hence, the corresponding plan actions or methods (i.e. the set of possible actions for a upcoming decision) needs to be mapped to human-understandable dialog information. As this mapping is potentially required for every plan action or method, and vice versa for every dialog information, coherent models between planner and DS become crucial for MIP systems. The thorough matching of both models would be an intricate and strenuous process, requiring constant maintenance, especially when models need to be updated. Thus, a more appropriate approach is the automatic generation or respectively extension of the respective models using one

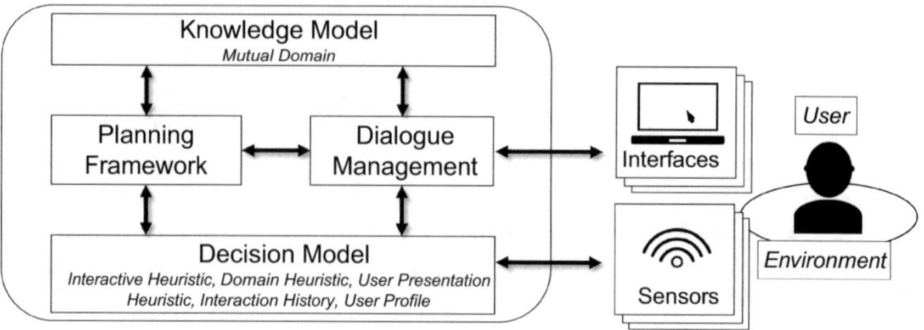

Fig. 1 Essential components of a MIP system (cf. [3])

mutual model as source, the *mutual knowledge model* (cf. [8]). From this model—in this case an OWL ontology [9]—the dialogs and their hierarchy can be derived, using the topmost elements as entry points for the dialog between user and machine. This is, for example, needed for the user to specify the objective for the planner, or to present the available *plan modifications* (i.e. the options in the decision-making), that have to be translated to a format understandable to the user (cf. [3]). The model is also used to extend the planning domain: hierarchical structures (i.e. decomposition methods) are derived using declarative background knowledge modeled in the ontology (cf. [8]). Using a mutual model addresses one of the challenges of MIP (cf. [3]), since translation problems between dialog and planner semantics can be prevented, even when updating the domain (e.g. by acquiring new knowledge, such as new workouts or rehabilitation methods). Another challenge related to the specific interaction between man and machine is *if*, *how*, and at *what* specific point in the dialog user-involvement is necessary or useful. This is one of the most essential challenges, as the integration process, and how the shift of initiative towards one of the parties is framed, affects how effective and user-friendly the MIP will be (Fig. 1).

Our multimodal MIP system was implemented using a knowledge-based cognitive architecture. The multimodal interface uses mainly speech, and graphical user interfaces as input and output. In addition, gestures, and sensory information such as the user location can be used as input. The use of multiple modalities enables us to vary the means of collaboration from uni- to multimodal interaction. The *Dialog Management* uses a modality-independent representation, communicating with the user via the *Fission* [10], *User Interface* [11], *Sensors* [12], and *Fusion* [13] modules.

3 Related Work

Initial work on combing dialog and planning in a mixed-initiative fashion has been done by George Ferguson and James Allen in their TRAINS [14] and TRIPS [15] systems. Their systems include the collaborative capabilities of reasoning, planning, execution, and communication and are based on the belief-desire-intention (BDI) model of agency [16]. Important work, approaching the problem from a different perspective, has been done by Rich et al. in COLLAGEN [17], aiming at applying collaborative discourse theory to human-computer interaction. Their work is based on the SharedPlans theory [18], and models the dialog state of the agents (i.e. user and system) as they interact and perform activities. More recent work involving mixed-initiative interaction has been done in various application domains (e.g. [4, 5, 19, 20]). One of the most well known is MAPGEN [5], applying a mixed-initiative planning and scheduling approach for the ground operations system for the Mars Exploration Rover of NASA. Abstract goals were planned by the user, yet the planner assured that all constraints, which is very complex in such a setting, are satisfied. Another example is DiamondHelp [21], a generic collaborative task-guidance system, which may also integrate the COLLAGEN system. DiamondHelp can be used for a multitude of tasks (e.g. help the user in programming a washing machine or thermostat).

What these work is missing is to investigate *how* the user's involvement should be framed. If the user is to be involved, the question arises how this should be rendered, i.e. what kind of integration is the most beneficial. In addition, if the user is not involved in the decision-making, it has to be decided if and how the user may be informed about the decisions the planner has made. The decision whether and how to involve the user into the planning process is not only controlled by a degree of necessity dependent on the current task and situation, but should also take into consideration the effects on the user's system experience. Usually, the user involvement is done by presenting a list of possible options for upcoming decisions to the user. If this form of user involvement is always necessary or simply best for the user experience is rather questionable. User involvement strategies may actually range from almost unrestricted decision-making (i.e. set of options), limited only by valid solution constraints, over explicit confirmations of system-preselected decisions, to only informing the user of made decisions. Hence, we designed the study examining the effects of different strategies of user involvement on the user experience.

4 User Study About User Involvement Strategies

For this study, we used our prototypical MIP system [3] and implemented several strategies to involve the user into the decision-making. This means that we evaluated different degrees of user involvement into a planning process, ranging from only informing the user of system-made decisions to explicitly requesting a user confirmation for the proposed system decision. In this scenario the user's task was to create individual strength training workouts. In each strength training workout at least three different muscle groups had to be trained and exercises chosen accordingly. The user was guided through the process by the system, which provided a selection of exercises for training each specific muscle group necessary for the workout. For example, when planning a strength training for the upper body, the user had to select exercises to train the chest. This selection is an involvement of the user into the MIP process. The decision how to refine the task of training the chest is not made by the system, but left to the user. The system decision was based on previously made selections by the user. This means that when in a previous interaction the same decision (i.e. the same situation with the exact same options) had do be done, this user-selected option was remembered for future interactions, and selected accordingly by the system. Of course, in a more complex scenario this decision would depend not only on the interaction history, but also on additional information (e.g. affective user states like overextension, interest, or engagement) stored in the user state. The system-made selection was presented in various ways, which were the following:

Explicit confirmation (EC) based on previous selections the choice was already made by the system and presented to the user, who had to explicitly confirm the choice by clicking "okay".

Implicit confirmation (IC) the system-made decision (i.e. the selection) was presented to the user, but the user could intervene, in a certain time frame, by clicking "Let me decide". Therefore, this is a form of implicit confirmation.

Information (INF) the system-made decision was presented to the user without the need of confirmation. Hence, the user was only informed of the system's decision-making, without the option to intervene.

Unsorted (US) the baseline was the usual unsorted selection task. No proactive behavior by the system was present, meaning that users had to select from a list.

In all conditions the system-made selection was explained by the system using a phrase similar to "For training this muscle group, you previously selected this exercise. Therefore, it was already selected for you." The participants were distributed by a random-function to the variants, resulting in 23 participants receiving the known unsorted selection, 25 asked for explicit confirmation, 30 with implicit confirmations, and 26 receiving only an information by the system.

4.1 Used Questionnaires

For the assessment of the study we chose two questionnaires. The AttrakDiff 2 questionnaire [22], which extends the assessment of technical systems or software in general from the limited view of usability, which represents mostly pragmatic qualities, to the integration of scales measuring hedonic qualities. It consists of four basic scales: *perceived pragmatic quality*, which measures the product's ability to achieve the user's goals efficiently and effectively without inducing a high mental load; *hedonic quality-stimulation*, which measures whether novel, interesting and inspiring qualities are present to increase the user's attention and foster the user's abilities and skills; *hedonic quality-identity*, which assesses the user's perceived identity of the subject at evaluation; and *perceived attractiveness*, which is a global rating based on the perceived qualities (Fig. 2).

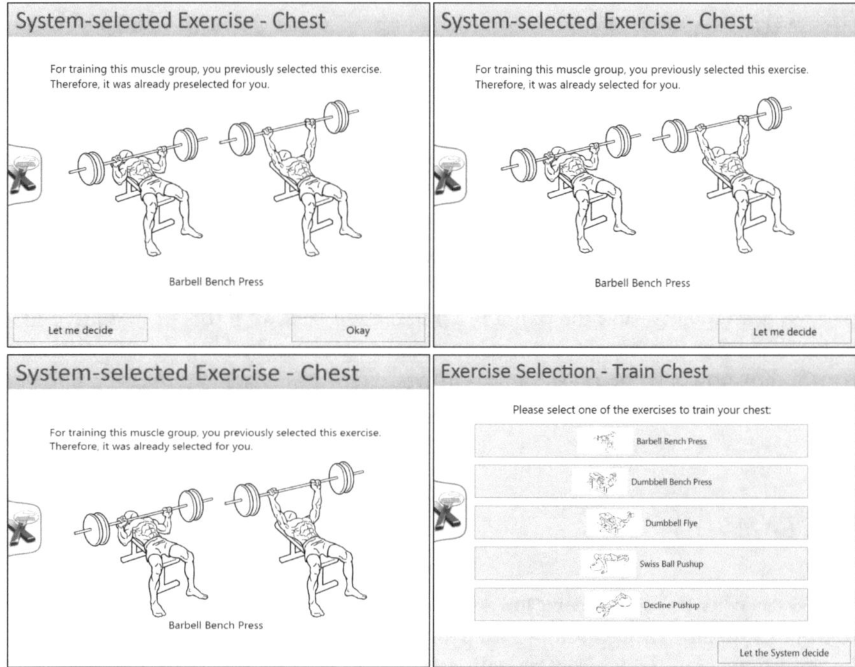

Fig. 2 The different degrees of user involvement: on the *top left* the proactive decision by the system has to be confirmed explicitly by the user (EC). On the *top right*, the decision is presented and the user may intervene in a given time frame, else it is confirmed (IC). On the *bottom left* the user is only informed of the system's decision-making (INF), and on the *bottom right* the usual selection is presented as baseline condition (US)

The other used questionnaire measured the cognitive load. Cognitive load, which consists of the three different types, should not exceed the working memory capacity [23]. One of the basic ideas of cognitive load theory is that a low extraneous load, resulting from a good instructional design, enhances the potential that users engage in cognitive processes (i.e. germane load) related to learning [24]. Hence, the better the instructional design, the greater potential for germane cognitive load and learning. We used an experimental questionnaire developed by [25] which measures all three types of cognitive load separately. The questionnaire consisted of 12 items, with four items each for every type of cognitive load: *intrinsic cognitive load*, which can be described as the inherent load induced by the content itself. This type of load can not be changed by the design of the learning material and is caused mainly by the difficulty of the task. In other words it results e.g. from the number of elements that must be simultaneously processed in the working memory; *extraneous cognitive load*, which is caused by the presentation form of the learning material and is considered to be manipulable by the design of the learning material; *germane cognitive load*, which is considered the load inflicted by the learning process. Germane cognitive load is "good" cognitive load, which helps in fostering the processes inherent in the construction and automation of schemas.

4.2 Hypotheses

Our hypotheses were that in general the various conditions will perform differently, especially regarding perceived cognitive load, pragmatic qualities and attractiveness of the system. We expected no significantly different influences on the human-computer trust relationship between human and machine for the conditions. The exclusion of the user from the decision making (i.e. only informing the user) was expected to reduce the hedonic quality compared to the use of explicit and implicit confirmations. The baseline was expected to perform worst for the perceived pragmatic system quality, and the explicit confirmation best. In terms of cognitive load we expected that when the system takes over the decision making (i.e. implicit confirmation or user information), the cognitive load for the user is reduced compared to the other conditions.

4.3 Results

The results were collected using the AttrakDiff, cognitive load, and human-computer trust questionnaire. In addition, we used an open questions form for user feedback. As the conditions would not affect objective measures like task completion or efficiency rate, they were neglected in this paper.
AttrakDiff Assessing the results of the AttrakDiff questionnaire, using a one-way ANOVA, we found marginal differences between the conditions for the dimensions (see Table 1) of perceived *hedonic quality-identity* ($F(3, 96) = 2.172$, $p = 0.096$) and the perceived overall *attractiveness* ($F(3, 96) = 2.420$, $p = 0.071$) of the system. Post-hoc comparisons using the Fisher Least significant difference (LSD) test indicated that the mean score of *hedonic quality-identity* for the US condition ($M = 3.71$, $SD = 0.705$) was significantly different, at the $p = 0.015$ level, than the INF condition ($M = 4.37$, $SD = 0.86$). For *attractiveness* the US condition ($M = 3.88$, $SD = 0.77$) was also significantly different ($p = 0.009$) than the INF condition ($M = 4.62$, $SD = 0.93$).

Table 1 This table shows the mean values of the AttrakDiff questionnaire dimensions

		PQ	HQ-I	HQ-S	ATT
US	M	4.27	3.71	3.77	3.88
	SD	0.91	0.71	1.02	0.77
EC	M	4.47	4.13	3.95	4.17
	SD	1.12	1.29	1.17	1.24
IC	M	4.31	3.97	3.97	4.19
	SD	1.12	0.68	0.77	0.84
INF	M	4.81	4.37	4.10	4.62
	SD	0.99	0.87	0.96	0.93

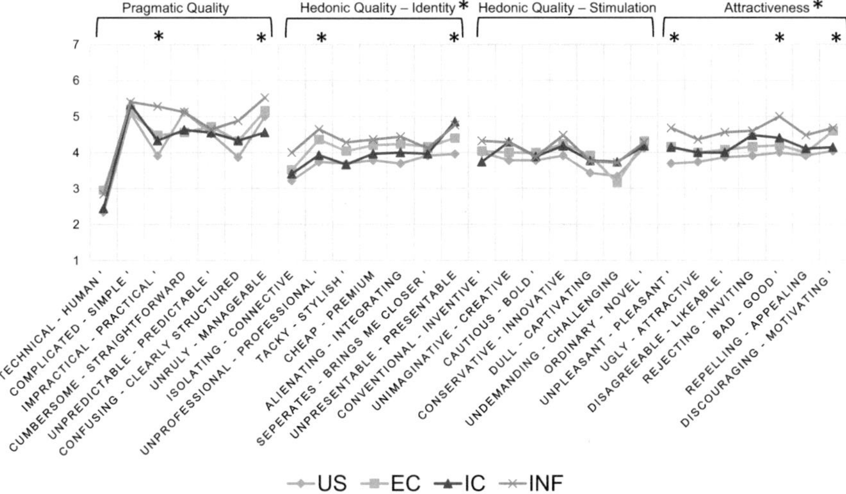

Fig. 3 This shows the average means of the AttrakDiff comparing the confirmation conditions on a 7-point Likert scale. US is a unsorted list of options, EC and IC explicit and implicit confirmations, and INF only informs the user of the system decision. The ' indicates inverted, for the sake of readability, scales. The * indicates significance

Looking further into the data and analyzing the single word pairs of the AttrakDiff questionnaire (see Fig. 3), to find the origin of the differences, we could find more detailed results. Using one-way ANOVA tests we found significant differences in the word pair *impractical-practical* and marginal significance for *unruly-manageable*, which both belong to the dimension of *pragmatic quality*. Post hoc comparisons using Fishers LSD showed that for *impractical-practical* the INF condition performed significantly better ($p = 0.002$) than the US condition and also significantly better ($p = 0.024$) than the IC condition. For *unruly-manageable* the INF condition performed significantly better ($p = 0.14$) than the IC condition.

For the dimension of *hedonic quality-identity* we found a marginal significant differences for the word pairs *unprofessional-professional* and *unpresentable-presentable*. Fishers LSD post hoc test revealed that INF performed significantly better ($p = 0.018$) than US for *unprofessional-professional*. For *unpresentable-presentable* the IC condition was significantly better ($p = 0.020$) than US, which was also significantly worse ($p = 0.040$) than the INF condition.

For *attractiveness* a marginal significant difference, using also a one-way ANOVA, was found in *unpleasant-pleasant* ($F(3, 96) = 2.211$, $p = 0.092$), *bad-good* ($F(3, 96) = 2.397$, $p = 0.073$), and *discouraging-motivating* ($F(3, 96) = 2.314$, $P = 0.081$). Post hoc tests revealed that the significant differences for *unpleasant-pleasant* were between INF and US at ($p = 0.012$). For *bad-good* the average mean of the INF condition was significantly better at ($p = 0.014$) than US, and also better at ($p = 0.043$) than the EC condition The third word pair showing significant

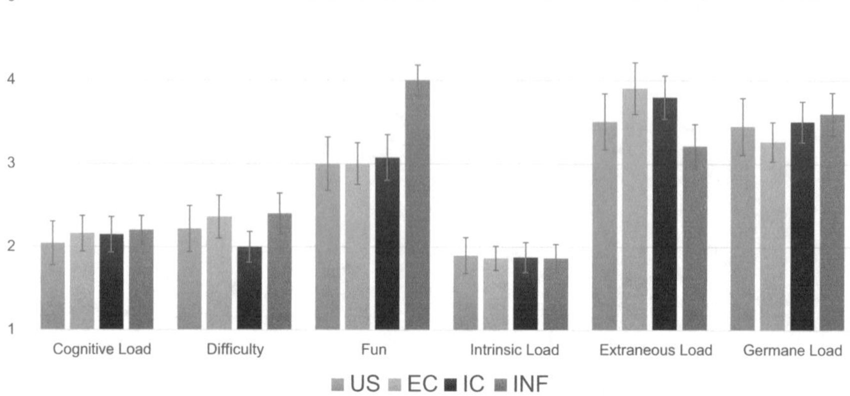

Fig. 4 This shows the average means of the cognitive load comparing the confirmation conditions on a 7-point Likert scale. Intrinsic, extraneous and germane load originate from the experimental questionnaire

results ($p = 0.037$) was *discouraging-motivating* with INF performing better than US. In addition, we found also that the user information condition was significantly performing better for *ugly-attractive* with US and INF at ($p = 0.037$), for *rejecting-inviting* with US and INF at ($p = 0.039$), and also for *repelling-appealing* with US ($M = 3.91$, $SD = 1.08$) and INF at ($p = 0.046$).

Cognitive Load Analyzing the cognitive load items (see Fig. 4) we found significant differences, using a one-way ANOVA, for *fun* with ($F(3, 96) = 3.488$, $p = 0.019$). Fishers LSD showed that the user information condition ($M = 4.00$, $SD = 0.91$) was significantly better than the rest. Compared to US ($M = 3.00$, $SD = 1.53$) at ($p = 0.009$), to EC ($M = 3.00$, $SD = 1.25$) at ($p = 0.008$), and to IC ($M = 3.07$, $SD = 1.43$) it was significant at ($p = 0.012$).

Open Questions The following comments were entered by the participants: 'carrying over previous made decisions should be confirmed by the user', and 'If the system selects an exercise, the system should notify why this exercise was thought to be the most fitting one'.

4.4 Discussion

Surprisingly it showed that only informing the user of the system-made selection, without any possibility to intervene, was performing best in almost any category. The pragmatic quality, the identification with the system (hedonic quality) and the overall attractiveness were best for the INF condition. The automatic selection of the system was perceived as very practical and increased the perception that the system is predictable and manageable. This goes along with the fact that the system behavior was explained to the user, considering earlier results on explanations and system acceptance. Additionally, the INF condition was experienced as the most enjoyable of all, along with reducing the extraneous load (cf. Fig. 4) of the task at hand. Even though the baseline condition of selecting from an unsorted list as before, was experienced before, and thus would require no additional cognitive load, the automatic selection and informing the user of this decision tends to be less demanding on the extraneous load. Also the technical competence of the system was perceived better than for the baseline condition.

The integration of the user into the decision making using explicit confirmations seem to perform second best for most dimensions and items. Though it seems to increase the extraneous load of the task, by requiring additional user input, the identification of the user with the system, measured by the *hedonic quality-identity* seems to be greater. The fact that the *implicit confirmation* condition performed that much worse than the *user information* actually seems odd to us. It appears to us that the combination of informing the user and presenting, for a defined time frame, the explicit interaction possibility to deny the automated selection, was confusing for the user. Maybe the button labeled ‘Let me decide’, or the definition of a restricted time frame was not clearly understandable, thus leading to a worse user-experience. These results show that the decision on how to frame the interaction dialog between user and system will affect the user experience and potentially the cognitive load of the user. As future DS will become more complex and evolve to collaborative intelligent assistants rather than simple problem solvers, the way of interaction between those two parties will be crucial.

5 Conclusion

Overall, it seems, that for decisions which are understandable and reasonable, informing the user of system-made decisions may contribute to a more practical, attractive, fun and less demanding dialog system. However, one must be careful to transfer these findings to other domains or more complex tasks. The positive experience of the user information condition might be due to the task at hand. Usually, workouts are planned, at least for inexperienced users, by experts (e.g. coaches). Addressing competences to a workout planner system, and therefore trusting its decisions, seems like a logical conclusion. For future evaluations it will be interesting to compare these results to automated system behavior for tasks, where usually the user is in charge and dictates the decision-making process. This might lead to a decrease of acceptance for user information conditions and to an increase for conditions, where the user has more control. Nevertheless, this work shows that it is important to investigate in the collaboration dialog (e.g. user involvement strategies) between user and system, which will be important for future more intelligent and assistive dialog systems.

References

1. Wendemuth, A., Biundo, S.: A companion technology for cognitive technical systems. In: Anna Esposito, Alessandro Vinciarelli, R.H.V.C.M. (ed.) Proceedings of the EUCogII-SSPNET-COST2102 International Conference, pp. 89-103 (2011). Lecture Notes in Computer Science, Springer Berlin Heidelberg (2012)
2. Sohrabi, S., Baier, J., McIlraith, S.A.: HTN planning with preferences. In: Proceedings of the 21st Int. Joint Conference on Artificial Intelligence (IJCAI 2009). pp. 1790-1797. AAAI Press (2009)
3. Nothdurft, F., Behnke, G., Bercher, P., Biundo, S., Minker, W.: The interplay of user-centered dialog systems and ai planning. In: Proceedings of the 16th Annual Meeting of the Special Interest Group on Discourse and Dialogue (SIGDIAL). pp. 344-353. Association for Computational Linguistics, Prague, Czech Republic (September 2015)
4. Myers, K.L., Tyson, W.M., Wolverton, M.J., Jarvis, P.A., Lee, T.J., desJardins, M.: PASSAT: a user-centric planning framework. In: Proceedings of the 3rd International NASA Workshop on Planning and Scheduling for Space, pp. 1-10 (2002)
5. Ai-Chang, M., Bresina, J., Charest, L., Chase, A., Hsu, J.J., Jonsson, A., Kanefsky, B., Morris, P., Rajan, K., Yglesias, J., et al.: Mapgen: mixed-initiative planning and scheduling for the mars exploration rover mission. IEEE Intell. Syst. **19**(1), 8–12 (2004)
6. Fernández-Olivares, J., Castillo, L.A., García-Pérez, Ó., Palao, F.: Bringing users and planning technology together. Experiences in SIADEX. In: Proceedings of the 16th International Conference on Automated Planning and Scheduling (ICAPS 2006). pp. 11-20. AAAI Press (2006)
7. Tecuci, G., Boicu, M., Cox, M.T.: Seven aspects of mixed-initiative reasoning: an introduction to this special issue on mixed-initiative assistants. AI Mag. **28**(2), 11 (2007)
8. Behnke, G., Ponomaryov, D., Schiller, M., Bercher, P., Nothdurft, F., Glimm, B., Biundo, S.: Coherence across components in cognitive systems–one ontology to rule them all. In: Proceedings of the 25th International Joint Conference on Artificial Intelligence (IJCAI 2015). AAAI Press (2015)
9. W3C OWL Working Group: OWL 2 Web Ontology Language: Document Overview (2009). http://www.w3.org/TR/owl2-overview/
10. Honold, F., Schüssel, F., Weber, M.: Adaptive probabilistic fission for multimodal systems. In: Proceedings of the 24th Australian Computer-Human Interaction Conference, pp. 222-231. OzCHI'12, ACM, New York, NY, USA (November, 26-30 2012)
11. Honold, F., Schussel, F., Weber, M., Nothdurft, F., Bertrand, G., Minker, W.: Context models for adaptive dialogs and multimodal interaction. In: Proceedings of the 9th International Conference on Intelligent Environments (IE 2013), pp. 57-64. IEEE (2013)
12. Glodek, M., Honold, F., Geier, T., Krell, G., Nothdurft, F., Reuter, S., Schüssel, F., Hoernle, T., Dietmayer, K., Minker, W., Biundo, S., Weber, M., Palm, G., Schwenker, F.: Fusion paradigms in cognitive technical systems for Human-Computer interaction. Neurocomputing **161**, 17–37 (2015)
13. Schüssel, F., Honold, F., Weber, M.: Using the transferable belief model for multimodal input fusion in companion systems. In: Multimodal Pattern Recognition of Social Signals in HCI, LNCS, vol. 7742, pp. 100-115. Springer (2013)
14. Allen, J.F., Schubert, L.K., Ferguson, G., Heeman, P., Hwang, C.H., Kato, T., Light, M., Martin, N., Miller, B., Poesio, M., et al.: The trains project: a case study in building a conversational planning agent. J. Exp. Theor. Artif. Intell. **7**(1), 7–48 (1995)
15. Ferguson, G., Allen, J.F., et al.: Trips: An integrated intelligent problem-solving assistant. In: Proceedings of the AAAI/IAAI, pp. 567–572 (1998)
16. Rao, A.S., Georgeff, M.P.: Modeling rational agents within a bdi-architecture. KR **91**, 473–484 (1991)
17. Rich, C., Sidner, C.L.: Collagen: a collaboration manager for software interface agents. User Model. User-Adapt. Interact. **8**(3–4), 315–350 (1998)

18. Grosz, B.J., Kraus, S.: Collaborative plans for complex group action. Artif. Intell. **86**(2), 269–357 (1996)
19. Myers, K.L., Jarvis, P.A., Tyson, W.M., Wolverton, M.J.: A mixed-initiative framework for robust plan sketching. In: Proceedings of the 13th International Conference on Automated Planning and Scheduling (ICAPS 2003). pp. 256-266. AAAI Press (2003)
20. de la Asunción, M., Castillo, L., Fdez-Olivares, J., García-Pérez, Ó., González, A., Palao, F.: Siadex: an interactive knowledge-based planner for decision support in forest fire fighting. AI Commun. **18**(4), 257 (2005)
21. Rich, C., Sidner, C.L.: Diamondhelp: a generic collaborative task guidance system. AI Mag. **28**(2), 33 (2007)
22. Hassenzahl, M., Burmester, M., Koller, F.: Attrakdiff: Ein fragebogen zur messung wahrgenommener hedonischer und pragmatischer qualitt. In: Szwillus, G., Ziegler, J. (eds.) Mensch & Computer 2003: Interaktion in Bewegung, pp. 187–196. B. G. Teubner, Stuttgart (2003)
23. Kirschner, P.A.: Cognitive load theory: implications of cognitive load theory on the design of learning. Learn. Instr. **12**(1), 1–10 (2002)
24. Paas, F., Van Merriënboer, J.: Variability of worked examples and transfer of geometrical problem-solving skills: a cognitive-load approach. J. Educ. Psychol. **86**(1), 122 (1994)
25. M. Klepsch, F.W., Seufert, T.: Differentiated Measurement of Cognitive Load: Possible or Not? (2015), in preparation

Part III
Socio-Cognitive Language Processing

Natural Language Dialog System Considering Speaker's Emotion Calculated from Acoustic Features

Takumi Takahashi, Kazuya Mera, Tang Ba Nhat, Yoshiaki Kurosawa and Toshiyuki Takezawa

Abstract With the development of Interactive Voice Response (IVR) systems, people can not only operate computer systems through task-oriented conversation but also enjoy non-task-oriented conversation with the computer. When an IVR system generates a response, it usually refers to just verbal information of the user's utterance. However, when a person gloomily says "I'm fine," people will respond not by saying "That's wonderful" but "Really?" or "Are you OK?" because we can consider both verbal and non-verbal information such as tone of voice, facial expressions, gestures, and so on. In this article, we propose an intelligent IVR system that considers not only verbal but also non-verbal information. To estimate a speaker's emotion (positive, negative, or neutral), 384 acoustic features extracted from the speaker's utterance are utilized to machine learning (SVM). Artificial Intelligence Markup Language (AIML)-based response generating rules are expanded to be able to consider the speaker's emotion. As a result of the experiment, subjects felt that the proposed dialog system was more likable, enjoyable, and did not give machine-like reactions.

Keywords Interactive Voice Response System (IVR) · Acoustic features · Emotion · Support Vector Machine (SVM) · Artificial Intelligence Markup Language (AIML)

T. Takahashi · K. Mera (✉) · Y. Kurosawa · T. Takezawa
Graduate School of Information Sciences, Hiroshima City University, 3-4-1, Ozuka-higashi, Asa-minami-ku, Hiroshima 731-3194, Japan
e-mail: mera@hiroshima-cu.ac.jp

T. Takahashi
e-mail: takahashi@ls.info.hiroshima-cu.ac.jp

Y. Kurosawa
e-mail: kurosawa@hiroshima-cu.ac.jp

T. Takezawa
e-mail: takezawa@hiroshima-cu.ac.jp

T.B. Nhat
FPT Software, Tokyo, Japan
e-mail: tang@ls.info.hiroshima-cu.ac.jp

K. Jokinen and G. Wilcock (eds.), *Dialogues with Social Robots*,
Lecture Notes in Electrical Engineering 427,
DOI 10.1007/978-981-10-2585-3_11

1 Introduction

Interactive Voice Response (IVR) systems are rapidly spreading in our daily life, e.g. Siri by Apple Inc. and some advanced car navigation systems. Although previous IVR systems can only deal with task-oriented dialog that responds to the user's demands such as answering a user's question, executing a user's order, and recommending useful items, recent IVR systems attempt non-task-oriented dialog like chat and small talk. These kinds of functions are useful for people to not only waste time but also increase their affinity with and tolerance of computer systems.

When standard IVR system generates a response to the speaker's utterance, only verbal information of the utterance is considered. However, when a person gloomily says "I'm fine," people will respond not by saying "That's wonderful" but "Really?" or "Are you OK?" because we can consider both verbal and non-verbal information such as tone of voice, facial expressions, gestures, and so on.

In this article, we propose an intelligent IVR system that considers not only verbal but also non-verbal information. To estimate a speaker's emotion (positive, negative, or neutral), 384 acoustic features extracted from the speaker's utterance are utilized to machine learning (SVM). Artificial Intelligence Markup Language (AIML) based response generating rules are expanded to be able to consider both text expressions and the speaker's emotion as condition parts of the rules.

Section 2 introduces the overview of our proposed method. Sections 3 and 4 explain an expansion of AIML-based response generating rules and an emotion estimating method from acoustic features, respectively. Section 5 discusses experimental results. The conclusion and future works are presented in Sect. 6.

2 Overview of the Proposed Method

The proposed method is based on a stimulus-response style response generating system. Response generating rules usually consist of a condition part that matches input utterance and a response part that generates response sentence. The general rule-based method compares the verbal expression of input utterance with the condition part of the rules. However, even if a speaker utters completely the same sentence, utterances with different non-verbal expressions often make different impressions as discussed in Sect. 1.

In this paper, we propose a method that can consider both verbal expressions and the speaker's emotion estimated from acoustic features when the method selects the most adequate response generating rule. Figure 1 shows the overview of our proposed method. To carry out pattern matching process, two types of information (i.e. verbal and non-verbal) are required. As verbal information, the text expression of the speaker's utterance is used. It is obtained by a speech recognition unit, and the text is analyzed morphologically because Japanese sentences are written without spaces between words.

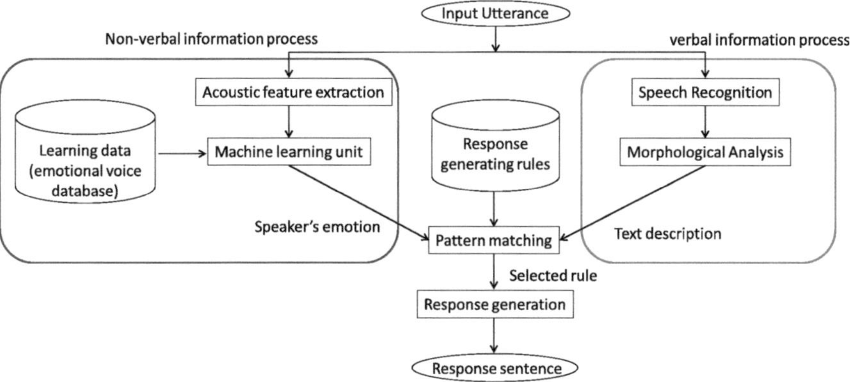

Fig. 1 Overview of our proposed method

On the other hand, the speaker's emotion is estimated from acoustic features of the utterance. First, acoustic features of the utterance are calculated. Next, the speaker's emotion is estimated from the acoustic features by using a machine learning unit that previously learned the relationship between emotion types and the acoustic features. The pattern matching process compares the condition part of all rules with the speaker's utterance. Condition of the rule specifies both text features and emotion state. Finally, response sentence is generated on the basis of the response part of the selected rule.

Section 3 introduces the expansion of AIML-based response generating rules, and Sect. 4 describes the emotion estimating method from acoustic features.

3 Expansion of AIML-based Response Generating Rule

There are response generating methods based on the stimulus-response model. One of the most famous models is Artificial Intelligence Markup Language (AIML), which is an XML-compliant language [1]. Artificial Linguistic Internet Computer Entity (A.L.I.C.E), which has won the annual Loebner Prize Competition in Artificial Intelligence [2] three times, is also constructed by using AIML.

The most important units of AIML are the follows:

- < *aiml* >: the tag that begins and ends an AIML document
- < *category* >: the tag that marks a "unit of knowledge" in an Alicebot's knowledge base
- < *pattern* >: used to contain a simple pattern that matches what a user may say or type to an Alicebot
- < *template* >: contains the response to a user input

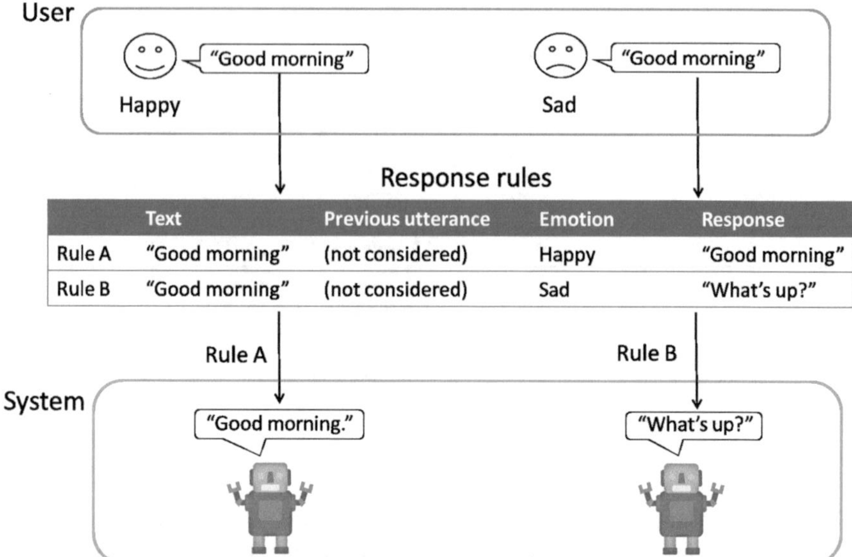

Fig. 2 Example of response generation

The AIML pattern language is consisting only of words, spaces, and the wildcard symbols _ and $*$. Template language is described by not only complete sentences but also "template" expressions that are filled up with additional words inherited from $< pattern >$ or other units.

Our proposed method improves $< pattern >$ unit of AIML in order to consider both text information and the speaker's emotion. The $< pattern >$ part of our proposed method consists of three conditions: "Text," "Previous utterance," and "Emotion." "Text" is the text of a speaker's utterance. It corresponds to the $< pattern >$ unit of AIML. "Previous utterance" refers to the robot's previous utterance. In AIML, the syntax $< that > ... < /that >$ encloses a pattern that matches the robot's previous utterance. "Emotion" is our original factor that indicates the speaker's emotion estimated from acoustic features of the input utterance. Figure 2 shows an example of the response rule matching process. When a speaker cheerfully utters "Good morning," the speaker's emotion is estimated to be "happy," and rule A is selected. As a result, the system responds with "Good morning." On the other hand, when the speaker gloomily utters "Good morning," the speaker's emotion is estimated "sad," and rule B is selected. As a result, the system responds "You don't sound happy. What's up?" As shown in the example, our proposed method can respond with an appropriate reaction by considering the speaker's emotion.

4 Emotion Estimation Method Based on Acoustic Features

The young field of emotion recognition from voices has recently gained considerable interest in human-machine communication, human-robot communication, and multimedia retrieval. Numerous studies have been seen in the last decade trying to improve on features and classifiers. Therefore, the "Emotion Challenge" competition, in which various emotion recognition methods competed, was held at INTERSPEECH 2009 [3]. In the competition, open source software (openSMILE [4]) was provided that can extract various features. Schuller exhibited baseline results for 2-class emotion recognition, NEGative (subsuming *angry, touchy, reprimanding, and emphatic*) and IDLe (consisting of all non-negative states), by using 384 features. As a result, weighted averages of the recall and precision were 0.73 and 0.71, resprectively.

Our proposed method also uses the feature set used at the INTERSPEECH 2009 Emotion Challenge that calculates 384 static feature values. The feature set refers to the following 16 low-level descriptors (contours):

- **pcm_RMSenergy:** Root-mean-square frame energy
- **mfcc:** Mel-Frequency cepstral coefficients 1–12
- **pcm_zcr:** Zero-crossing rate of time signal (frame-based)
- **voiceProb:** The voicing probability computed from the ACF (Autocorrelation Function)
- **F0:** The fundamental frequency computed from the Cepstrum

Because a 1st order delta coefficient (differential) of each contour is calculated for each contour, the number of contours becomes 32. The following 12 features are calculated for each contour:

- **max:** The maximum value of the contour
- **min:** The minimum value of the contour
- **range** = max − min
- **maxPos:** The absolute position of the maximum value (in frames)
- **minPos:** The absolute position of the minimum value (in frames)
- **amean:** The arithmetic mean of the contour
- **linregc1:** The slope (m) of a linear approximation of the contour
- **linregc2:** The offset (t) of a linear approximation of the contour
- **linregerrQ:** The quadratic error computed as the difference in the linear approximation and the actual contour
- **stddev:** The standard deviation of the values in the contour
- **skewness:** The skewness (3rd order moment)
- **kurtosis:** The kurtosis (4th order moment)

Our proposed method uses a Support Vector Machine (SVM) as a machine learning unit. For machine learning, 26 positive voices, 29 negative voices, and 30 neutral voices were prepared. They were collected from TV dramas and animations and were

Table 1 Confusion matrix of emotion estimation result

		Predict			
		Positive	Neutral	Negative	Recall
Actual	Positive	15	0	11	0.58
	Neutral	0	27	3	0.90
	Negative	11	0	18	0.62
	Precision	0.58	1.00	0.56	

annotated with one type of emotion (positive, negative, and neutral) by five university students. Only the data in which all annotators annotated the same emotion type were used.

The performance of the trained classifier was measured by leave-one-out cross validation. Table 1 is a confusion matrix of the emotion estimation result. Although accuracy and precision of "neutral" were good, "positive" and "negative" sometimes confused. The total accuracy was 0.71. Although two types of pre-processing strategies (standardization after balancing by SMOTE and balancing after standardization) were also applied, both accuracies (0.71) were the same as that of the non-pre-processing strategies.

5 Experiment

A dialog system that can consider speaker's emotion on the basis of the proposed method was constructed to evaluate the impression for the dialog agent. Julius [5] and openSMILE [4] were utilized for speech recognition and acoustic feature extraction, respectively. SVM in a statistical computing environment R [6] was utilized for machine learning. Conversation logs by the previous method, which considers text expression only, and by the proposed method, which can consider both text expression and speaker's emotion, were evaluated from the point of view of likeability, machine-creature likeness, and feeling of using. Table 2 contains two conversation logs by

Table 2 Conversation logs by previous method

User:	Hello.
System:	Hello. Do you have plans for today?
User:	Not much. (*Said happily*)
System:	I understand.
User:	Will it be fine tomorrow?
System:	It will rain.
User:	Rain... (*Said gloomily*)
System:	You should take an umbrella.

Table 3 Conversation logs by proposed method (correct estimation)

User:	Hello.
System:	Hello. Do you have plans for today?
User:	Not much. (*Said happily*)
System:	Really? Your voice is bouncy!
User:	Will it be fine tomorrow?
System:	It will rain.
User:	Rain... (*Said glooomily*)
System:	Is it bad for you?
User:	I wanted to go on a picnic tomorrow.
System:	It's so pity. But, the weather will get better from afternoon.

Table 4 Conversation logs by proposed method (inadequate estimation)

User:	Hello.
System:	Hello. Do you have plans for today?
User:	Not much. (*Said gloomily*)
System:	Really? Your voice is bouncy!
User:	Will it be fine tomorrow?
System:	It will rain.
User:	Rain... (*Said happily*)
System:	Is it bad for you?
User:	Not at all.

Table 5 A conversation log by proposed method (including both correct and inadequate estimations)

User:	Please recommend a movie.
System:	How about "Friday the 13th?"
User:	What kind of movie is it ?
System:	It's a horror movie.
User:	Hmm, Horror movie... (*Said gloomily*)
System:	Don't you like horror movies? Then, how about "Stand by me?"
User:	Hmm, Stand by me... (*Said acceptably*)
System:	Don't you like "Stand by me," too?
User:	Yes, I like it. Thank you. (*Said happily*)
System:	You're welcome.

the previous method. Tables 3 and 4 are conversation logs that proposed method estimated speaker's emotion correctly and inadequately, resprectively. Table 5 is a conversation log that proposed method succeeded and failed to estimate speaker's emotion in a conversation.

Table 6 Added AIML rules especially for this experiment

< *pattern* >			< *template* >
Text	Previous utterance	Emotion	Response
Hello.	*	*	Hello. Do you have plans for today?
Not much.	Hello. Do you have plans for today?	Positive	Really? Your voice is bouncy!
Not much.	Hello. Do you have plans for today?	Neutral	I understand.
Not much.	Hello. Do you have plans for today?	Negative	May I recommend something amusing?
Will it be fine tomorrow?	*	*	It will rain.
Rain.	It will rain.	Positive	Is it happy for you?
Rain.	It will rain.	Neutral	You should take an umbrella.
Rain.	It will rain.	Negative	Is it bad for you?
I wanted to go on a picnic tomorrow.	Is it bad for you?	*	It's so pity. But, the weather will get better from afternoon.
Please recommend a movie.	*	*	How about "Friday the 13th?"
What kind of movie is it?	How about "Friday the 13th?"	*	It's a horror movie.
Horror movie.	It's a horror movie.	Positive	Yes! Do you like it?
Horror movie.	It's a horror movie.	Neutral	Yes.
Horror movie.	It's a horror movie.	Negative	Don't you like horror movies? Then, how about "Stand by me?"
Stand by me.	How about "Stand by me?"	Positive	Yes! Do you like it?
Stand by me.	How about "Stand by me?"	Neutral	Yes.
Stand by me.	How about "Stand by me?"	Negative	Don't you like "Stand by me," too?
Thank you.	*	Positive	You're welcome.
Thank you.	*	Neutral	You're welcome.
Thank you.	*	Negative	Were you really satisfied?

In order to simulate these dialog examples, 20 improved AIML rules as shown in Table 6 were added into our proposed dialog system. Format of the rules is explained in Sect. 3.

Table 7 Evaluation factors for computer dialog system

Factors	Adjectives
Likeability factor	Diplomatic, want to mimic, cool, pleasing, want to be friends, want to see again, gentle, arouse sympathy, sensible
Personality factor	Faithfully, calm, mischievous, likely unfaithfully
Machine-creature likeness	Machine-like, creature-like
Feeling of using	Enjoyable, annoying, systematic

All of the conversation scenes were presented in videos to subjects (nine males and a female; 22–24 years old). All of the videos were about a minute long. The subjects gave their impressions for each conversation scene after they finished watching each video.

In this experiment, the conversations were evaluated by a "likeability factor" with 15 adjectives, "personality factor" with four adjectives, and "machine-creature likeness" with two adjectives as shown in Table 7. These factors were selected with reference to the work Takayoshi and Tanaka [7], who evaluated impressions of a robot. Furthermore, "feeling of using" was also evaluated by using three adjectives.

The subjects evaluated all conversations from the point of view of each adjective by a five-grade evaluation (+2:think so, +1:slightly think so, 0:neither, −1:do not much think so, −2:do not think so).

The mean value and the standard deviation for each adjective are shown in Table 8. In addition, the list of significant different adjectives are shown in Tables 9, 10, 11, 12, and 13.

About likeability factor: Likeability of a dialog system directly relates to being personable. The proposed method that could estimate the speaker's emotions correctly had significantly higher mean values of adjectives about likeability than the previous method, especially "pleasing ($p < 0.001$)" and "want to be friends ($p < 0.001$)." On the other hand, the proposed method that estimated inadequate emotions had significantly lower mean values of "want to mimic ($p < 0.05$)" and "cool ($p < 0.01$)" than the previous method.

About machine-creature likeness factor: The result denoted that the subjects felt that the previous method was more "machine-like ($p < 0.001$)" and the proposed method was more "creature-like ($p < 0.001$)" even if the proposed method estimated inadequate emotion (machine-like: $p < 0.001$, creature-like: $p < 0.001$). Although the previous method always responded with the same response to the same text expression, the proposed method changed responses in accordance with the speaker's emotion. This provides two impressions: the previous method is monotonous, and the proposed method is not machine-like because it can deal with emotion.

Table 8 Mean value and standard deviation for each factor and adjective

			Previous method (Table 2)	Proposed method		
				Correct estimation (Table 3)	Inadequate estimation (Table 4)	Imperfect estimation (Table 5)
		N	10	10	10	10
Likeability factor	Diplomatic	Mean	0.35	1.60	0.20	1.90
		SD	1.15	0.58	1.12	0.30
	Want to mimic	Mean	0.05	0.70	−0.50	1.00
		SD	0.86	0.78	0.59	0.89
	Cool	Mean	0.10	0.20	−0.50	0.80
		SD	0.70	0.68	0.50	0.87
	Pleasing	Mean	−0.05	1.35	−0.5	1.30
		SD	1.12	0.96	1.28	1.19
	Want to be friends	Mean	−0.35	0.95	−0.30	1.20
		SD	1.24	1.07	1.05	1.17
	Want to see again	Mean	−0.40	0.85	−0.40	1.20
		SD	0.97	1.11	0.80	1.17
	Gentle	Mean	0.20	1.35	0.00	1.70
		SD	1.08	0.65	0.89	0.46
	Arouse sympathy	Mean	−0.05	0.95	−0.45	1.00
		SD	0.97	0.97	0.80	1.10
	Sensible	Mean	0.10	1.05	−0.55	1.90
		SD	1.48	1.24	1.36	0.30
Personality factor	Faithfully	Mean	1.05	0.40	0.25	0.40
		SD	0.74	0.80	0.99	0.80
	Calm	Mean	0.40	−0.55	0.10	−0.20
		SD	0.80	0.74	0.94	0.60
	Mischievous	Mean	−0.80	−0.20	−0.20	−0.10
		SD	0.93	0.93	0.93	0.54
	Unfaithfully	Mean	−0.75	−0.80	−0.30	−1.00
		SD	0.83	0.60	1.10	0.63

(continued)

Table 8 (continued)

			Previous method (Table 2)	Proposed method		
				Correct estimation (Table 3)	Inadequate estimation (Table 4)	Imperfect estimation (Table 5)
		N	10	10	10	10
Machine-creature likeness	Machine-like	Mean	1.35	−1.15	0.50	−1.00
		SD	0.79	0.91	0.97	1.10
	Creature-like	Mean	−0.90	1.30	−0.15	1.20
		SD	−0.89	0.64	0.96	0.98
Feeling of using	enjoyable	Mean	−0.50	1.20	−0.40	1.30
		SD	0.87	0.40	0.80	0.64
	Be annoying	Mean	−0.10	−1.00	0.30	−0.60
		SD	0.83	1.41	1.00	1.56
	Systematic	Mean	0.90	−0.90	0.65	−1.00
		SD	0.89	0.77	0.73	0.89

N number of subjects

Table 9 A list of significant different adjectives between previous method and proposed method (correct estimation)

Previous method	Proposed method (correct estimation)
Faithfully*, calm**, machine-likeness***, be annoying**, systematic***	Diplomatic***, want to mimic*, pleasing***, want to be friends***, want to see again***, gentle**, arouse sympathy**, sensible*, creature-likeness***, enjoyable***

$^{***}p < 0.001$ $^{**}p < 0.01$ $^{*}p < 0.05$

Table 10 A list of significant different adjectives between previous method and proposed method (inadequate estimation)

Previous method	Proposed method (inadequate method)
Want to mimic*, cool**, faithfully*, machine-likeness***	Unfaithfully*, creature-likeness**

$^{***}p < 0.001$ $^{**}p < 0.01$ $^{*}p < 0.05$

About feeling of using factor: The proposed method that could estimate the speaker's emotions correctly had a significantly higher mean value of "enjoyable ($p < 0.001$)" than the previous method. On the other hand, the previous method had a significantly higher mean value of "systematic ($p < 0.001$)" than the proposed method. Subjects could enjoy the reaction of the proposed method because changing response patterns in accordance with acoustic features of utterance was novel. However, the proposed method that estimated inadequate emotions had a

Table 11 A list of significant different adjectives between previous method and proposed method (imperfect estimation)

Previous method	Proposed method (imperfect estimation)
Faithfully*, machine-likeness***, systematic***	Diplomatic***, want to mimic*, cool*, pleasing**, want to be friends**, want to see again***, gentle***, arouse sympathy*, sensible***, mischievous*, creature-likeness***, enjoyable***

$^{***}p < 0.001$ $^{**}p < 0.01$ $^{*}p < 0.05$

Table 12 A list of significant different adjectives between proposed method (correct estimation) and proposed method (inadequate estimation)

Proposed method (correct estimation)	Proposed method (inadequate estimation)
Diplomatic***, want to mimic***, cool**, pleasing***, want to be friends***, want to see again***, gentle**, arouse sympathy**, sensible*, creature-likeness***, enjoyable***	Calm*, unfaithfully* machine-likeness***, be annoying***, systematic***

$^{***}p < 0.001$ $^{**}p < 0.01$ $^{*}p < 0.05$

Table 13 A list of significant different adjectives between proposed method (correct estimation) and proposed method (imperfect estimation)

Proposed method (correct estimation)	Proposed method (imperfect estimation)
–	Sensible**

$^{***}p < 0.001$ $^{**}p < 0.01$ $^{*}p < 0.05$

significantly higher mean value of "be annoying ($p < 0.001$)" than that in succeed situation.

Comparison between perfect and imperfect emotion estimation: The impressions between the system which estimated the speaker's emotions perfectly and that sometimes estimated inadequate emotions were almost the same except "sensible ($p < 0.01$)." It denotes that the likeability of the dialog system does not decrease so much even if the system sometimes fail to estimate the speaker's emotion.

6 Conclusion

In this article, we proposed a natural language dialog method that can consider both text expressions and the speaker's emotion estimated from acoustic features. The speaker's emotion was estimated by using SVM based on 384 acoustic features extracted by openSMILE. Response generating rules were expanded from the AIML format. The $< pattern >$ part of the rule consisted of three conditions: "Text," "Previous utterance," and "Emotion."

The impressions of conversations by the previous method and the proposed method were evaluated from the point of view of "likeability factor," "personality factor," "machine-creature likeness," and "feeling of using." As a result, subjects felt the proposed method was more personable, creature-like, and enjoyable than previous method. Furthermore, the positive impression was kept up even if the emotion estimation result was not perfect.

For the future work, the number of response generating rules should be increased in large quantities to realize an intellectual and flexible dialog system. Furthermore, not only acoustic feature but also other non-verbal features like facial expressions should be considered to improve the accuracy of emotion estimation.

Acknowledgements This research is supported by JSPS KAKENHI Grant Number 26330313 and the Center of Innovation Program from Japan Science and Technology Agency, JST.

References

1. AIML—The Artificial Intelligence Markup Language. http://www.alicebot.org/aiml.html. Accessed 9 May 2016
2. Home Page of the Loebner Prize. http://www.loebner.net/Prizef/loebner-prize.html. Accessed 9 May 2016
3. Emotion Challenge—AAAC emotion-research.net—Association for the Advancement of Affective Computing. http://emotion-research.net/sigs/speech-sig/emotion-challenge. Accessed 9 May 2016
4. Eyben, F., Wöllmer, M., Schuller, B.: openSMILE: The Munich versatile and fast open-source audio feature extractor. In: Proceedings of the International Conference on Multimedia (2010)
5. Lee, A., Kawahara, T., Shikano, K.: Real-time confidence scoring based on word posterior probability on two-pass search algorithm. Tech. Rep. IEICE **103**(520), 35–40 (2003). (in Japanese)
6. Ihaka, R., Gentleman, R.: A language for data analysis and graphics. J. Comput. Graph. Stat. **5**, 299–314 (1996)
7. Takayoshi, K., Tanaka, T.: The relationship between the behavior of kuwata and impression of his intelligence & personality. IPSJ SIG Tech. Rep. **160**, 43–48 (2007). (in Japanese)

Salient Cross-Lingual Acoustic and Prosodic Features for English and German Emotion Recognition

Maxim Sidorov, Christina Brester, Stefan Ultes and Alexander Schmitt

Abstract While approaches on automatic recognition of human emotion from speech have already achieved reasonable results, a lot of room for improvement still remains there. In our research, we select the most essential features by applying a self-adaptive multi-objective genetic algorithm. The proposed approach is evaluated using data from different languages (English and German) with two different feature sets consisting of 37 and 384 dimensions, respectively. The obtained results of the developed technique have increased the emotion recognition performance by up to 49.8 % relative improvement in accuracy. Furthermore, in order to identify salient features across speech data from different languages, we analysed the selection count of the features to generate a feature ranking. Based on this, a feature set for speech-based emotion recognition based on the most salient features has been created. By applying this feature set, we achieve a relative improvement of up to 37.3 % without the need of time-consuming feature selection using a genetic algorithm.

Keywords Emotion recognition · Speech analysis · Dimensionality reduction · Salient features

M. Sidorov (✉) · S. Ultes · A. Schmitt
Ulm University, Ulm, Germany
e-mail: maxim.sidorov@uni-ulm.de

S. Ultes
e-mail: stefan.ultes@uni-ulm.de

A. Schmitt
e-mail: alexander.schmitt@uni-ulm.de

C. Brester
Siberian State Aerospace University, Krasnoyarsk, Russia
e-mail: christina.brester@sibsau.ru

K. Jokinen and G. Wilcock (eds.), *Dialogues with Social Robots*,
Lecture Notes in Electrical Engineering 427,
DOI 10.1007/978-981-10-2585-3_12

1 Introduction

Automatic recognition of human emotions based on the speech signal is in the focus of research groups all over the world. While enabling machines to recognize human emotions may be useful in various applications, e.g., improvement of Spoken Dialogue Systems (SDSs) or monitoring agents in call-centers, the recognition performance is still not satisfying.

Emotion recognition (ER) rendered as a classification problem may be solved with supervised learning approaches by extracting a huge amount of numerical features out of the speech signal. However, the curse of dimensionality [1], i.e., having more features results in worse performance, poses a critical issue. Some of the features may be highly-correlated or their level of variability may be dramatically low. Therefore, some attributes might not bring a beneficial impact to the system—or even decrease its performance. Hence, feature selection techniques are applied which not simply results in a trade-off between time-consuming feature extraction and the accuracy of the model. Moreover, selecting an optimal feature set potentially increases the overall performance of the system. Furthermore, as feature selection is a time-consuming procedure, it is highly desirable to have a set of salient features which is known to result in good emotion recognition performance for different settings, e.g., for different languages.

However, a suitable feature set should be both, representative and compact, i.e., result in good performance while being as small as possible. Hence, in this contribution, we propose the usage of a multi-objective genetic algorithm (MOGA), which is a heuristic algorithm of pseudo-boolean optimization, in order to maximize ER accuracy and minimize the size of the feature set simultaneously. Furthermore, we proposed a self-adaptive scheme of MOGA which exempts from the necessity of choosing the parameters of the algorithm. Eventually, our main focus lies on identifying salient cross-lingual features for emotion recognition. Hence, we analysed the distributions of selected features of three different databases (English and German) in order to identify the most salient features across different databases and languages. Finally, the resulting feature set is evaluated for each of the databases used for creating the feature set.

The rest of the article is organized as follows: Significant related work is presented in Sect. 2. Section 3 describes the applied corpora and renders their differences. Our approach to automatic emotion recognition using MOGA-based feature selection is presented in Sect. 4 having its results of numerical evaluations in Sect. 5. In Sect. 6, we analyze the cross-lingual feature set and present the list of salient features. Conclusion and future work are described in Sect. 7.

2 Significant Related Work

One of the pilot experiments which deals with speech-based emotion recognition has been presented by Kwon et al. [2]. The authors compared emotion recognition performance of various classifiers: support vector machine, linear discriminant analysis, quadratic discriminant analysis and hidden Markov model on SUSAS [3] and AIBO [4] databases of emotional speech. The following set of speech signal features have been used in the study: pitch, log energy, formant, mel-band energies, and mel frequency cepstral coefficients (MFCCs). The authors have managed to achieve the highest value of accuracy (70.1 and 42.3 % on the databases, correspondingly) using Gaussian support vector machine.

The authors in [5] highlighted the importance of feature selection for the ER manifested by determining an efficient feature subset was using the fast correlation-based filter feature selection method. A fuzzy ARTMAP neural network [6] was used as an algorithm for emotion modelling. The authors have achieved an accuracy of over 87.52 % for emotion recognition on the FARSDAT speech corpus [7].

While our research of identifying cross-lingual salient features includes four different emotions, Polzehl et al. [8] only focused on the emotion anger. The authors analysed two different anger corpora of German and American English to determine the optimal feature set for anger recognition. The German database contains 21 h of recordings from a German Interactive Voice Response (IVR) portal offering assistance troubleshooting. For each utterance, three annotators assigned one of the following labels: not angry, not sure, slightly angry, clear anger, clear rage, and garbage. Garbage marked utterances are non applicable, e.g., contain silence or critical noise. The English corpus originates from an US-American IVR portal capable of fixing Internet-related problems. Three labelers divided the corpus into angry, annoyed, and non-angry turns. A total of 1,450 acoustic features and their statistical description (e.g., means, moments of first to fourth order, the standard deviation) have been extracted from the speech signal. The features are divided into seven general groups: pitch, loudness, MFCC, spectrals, formants, intensity, and other (e.g., harmonics-to-noise). Analyzing each feature group separately, the authors achieve in a baseline approach without further feature selection a maximal $f1$ score of 68.6 with 612 MFCC-based features of the German corpus and a maximal $f1$ score of 73.5 with 171 intensity-based features of the English corpus.

3 Corpora

For identifying salient features for emotion recognition, the following three different speech databases have been used:

Berlin DB The Berlin emotional database [9] was recorded at the Technical University of Berlin and consists of labeled emotional German utterances which were spoken by 10 actors (5 f). Each utterance has been assigned one of the following emotion labels: neutral, anger, fear, joy, sadness, boredom or disgust.

SAVEE The SAVEE (Surrey Audio-Visual Expressed Emotion) corpus [10] was initially recorded for research on audio-visual emotion classification containing four native English male speakers. One emotion label for each utterance has been applied using the standard set of emotions (anger, disgust, fear, happiness, sadness, surprise and neutral).

Table 1 Databases description

Database	Language	Length (min)	# of emotions	File level duration		Emotion level duration	
				Mean (s)	Std. (s)	Mean (s)	Std. (s)
Berlin	German	24.7	7	2.7	1.02	212.4	64.8
SAVEE	English	30.7	7	3.8	1.07	263.2	76.3
VAM	German	47.8	4	3.02	2.1	717.1	726.3

VAM The VAM database [11] was created at the University of Karlsruhe and consists of utterances extracted from the popular German talk-show "Vera am Mittag" (Vera in the afternoon). The emotion annotation of the first part of the corpus (speakers 1–19) were given by 17 human evaluators and the rest of the utterances (speakers 20–47) were annotated by 6 raters, all on the 3-dimensional emotional basis (valence, activation and dominance). For this work, only pleasantness (or evaluation) and the arousal axis are used. The quadrants (counterclockwise, starting in positive quadrant, assuming arousal as abscissa) are then assigned to emotional labels happy-exciting, angry-anxious, sad-bored, and relaxed-serene (cf. [12]).

A statistical description of the used corpora is depicted in Table 1. Both, the Berlin and the SAVEE database consist of acted emotions while VAM comprises real emotions. Furthermore, the VAM database is highly unbalanced (see Emotion level duration columns in Table 1).

4 Feature Selection with MOGA

For our main contribution of applying feature selection using an adaptive multi-objective genetic algorithm (MOGA) to identify cross-language salient features, a probabilistic neural network (PNN) [13] has been chosen arbitrarily as a classification algorithm for building emotion recognition models as it has shown to be a fast classification algorithm providing good results.

A MOGA, being a genetic algorithm (GA) implementing an effective pseudo-boolean optimization procedure, is used to solve the multi-objective optimization problem. A multi-objective GA operates with a set of binary vectors coding the subsets of informative features, where *false* corresponds to non-essential attributes and *true* corresponds to essential ones.

In this work, the applied MOGA is based on the Strength Pareto Evolutionary Algorithm (SPEA) [14], where non-dominated points are stored in the limited capacity archive named *outer set*. The content of this set is updated throughout the algorithm execution and as a result we have an approximation of the Pareto set.

The scheme of the SPEA method includes several steps:

1. Determination of the initial population $P_t(t = 0)$.
2. Copying of the individuals not dominated by P_t into the intermediate outer set $(\bar{P}')$.
3. Deletion of the individuals dominated by $\bar{P}'$ from the intermediate outer set.
3. Clustering of the outer set $\bar{P}_{t+1}$ (if the capacity of the set $\bar{P}'$ is more than the fixed limit).
4. Compilation of the outer set $\bar{P}_{t+1}$ with the set $\bar{P}'$ individuals.
5. Application of all of the genetic operators: selection, crossover, mutation.
6. Test of the stop-criterion: If it is true, then the GA is completed. Otherwise, continue from the second step.

It is well-known that the performance of conventional GAs completely depends on the settings of the genetic operators (selection, crossover, mutation). Consequently, to achieve a good performance level with SPEA, its parameter settings have to bee adjusted carefully (step 6). Therefore, SPEA-modification based on the idea of self-adaptation has been developed [15].

In this self-adaption version of SPEA, tournament selection is applied: individuals can be selected both from the current population and from the outer set. Hence, only recombination and mutation operators require adjustment (tuning or control).

The mutation probability p_m can be determined according to one of the rules developed by Daridi et al. [16]. As parametrized by Daridi et al., the following rule was used for the proposed self-adaptive SPEA-modification:

$$p_m = \frac{1}{240} + \frac{0.11375}{2^t}, \tag{1}$$

where t is the current generation number.

The self-configurable recombination operator is based on the *co-evolution* idea [17]: the population is divided into groups and each group is generated with a particular type of recombination (it may be *one-point*, *two-point* or *uniform* crossover). The size of the subpopulation depends on the fitness value of the corresponding recombination type, where the fitness value q_i of the i-th recombination operator is determined by

$$q_i = \sum_{l=0}^{T-1} \frac{T-l}{l+1} \cdot b_i, \tag{2}$$

where T is the adaptation interval, $l = 0$ corresponds to the latest generation in the adaptation interval, $l = 1$ corresponds to the previous generation, etc. b_i, indicating the effectiveness of the individuals, is defined as

$$b_i = \frac{p_i}{|\bar{P}|} \cdot \frac{N}{n_i}, \tag{3}$$

where p_i is the amount of individuals in the current outer set generated with the i-th type of recombination operator, $|\bar{P}|$ is the outer set size, n_i is the amount of individuals in the current population generated with the i-th type of crossover, and N is the population size.

The efficiency of the operators is compared in pairs in every T-th generation to reallocate resources on the basis of the fitness values. Hence, the maximum number of applications $f(n)$ of each individual is defined by

$$f(n) = \begin{cases} 0, & \text{if } n_i \leq social\_card \\ int\left(\frac{n_i - social\_card}{n_i}\right), & \text{if } (n_i - h_i \cdot penalty) \\ & \quad \leq social\_card \\ penalty, & \text{otherwise} \end{cases} \tag{4}$$

Here, s_i is the size of a resource given by the i-th operator to those which won, h_i is the number of losses of the i-th operator in paired comparisons, the $social\_card$ is the minimum allowable size of a group, the $penalty$ is a negative score for defeated operators.

The effectiveness of the proposed approach was evaluated on the set of test problems [18] developed by the scientific community for the comparison of evolutionary algorithms: the effectiveness of the adaptive SPEA outperformed the effectiveness of the average conventional SPEA proving that self-adaptation is an alternative to the random choice of genetic operators or multiple runs of the GA for each variant of settings.

5 Evaluation and Results

To investigate the performance of the MOGA-based feature selection for emotion recognition, two experiments using two different feature sets have been conducted. The first feature set consists of the 37 most common acoustic features and is used as a baseline. Average values of the following speech signal features are included in the baseline feature set: power, mean, root mean square, jitter, shimmer, 12 MFCCs and 5 formants. Mean, minimum, maximum, range and deviation of the following features have also been used: pitch, intensity and harmonicity. All of the 37 features have been extracted for each speech signal file.

The extended feature set consists of 384 features and taken from the Interspeech 2009 Emotion Challenge [19]. It is an extension to the baseline feature set containing additional features and so called functionals. In contrast to the baseline feature set, it is not only applied completely but also using the previously presented MOGA feature selection procedure.

Table 2 Evaluation result of emotion recognition: accuracy with the 37-dimensional feature set (Baseline), the extended feature set (IS'09), and the reduced feature set (GA) having the number of features in parentheses (Num.) and relative improvement of the feature selection approach (Gain)

Database	Baseline	IS'09	GA (Num.)	Gain (%)
Berlin	56.7	58.9	71.5 (68.4)	26.1
VAM	68.0	67.1	70.6 (64.8)	3.9
SAVEE	41.6	47.3	48.4 (84.1)	16.3

Features of the baseline feature set have been extracted using Praat [20]. The features of the extended feature set have been extracted for the audio signal using the openSMILE toolkit [21].

For evaluating the emotion recognition performance, the data is divided into training and testing sets with a ratio of 0.7 for training and 0.3 for testing. The training set is used for creating and training the PNN-based emotion model while the testing set is used for measuring the model's performance. In order to get representative results, this procedure has been repeated fifteen times. The achieved accuracy of emotion recognition with the baseline and the extended feature set is shown in Table 2 (Columns Baseline and IS'09).

For applying feature selection and creating the corresponding emotion model, the training set is used as well. For the feature selection process, the feature set was coded with boolean vectors (*true* corresponds to an essential attribute, *false* represents an unessential one) each representing one individual of the self-adaptive MOGA. To test the fitness of the individuals, the training set was in turn divided into two sets of which 80 % were used to train a PNN-based emotion model and 20 % to evaluate it. As fitness, the average accuracy computed out of the accuracies of fifteen evaluation cycles for each individual was used. After 100 iterations of the MOGA, the optimal feature set, i.e., the fittest individual, was used for creating an emotion model whose performance was then evaluated using the testing set. This complete procedure has also been applied fifeteen times. The results depicted in Table 2 show significant improvement of applying feature selection over the baseline and the extended feature set. As described before, feature selection was only applied for the extended feature set.

6 Salient Features for Cross-Lingual Emotion Recognition

While feature selection poses a good way of improving the performance, GA-based feature selection is resource-consuming and thus hard to deploy in real-world applications. Hence, in this contribution, we derive the most appropriate feature set based on information about the selected features. The resulting feature set may then be applied to unseen data (language and domain) without the need of running through the complete process anew.

For the creation of optimal feature set for each database, a ranking of the selected feature set has been created for each database separately (Experiment 1). Here, the rank of each feature is based on the number of its participation in the individuals of the GA (feature counts). Then, in order to obtain an optimal feature set for ER in general, a combined ranking of the selected features has been created by combining the feature counts of each database.

For all databases, the information about extracted features was accumulated through multiple executions of the algorithm. Based on these, a distribution of feature selection frequencies (FSF), i.e. the relative number of cases in which a particular feature was selected, was created. Then, the list of features was ranked in the descending order of their FSF-values. During this iterative procedure, the optimal number of features was defined for each database separately and a learning curve was generated by subsequently adding the top-ranked feature to the applied feature set. For providing statistically significant results, the classification procedure was executed 25 times for each set of feature sets. At Fig. 1, the accuracy of emotion recognition for all databases is illustrated depending on the number of involved features. It may be noticed that every obtained curve has only one maximum and, moreover, the maximums are located in a similar range for all databases.

Based on individual distribution of FSF-values for each database, an combined distribution was derived by calculating the average frequencies. The next steps are similar to Experiment 1: based on the new ranking, the iterative procedure was launched for each database. Figure 2 presents the learning curve, i.e., accuracy of emotion recognition for each iteration. Furthermore, the figure also contains an average learning curve also indicating the combined maximum accuracy with 147 features constituting the common feature set. Again, all values have been determined by taking the average of 25 interactions of the experiment.

Table 3 shows the results for individual and common ranking as well as by applying the common feature set. Individual ranking clearly achieves the best results increasing the performance up to 49.8 % relative to the baseline. However, applying

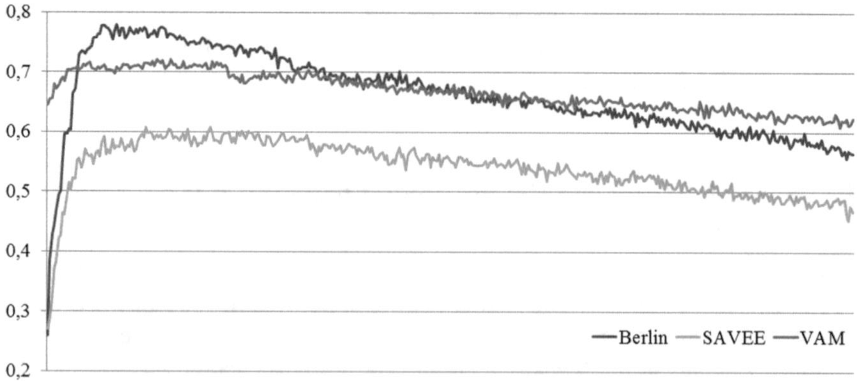

Fig. 1 Accuracy of emotion recognition system A: usage of individual distribution

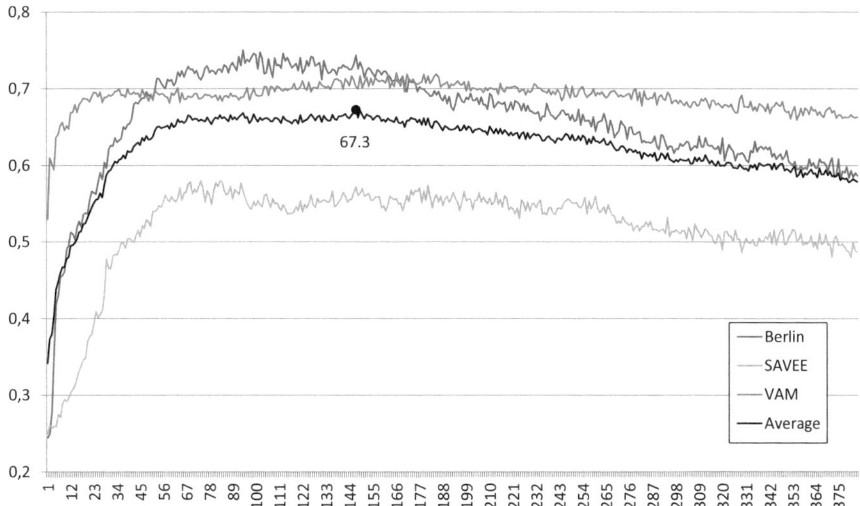

Fig. 2 Accuracy of emotion recognition system B: usage of average distribution

Table 3 Evaluation results of the individual and combined ranking (Number of features in parentheses) of the extended feature set

Corpus	Individual ranking	Common ranking	Common set	Gain (%)
Berlin	79.7 (86)	75.1 (94)	74.37	31.2
VAM	73.6 (81)	72.1 (166)	70.24	3.3
SAVEE	62.3 (148)	58.0 (74)	57.17	37.3

Furthermore, results for the common feature set plus relative improvement (Gain)

Table 4 Top 20 ranked features using self-adaptive MOGA feature selection

Rank	Feature group	Rank	Feature group
1	MFCC	11	MFCC
2	Energy	12	Energy
3	ZCR	13	Energy
4	MFCC	14	Energy
5	MFCC	15	MFCC
6	MFCC	16	MFCC
7	MFCC	17	MFCC
8	MFCC	18	MFCC
9	MFCC	19	MFCC
10	MFCC	20	MFCC

common ranking still provides results clearly above the baseline. Moreover, both results outperform the results of simply applying MOGA-based feature selection as in Table 2. Finally, applying the derived common feature set consisting of the 147 most salient features also results in improvement of accuracy by up to 37.3 % relative to the baseline. In Table 4, the top twenty rankend features of the common feature set are shown along with their feature group. When comparing these to the results of Polzehl et al. [8], they belong to similar groups confirming their results for general emotion recongition.

7 Conclusion and Future Work

In this work, we presented the application of a PNN-MOGA hybrid emotion recognition approach where MOGA is used for feature selection. By applying the algorithm to three different corpora containing English or German speech, the overall accuracy could be improved for all data by up to 49.8 %. Finally, we created a common feature set out of the individual rankings consisting of the 147 most salient features improving the performance on all corpora by up to 37.3 %.

While a PNN has already provided reasonable results for emotion recognition, we still examine its general appropriateness. The usage of other possibly more accurate classifiers may improve the performance of this system. Furthermore, dialogues may not only consist of speech, but also of a visual representation. Hence, an analysis of video recordings may also improve the ER performance.

References

1. Bellman, R.: Dynamic Programming. Princeton University Press (1957)
2. Kwon, O.W., Chan, K., Hao, J., Lee, T.W.: Emotion recognition by speech signals. In: Proceedings of the INTERSPEECH (2003)
3. Hansen, J.H., Bou-Ghazale, S.E., Sarikaya, R., Pellom, B.: Getting started with SUSAS: a speech under simulated and actual stress database. In: Proceedings of the EUROSPEECH **97**, 1743–1746 (1997)
4. Batliner, A., Hacker, C., Steidl, S., Nöth, E., D'Arcy, S., Russell, M.J., Wong, M.: "you stupid tin box"—children interacting with the Aibo robot: A cross-linguistic emotional speech corpus. In: Proceedings of LREC (2004)
5. Gharavian, D., Sheikhan, M., Nazerieh, A., Garoucy, S.: Speech emotion recognition using fcbf feature selection method and ga-optimized fuzzy artmap neural network. Neural Comput. Appl. **21**(8), 2115–2126 (2012)
6. Carpenter, G.A., Grossberg, S., Markuzon, N., Reynolds, J.H., Rosen, D.B.: Fuzzy artmap: a neural network architecture for incremental supervised learning of analog multidimensional maps. IEEE Trans. Neural Netw. **3**(5), 698–713 (1992)
7. Bijankhan, M., Sheikhzadegan, J., Roohani, M., Samareh, Y., Lucas, C., Tebyani, M.: Farsdat-the speech database of Farsi spoken language. In: Proceedings of the Australian Conference on Speech Science and Technology. vol. 2, pp. 826–830 (1994)

8. Polzehl, T., Schmitt, A., Metze, F.: Salient features for anger recognition in German and English IVR portals. In: Minker, W., Lee, G.G., Nakamura, S., Mariani, J. (eds.) Spoken Dialogue Systems Technology and Design, pp. 83–105. Springer, New York (2011). doi:10.1007/978-1-4419-7934-6_4
9. Burkhardt, F., Paeschke, A., Rolfes, M., Sendlmeier, W.F., Weiss, B.: A database of German emotional speech. In: Proceeding of Interspeech. pp. 1517–1520 (2005)
10. Haq, S., Jackson, P.: Machine Audition: Principles, Algorithms and Systems, chap. Multimodal Emotion Recognition, pp. 398–423. IGI Global, Hershey PA (Aug 2010)
11. Grimm, M., Kroschel, K., Narayanan, S.: The Vera am Mittag German audio-visual emotional speech database. In: Proceedings of the 2008 IEEE International Conference on Multimedia and Expo, pp. 865–868. IEEE (2008)
12. Schuller, B., Vlasenko, B., Eyben, F., Rigoll, G., Wendemuth, A.: Acoustic emotion recognition: a benchmark comparison of performances. In: Proceedings of the IEEE Workshop on Automatic Speech Recognition & Understanding, 2009, ASRU 2009, pp. 552–557. IEEE (2009)
13. Specht, D.F.: Probabilistic neural networks. Neural Netw. **3**(1), 109–118 (1990)
14. Zitzler, E., Thiele, L.: Multiobjective evolutionary algorithms: a comparative case study and the strength pareto approach. IEEE Trans. Evol. Comput. **3**(4), 257–271 (1999)
15. Eiben, A.E., Hinterding, R., Michalewicz, Z.: Parameter control in evolutionary algorithms. IEEE Trans. Evol. Comput. **3**(2), 124–141 (1999)
16. Daridi, F., Kharma, N., Salik, J.: Parameterless genetic algorithms: review and innovation. IEEE Can. Rev. **47**, 19–23 (2004)
17. Potter, M.A., De Jong, K.A.: A cooperative coevolutionary approach to function optimization. In: Parallel Problem Solving from Nature–PPSN III, pp. 249–257. Springer (1994)
18. Zhang, Q., Zhou, A., Zhao, S., Suganthan, P.N., Liu, W., Tiwari, S.: Multiobjective optimization test instances for the CEC 2009 special session and competition. University of Essex, Colchester, UK and Nanyang Technological University, Singapore, Special Session on Performance Assessment of Multi-Objective Optimization Algorithms, Technical Report (2008)
19. Kockmann, M., Burget, L., Černockỳ, J.: Brno University of Technology system for Interspeech 2009 emotion challenge. In: Proceedings of the Tenth Annual Conference of the International Speech Communication Association (2009)
20. Boersma, P.: Praat, a system for doing phonetics by computer. Glot Int. **5**(9/10), 341–345 (2002)
21. Eyben, F., Wöllmer, M., Schuller, B.: Opensmile: the Munich versatile and fast open-source audio feature extractor. In: Proceedings of the international conference on Multimedia, pp. 1459–1462. ACM (2010)

Entropy-Driven Dialog for Topic Classification: Detecting and Tackling Uncertainty

Manex Serras, Naiara Perez, María Inés Torres and Arantza del Pozo

Abstract A frequent difficulty faced by developers of Dialog Systems is the absence of a corpus of conversations to model the dialog statistically. Even when such a corpus is available, neither an agenda nor a statistically-based dialog control logic are options if the domain knowledge is broad. This article presents a module that automatically generates system-turn utterances to guide the user through the dialog. These system-turns are not established beforehand, and vary with each dialog. In particular, the task defined in this paper is the automation of a call-routing service. The proposed module is used when the user has not given enough information to route the call with high confidence. Doing so, and using the generated system-turns, the obtained information is improved through the dialog. The article focuses on the development and operation of this module, which is valid for agenda-based and statistical approaches, being applicable in both types of corpora.

Keywords Dialog system · System turn generation · Uncertainty detection · Information recovery

M. Serras (✉) · N. Perez · A. del Pozo
Vicomtech-IK4, Mikeletegi Pasealekua 57, Donostia-San Sebastian, Spain
e-mail: mserras@vicomtech.org
URL: http://www.vicomtech.org

N. Perez
e-mail: nperez@vicomtech.org

A. del Pozo
e-mail: adelpozo@vicomtech.org

M.I. Torres
SPIN Research Group, Universidad del País Vasco UPV/EHU, Bilbao, Spain
e-mail: manes.torres@ehu.es
URL: http://www.ehu.eus/en/web/speech-interactive

K. Jokinen and G. Wilcock (eds.), *Dialogues with Social Robots*,
Lecture Notes in Electrical Engineering 427,
DOI 10.1007/978-981-10-2585-3_13

1 Introduction

Developing a Dialog System (DS) may be a hard task, especially if there is a lack of dialog-based corpus and/or the domain to model within the dialog is too broad. The agenda-based approach is commonly adopted for dialog modeling [1] in the first case. However, devising plan-based conversation streams requires that developers have full knowledge of the domain [2], and/or dialog streams defined beforehand for every possible situation. For this reason, this approach has been mostly used for fixed domains and formulaic tasks, such as booking airline tickets and consulting bus schedules or fares [3–5]. But, even when it is possible to use a Markov Decision Process (MDP) or Partially Observable MDP-based approach, developers may find problems when the application domain is too vast to cover.

This article proposes a novel module to enhance the common Dialog Manager (DM) logic: the Suggestion Generation (SG) module. SG generates a dialog system turn to retrieve more information from the user to achieve the task set when the information given by the user is not enough. In order to measure the amount of information, the entropy score is used in a similar way as [6], but to quantify a Topic Classifier's (TC) decision quality. TC bases its decisions in language units extracted from the user turn by the Natural Language Understanding (NLU) module.

The specific project task of this article is the automation of a call-routing service. The corpus available consists of written records and customer e-mails, which describe a vast range of problems—and not necessarily all the possible problems: new ones are arising all the time. The main contribution of the article lies in embedding within a Dialog System (DS) a module that allows the user-machine interaction to be guided through questions not established beforehand, generating a dialog stream automatically without a corpus of dialogs nor having to design manually an agenda-based strategy for each possible situation. SG grants the system enough flexibility to adapt to and route unseen situations despite the domain being too broad.

The article is structured as follows: Sect. 2 provides an overview of the DS in terms of architecture and interaction between modules; Sect. 3 explains how SG and its algorithm work; Sect. 4 describes the task corpus, and explains the automatic selection of language units for SG and TC; Section 5 evaluates SG and its impact on uncertainty detection; finally, in Sect. 6, the conclusions reached and future guidelines are presented.

2 Overview: The Dialog System

The system consists currently of 5 modules, as shown in Fig. 1: a NLU module for semantic parsing and language unit extraction, an agenda-based DM, the Topic Classifier (TC), the Suggestion Generation (SG) module, and the Natural Language Generator (NLG). Automatic Speech Recognition (ASR) and Text-to-Speech (TTS) systems have not been implemented still. An internal server is responsible for the communication between the existing modules.

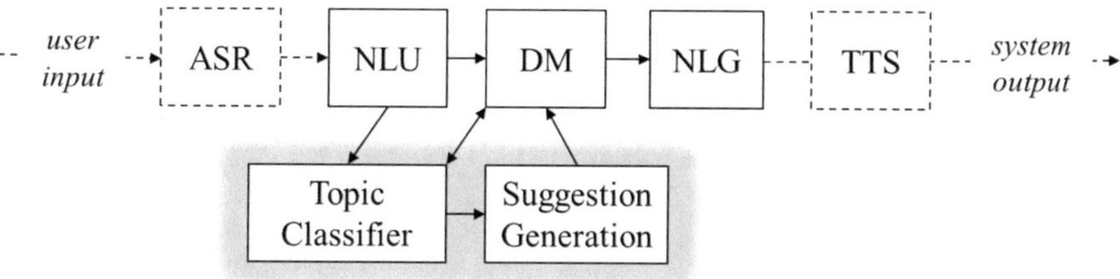

Fig. 1 Proposed system architecture

The dialog designed is plan-based in the general thread and most simple subtasks controlled by the DM. When the point of the conversation is reached when the user issue must be categorized, however, the DM delegates most of the decision making burden to the statistical components of the system: TC and SG. In very few words, this is how the two modules come into play:

1. The user turn is processed with the parser. Its output is passed to TC.
2. TC assigns the class of highest probability to the observed output. The set of classes is composed of the departments of the company providing the service.
3. The entropy of the taken decision is measured. If uncertainty remains low, TC's decision is accepted. Otherwise, the decision is revoked and SG suggest a language unit for the NLG to generate a system turn, in an attempt to retrieve more information from the user. TC is re-evaluated with the user's answer.

3 The Suggestion Generation Module

The Suggestion Generation (SG) module serves the purpose of retrieving information from the user when the DM considers that the classification confidence is not high enough.

The call-routing procedure is explained as follows: first, the utterance representation $Y = (u_{Y_1}, u_{Y_2}, \cdots, u_{Y_{|Y|}})$ is extracted from the user turn by the NLU module, where each u_{Y_i} corresponds to a language unit (a concept, word, lemma, …). Y is passed to the Topic Classifier (TC). Then, having defined the set of departments as the class set Ω, where each class ω_j is one department, the call is routed to the assigned department if the classification confidence is high. Otherwise—if the confidence is low—, the DM uses SG to choose a language unit unseen in Y, $\hat{u}$, about which to ask the user in the next system turn. SG obtains $\hat{u}$ from a graph structure that contains the informational relationships of the corpus, the Information Graph (IG), exploring it with the Suggestion Generation Algorithm (SGA) designed. When $\hat{u}$ is chosen, the NLG module generates the question with framed prompts prepared beforehand. It picks one of the prompts, and merges $\hat{u}$ in the empty frame.

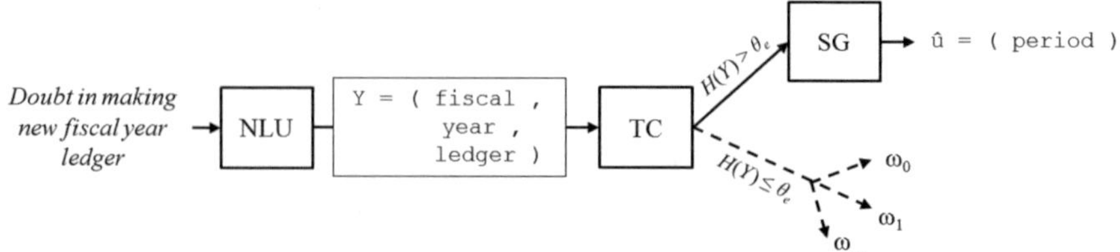

Fig. 2 Processing of a low information user utterance through the system

The classification confidence to determine whether SG has to be activated is obtained using the entropy measure when the TC assigns a class ω_j to the utterance representation Y:

$$H(Y) = -\sum_{j=1}^{|\Omega|} P(Y\,|\omega_j) log_{|\Omega|}(P(Y\,|\omega_j)) \tag{1}$$

where $P(Y\,|\omega_j)$ is the probability of Y given the class ω_j. In this stage, the uncertainty evaluation is done: an utterance representation Y is labeled as an Uncertain Case (UC) if and only if:

$$H(Y) \geq \theta_e \tag{2}$$

where θ_e is the entropy threshold. When Y is labeled as UC, the DM activates the SG module.

Figure 2 depicts an example where the user does not grant enough information and the system draws on SG to choose the language unit for the next system turn. This section explains how IG is modeled and explored by the SGA.

3.1 *The Information Graph*

The Information Graph is the structure that captures the knowledge of the corpus. In this structure, those combinations of language units that grant information to achieve the proposed goal (here, call-routing) are represented as directed edges. The entropy measure of the language units is adopted as the graph connection criterion.

Being U the set of selected language units from the training set S, $H(u_i)$ is the entropy of the language unit $u_i \in U$:

$$H(u_i) = -\sum_{j=1}^{|\Omega|} \hat{P}(u_i|\omega_j) log_{|\Omega|}(\hat{P}(u_i|\omega_j)) \tag{3}$$

$H(u_i)$ measures the amount of information given by u_i in Ω: the lower $H(u_i)$ is, the higher information it grants.

For each $u_i,\ u_m \in U : u_i \neq u_m$ the joint probability given the class $\hat{P}(u_i, u_m|\omega_j)$ is estimated for each class, in order to calculate the joint entropy measure:

$$H(u_i, u_m) = -\sum_{j=1}^{|\Omega|} \hat{P}(u_i, u_m|\omega_j) log_{|\Omega|}(\hat{P}(u_i, u_m|\omega_j)) \tag{4}$$

When the entropy for each u_i and $(u_i, u_j) : u_i \neq u_j$ in U is calculated, the Information Graph of S is defined. Let IG be a directed graph represented as a set of nodes and edges, $IG = (U, E)$, where each language unit $u \in U$ is a node of the graph. Two units $u_i, u_m \in U$ are connected by a directed edge $e_{u_i,u_m} \in E$ starting from u_i and heading to u_m, if and only if

$$H(u_i) \geq H(u_i, u_m) + \theta_1 \tag{5}$$

that is, if the entropy measure is reduced above a threshold θ_1—set in a tuning phase—when combining these two units.

Definition 1 A language unit u_j is a child of another unit u_i in $IG = (U, E)$ if and only if $e_{u_i,u_j} \in E$ exists.

Definition 2 The set of children of u_i in the $IG = (U, E)$ is the union of all the children of u_i, and is denoted as $ch(u_i) = \{u_j \in U : \exists\, e_{u_i,u_j} \in E\}$.

3.2 The Suggestion Generation Algorithm

The Suggestion Generation Algorithm (SGA) is responsible for picking a language unit $\hat{u}$ about which to ask the user in the next system turn. In order to achieve the goal defined in the DM, the suggested unit $\hat{u}$ must provide additional information regarding the task. To this end, SGA explores IG to choose $\hat{u}$.

3.2.1 Association Rules

The graph is explored using Association Rules, common in Data Mining [7, 8] because of their flexibility and real-time results, even in domains with a huge amount of features. As the suggestion generated needs to be related with the user's utterance representation Y and, at the same time, have the highest occurrence rate possible, the association rules support and confidence are used. Subsets are used instead of sequences in these rules, so let R be the set of units observed in the sequence Y where $R \subseteq U$.

$$Y = (u_{Y_1}, \cdots, u_{Y_{|Y|}}) \Rightarrow R = \{u_{Y_i}\} : i = 1, \cdots, |Y| \tag{6}$$

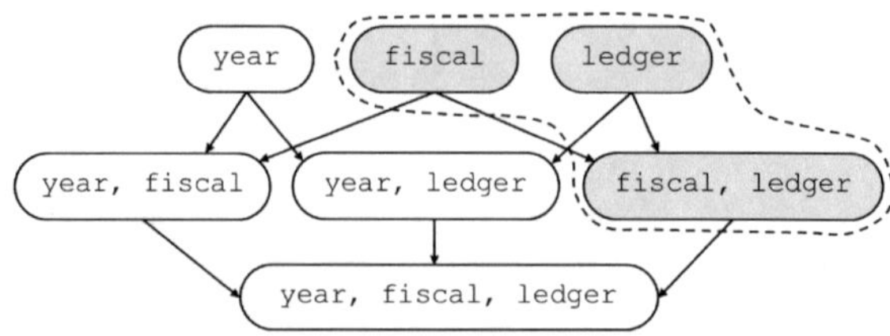

Fig. 3 A-Priori pruning with K=2

Being $S = \{s_j\}\ j = 1, \cdots, |S|$ the corpus where each s_j is a sample represented as a set of language units, the support count of R is:

$$\sigma(R) = |\{s_j \in S \mid R \subseteq s_j\}| \tag{7}$$

Being Z another set of language units, the support and confidence rules are defined as follows:

$$Support:\ s(R \rightarrow Z) = \frac{\sigma(R \cup Z)}{|S|} \quad Confidence:\ c(R \rightarrow Z) = \frac{\sigma(R \cup Z)}{\sigma(R)} \tag{8}$$

The support represents the probability of observing the sets R and Z together in the training set. The confidence rule measures how likely subset Z is to appear once R has been observed.

3.2.2 Graph Exploration Algorithm

The SGA obtains a language unit $\hat{u} \notin R$ that maximizes the confidence score $c(R \rightarrow \hat{u})$ using the connections defined in the IG.

When searching for the new language unit $\hat{u}$ related with R, it may happen that the set R or $(R \cup \hat{u})$ have not been seen in the training set S. This is why the search is done through the subsets $r' \subseteq R$, and the aim is reformulated as obtaining the $\hat{u} \notin R$ which maximizes $c(r' \rightarrow \hat{u})$ being r' the biggest subset possible.

Because the number of subsets of R are $\sum_{k=1}^{|R|} \binom{|R|}{k}$ and they grow exponentially depending on $|R|$, the A-Priori pruning algorithm is used to reduce the search space.

Theorem 1 ***A-Priori principle:*** *If a set of elements is frequent, then all of its subsets must also be frequent.*

Setting a K pruning-parameter, K amount of language units of highest support are chosen from R. Using these units, the subsets $r' \subseteq R$ are generated. The set of these subsets r' is denoted as R'. As depicted in Fig. 3, when choosing the language units `fiscal` and `ledger`, the number of subsets to search through is decreased to $|R'| = \sum_{k=1}^{K} \binom{K}{k}$.

When the most frequent subsets $r' \in R'$ are generated, the Information Graph is used to retrieve the semantic unit $\hat{u}$ which maximizes $c(R \rightarrow \hat{u})$ for the biggest r' where $\sigma(r' \cup \hat{u}) \neq 0$.

Summarizing, the suggested unit $\hat{u}$ has to meet the next conditions:

1. $\hat{u}$ must be a child of every element of a set $r' \in R'$ in the IG.
2. The support count of $(r' \cup \hat{u})$ must be non-zero: $\sigma(r' \cup \hat{u}) \neq 0$.
3. The set r' must contain as many language units as possible, up to K.

This is the actual algorithm designed to explore the graph; R' is sorted from the largest to the smallest subset due to condition 3:

Algorithm 1: Suggestion Generation Algorithm

```
input          : R', IG, K, R
initialization: size = K, valid_units = {}, c_score = {}
for r' in R' do
    for each unit u_i in r' do
        Obtain the children of u_i in the IG, ch(u_i);
    Set ch(r') = ∩_{i=1}^{|r'|} ch(u_i);
    if ch(r') ≠ ∅ and size = |r'| then
        Append all the u_i ∈ ch(r') to valid_units;
        Append confidence c(r' → u_i) to c_score;
    if valid_units = ∅ and |r'| = size − 1 then
        size ← |r'|;
if valid_units ≠ ∅ then
    Set û as the valid unit with highest confidence score unseen in R;
```

In short, the algorithm finds through the IG those children of the largest possible sets of R' and tag them as valid units. From those valid units, the one with highest confidence score and unseen in R is chosen.

3.3 Answer Evaluation

Once the unit $\hat{u}$ is suggested through a system turn, three scenarios have been defined for the next user turn:

1. If the user confirms that the problem is related to $\hat{u}$, the classifier is re-evaluated with the updated sequence $Y_{new} = (Y, \hat{u})$.
2. If the user rejects $\hat{u}$, the SGA is re-evaluated omitting from the training set every sample $s_j \in S$ that contains $\hat{u}$.
3. If the user does not know whether $\hat{u}$ is related with the issue, the next valid unit with highest confidence is used to generate an additional system turn.

Those user calls that remain unclassified after some suggestions of SG are routed to a technician for further analysis. When this cannot be made, TC is forced to route the call to a department by setting $\theta_e = 1$.

4 The Task and Feature Selection

The input of the Topic Classifier (TC) and the Suggestion Generator (SG) module is the utterance representation of the user turn, Y, as extracted by the NLU module. The representation is a sequence of language units: $Y = (u_{Y_1}, u_{Y_2}, \cdots, u_{Y_{|Y|}})$ (see Fig. 2), where $u_{Y_j} \in U$, U being the set of language units selected from the training set S. The units chosen to evaluate the system proposed in this paper are lemmas automatically selected from the corpus available. As proven in [9], words tend to overfit the task and are computationally too demanding, whereas concepts are lighter but do not lend that easily to automatic selection and update. Lemmas seek the balance between the two.

In the best-case scenario, the corpus from which to extract the lemmas should consist of recorded calls of the service to be automatized. However, the corpus actually available is composed of 25568 written technical records of issues consulted. They are in Spanish and organized in 5 classes, each corresponding to one of the departments to which calls must be routed: ω_0-Finance, ω_1-Human Resources, ω_2-Information Technologies, ω_3-Logistics, and ω_4-Software & Hardware. They include e-mails sent by customers explaining their problems and/or quick notes taken by the technicians that attended to them. Thus, a normalization pre-processing has been necessary to make records as similar as possible to the output of an automatic speech recognizer: typographic and orthographic errors have been corrected, shortened forms expanded, and digits converted to character sequences.

The proposed architecture requires that the lemmas selected from the corpus be encoded in grammar-rules for the NLU module to extract the language units from user turns. In this light, automatic selection of lemmas comes down to automatic generation of lemma-grammars. The procedure designed for this purpose draws on the language analyzer toolkit Freeling [10] and consists on the following steps: (*a*) tokenize the corpus and discard stopwords; (*b*) lemmatize and analyze morphologically the words in the resulting list; (*c*) keep the words labeled Noun, Verb or Adjective with a confidence score higher than 0.3—a threshold fixed in a tuning phase—; and, (*d*) group together the words with the same lemma. Each resulting cluster <lemma + words> is a rule of the grammar that tells the parser to return the lemma whenever the user turn contains one of the words associated to it. Figure 4 shows part of a real example of a grammar-rule automatically generated.

The grammar generated applying the procedure explained to the entire normalized corpus has 5882 rules (i.e., lemmas), reducing the dimension of U by 60.96 % as compared to using the whole vocabulary of the corpus as language units (see Table 1). More significantly, parsing the corpus with this grammar has revealed that the utterance representation dimension diminishes 76.69 % on average (see Table 2).

```
[Recuperar]              \\ Lemma      To recover
  (recuperar)            \\ Word 1     to recover
  (recuperados)          \\ Word 2     recovered
  (recuperas)            \\ Word 3     you recover
  ...
;
```

Fig. 4 Partial lemma-grammar rule example

Table 1 Language unit set (U) dimension reduction

Vocabulary size (i.e., amount of distinct words in the corpus)	15070
Amount of lemmas selected	5882

Table 2 Utterance representation (Y) dimension reduction

Amount of distinct words per record	26.26 {820.40 ±}
Amount of lemmas extracted per record	6.12 {8.60 ±}

5 Experimental Evaluation

The experimental section is divided in two sections. The first one evaluates the impact of the entropy threshold θ_e in the classifier's metrics. In the second section, the SGA is used to recover information for Uncertain Cases, making suggestions throughout the dialog.

5.1 *Classifier Evaluation*

Once the user has explained their issue, the utterance representation Y is passed to TC to choose the target department of call-routing. As stated in Sect. 3, the entropy of Y is used to determine whether the user is giving enough information to route the call with high confidence or not. The call is routed if:

$$H(Y) \leq \theta_e \tag{9}$$

The representations Y which are above the threshold θ_e should not to be routed, so the classifier's decision is revoked.

In our particular case, the classification algorithm Multinomial Naive Bayes (M-NB)[1] is used in the Dialog System, as it has proven to be well-suited for the

[1]The implementation can be found in the Scikit-learn Python package [11].

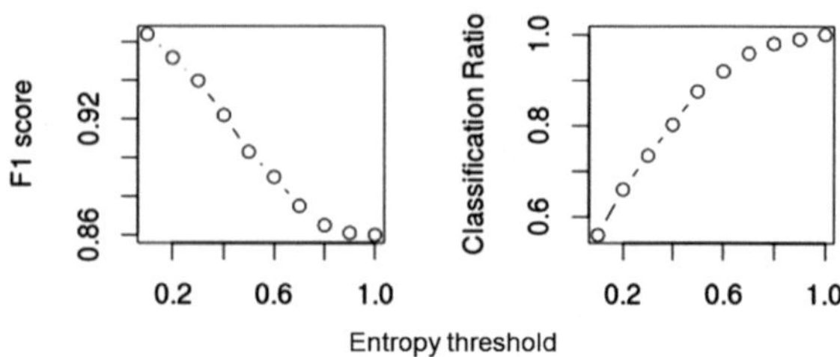

Fig. 5 Entropy threshold adjustment

Table 3 Results of MNB classifier with θ_e threshold set to 1 and 0.5

MNB	Precision	Recall	F_1	UC %
$\theta_e = 1$	0.86 ± 0.01	0.86 ± 0.01	0.86 ± 0.01	0.0
$\theta_e = 0.5$	**0.90** $\pm$ 0.01	**0.90** $\pm$ 0.01	**0.90** $\pm$ 0.01	12.5

task [9]. In order to observe the impact of the threshold in the classification metrics, the obtained F_1 score is depicted along with the ratio of classified instances in Fig. 5.

As can be observed, the lower θ_e is set, the lower is the ratio of calls routed—they are labeled as UC—and the higher is the F_1 score. Since an automatic algorithm has been designed to retrieve more information from the user—the SGA—, the main focus is set on seeking balance between the scores and the percentage of UC samples. For this reason, the threshold θ_e has been set to 0.5.

Table 3 shows the results of classification excluding the UC records obtained with the new threshold, as compared to having $\theta_e = 1$: excluding the 12.5 % UC from classification improves the metrics of call-routing by 4 points; these results suggest that entropy is useful to detect the cases at highest risk of being wrongly categorized and, thus, routed to an incorrect department.

5.2 *Recovering from Uncertainty*

A second set of experiments has been carried out in order to evaluate SGA and its reliability at recovering from Uncertain Cases (UC). First, an assessment has been done in terms of utterance appropriateness, as defined in [12]:

> An utterance is considered appropriate when [...] it asks for additional information which is essential to respond to the user's request [...]. A[ppropriate] U[tterance] evaluates whether the DM provides a coherent response at each turn according to its input.

We have measured the metric manually, taking the 12.5 % UC records of one of the folds in the previous experiments—322 in total—to emulate user turns as input for SG module. K is set to 5. Table 4 shows the results of the evaluation locally (i.e., taking as reference the amount of suggestions generated) and globally (taking as reference all the instances labeled as UC).

Table 4 Appropriateness of system turns generated with SG (K = 5)

	Total instances	Total suggestions	AU	Non-AU
SGA	322	318	267	51
Global %	–	**98.7**	**83.0**	15.7
Local %	–	–	84.0	16.0

Table 5 Classification results using SG to resolve low information user turns

Suggestions		Turn					Total Instances = 322	
		1	2	3	4	5		
SGA	318	159	68	33	17	17	Local	Global
Classified %							92.5	**91.3**
Not Classified %							7.5	**8.7**

The second experiment aims at checking that the information retrieved with SG helps improve the classification confidence. To this end, the same 322 UC-input/system turn pairs of the previous experiment have been used as the initial state. A user turn has been emulated that always accepts the suggestion offered by the system (the first scenario of answer evaluation, Sect. 3.3). Up to 4 additional suggestion and acceptance rounds have been emulated whenever the classifier returned UC after its update. Table 5 shows the results.

The results obtained indicate that the suggested language unit $\hat{u}$ reduces the entropy of the sample, improving the classification confidence and helping in the intended call-routing task.

6 Conclusions and Future Work

This article proposes a strategy that reinforces the common Dialog Manager logic. More specifically, it generates dialog systems turns automatically to retrieve information from the user when the latter does not provide enough information to complete the task set. This is done introducing an entropy measure criterion to detect uncertainty and low classification confidences, plus the the Suggestion Generation (SG) module, which chooses, using a graph representation of the domain knowledge—the Information Graph (IG)—, a language unit related with high confidence to the current dialog.

The proposed SG Algorithm has proven to be effective, rendering an appropriate suggestion 82.91 % of the times. The suggestions generated are meaningful for the classification task 91.3 % of the times. These results prove that IG is a valid structure to represent the informational relationships of the task domain. In conclusion, the inclusion of the entropy-based uncertainty detection and the SG module in a Spoken Dialog System improves the Dialog Manager logic, obtaining a more flexible and adaptive system.

Regarding future work, full implementation of the Dialog System in a real-user environment is planned, to face real and new situations, and thus evaluating SG in a wider range of situations and obtaining new data directly from the users. Also, other ways to explore the IG are being developed, taking into account the amount of information provided by each language unit of the graph.

References

1. Bohus, D., Rudnicky, A.: The RavenClaw dialog management framework: architecture and systems. Comput. Speech Lang. XXII **I**(3), 257–406 (2008)
2. Denecke, M., Waibel, A.: Dialogue strategies guiding users to their communicative goals. Eurospeech **3**(November), 1339–1342 (1997)
3. Chu-Carroll, J., Carpenter, B.: Vector-based natural language call routing. Comput. Linguist. **25**(3), 361–388 (1999)
4. Griol, D., Torres, F., Hurtado, L.F., Grau, S., Sanchis, E., Segarra, E.: Development and evaluation of the DIHANA project dialog system. In: Proceedings of Interspeech-06 Satellite Workshop Dialogue on Dialogues. Multidisciplinary Evaluation of Advanced Speech-based Interactive Systems. Pittsburgh (2006)
5. Gupta, N., Tür, G., Hakkani-Tür, D., Bangalore, S., Riccardi, G., Gilbert, M.: The AT&T spoken language understanding system. IEEE Trans. Audio Speech Lang. Process. **14**(1), 213–222 (2006)
6. Misu, T., Kawahara, T.: Dialogue strategy to clarify users queries for document retrieval system with speech interface. Speech Commun. **48**(9), 1137–1150 (2006)
7. Gosain, A., Bhugra, M.: A comprehensive survey of association rules on quantitative data in data mining. In: Information and Communication Technology (ICT 2013) pp. 1003–1008 (2013), http://ieeexplore.ieee.org/xpls/abs_all.jsp?arnumber=6558244
8. Yen, S.J., Chen, A.L.P.: An efficient data mining technique for discovering interesting association rules. In: Proceedings of the Eighth International Workshop on Database and Expert Systems Applications, Sep 1997, pp. 664–669 (1997)
9. Serras, M., Perez, N., Torres, M.I., Del Pozo, A., Justo, R.: Topic classifier for customer service dialog systems. In: Proceedings of the 18th International Conference on Text, Speech and Dialogue. Springer (2015)
10. Carreras, X., Chao, I., Padró, L., Padró, M.: Freeling: An open-source suite of language analyzers. In: Proceedings of the 4th Language Resources and Evaluation Conference (LREC 2004) IV, 239–242 (2004). http://hnk.ffzg.hr/bibl/lrec2004/pdf/271.pdf
11. Pedregosa, F., Varoquaux, G., Gramfort, A., Michel, V., Thirion, B., Grisel, O., Blondel, M., Prettenhofer, P., Weiss, R., Dubourg, V., Vanderplas, J., Passos, A., Cournapeau, D., Brucher, M., Perrot, M., Duchesnay, E.: Scikit-learn: machine learning in Python. J. Mach. Learn. Res. XII, 2825–2830 (2011)
12. Ghigi, F., Torres, M.I., Justo, R., Benedí, J.M.: Evaluating spoken dialogue models under the interactive pattern recognition framework. In: Proceedings of the INTERSPEECH 2013. pp. 480–484. Lyon (2013)

Evaluation of Question-Answering System About Conversational Agent's Personality

Hiroaki Sugiyama, Toyomi Meguro and Ryuichiro Higashinaka

Abstract We develop a question-answering system for questions that ask about a conversational agent's personality based on large-scale question-answer pairs created by hand. In casual dialogues, the speaker sometimes asks his conversation partner questions about favorites or experiences. Since this behavior also appears in conversational dialogues with a dialogue system, systems must be developed to respond to such questions. However, the effectiveness of personality-question-answering for conversational agents has not been investigated. Our user-machine chat experiments show that our question-answering system, which estimates appropriate answers with 60.7 % accuracy for the personality questions in our conversation corpus, significantly improves user's subjective evaluations.

Keywords Conversational systems · Question answering · Agent's personality

1 Introduction

Recent research on dialogue agents has actively investigated casual dialogues [1–5], because conversational agents are useful not only for entertainment or counseling purposes but also for improving performance in task-oriented dialogues [6]. In conversations, people often ask questions related to the specific personality of the person with whom they are talking, such as favorite foods and experience playing sports [7].

H. Sugiyama (✉) · T. Meguro · R. Higashinaka
NTT Communication Science Laboratories, 2-4, Hikari-dai, Seika-cho, Souraku-gun, Kyoto, Japan
e-mail: sugiyama.hiroaki@lab.ntt.co.jp
URL: http://www.kecl.ntt.co.jp/icl/ce/member/sugiyama/

T. Meguro
e-mail: meguro.toyomi@lab.ntt.co.jp

R. Higashinaka
NTT Media Intelligence Laboratories, 1-1, Hikari-no-oka, Yokosuka-shi, Kanagawa, Japan
e-mail: higashinaka.ryuichiro@lab.ntt.co.jp

K. Jokinen and G. Wilcock (eds.), *Dialogues with Social Robots*, Lecture Notes in Electrical Engineering 427,
DOI 10.1007/978-981-10-2585-3_14

Nishimura et al. showed that such personality questions also appeared in conversations with conversational agents [8]. If an agent avoids answering personality questions like ELIZA [9] which repeats almost the same questions or asks the talker back the question why the talker asks such questions, people will be disappointed with the agent and will stop talking; therefore, the capability to answer personality questions is one important function in the development of conversational agents.

Most previous research on the personality of conversational agents has investigated the agent's personality using roughly-grained categories, such as the Big-Five [10–12]. All of these studies parameterized the personalities, but they did not deal with specific subjects of the personalities, which are required to answer personal questions. To answer such personality questions, Batacharia et al. developed the Person DataBase (PDB), which consists of question-answer pairs (*QA pairs*) evoked by a pre-defined persona named *Catherine*, a 26-year-old female living in New York [13]. Their approach retrieves a question sentence that resembles the user's question utterance from the PDB and returns an answer sentence associated with the retrieved question. However, they did not evaluate the effectiveness for improving user satisfaction of their question-answering system (*QA system*), which is based on PDB. Traum et al. recently proposed and objectively evaluated time-offset interaction [14, 15], but they only evaluated the adequacy of each generated response without examining the effectiveness of the personality QA.

In this study, we examine the effectiveness of personality QA in casual dialogues between users and conversational agents. To develop our QA system that can answer specific personality questions, such as favorite asian foods or high scool athletic participation, we utilized the previously developed Person DataBase (PDB) [16] that contains 26,595 question-answer pairs about specific personality traits, which are classified into 10,082 *question categories*, where each category represents the identical subject. Our personality QA system estimates a question category from a personality question sentence and generates an answer associated with the estimated question category. Since a wide range of topics appear in casual dialogues, our large-scale QA pairs will be useful to cover a range of questions and generate appropriate answers. First we objectively examine the estimation accuracy of the question categories for personality inquiries that appeared in a human to human conversation corpus. After that, we combine our personality QA system with a conversational system [17] and investigate the former's effectiveness through chat experiments.

2 Person DataBase: PDB

We developed a personality QA system that can answer specific personality questions. To answer personality questions, we utilized our previously developed Person DataBase (PDB) [16]. In this section, we explain its details.

Figure 1 shows the procedure for collecting QA pairs. First, we collected 26,595 question sentences about the six personas shown in Fig. 1 from 42 Japanese-speaking participants (*questioners*). Each questioner created sentences under the following five

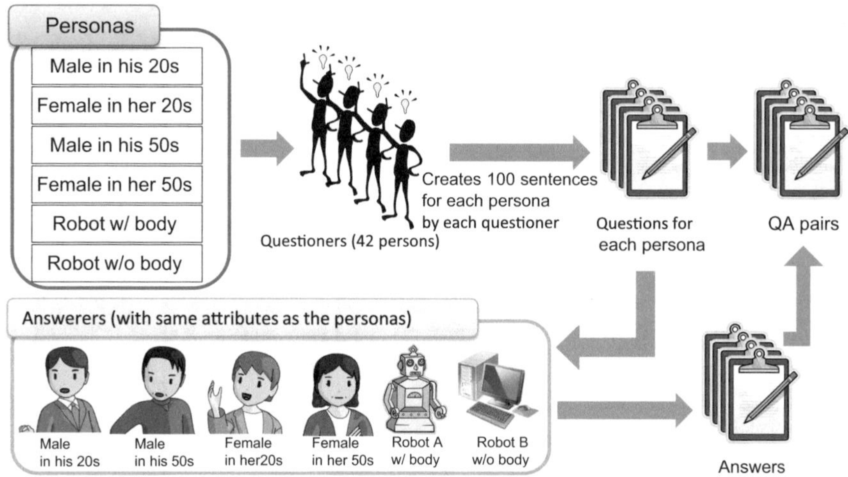

Fig. 1 Overview of collection of QA pairs [16]

rules: (A) create sentences that reflect what you want to know; (B) create grammatically complete sentences; (C) create exactly one sentence for each question; (D) do not create duplicate questions for a persona (e.g., *Where do you live?* and *What's your current address?*); and (E) do not copy questions from other sources like the web.

Then a participant called *an answerer* (not the questioner) with the same personality attribute as one of the personas created answers for the questions associated with the persona based on the following instructions: (a) create answers based on your own experiences or favorites; (b) create the same answers to the questions that represent identical subjects; and (c) create as many *Yes/No* answers as possible (called *Yes/No* restrictions). For the robot personas, the answerers created answers based on a robot character they imagined themselves.

After the collection stage, the answerers classified the question-answer sentence pairs (*QA pairs*) into question categories, where each represents an identical subject. This approach enables us to identify frequently asked question subjects based on the number of question sentences in each question category. Through this classification, we classified the question sentences into 10,082 question categories. Their distribution is long-tailed (half of the question sentences belong to the top 11 % (1110) categories), and 65.1 % (6568) have only one sentence. We call the top 11 % frequently-appearing question categories *frequent question categories*, and the others are *infrequent question categories*.

Finally, another participant (not the author, not the questioner, and not the answerer) called *an information annotator* annotated the collected QA pairs with answer types. We called the collected QA pairs with such information into a PDB. Table 1 illustrates examples of the collected PDBs.

Table 1 PDB examples (all columns translated by authors)

Question sentences	Question categories	Answer sentences	Answer types	Persona
How accurately can you understand what people say?	How accurately you can understand what people say	98 %	Quantity: Other	Robot B
Do you want to hold a wedding ceremony overseas?	Whether you want to hold a wedding ceremony overseas	Yes	Yes/No	20s Male
When is your birthday?	Birthday	Sept. 10, 1986.	Quantity: Date	20s Male
Do you usually eat pancakes?	Whether you usually eat pancakes	No	Yes/No	20s Male
Do you buy groceries by yourself?	Whether you buy groceries by yourself	No	Yes/No	50s Female
Do you have any persimmon trees in your garden?	Whether you have persimmon trees in your garden	No	Yes/No	50s Female
Do you have any pets?	Whether you have some pets	No	Yes/No	20s Female
What manufacturer of camera do you have?	Manufacturer of your camera	Canon	Company	20s Female

3 Experiments

Our QA system estimates question categories to generate answers for input question sentences. In this section, first we objectively evaluate the estimation accuracy of the question categories. After that, we investigate the effectiveness of our personality QA system by examining response appropriateness and user-machine chat experiments.

3.1 *Objective Evaluation: Estimation Accuracy of Question Categories*

3.1.1 Experiment Settings

In this section, we compare the estimation accuracy among a combination of estimation methods and features. For the estimation methods, we adopt the following three estimators: the RBF-kernel SVM, the linear SVM, and cosine-based retrieval. The former two SVM approaches directly estimate question categories, and the latter finds the most similar question sentence to an input question sentence. When we calculate estimation accuracy, we use the question category associated with the retrieved question sentence. We experimentally determined each estimator's parameters. We also compared the estimation accuracy among the following typical classification and retrieval features: bag-of-words (unigram or uni+bigram) vectors extracted from the question sentences, and those weighted with TF-IDF, which are calculated using words contained in the PDB and the conversation corpus, which treats each sentence as a document.

To train the estimators, we limited the estimation target to frequent question categories (1,110 categories and 13,917 sentences) that appear five or more times in PDB without considering the personas, since the estimators require at least a few training examples to estimate the question categories.

We evaluated the estimation accuracy of the question categories using two types of question sentences: those contained in the PDB and those sampled from the conversation corpus gathered by Higashinaka et al. [5], which contains 3,680 conversations based on text chats (134 K sentences). We evaluated the estimation accuracy of the PDB sentences with 5-fold cross-validation with 13,917 sentences. For the sentences in the conversation corpus, first we sampled 166 personality question sentences from the corpus and two annotators (not the authors) annotated the question categories to the sentences (Cohen's κ: 0.816). In the 166 sentences, 51 personality question sentences are associated with the frequent question categories by both the annotators. We trained the classifier with all the PDB sentences and evaluated theie accuracy with the 51 personality question sentences.

3.1.2 Result and Discussion

Figures 2a, b show the estimation accuracy of the question categories for the PDB and the conversation corpus. The linear SVM with unigrams showed the highest accuracy with 0.841 for PDB (without TF-IDF) and 0.607 for the conversation corpus (with TF-IDF). The RBF-kernel SVM and cosine-based retrieval showed lower accuracy probably because of a lack of data for each category. In the comparison between the raw bag-of-words and the TF-IDF weighted ones, the latter outperformed the former in most of the settings, suggesting that TF-IDF compensates for the lack of data.

Figure 2 also shows that the estimation accuracy for the conversation corpus is lower than that for the PDB. This is because the personal question sentences sampled from the conversation corpus have many words that are not contained in the PDB. Besides, contrary to the PDB where question sentences contain only the necessary information to represent the questions, the sentences in the conversation corpus sometimes contain words unrelated to the questions, such as the introduction of question topics.

Next, we analyzed the incorrect estimations based on one trial of 5-fold cross-validation. Table 2 shows the error-categories, their ratios, and their examples. 29.8 %

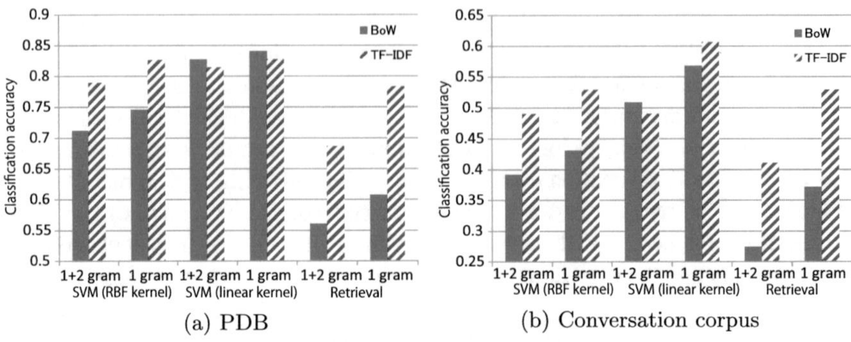

Fig. 2 Accuracy of question categories: majority baseline in **a** PDB is 0.011 (155/13917)

Table 2 Error-categories, their ratios, and their examples (translated by authors)

Almost identical meaning (Categories differed by tense or modalities)	29.8 % 135/453	Q sent.: Have you ever ridden a bicycle? Correct: Whether you can ride a bicycle Est.: Whether you usually ride a bicycle
Different grain size of topics	9.1 % 41/453	Q sent.: Are you interested in fashion? Correct: Whether you pay attention to clothes Est.: Whether you have favorite brands of clothing or bags
Different topics	35.8 % 162/453	Q sent.: Do you color your hair? Correct: Whether you dye (have ever dyed) your hair Est.: Whether you have a job now
Answer-type mismatch	10.3 % 47/453	Q sent.: How often do you drink a week? Correct: Frequency of drinking in a week Est.: Favorite kind of alcohol
Annotation error (Existence of more appropriate category)	7.7 % 35/453	Q sent.: Do you usually drive? Correct: Whether you can drive a car/whether you have a driver's license Est.: Whether you drive a car frequently
Character mismatch (Categories varied by character)	1.5 % 7/453	Q sent.: What is your favorite opposite sex type? Correct: Favorite female type Est.: Favorite male type

of the errors can be categorized as *almost identical meaning*, where the question sentences were correctly answered with the estimated question categories, such as *whether you can ride a bicycle* as the correct category and *whether you usually ride a bicycle* as the estimated category. The most frequent (35.8 %) error type was *different topics*, which unfortunately cannot be answered with the estimated question categories, such as *whether you are attentive to clothes* as the correct category and *whether you have work now* as the estimated category. Such errors were caused when important words in the question sentences were not covered in the estimator's training data. We assume that thesauri like WordNet or word-clustering methods like brown-clustering [18] or word2vec [19] can improve the estimation accuracy of such question sentences. Answer-type mismatch explains about 10.3 % of the errors. This error can be reduced by using a separate classifier for the answer-types.

3.2 Subjective Evaluation 1: Response Appropriateness

As a subjective evaluation, first we examined the effectiveness of a personality QA system through module-based experiments that compared the appropriateness of the sentences generated by our QA system and a conventional approach (without our personality QA system) on input personality questions.

3.2.1 Experiment Settings

Based on the objective evaluation results, we adopted linear SVM with TF-IDF weighted unigrams as features of the estimators of the question category. When a personality question sentence is given, our personality QA system estimates the question category that corresponds to the question sentence and returns an answer that is randomly selected from answer sentences associated with the estimated question category.

For the conventional system, we adopted an open-domain conversational system proposed by Sugiyama et al. [17]. Despite the wide range of topics that appear in casual dialogues, this system generates system utterances related to the user utterances by assembling two phrase-pairs: one extracted from user utterances and another from a Twitter corpus that has dependency relations with the former pair. Since Sugiyama's system outperforms other rule-based and retrieval-based systems and has no question-answering function, it is reasonable to examine the effectiveness of our personality QA system.

We randomly selected 52 personality questions from the conversational corpus (Sect. 3.1), and our QA and conversational systems generated responses for each sentence. We shuffled the sentence orders for each input sentence, and two annotators (not the authors) subjectively labeled them on a 7-point Likert scale in terms of response naturalness. The Spearman's correlation coefficients of the two annotators were 0.71. We used the mean labeled scores for our evaluation.

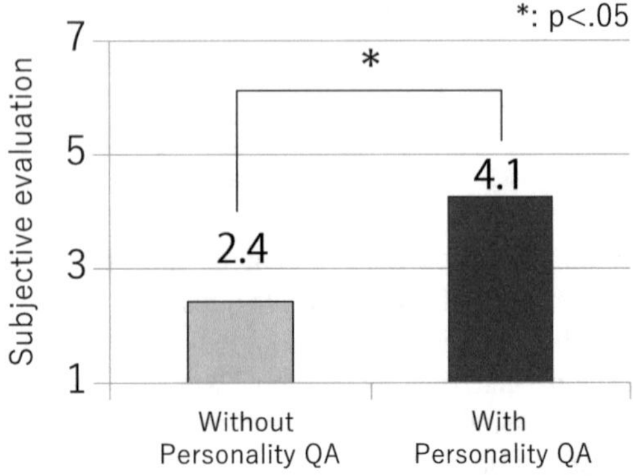

Fig. 3 Comparison of appropriateness of one-turn responses for personality questions with 7-point Likert scale

Table 3 Examples of question sentences, answer sentences, and their evaluation scores (translated by authors)

Input questions	Answer sentences	Scores
What are you going to do today?	With PQA: I'm going to watch a movie Without PQA: Yes. What are your plans?	7 1
Do you like wasting time?	With PQA: I like drawing postcards. How about you? Without PQA: I got it. What do you like to collect?	2 1

3.2.2 Results

Figure 3 shows that the response sentences generated by our personality QA system acquired significantly higher scores (4.1) than the conventional system (2.4) ($p < 0.05$ with a Wilcoxon signed-rank test). This indicates that the personality QA's capability adequately responds to personality questions. However, its evaluation scores are less satisfactory, since some questions are too specific and could not appropriately be classified into our question categories. For example, the question, *What are you going to do today?*, in Table 3 is common, so our QA system's response was appropriate. On the other hand, since the question, *Do you like wasting time?*, is not so common, neither system generated appropriate answers, which are evaluated with scores 1 or 2 that decreased the mean of the evaluation scores.

3.3 *Subjective Evaluation 2: Online-Chat Experiments*

We also examined the effectiveness of our personality QA system through user-machine chat experiments.

Table 4 Objective evaluation scores on 7-point Likert scale ($*$: $p < 0.1$, $**$: $p < 0.05$)

	Without QA	With QA
(1) Naturalness of dialogue flow	2.96	3.52**
(2) Grammatical correctness	3.68	4.02**
(3) Appropriateness as one-turn response	3.12	3.73**
(4) Semantic consistency	2.66	3.50**
(5) Dialogue usefulness	2.50	3.27**
(6) Ease of considering next utterance	3.04	3.75**
(7) Variety of system utterances	3.02	3.64**
(8) User motivation	3.96	4.50**
(9) System motivation felt by user	3.79	4.14**
(10) Desire to chat again	2.79	3.27*

3.3.1 Experiment Settings

Since our personality QA system is not designed to respond to user utterances except for personality questions, we combined our personality QA system with another conversational system that can handle such utterances to perform chat experiments. In this study, we used aforementioned open-domain conversational system [17]. When a user's utterance is estimated to be a personality question, the personality QA system generates system utterances; if it is estimated to be a normal utterance (not a personality question), the conversational system generates a system utterance.

We recruited 30 native Japanese-speaking participants whose ages ranged from 20 to 50 (controlled by gender and age) from outside the authors' organization who have experience using chat systems. Each participant talked with the conversational systems with and without the personality QA system and provided subjective evaluation scores for both systems for each of the ten criteria shown in Table 4 using 7-point Likert scales.

Each ten-minute dialogue was conducted through a text chat interface; the presentation orders of the systems were randomized. Since it takes about 30 to 60 s for the system to generate responses, about 20 utterances were performed in a dialogue.

3.3.2 Results

Table 4 shows that personality QA system outperformed that without the QA system in all the criteria. The dialogue examples in Table 5 show that the users sometimes asked the system personality questions, which it couldn't answer without the personality QA. This failure caused the entire dialogue to collapse. On the other hand, the personality QA system answered the personality questions. In this example, although the user's factoid questions couldn't be answered, the dialogue's thread was recaptured by other personality questions. These results indicate that the capability of

Table 5 Example dialogues translated by authors

(a) Without personality QA	
User	What are you doing?
System	OK. Do you usually watch TV?
User	I don't watch TV because I don't even have one.
System	Oh, it's a TV. Please watch TV when you get back home.
User	But I don't have a TV!
System	Well, you're at home. It's dangerous after you get back home.
User	What's dangerous?
System	That's right.
User	What are you talking about? What's your favorite TV program?
System	That's right.
User	Please tell me about your favorite TV program.
System	That's right.
User	What's your hobby?
System	Well, what's your hobby?
User	Tell me first.
System	Yes.
User	What do you like?
System	Yeah. What do you like?
User	You're playing strange games with me. I like fish paste.
System	Let's come to like fish paste parfait.
User	What's that?
(b) With personality QA	
User	Hello!
System	Yeah. What's your hobby?
User	Hobby? Well, I've been doing lots of bowling recently.
System	OK.
User	What's your hobby?
System	Reading.
User	What have you read recently?
System	*Do Androids Dream of Electric Sheep?*.
User	Who is the author?
System	Oh, who is it?
User	You seem to answer without really answering...
System	I feel sick when it pushes and is pushed.
User	Well, what book do you recommend?
System	*The Brothers Karamazov*.
User	That's really long, isn't it.
System	That's right. Please recommend some more books since autumn is my favorite time for reading.
User	Well, I haven't read any books recently.
System	Oh, reading? It's like autumn and I like it.
User	Speaking of autumn, sauries are good.
System	It is like autumn and I like it

Underlined sentences are personality questions

the personality QA is both required and useful for conversational agents to prevent dialogues from breaking down.

4 Conclusion

We developed a question-answering system for questions that ask about a conversational agent's specific personality using large-scale question-answer pairs called a Person DataBase (PDB). Our objective evaluation experiment showed that our personality QA system estimated appropriate question categories with 84.1 % accuracy for question sentences in PDB and 60.7 % for a conversation corpus. We also analyzed the incorrect estimations and determined that 29.8 % of the errors are not critical; other errors can be improved. Furthermore, we examined our personality QA system's effectiveness through subjective evaluation experiments, which indicated that it is useful to prevent the breakdown of conversation.

References

1. Shibata, M., Nishiguchi, T., Tomiura, Y.: Dialog system for open-ended conversation using web documents. Informatica **33**, 277–284 (2009), http://www.informatica.si/PDF/Informatica_2009_3.pdf#page=37
2. Ritter, A., Cherry, C., Dolan, W.: Data-driven response generation in social media. In: Proceedings of the 2011 Conference on Empirical Methods in Natural Language Processing, pp. 583–593 (2011). http://ai.cs.washington.edu/www/media/papers/paper2.pdf
3. Wong, W., Cavedon, L., Thangarajah, J., Padgham, L.: Strategies for mixed-initiative conversation management using question-answer pairs. In: Proceedings of the 24th International Conference on Computational Linguistics. pp. 2821–2834 (2012). https://www.aclweb.org/anthology-new/C/C12/C12-1172.pdf
4. Meguro, T., Higashinaka, R., Minami, Y., Dohsaka, K.: Controlling listening-oriented dialogue using partially observable markov decision processes. In: Proceedings of the 23rd International Conference on Computational Linguistics, pp. 761–769 (2010)
5. Higashinaka, R., Imamura, K., Meguro, T., Miyazaki, C., Kobayashi, N., Sugiyama, H., Hirano, T., Makino, T., Matsuo, Y.: Towards an open-domain conversational system fully based on natural language processing. In: Proceedings of the 25th International Conference on Computational Linguistics, pp. 928–939 (2014)
6. Bickmore, T., Cassell, J.: Relational agents: a model and implementation of building user trust. In: Proceedings of the SIGCHI Conference on Human Factors in Computing Systems, pp. 396–403 (2001). http://dl.acm.org/citation.cfm?id=365304
7. Tidwell, L.C., Walther, J.B.: Computer-mediated communication effects on disclosure, impressions, and interpersonal evaluations. Hum. Commun. Res. **28**(3), 317–348 (2002)
8. Nisimura, R., Nishihara, Y., Tsurumi, R., Lee, A., Saruwatari, H., Shikano, K.: Takemaru-kun.: Speech-oriented information system for real world research platform. In: Proceedings of the First International Workshop on Language Understanding and Agents for Real World Interaction, pp. 70–78 (2003). http://library.naist.jp:12180/dspace/handle/10061/8226
9. Weizenbaum, J.: ELIZA—a computer program for the study of natural language communication between man and machine. Commun. ACM **9**(1), 36–45 (Jan 1966). http://portal.acm.org/citation.cfm?doid=365153.365168

10. Caspi, A., Roberts, B.W., Shiner, R.L.: Personality development: stability and change. Ann. Rev. Psychol. **56**, 453–484 (Jan 2005). http://www.ncbi.nlm.nih.gov/pubmed/15709943
11. John, O.P., Srivastava, S.: The Big Five trait taxonomy: history, measurement, and theoretical perspectives. No. 510 (1999). http://scholar.google.com/scholar?hl=en&btnG=Search&q=intitle:The+Big-Five+Trait+Taxonomy:+History,+Measurement,+and+Theoretical+Perspectives#0
12. Mairesse, F., Walker, M.: PERSONAGE: Personality generation for dialogue. In: Proceedings of the Annual Meeting of the Association For Computational Linguistics, pp. 496–503 (2007). http://acl.ldc.upenn.edu/P/P07/P07-1063.pdf
13. Batacharia, B., Levy, D., Catizone, R., Krotov, A., Wilks, Y.: CONVERSE: a conversational companion. Machine conversations, pp. 205–215 (1999). http://link.springer.com/chapter/10.1007/978-1-4757-5687-6_17
14. Traum, D., Georgila, K., Artstein, R., Leuski, A.: Evaluating spoken dialogue processing for time-offset interaction. In: Proceedings of the 16th Annual SIGdial Meeting on Discourse and Dialogue, pp. 199–208 (2015)
15. Leuski, A., Traum, D.: Creating virtual human dialogue using information retrieval techniques. AI Mag. **32**(2), 42–56 (2011). http://www.aaai.org/ojs/index.php/aimagazine/article/view/2347
16. Sugiyama, H., Meguro, T., Higashinaka, R.: Large-scale collection and analysis of personal question-answer pairs for conversational agents. In: Proceedings of Intelligent Virtual Agents, pp. 420–433 (2014)
17. Sugiyama, H., Meguro, T., Higashinaka, R., Minami, Y.: Open-domain utterance generation using phrase pairs based on dependency relations. In: Proceedings of Spoken Language Technology Workshop, pp. 60–65 (2014)
18. Turian, J., Ratinov, L., Bengio, Y.: Word representations: A simple and general method for semi-supervised learning. In: Proceedings of the 48th Annual Meeting of the Association for Computational Linquistics. pp. 384–394. July (2010)
19. Mikolov, T., Chen, K., Corrado, G., Dean, J.: Distributed representations of words and phrases and their compositionality. In: Proceedings of the 27th Annual Conference on Neural Information Processing Systems, pp. 1–9 (2013)

Fisher Kernels on Phase-Based Features for Speech Emotion Recognition

Jun Deng, Xinzhou Xu, Zixing Zhang, Sascha Frühholz, Didier Grandjean and Björn Schuller

Abstract The involvement of affect information in a spoken dialogue system can increase the user-friendliness and provide a more natural way for the interaction experience. This can be reached by speech emotion recognition, where the features are usually dominated by the spectral amplitude information while they ignore the use of the phase spectrum. In this chapter, we propose to use phase-based features to build up such an emotion recognition system. To exploit these features, we employ Fisher kernels. The according technique encodes the phase-based features by their deviation from a generative Gaussian mixture model. The resulting representation is fed to train a classification model with a linear kernel classifier. Experimental results on the GeWEC database including 'normal' and whispered phonation demonstrate the effectiveness of our method.

Keywords Speech emotion recognition · Phase-based features · Fisher kernels · Modified group delay features

1 Introduction

For a spoken dialogue systems, a recent trend is to consider the integration of emotion recognition in order to increase the user-friendliness and provide a more natural interaction experience [1–6]. In fact, this may be particularly relevant for systems

J. Deng (✉) · Z. Zhang · B. Schuller
Chair of Complex & Intelligent Systems, University of Passau, Passau, Germany
e-mail: jun.deng@uni-passau.de

X. Xu
Machine Intelligence & Signal Processing Group, MMK,
Technische Universität München, Munich, Germany

S. Frühholz · D. Grandjean · B. Schuller
Swiss Center for Affective Sciences, University of Geneva, Geneva, Switzerland

B. Schuller
Department of Computing, Imperial College London, London, UK

K. Jokinen and G. Wilcock (eds.), *Dialogues with Social Robots*,
Lecture Notes in Electrical Engineering 427,
DOI 10.1007/978-981-10-2585-3_15

that accept whispered speech as input given the social and emotional implications of whispering. At present, acoustic features used for speech emotion recognition are dominated by the conventional Fourier transformation magnitude part of a signal, such as in Mel-frequency cepstral coefficients (MFCCs) [7–10]. In general, the phase-based representation of the signal has been neglected mainly because of the difficulties in phase wrapping [11, 12]. In spite of this, the phase spectrum is capable of summarising the signal. Recent work has proved the effectiveness of using phase spectrum in different speech audio processing applications, including speech recognition [13, 14], source separation [15], and speaker recognition [16]. However, there exists little research, which applies phase-based features for speech emotion recognition. Recently, the phase distortion, which is the derivative of the relative phase shift, has been investigated for emotional valence recognition [17]. In this short chapter, the key objective is to demonstrate the usefulness of the phased-based features for speech emotion recognition. In particular, this chapter investigates whether the *modified group delay feature* is capable of improving the performance of an emotion recogniser because it has not yet been applied for speech emotion recognition. Besides, we propose to use *Fisher kernels* to encode the varied length series of the modified group delay features into a fixed length Fisher vector. The Fisher kernel is a powerful framework, which enjoys the benefits of generative and discriminative approaches to pattern classification [18]. Eventually, a linear kernel support vector machine (SVM) is adopted to train the emotion recognition model with the resulting Fisher vectors.

The organization of this chapter is as follows. Section 2 first introduces the proposed methods, including the selected phase-based features and Fisher kernels. Next, Sect. 3.2 presents the experimental results on the GeWEC database. In Sect. 4, finally, we conclude this chapter and point out the future work.

2 Methods

In this section, we theoretically brief on the modified group delay feature as well the Fisher kernels. Afterwards, we give the proposed system for speech emotion recognition based on the modified group delay feature and the Fisher kernel, which is illustrated in Fig. 1.

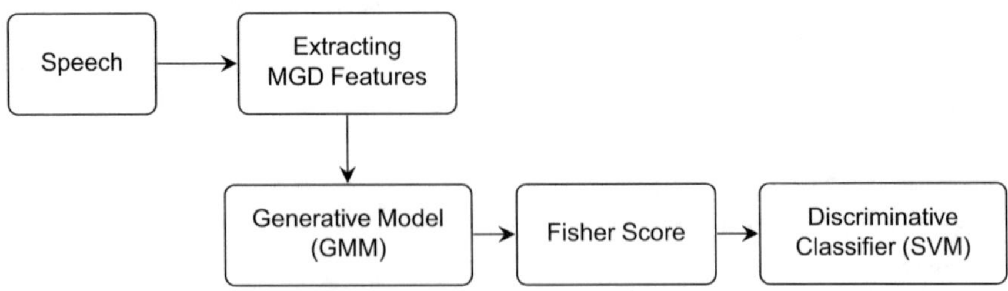

Fig. 1 Block scheme of the proposed speech emotion recognition system, using phase-based features, Fisher kernels, and SVMs

2.1 Modified Group Delay Feature

The Fourier transform of a discrete time digital signal $x(n)$ can be computed as

$$X(\omega) = |X(\omega)|e^{j\phi(\omega)} \tag{1}$$

where $|X(\omega)|$ and $\phi(\omega)$ are the magnitude and phase spectrum.

The common spectral features for speech processing only make use of the magnitude spectrum whereas often rejecting the use of the phase spectrum. Extracting useful features from the phase spectrum is a challenging task because of wrapping of the phase spectrum and its dependency on the window position [19, 20]. In the community, however, a large body of previous studies have shown that extracting the phase information of a signal is applicable and systems using the extracted phase information deliver promising performance. Inspired by the big success of them, we explore the Modified Group Delay feature (referred to as MGD) to build up a speech emotion recognition system in this work.

Given a discrete time signal $x(n)$, the group delay feature is written as follows

$$\tau_g(\omega) = \frac{X_R(\omega)Y_R(\omega) + X_I(\omega)Y_I(\omega)}{|X(\omega)|^2}, \tag{2}$$

where the angular frequency ω is limited in $[0, 2\pi]$, n is an integer, $|X(\omega)|$ is the magnitude of the Fourier transforms of $x(n)$, $Y(\omega)$ is the Fourier transforms of the signal $y = nx(n)$, and the subscripts R and I indicate real and imaginary parts, respectively. Although the group delay feature is discriminative and additive, it is ill behaved if the zeros of the system transfer function are close to the unit circle [13]. For this reason, the group delay function at frequency bins near these zeros inevitably results in spurious spikes and becomes ill-behaved although it is able to produce a meaningful representation of a signal to a certain extent.

To address this issue, a modification of the group delay function is proposed in [13]. The modified feature is computed as

$$\tau_m(\omega) = \frac{\tau_p(\omega)}{|\tau_p(\omega)|}|\tau_p(\omega)|^{\alpha}, \tag{3}$$

where

$$\tau_p(\omega) = \frac{X_R(\omega)Y_R(\omega) + X_I(\omega)Y_I(\omega)}{|S(\omega)|^{2\gamma}}, \tag{4}$$

and $S(\omega)$ is a smoothed form of $|X(\omega)|$. The two tuning parameters γ and α, which control the range dynamics of the MGD spectrum, are set to 1.2 and 0.4 in this work based on previous work [21, 22]. Note that $P(\omega) = X_R(\omega)Y_R(\omega) + X_I(\omega)Y_I(\omega)$, called the product spectra, includes information from both the magnitude and phase spectrum [23]. In practice, the cepstrally smoothed form $|S(\omega)|$ is commonly derived as follows:

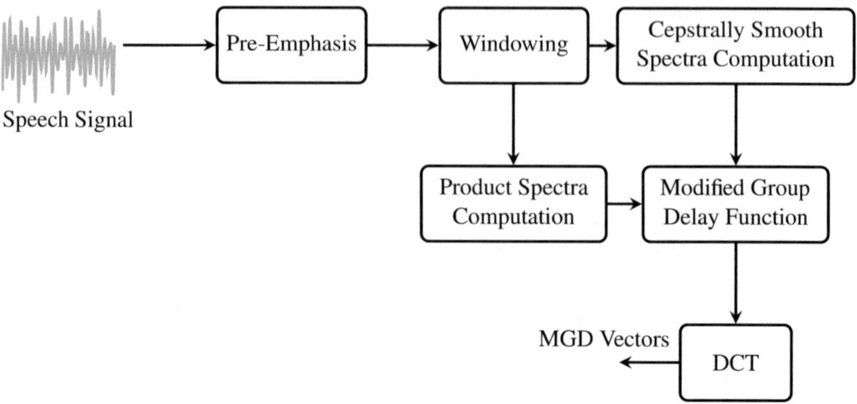

Fig. 2 *MGD* computation process

1. Take the log-magnitude Fourier spectra of $X(\omega)$ and smooth the log spectra by applying the 5 order of the median filter.
2. Take the Discrete Cosine Transform (DCT) of the log spectra and retain the first 30 coefficients.
3. Take the inverse DCT of the cepstral coefficients to result in the smoothed spectra $|S(\omega)|$.

Figure 2 shows the complete computation process of the MGD-based feature extraction. In a manner similar to the computation of MFCCs, the speech signal is first been pre-emphasised and then framed by a Hamming window. Afterwards, the MGD features are computed by Eq. (3). Finally, the DCT is applied on the MGD features so as to perform a decorrelation. In general, the first coefficient obtained by the DCT is excluded to avoid the effects of the average value. The first 12 coefficients are retained (excluding the 0th coefficient).

2.2 *Fisher Kernels*

The Fisher kernel was firstly introduced in [18] to exploit generative probability models in discriminative classifiers such as SVMs. Moreno and Rifkin [24] used the Fisher kernel method for large Web audio classification, where a Gaussian mixture model is used as a probabilistic generative model to produce appropriate feature vectors. Furthermore, based on the Fisher kernel method, [25] proposed a hybrid system for a text-independent speaker recognition task.

Recently, the Fisher kernel is popularly used in large scale image classification and image retrieval [26, 27]. The basic idea is to look at how the low level descriptors (e.g., a sequence of the phase-based features) affect the learnt generative model, which is typically a Gaussian mixture model. The effect is obtained by computing the derivative of the log-likelihood with respect to the model parameters. Formally, given a parametric generative model p_λ with parameters λ, the Fisher score function of a given example X is given by computing the first derivative of the log-likelihood function

$$\Phi(X) = \nabla_\lambda \log p_\lambda(X). \tag{5}$$

The Fisher score allows us to embed a sequence of low-level descriptors into a fixed-length vector, whose dimensionality depends on the size of the model parameters, not on the length of X.

The Fisher kernel is then defined as

$$K(X_i, X_j) = \Phi(X_i)^T I^{-1} \Phi(X_j) \tag{6}$$

where I is the Fisher information matrix. The Fisher kernel leads to a measure of similarity between two examples X_i and X_j by computing a distance between them, and can be used with any kernel classifier such as SVMs.

3 Experiments

3.1 Geneva Whispered Emotion Corpus

We employ the Geneva Whispered Emotion Corpus (GeWEC) to evaluate the effectiveness of the proposed system. The corpus provides normal and whispered paired utterances. Two male and two female professional French-speaking actors in Geneva were recruited to speak eight predefined French pseudo-words ("*belam*", "*molen*", "*namil*", "*nodag*", "*lagod*", "*minad*", and "*nolan*") with a given emotional state in both normal and whispered speech modes as in the GEMEPS-corpus that was used in the Interspeech 2013 Computational Paralinguistics Challenge [28]. Speech was expressed in four emotional states: *angry*, *fear*, *happiness*, and *neutral*. The actors were requested to express each word in all four emotional states five times. The utterances were labelled based on the state they should be expressed in, i.e., one emotion label was assigned to each utterance. As a result, GeWEC consists of 1280 instances in total. In the experiments, cross-speech-mode evaluation is considered. That is, one speech mode of the GeWEC data is used for training while the other speech mode data is used for testing.

Recording was done in a sound proof chamber using professional recording equipment. All recordings were recorded with a 16 bit PCM encoded single channel at a sampling rate of 44.1 kHz. The distance from the microphone was about 0.5 m during recording. Recordings were accompanied by visual cues on a screen, which indicated which word has to be vocalized and which emotional state needs to be expressed. Cues were on the screen for 1 s length, separated by a blank screen of 2 s. The cue duration of 1 s was chosen such that the actors were guided to vocalize each word with a duration of about 1 s, which ensures that the vocalizations were comparable in length.

Pre-processing steps were applied to each utterance before feature extraction, in which all utterances were normalized to mean energy, as well as scaled to a mean of 70 dB sound pressure level (SPL) and added manually a fade-in/fade-out duration of 15 ms.

3.2 Experimental Setup

To extract the phase-based features, frame windowing is performed using a Hamming window with a frame-length of 25 ms and a frame-shift of 20 ms. Pre-emphasis is conducted. Then, we compute the phase-based features based on Eq. (3), which ends up 12-dimensional features for each frame.

To generate Fisher vectors on the low level descriptors, we use a K-component GMM with diagonal covariances as the generative model p_λ. As suggested in [26], only gradients of the means and covariances are taken into account, leading to a $2 \times d \times K$ dimensional vector, where $d = 12$ for the phase-based features. The number of components for GMMs is chosen in a range $K = \{2, 4, \ldots, 30\}$ via cross-validation. As for the basic supervised learner in the classification step, we use linear SVMs implemented in LIBLINEAR [29].

As for the baselines, we chose to use SVMs with various state-of-the-art and publicly available feature sets, provided by the open source openSMILE toolkit [30, 31]. The feature sets used include Interspeech Challenges on Emotion in 2009 [32] (IS09), Level of Interest in 2010 [33] (IS10), Speaker States in 2011 [34] (IS11), Speaker Traits in 2012 [35] (IS12), Emotion in 2013 [28] (IS13), and the most recently proposed Geneva minimalistic acoustic parameter set (GeMAPS) and the extended Geneva minimalistic acoustic parameter set (eGeMAPS) [36]. Unweighted average recall (UAR) is used as a performance metric as in these challenges. It equals the sum of the recalls per class divided by the number of the classes, and appears more meaningful than overall accuracy in the given case of presence of class imbalance. Besides, significance tests are conducted by computing a one-sided z-test, which has also been the competition measure of all Interspeech Challenges on Emotion and Paralinguistics.

3.3 Results

Table 1 presents the experimental results on the GeWEC data. As can be seen from Table 1, the proposed method always performs best for the two experimental settings. For the first one, where the model is trained on 'normal' while tested on whispered speech, the phase-based features in conjunction with the Fisher kernel reaches 50.3 %

Table 1 UAR for four-way cross-mode emotion recognition on GeWEC: When one speech mode of GeWEC (normal speech (norm.) or whispered speech (whisp.)) is used for training, the other one is used for testing

UAR [%]	IS09	IS10	IS11	IS12	IS13	GeMAPS	eGeMAPS	Proposed
Norm. (train), Whisp. (test)	35.5	39.5	40.3	33.3	36.4	34.1	41.9	∗**50.3**
Whisp. (train), Norm. (test)	53.4	52.3	52.8	46.4	48.4	32.0	38.9	**54.8**

Different feature sets are considered. Significant results (p-value < 0.05, one-sided z-test) are marked with an asterisk. The proposed method compares with the state-of-the-art recognition systems using brute-forced features sets, including the INTERSPEECH (IS) Challenge feature sets in 2009–2013, the Geneva minimalistic acoustic parameter set (GeMAPS), and the extended Geneva minimalistic acoustic parameter set (eGeMAPS)

UAR, which outperforms the other considered methods significantly by a large margin. As for the second setting, the proposed method achieves 54.8 % UAR, which is as competitive as the other approaches.

4 Conclusions

We focused on improving speech emotion recognition, e. g., for the embedding in spoken dialogue systems, in a challenging whispered versus non-whispered speech and vice-versa cross-mode setting. Specifically, we presented a novel framework, using phased-based features (i.e., modified group delay features) with a Fisher kernel. Cross-speech-mode experiments on the GeWEC data were conducted, demonstrating that the present framework is competitive with other modern emotion recognition models. Further evaluations of alternative types of phase-based features are to be considered. Other future work includes to incorporate different normalisation techniques such as L2 normalisation [27] into the recognition framework.

Acknowledgements This work has been partially supported by the BMBF IKT2020-Grant under grant agreement No. 16SV7213 (EmotAsS) and the European Communitys Seventh Framework Programme through the ERC Starting Grant No. 338164 (iHEARu).

References

1. Andre, E., Rehm, M., Minker, W., Bühler, D.: Endowing spoken language dialogue systems with emotional intelligence. In: Affective Dialogue Systems, pp. 178–187. Springer (2004)
2. Acosta, J.C.: Using emotion to gain rapport in a spoken dialog system. In: Proceedings of NAACL HLT, pp. 49–54. Boulder, CO (2009)
3. Pittermann, J., Pittermann, A., Minker, W.: Emotion recognition and adaptation in spoken dialogue systems. Int. J. Speech Technol. **13**(1), 49–60 (2010)

4. Callejas, Z., Griol, D., López-Cózar, R.: Predicting user mental states in spoken dialogue systems. EURASIP J. Adv. Sign. Process. **2011**, 6 (2011)
5. Vinciarelli, A., Pantic, M., Heylen, D., Pelachaud, C., Poggi, I., D'Errico, F., Schröder, M.: Bridging the gap between social animal and unsocial machine: a survey of social signal processing. IEEE Trans. Affect. Comput. **3**(1), 69–87 (2012)
6. Benyon, D., Gamback, B., Hansen, P., Mival, O., Webb, N.: How was your day? Evaluating a conversational companion. IEEE Trans. Affect. Comput. **4**(3), 299–311 (2013)
7. Dumouchel, P., Dehak, N., Attabi, Y., Dehak, R., Boufaden, N.: Cepstral and long-term features for emotion recognition. In: Proceedings of INTERSPEECH, pp. 344–347. Brighton, UK (2009)
8. Schuller, B.: Intelligent Audio Analysis. Signals and Communication Technology, Springer (2013), 350 p
9. Anagnostopoulos, C.N., Iliou, T., Giannoukos, I.: Features and classifiers for emotion recognition from speech: a survey from 2000 to 2011. Artif. Intell. Rev. pp. 1–23 (2012)
10. Attabi, Y., Alam, M.J., Dumouchel, P., Kenny, P., O'Shaughnessy, D.: Multiple windowed spectral features for emotion recognition. In: Proceedings of ICASSP, pp. 7527–7531. IEEE, Vancouver, BC (2013)
11. Mowlaee, P., Saeidi, R., Stylanou, Y.: INTERSPEECH 2014 special session: phase importance in speech processing applications. In: Proceedings of INTERSPEECH. Singapore (2014), 5 p
12. Yegnanarayana, B., Sreekanth, J., Rangarajan, A.: Waveform estimation using group delay processing. IEEE Trans. Acoust. Speech Sign. Process. **33**(4), 832–836 (1985)
13. Murthy, H., Gadde, V., et al.: The modified group delay function and its application to phoneme recognition. In: Proceedings of ICASSP, vol. 1, pp. I–68. Hong Kong, China (2003)
14. Hegde, R., Murthy, H., Gadde, V.: Significance of the modified group delay feature in speech recognition. IEEE Trans. Audio Speech Lang. Process. **15**(1), 190–202 (2007)
15. Mowlaee, P., Saiedi, R., Martin, R.: Phase estimation for signal reconstruction in single-channel speech separation. In: Proceedings of ICSLP, pp. 1–4. Hong Kong, China (2012)
16. Hernáez, I., Saratxaga, I., Sanchez, J., Navas, E., Luengo, I.: Use of the harmonic phase in speaker recognition. In: Proceedings of INTERSPEECH, pp. 2757–2760. Florence, Italy (2011)
17. Tahon, M., Degottex, G., Devillers, L.: Usual voice quality features and glottal features for emotional valence detection. In: Proceedings of ICSP, pp. 693–696. Beijing, China (2012)
18. Jaakkola, T., Haussler, D.: Exploiting generative models in discriminative classifiers. In: Proceedings of NIPS, pp. 487–493. Denver, CO (1999)
19. McCowan, I., Dean, D., McLaren, M., Vogt, R., Sridharan, S.: The delta-phase spectrum with application to voice activity detection and speaker recognition. IEEE Trans. Audio Speech Lang. Process. **19**(7), 2026–2038 (2011)
20. Diment, A., Rajan, P., Heittola, T., Virtanen, T.: Modified group delay feature for musical instrument recognition. In: Proceedings of CMMR, pp. 431–438. Marseille, France (2013)
21. Wu, Z., Siong, C.E., Li, H.: Detecting converted speech and natural speech for anti-spoofing attack in speaker recognition. In: Proceedings of INTERSPEECH. Portland, OR (2012), 4 p
22. Xiao, X., Tian, X., Du, S., Xu, H., Chng, E.S., Li, H.: Spoofing speech detection using high dimensional magnitude and phase features: The NTU approach for ASVspoof 2015 challenge. In: Proceedings of INTERSPEECH, pp. 2052–2056. Dresden, Germany (2015)
23. Zhu, D., Paliwal, K.K.: Product of power spectrum and group delay function for speech recognition. In: Proceedings of ICASSP, pp. 125–128. Montreal, Canada (2004)
24. Moreno, P.J., Rifkin, R.: Using the Fisher kernel method for web audio classification. In: Proceedings of ICASSP, pp. 2417–2420. Istanbul, Turkey (2000)
25. Fine, S., Navrátil, J., Gopinath, R.A.: A hybrid GMM/SVM approach to speaker identification. In: Proceedings of ICASSP, pp. 417–420. Utah, USA (2001)
26. Perronnin, F., Dance, C.: Fisher kernels on visual vocabularies for image categorization. In: Proceedings of CVPR, pp. 1–8. Minneapolis, MN (2007)
27. Perronnin, F., Sánchez, J., Mensink, T.: Improving the Fisher kernel for large-scale image classification. In: Proceedings of ECCV, pp. 143–156. Crete, Greece (2010)

28. Schuller, B., Steidl, S., Batliner, A., Vinciarelli, A., Scherer, K., Ringeval, F., Chetouani, M., Weninger, F., Eyben, F., Marchi, E., Mortillaro, M., Salamin, H., Polychroniou, A., Valente, F., Kim, S.: The INTERSPEECH 2013 computational paralinguistics challenge: social signals, conflict, emotion, autism. In: Proceedings of INTERSPEECH, pp. 148–152. Lyon, France (2013)
29. Fan, R.E., Chang, K.W., Hsieh, C.J., Wang, X.R., Lin, C.J.: LIBLINEAR: a library for large linear classification. J. Mach. Learn. Res. **9**, 1871–1874 (2008)
30. Eyben, F., Wöllmer, M., Schuller, B.: openSMILE—the Munich versatile and fast open-source audio feature extractor. In: Proceedings of MM, pp. 1459–1462. Florence, Italy (2010)
31. Eyben, F., Weninger, F., Groß, F., Schuller, B.: Recent developments in openSMILE, the Munich open-source multimedia feature extractor. In: Proceedings of MM, pp. 835–838. Barcelona, Spain (2013)
32. Schuller, B., Steidl, S., Batliner, A.: The INTERSPEECH 2009 emotion challenge. In: Proceedings of INTERSPEECH, pp. 312–315. Brighton, UK (2009)
33. Schuller, B., Steidl, S., Batliner, A., Burkhardt, F., Devillers, L., Müller, C., Narayanan, S.: The INTERSPEECH 2010 paralinguistic challenge. In: Proceedings of INTERSPEECH, pp. 2794–2797. Makuhari, Japan (2010)
34. Schuller, B., Batliner, A., Steidl, S., Schiel, F., Krajewski, J.: The INTERSPEECH 2011 speaker state challenge. In: Proceedings of INTERSPEECH, pp. 3201–3204. Florence, Italy (2011)
35. Schuller, B., Steidl, S., Batliner, A., Nöth, E., Vinciarelli, A., Burkhardt, F., van Son, R., Weninger, F., Eyben, F., Bocklet, T., Mohammadi, G., Weiss, B.: The INTERSPEECH 2012 speaker trait challenge. In: Proceedings of INTERSPEECH, Portland, OR (2012)
36. Eyben, F., Scherer, K., Schuller, B., Sundberg, J., André, E., Busso, C., Devillers, L., Epps, J., Laukka, P., Narayanan, S., Truong, K.: The Geneva minimalistic acoustic parameter set (GeMAPS) for voice research and affective computing. IEEE Trans. Affect. Comput. **7**(2), 190–202 (2016)

Part IV
Towards Multilingual, Multimodal, Open Domain Spoken Dialogue Systems

Internationalisation and Localisation of Spoken Dialogue Systems

Niklas Laxström, Graham Wilcock and Kristiina Jokinen

Abstract In modern software development, localisation is a straightforward process–assuming internationalisation has been considered during development. The localisation of spoken dialogue systems is less mature, possibly because they differ from common software in that interaction with them is situated and uses multiple modalities. We claim that it is possible to apply software internationalisation practices to spoken dialogue systems and that this helps the rapid localisation of such systems to new languages. We internationalised and localised the WikiTalk spoken dialogue system. During the process we identified needs relevant to spoken dialogue systems that will benefit from further research and engineering efforts.

Keywords Spoken dialogue systems · Internationalisation · Localisation · Wikipedia · Language selection

1 Introduction

A spoken dialogue system is a software application that is able to converse in speech. It accepts input from the speech recogniser, interacts with external knowledge sources, and produces messages as output to the user [1]. An important component is the dialogue manager which enables interaction with the user, and generally controls the dialogue flow.

Speech interfaces are useful in situations where the hands are not free. They provide access to digital databases and allow the user to search for information using natural language. Since the 1990s, a large number of speech-based systems have been

N. Laxström (✉) · G. Wilcock · K. Jokinen
University of Helsinki, Helsinki, Finland
e-mail: niklas.laxstrom@helsinki.fi

G. Wilcock
e-mail: graham.wilcock@helsinki.fi

K. Jokinen
e-mail: kristiina.jokinen@helsinki.fi

K. Jokinen and G. Wilcock (eds.), *Dialogues with Social Robots*,
Lecture Notes in Electrical Engineering 427,
DOI 10.1007/978-981-10-2585-3_16

developed that support mobile services, car navigation, and various phone-based voice applications. Personal assistants with speech interfaces on smart phones are becoming commonplace, for example Siri and Cortana. In addition, spoken dialogue systems are now essential for humanoid robots which are expected to move rapidly into everyday use.

As smart phones propagate around the world, often being people's only way to access the Internet, localisation of personal assistants is as important as for other software. A particularly challenging area is robotics, where spoken dialogue systems need to maintain a two-way conversation by listening as well as talking with the situated agent, and focus on the physical environment as well.

The importance of localisation is widely acknowledged in the software industry: localisation is considered to be necessary for the software developers to remain competitive [2, 3]. Localisation removes the language barrier, making the software usable for more people. It also allows people to work using their most fluent language.

Localisation helps preserve non-dominant languages by encouraging their use and making them visible in the digital world. Localisation can also serve to revitalise endangered languages by motivating their speakers to use them. For instance, [4] points out that, to survive digitalisation, a language has to have a digitally performed function, i.e. it has to produce new, publicly available digital material.

The DigiSami project[1] focuses on the viability of the North Sami language (one of the small Finno-Ugric languages spoken in the Northern Europe) by supporting digital content generation and experimenting with language technology applications that can strengthen the user's interest in using the language in various interactive contexts. For the latter purpose it uses WikiTalk [5].

We investigated whether common software internationalisation and localisation methods are applicable to spoken dialogue systems. Our hypotheses are:

1. existing software internationalisation practices can be applied to spoken dialogue systems;
2. once internationalisation is in place, localisation of a spoken dialogue system to a new language is rapid.

To test the hypotheses we performed a practical case study using the WikiTalk spoken dialogue system described in Sect. 3. WikiTalk was initially developed without consideration for internationalisation and localisation. For this study we first created an internationalised version of WikiTalk to test the first hypothesis.

We also localised WikiTalk to English and Finnish. We then involved other people to localise WikiTalk to Japanese in order to test the second hypothesis. Although in this study we used WikiTalk with a Nao robot, most of the observations apply to spoken dialogue systems in general.

The article is structured as follows. Section 2 summarises previous work on the theory and methods of internationalisation and localisation of spoken dialogue systems and of software. Section 3 describes WikiTalk. Section 4 contains the results of the study and Sect. 5 discusses the limitations of our internationalisation. Section 5.2

[1] http://www.helsinki.fi/digisami/.

addresses issues specifically related to the Nao robot and multimodality. Finally, we discuss the results and conclude with recommendations for future improvements to the internationalisation of spoken dialogue systems.

2 Previous Work and Methods

Software internationalisation and localisation has a long history [6]. We can therefore adopt long-standing software internationalisation practices and apply them to spoken dialogue system software.

While there is much standardisation in software internationalisation, there is great variance in its methods and quality. Hence there are no unified practices [7]. However, we consider libraries such as GNU Gettext [8], in use for two decades, to encapsulate a stable core of practices. The Gettext ecosystem is well documented [9]. In addition we rely on our experience working with internationalisation and localisation of many open source projects which use various libraries.

2.1 Software Internationalisation and Localisation

Internationalisation is the process of designing software so that it can be adapted to different languages or regions. Localisation is the process of adapting the software to a specific locale. World Wide Web Consortium (W3C) provides a concise summary of the concepts [10].

Internationalisation and localisation are closely related. Internationalisation is done by the software developer to make localisation possible and easy. Localisation is usually done by an expert in a language, often known as a translator even though the localisation work is not only translation.

The goal of internationalisation is to facilitate the best software localisation while minimising changes required to the software source code. Poor implementation of internationalisation restricts the quality of localisation. For example, if the software places restrictions on the word order of an interface string, it is impossible to produce good localisation in languages where the natural word order differs. This issue is solved in Ex. 1, which shows a string in English (a) translated into German (b). A good German translation would be impossible if the translator was only able to translate the word *Show*. Because in the example we have a placeholder (`$1`) that can be moved, a good translation is possible.

(1) a. Show $1?
    b. $1 anzeigen?

It is common to implement internationalisation with programming libraries. Such libraries are beneficial in many ways: they reduce implementation costs and increase interoperability with various tools used by the translators. In practice, however, there are many different internationalisation libraries that are not completely compatible with each other. Interoperability between libraries and translation tools is most often achieved only through the file format of the interface strings. For example, web-based translation platforms such as Pootle or translatewiki.net are able to provide translators with an easy interface and various translation aids (e.g. translation memory), as long as they support the given file format.

2.2 Dialogue System Localisation

A common way to localise spoken dialogue systems is to have a general dialogue model with well-defined dialogue acts which are considered universally valid, and then realise the communicative intentions with respect to different languages. The system's presentation language is localised with respect to the user's actual language. The dialogue model approach has often been considered one of the main benefits for building separate and hierarchical modules of a more general model of interaction, while leaving realisation of the dialogue acts to separate language-specific components, so that language generation is a task for localisation [11].

There has been work on standardising dialogue acts and interaction management, and this has resulted in ISO standards [12] which aim at a general and scalable set of dialogue acts which could be used amongst different tasks, activities and languages. However, very few dialogue systems are actually built on these theoretically and practically well-grounded acts. The reason may be that for a simple voice interface such a set of dialogue acts is considered too elaborate, or maybe the acts are considered too theoretical for an engineering task of building practical interfaces.

Dialogue strategies are usually designed for a user who is a generic mean of the possible users. Adaptation to different users is done via user modelling which takes the various users' individual preferences into account: this can range from a list of preferences related to language, speed, etc. to a more elaborated modelling of the user's likes and habits as in recommendation systems. The system can use the user model to provide suitably tailored help [13] or proactively take the user's needs into account [14].

Another problem in localising spoken dialogue systems is that interaction management also requires localisation. It is much more difficult to standardise spoken interaction features such as turn-taking practices, politeness codes, and feedback realisation with the help of standard interaction management protocols. The common patterns of interaction with system initiative dialogues ("Please give the name of the arrival city") and clarification question ("Did I hear you say...") are common practices, but render the systems rather clumsy and unnatural. Natural interaction features vary depending on the culture and language, and also vary for each individual, so various large dialogue modelling studies are conducted, and also intercultural studies are necessary to model the differences. It may not be easy or even possible to have a standard set of interaction patterns which can then be localised for interaction management in various tasks, activities, contexts and cultures.

3 WikiTalk

WikiTalk [5] is a spoken information access system that gets its information from Wikipedia. The system talks about a given topic using sentences extracted from the Wikipedia article about the topic. The sentences are processed to make them more suitable for spoken dialogue, mainly by removing large amounts of textual details (parenthetical information, footnotes, infoboxes and so on). The hyperlinks in the Wikipedia text are also extracted and are used to predict the user's utterances in order to support smooth shifts to related topics. Example dialogues with WikiTalk are given by [5, 15].

WikiTalk is bound by the limitations of the automatic speech recognition (ASR) and text-to-speech (TTS) systems available on the robot. The Nao robot currently supports ASR and TTS for 19 different languages (which must be purchased separately, apart from English and one other language). An important limitation is that ASR on Nao can process only one language at a time. This makes it difficult to handle language switching in multilingual situations and rules out multilingual conversations.

Nao does not provide language identification of speech. Such technology exists [16] but is not readily available. Language identification of text has become a common tool in translation agencies and machine translation services. In the case of Nao, implementing spoken language identification is expected to be complicated due to limited processing power on the robot. Spoken language systems would benefit from such a technology: in WikiTalk it would allow more fluent language selection. Our language selection mechanisms are described in Sects. 4.1 and 4.3.

In WikiTalk, the interface language, the language the user uses to communicate with the application, is linked to the Wikipedia in the same language. This means that to have WikiTalk in a particular language, we must have all of the following: localisation of WikiTalk in that language; ASR and TTS for that language; and a Wikipedia for that language.

4 WikiTalk Internationalisation

We start with the results regarding the first hypothesis. Section 4.4 discusses results regarding the second hypothesis of rapid localisation.

Generally the software internationalisation process goes as follows: First choose an internationalisation library. Using the library, make the interface strings translatable; adapt the code to handle other internationalisation features (character encoding, number formatting, calendar systems, etc.) depending on the needs of the application; and implement language selection.

We implemented the internationalisation step by step: after each feature we decided what to add next by seeing what would make the biggest difference in the quality of the localisation. The following sections contain more details of each step.

4.1 String Extraction, Variables and Language Selection

First, we had to identify the software interface strings. In WikiTalk, those are the available user commands and the interaction management utterances by the robot. For example, when the user interrupts the speaking robot, the robot says in English *oh sorry*, in Finnish *anteeksi* and in Japanese *sumimasen*. This process, known as string extraction, is a typical step when internationalisation is done after development [17, 18]. The aim is to move the strings to be translated into a file separate from the application code, by replacing the string in the code with a function call. This function needs to know which translation it should fetch. Therefore, an identifier was assigned to each extracted string. We stored the map of identifiers and string contents in a JSON file format, one file per language.

Further work was necessary on these strings as some translated utterances were awkward or even ungrammatical due to differences in word order. Some utterances of the robot were constructed from multiple segments which were concatenated together. We solved this issue by adopting variables, which act as movable placeholders in the translatable strings (see Ex. 1 in Sect. 2.1).

The keywords which are listened to in order to change topics are dynamic, but they come from Wikipedia content and hence do not need to be localised.

Once localised, every software needs a way to select what locale to use. Given the limitations in language identification (see Sect. 3), we added a configuration option that sets the default language to use. WikiTalk, when it starts up, announces the availability of other languages. The robot says *say Finnish if you want to use Finnish* in Finnish, and continues similarly for other available languages. This approach does not scale above a few languages: it would take a considerable amount of time to list even ten languages, not to mention hundreds of languages supported by some software applications.

4.2 Wikipedia Content Processing

A common practice in software internationalisation is to make the code generic enough to work in any language. Programming interfaces are used to abstract the differences behind a common interface.

Natural language processing is one WikiTalk component for which this principle is important. In order to avoid the difficulties of open-vocabulary speech recognition, WikiTalk predicts the user's next utterance and only listens for a specified vocabulary of words and phrases. The recognition vocabulary is dynamically generated according to the dialogue context, including the links extracted from the current Wikipedia article.

Wikipedia content processing in WikiTalk is summarised by [19]. We used the Python Beautiful Soup library to extract the plain text of Wikipedia articles and the Natural Language Toolkit library to tokenize the text into sentences. Information about links is used both to synchronise gestures (see Sect. 5.2) and to define the keywords the robot listens for to make topic shifts.

Internally we use a simplified version of data presentation called linear document model. In this model, text is represented as continuous multilayer chunks. One layer contains the plain text, while additional layers contain annotations, such as a link in this case. A chunk consists of all the continuous text having the same annotations.

4.3 Finnish WikiTalk

WikiTalk listens for language autonyms like "suomi" (Finnish) using English language ASR. We used this as an additional mechanism for language selection. This solution only works in this particular context. As an experiment, we tried to decouple the *interface language* from the *Wikipedia language*. We accessed Swedish Wikipedia using Finnish WikiTalk and found the quality of Swedish using Finnish TTS understandable but unnatural. The experience was comparable to using machine translation when proper translation is not available.

We did not use an internationalisation library to support Finnish localisation in WikiTalk as it would be faster and easier to adapt our own implementation to any special needs. Many internationalisation libraries are available for Python [9], the programming language of WikiTalk: a comparison shows that our development reimplemented many of their common features, supporting our first hypothesis.

4.4 Japanese WikiTalk

Following the internationalisation of WikiTalk and the creation of English and Finnish localisations, the next step was to make a Japanese localisation to test the second hypothesis. While English and Finnish use variants of the Latin writing system, Japanese uses a quite different writing system and is a good example for testing whether the language support is generic enough. The Japanese localisation is described by [20].

Due to the previous internationalisation of WikiTalk, the effort required to create the Japanese localisation was greatly reduced. The differences between the English and Finnish alphabetic characters and the Japanese kanji, hiragana and katakana characters were handled by using Unicode. The main problem was caused by the fact that both English and Finnish separate words by spaces unlike Japanese. For the Japanese localisation the code was rewritten so that it does not assume spaces between words.

The English WikiTalk uses the *Did you know?* section from the main page of English Wikipedia in order to suggest new topics that might be interesting. The topics in this section are new every day. Japanese Wikipedia does not have a *Did you know?* section, so this method could not be used unchanged. However, Japanese Wikipedia does have a list of new articles on the main page, and this list was used to suggest new topics in a similar way.

For language-switching, we found that the workaround mentioned in Sect. 4.3 also worked for Japanese. English ASR recognises *Nihongo* as a request to switch to Japanese, and Japanese ASR recognises *Ingirishu* (English) as a request to switch to English. Language-switching between English and Japanese with multilingual WikiTalk was demonstrated at SIGDIAL 2015 [21].

5 Limitations on Internationalisation

The previous section describes an implementation of internationalisation that makes usable localisations possible. Existing and future localisations, however, could benefit from additional internationalisation support. In this section we describe additional requirements which are not universally adopted in standard software internationalisation and hence were not provided in our study. Solving these issues would be the next steps for increasing the quality of internationalisation support for spoken dialogue systems such as WikiTalk.

5.1 Grammatical Features in Text Interfaces

Often, the translated string needs to be different depending on the actual value that will replace a placeholder. This is still an on-going engineering problem in software development. For number agreement this is a solved problem, however, as most internationalisation libraries let translators specify different versions depending on a numerical value. The rules for this are standardised and available in the Unicode Common Locale Data Repository [22]. Ex. 2 demonstrates one such system. An interface string (a) shows the syntax which translators use to specify different versions: with value 1 the chosen version is (b) and with 5 (c).

(2) a. $1 {{PLURAL:$1|day|days}} ago
    b. 1 day ago
    c. 5 days ago

Besides grammatical number, there are many other types of agreements that cannot be handled by common internationalisation libraries. In some software, such as those used by Wikipedia and Facebook, it is possible to alter the string based on the gender of the placeholders.[2] In fact there was very little need for grammatical features for WikiTalk.

5.2 Issues in Multimodal Interfaces

The approach to internationalisation and localisation for WikiTalk treats the system like a normal software project, but the speech and multimodal interfaces raise additional issues. ASR and TTS systems are clearly less mature in terms of internationalisation than text-based interfaces. While in most parts of the world people can use computers where all the necessary fonts and input methods are available for text, the same is not true for speech. ASR and TTS systems in many languages are either missing or low quality.

The technology for multimodal interaction (including face tracking, nodding and gesturing) is even less mature than for speech. Existing internationalisation libraries provide no support for issues related to multimodal interaction, although research on multimodal interaction clearly shows that there are significant differences between languages for all these modalities.

Some of these differences are well-known, for example gesturing with hands and arms varies widely between cultures. The Nao robot provides a large collection of predefined gestures, but they are not language-dependent. These predefined gestures are intended to express emotions such as surprise and embarrassment, but the way these emotions are actually expressed in different communities varies greatly and in subtle ways. In addition to the predefined gestures, applications can develop their own sets of gestures. WikiTalk uses its own gestures which are classified into gesture families according to their communicative functions, as discussed in [23].

Other differences in multimodal behaviours are less well-known. For example, head nodding can vary even between language communities that are geographically close to each other. The NOMCO multimodal Nordic video corpus [24] shows significant differences in head nodding between Danes, Swedes and Finns. Danes tend to make multiple rapid down-nods, Swedes tend to make up-nods relatively more often, and Finns tend to make a single extremely slow down-nod [25, 26].

Face tracking also needs to be localised. In some cultures it is normal to look straight at the eyes of the dialogue partner, in other cultures it is normal to avoid direct eye-contact, and in many cultures it depends on parameters of the situation which can be quite complex (gender, age, social status, and so on). The Nao robot performs face tracking rather well, but the only options are to switch face tracking on or switch it

---

[2] Support for gender is complicated given that natural gender and grammatical gender might be different in some languages. It becomes even more complex when a user of the software uses a different language than the user who is being referred to.

off, and when it is on the robot looks straight at the detected face. Any parametrisation of face tracking in order to support localisation would require replacing the standard face tracking module with custom software. WikiTalk currently includes only the option of running with face tracking on or off.

The proximity between the robot and the user is another factor in multimodal interaction. Unlike a computer, a humanoid robot can walk and can move closer to the user or away from the user. Initial measurements of users' preferred distance from the robot are reported by [27], but it is well-known that preferred proximity between human speakers varies between cultures. This suggests that proximity is another multimodal feature of human-robot interaction that requires localisation.

Apart from localisation of individual modalities, there is the further problem of integration across modalities. In WikiTalk, one of the most difficult challenges is the synchronisation of speech and gestures. The utterances spoken by the robot include some words and phrases that should be emphasised by specific gestures. In WikiTalk, *NewInfos* are pieces of new information that typically correspond to hyperlinks in the Wikipedia text, and these NewInfos should be emphasised by a *beat* gesture [28]. Like most gestures, beat gestures have a preparation phase before the main beat phase. Synchronisation of the beat gesture with the NewInfo therefore requires estimating both the time required for the preparation phase of the gesture and the time required for uttering the words that precede the NewInfo word or phrase. The problem for localisation is that the number of words preceding the NewInfo may vary greatly between languages, and it remains to be confirmed whether the methods for synchronising speech and gestures described by [23] can be made to work successfully across different languages.

6 Conclusions and Future Work

While this study is based on only one spoken dialogue system, the results are promising and support our hypotheses. We found that many existing software internationalisation practices and libraries are applicable for spoken dialogue systems. We also found interesting special internationalisation needs that are not present in common software applications.

The Japanese WikiTalk localisation was completed rapidly, proving that our internationalisation implementation for the WikiTalk spoken dialogue system enables rapid localisation. We have not evaluated the localised versions due to our assumption that localisation is not the bottleneck in naturalness of the conversations. The original WikiTalk was evaluated as described in [29].

We plan to add further locales. For example in the DigiSami project [30, 31] we plan to create a version of WikiTalk in Sami language (named SamiTalk). This work is described further in [32, 33].

This study has revealed special needs in internationalisation of spoken dialogue systems which are not yet adequately addressed in the common internationalisation libraries and frameworks. These special needs include identification of spoken language and many issues in multimodal interaction.

Computer gaming is probably the closest industry where related issues are found, together with localisation challenges presented by puns and sounds [34]. Future work should investigate whether spoken dialogue systems can borrow some ideas from lessons learnt by gaming. It might also be worth investigating the solutions for subtitle localisation which deal with multiple modalities of video, speech and text.

To understand how to best support these needs, further research and practical work should be conducted. For example, as it is currently impractical to provide a robot for each translator, further work could integrate robot emulators in tools used by translators. The aim is that translators would (in addition to translating the strings) be able to create appropriate localised gestures and tweak timings so that all modalities are properly synchronised.

Further development would be to make the robot's speech sound more familiar to the user by making it recognise the person it's talking with and address them by their name. In this case, WikiTalk would encounter another typical issue, the internationalisation of people names [35].

Even new products targeting mass markets do not always support localisation from the beginning. This kind of development makes well-resourced languages even more privileged and increases the risks that less-resourced languages will be absent from new technologies. We recommend the developers of spoken dialogue systems to consider internationalisation from the beginning of the development and to take advantage of existing libraries, in order to avoid wasted effort later. For developers of internationalisation frameworks, we recommend providing support for localisation of multimodal interaction. Translation of strings is not sufficient to ensure a good experience for users of spoken dialogue systems. Developers of devices used for spoken dialogue systems should provide spoken language identification.

References

1. Jokinen, K., McTear, M.: Spoken Dialogue Systems. Morgan & Claypool (2009)
2. Reina, L.A., Robles, G., González-Barahona, J.M.: A preliminary analysis of localization in free software: how translations are performed. In: Open Source Software: Quality Verification, pp. 153–167. Springer (2013)
3. Murtaza, M., Shwan, A.: Guidelines for Multilingual Software Development. Chalmers University of Technology, University of Gothenburg, Department of Computer Science and Engineering, Gothenburg, Sweden (November 2009)
4. Kornai, A.: Language death in the digital age. Language (1/8) (2012)
5. Wilcock, G.: WikiTalk: a spoken Wikipedia.based open-domain knowledge access system. In: Proceedings of the COLING 2012 Workshop on Question Answering for Complex Domains, pp. 57–69. Mumbai, India (2012)
6. Esselink, B.: The evolution of localization. The Guide from Multilingual Computing & Technology: Localization **14**(5), 4–7 (2003)
7. Cos, J., Toval, R., Toval, A., Fernandez-Aleman, J., Carrillo-de Gea, J., Nicolas, J.: Internationalization requirements for e-learning audit purposes. In: Global Engineering Education Conference (EDUCON), 2012 IEEE, pp. 1–6 (April 2012)

8. GNU Gettext. https://www.gnu.org/software/gettext/. Accessed 18 Sept 2015
9. Roturier, J.: Localizing Apps: A Practical Guide for Translators and Translation Students. Translation Practices Explained, Taylor & Francis (2015)
10. World Wide Web Consortium: Localization versus Internationalization, http://www.w3.org/International/questions/qa-i18n. Accessed 18 Sept 2015
11. Jokinen, K.: Three challenges for dialogue management: the constructive dialogue model approach. Papers from the 16th Scandinavian Conference of Linguistics, pp. 221–234 (1996)
12. Bunt, H., Alexandersson, J., Choe, J.W., Fang, A.C., Hasida, K., Petukhova, V., Popescu-Belis, A., Traum, D.R.: ISO 24617-2: a semantically-based standard for dialogue annotation. In: Proceedings of Eighth International Conference on Language Resources and Evaluation (LREC 2012), pp. 430–437. Istanbul (2012)
13. Jokinen, K.: Adaptation and user expertise modelling in AthosMail. Univ. Access Inf. Soc. **4**(4), 374–392 (2006)
14. Nothdurft, F., Ultes, S., Minker, W.: Finding appropriate interaction strategies for proactive dialogue systems—an open quest. In: Proceedings of the 2nd European and the 5th Nordic Symposium on Multimodal Communication 2014, pp. 73–80. LiU Electronic Press, Tartu, Estonia (2015)
15. Jokinen, K., Wilcock, G.: Constructive interaction for talking about interesting topics. In: Proceedings of Eighth International Conference on Language Resources and Evaluation (LREC 2012). Istanbul (2012)
16. Li, H., Ma, B., Lee, K.A.: Spoken language recognition: from fundamentals to practice. Proceedings of IEEE **101**(5), 1136–1159 (2013)
17. Cardenosa, J., Gallardo, C., Martin, A.: Internationalization and localization after system development: a practical case. In: Proceedings of the 4th International Conference on Information Research, Applications and Education, vol. 1, pp. 207–214 (2007)
18. Wang, X., Zhang, L., Xie, T., Mei, H., Sun, J.: Locating need-to-externalize constant strings for software internationalization with generalized string-taint analysis. IEEE Trans. Softw. Eng. **39**(4), 516–536 (2013)
19. Laxström, N., Jokinen, K., Wilcock, G.: Situated interaction in a multilingual spoken information access framework. In: Proceedings of the Fifth International Workshop on Spoken Dialog Systems (IWSDS 2014), pp. 161–171. Napa, California (2014)
20. Okonogi, K., Wilcock, G., Yamamoto, S.: Nihongo WikiTalk no kaihatsu (Development of Japanese WikiTalk). In: Forum on Information Technology (FIT 2015). Matsuyama, Japan (2015), (in Japanese)
21. Wilcock, G., Jokinen, K.: Multilingual WikiTalk: Wikipedia-based talking robots that switch languages. In: Proceedings of the 16th Annual SIGdial Meeting on Discourse and Dialogue. Prague (2015)
22. Unicode Consortium: Unicode Common Locale Data Repository. http://cldr.unicode.org/. Accessed 18 Sept 2015
23. Meena, R., Jokinen, K., Wilcock, G.: Integration of gestures and speech in human-robot interaction. In: Proceedings of the 3rd IEEE International Conference on Cognitive Infocommunications (CogInfoCom 2012), pp. 673–678. Kosice (2012)
24. Paggio, P., Allwood, J., Ahlsén, E., Jokinen, K., Navarretta, C.: The NOMCO multimodal nordic resource—goals and characteristics. In: Proceedings of the Seventh International Conference on Language Resources and Evaluation (LREC 2010), pp. 2968–2973. Valletta, Malta (2010)
25. Navarretta, C., Ahlsén, E., Allwood, J., Jokinen, K., Paggio, P.: Feedback in nordic first-encounters: a comparative study. In: Proceedings of the Eight International Conference on Language Resources and Evaluation Conference (LREC 2012). Istanbul, Turkey (2012)
26. Toivio, E., Jokinen, K.: Multimodal feedback signaling in Finnish. In: Proceedings of the Fifth International Conference: Human Language Technologies—The Baltic Perspective, pp. 247–255. Tartu, Estonia (2012)
27. Han, J., Campbell, N., Jokinen, K., Wilcock, G.: Investigating the use of non-verbal cues in human-robot interaction with a Nao robot. In: Proceedings of the 3rd IEEE International Conference on Cognitive Infocommunications (CogInfoCom 2012), pp. 679–683. Kosice (2012)

28. Cassell, J.: Embodied conversation: integrating face and gesture into automatic spoken dialogue systems (1999)
29. Anastasiou, D., Jokinen, K., Wilcock, G.: Evaluation of WikiTalk—user studies of human-robot interaction. In: Proceedings of the 15th International Conference on Human-Computer Interaction (HCII 2013). Las Vegas (2013)
30. Jokinen, K.: Open-domain interaction and online content in the Sami language. In: Proceedings of the Ninth International Conference on Language Resources and Evaluation Conference (LREC 2014) (2014)
31. Jokinen, K., Wilcock, G.: Community-based resource building and data collection. In: Proceedings of the 4th International Workshop on Spoken Language Technologies for Under-resourced Languages (SLTU'14). St. Petersburg, Russia (2014)
32. Wilcock, G., Laxström, N., Leinonen, J., Smit, P., Kurimo, M., Jokinen, K.: Towards SamiTalk: a Sami-speaking Robot linked to Sami Wikipedia. In: Jokinen, K. and Wilcock, G. (Eds.) Dialogues with Social Robots, Springer (2016) pp. 343–351 (this volume)
33. Jokinen, K., Hiovain, K., Laxström, N., Rauhala, I., Wilcock, G.: DigiSami and digital natives: interaction technology for the North Sami language. In: Jokinen, K. and Wilcock, G. (Eds.) Dialogues with Social Robots, Springer (2016) pp. 3–19 (this volume)
34. Bernal-Merino, M.: Translation and localisation in video games: making entertainment software global. Routledge Adv. Transl. Stud. Taylor & Francis (2015)
35. Lefman, G.: Internationalisation of people names. Master's thesis, University of Limerick (2013). http://hdl.handle.net/10344/3450

A Multi-lingual Evaluation of the vAssist Spoken Dialog System. Comparing Disco and RavenClaw

Javier Mikel Olaso, Pierrick Milhorat, Julia Himmelsbach, Jérôme Boudy, Gérard Chollet, Stephan Schlögl and María Inés Torres

Abstract vAssist (Voice Controlled Assistive Care and Communication Services for the Home) is a European project for which several research institutes and companies have been working on the development of adapted spoken interfaces to support home care and communication services. This paper describes the spoken dialog system that has been built. Its natural language understanding module includes a novel reference resolver and it introduces a new hierarchical paradigm to model dialog tasks. The user-centered approach applied to the whole development process led to the setup of several experiment sessions with real users. Multilingual experiments carried out in Austria, France and Spain are described along with their analyses and results in terms of both system performance and user experience. An additional experimental comparison of the RavenClaw and Disco-LFF dialog managers built into the vAssist spoken dialog system highlighted similar performance and user acceptance.

J.M. Olaso (✉) · M.I. Torres
Universidad del País Vasco UPV/EHU, Leioa, Spain
e-mail: javiermikel.olaso@ehu.es

M.I. Torres
e-mail: manes.torres@ehu.es

P. Milhorat · G. Chollet
Télécom ParisTech, Paris, France
e-mail: milhorat@telecom-paristech.fr

G. Chollet
e-mail: chollet@telecom-paristech.fr

J. Himmelsbach
AIT Austrian Institute of Technology GmbH, Vienna, Austria
e-mail: julia.himmelsbach@ait.ac.at

J. Boudy
Télécom SudParis, Évry, France
e-mail: jerome.boudy@telecom-sudparis.eu

S. Schlögl
MCI Management Center Innsbruck, Innsbruck, Austria
e-mail: stephan.schloegl@mci.edu

K. Jokinen and G. Wilcock (eds.), *Dialogues with Social Robots*,
Lecture Notes in Electrical Engineering 427,
DOI 10.1007/978-981-10-2585-3_17

Keywords Spoken dialog systems · Dialog management · Real-user experiments

1 Introduction

The vAssist project [6] aims at providing specific voice controlled home care and communication services for two target groups of older persons: seniors living with chronic diseases and persons living with (fine) motor skills impairments. The main goal is the development of simplified and adapted interface variants for tele-medical and communication applications using multilingual natural speech and voice interaction (and supportive graphical user interfaces where necessary) [19, 22]. The vAssist consortium consists of research institutes and companies from Austria, France and Italy. Toward the end of the project, the University of the Basque Country was included so as to expand the perimeter to Spanish speaking users.

2 Related Work

A Spoken Dialog System (SDS) is a system providing an interface to a service or an application through a dialog. An interaction qualifies as dialog when it exceeds one turn. It requires to keep track of the dialog state, including the history of past turns, in order to select the appropriate next step.

Those systems do not usually consist of a single component but comprise several specialized programs combined in order to recognize the speech, extract the relevant information in the transcriptions, act on back-end services, decide on the best next step, generate system responses and synthesize speech.

JUPITER [31] was one of the first SDSs released to the public. The phone-based weather information conversational interface has received over 30,000 calls between 1997 and 1999. Earlier, researchers from Philips implemented an automatic train timetable information desk for Germany [1]. More recently, Carnegie Mellon University provided Olympus [2], which has been used to build systems like the Let's Go! Bus Information System [20], leading to the biggest corpus of man-machine dialogs with real users publicly available today. Recent platforms for developing spoken dialog systems include the Opendial toolkit [17] and the architecture developed by the University of Cambridge [30] for its startup VocalIQ.

ELIZA [27] is considered by many as the first dialog system. The core of the system was based on scripts which associated a system's response by looking for a pattern in the input. Larsson and Traum argued that the state of the dialog, including its history, may be represented as the sum of the so far exchanged information [14]. An Information State (IS) designer defines the elements of the information relevant to a dialog, a set of update rules and an update strategy. An example-based dialog manager (DM) [15] constructs a request to a database from the annotated input Dialog Act (DA). The database stores examples seen in the interaction data so that the algorithm looks for the most similar entry and then executes the system's associated action. On the other hand, plan-based DMs [3, 21] require a pre-programmed task model.

On the stochastic side, Markov Decision Processes (MDPs) represent a statistical decision framework to manage dialogs [16, 29]. Here, the dialog state space contains all the states the dialog may be in and the transitions dependent on the user inputs. The behavior of a DM based on MDPs is defined by a strategy which associates each state to an executable action. Statistical methods used for dialog management also include Stochastic Finite-State models [9, 10, 25] and SemiMDPs [8]. Finally the state-of-the-art POMDP [28] extends the MDPs hiding the states which emit observations according to a probabilistic distribution [13, 28, 30]. This additional layer encodes the uncertainty about the Natural Language Understanding (NLU) and, in the case of SDSs, the Automatic Speech Recognition (ASR). Within a theoretical framework the proposal of a global statistical framework, allowing for optimization, is highlighted by POMDP. However, practical POMDP-based DMs are currently limited in the number of variables and by the intractability of the computing power required to find an optimal strategy [7, 13, 30].

In vAssist the development context, along with the difficulty to collect 'real' training dialogs, favored the use of a deterministic control formalism. This was also motivated by the overall requirements of the system it had to be integrated with.

3 Main Goals and Contributions

This article describes the vAssist SDS and presents the results of it's final system evaluation. The vAssist DM system is based on an open and adaptative software architecture that allows for an easy configuration of DMs according to a given target scenario, user requirements and context. In accordance with [18], the novelties of the vAssist SDS are the Semantic Unifier and Reference Resolver (SURR) defined in the natural understanding module and the Link-Form Filling (LFF) concept proposed to model the task (for both cf. Sect. 4.6). The vAssist prototype is based on the Disco plan-based DM and the LFF task model. For comparison purposes we have also integrated an alternative, plan-based DM, i.e. Ravenclaw (cf. Sect. 4.7).

The main contribution of this work is therefore a multilingual lab evaluation of the final vAssist assistive home care and communication service applications running on a smart-phone. Such was carried out with real users in Austria, France and Spain (cf. Sect. 5). As an additional contribution the evaluation has been carried out in terms of system performance and user experience (cf. Sect. 6). The final contribution of this work is the experimental comparison of the Disco-LFF DM and the Ravenclaw DM working within the same SDS architecture, dealing with the same task and language (i.e. Spanish), and interacting with the same users (also cf. Sect. 6).

4 System Description

The vAssist SDS extends the usual chained design (i.e. ASR + NLU + DM + NLG + TTS). Components were split into modified sub-modules and new processes were integrated into a state-of-the-art workflow chain. Figure 1 shows the resulting SDS architecture.

4.1 Speech Recognition

The system uses the Google Speech API where an HTTP POST request transmits the signal segment to be recognized. The API returns the n-best hypotheses, being n a parameter of the request, as well as the confidence score for the best one. An empty result is returned when the segment cannot be recognized with enough confidence, i.e. when it does not contain speech.

4.2 Natural Language Generation and Text-to-Speech

A simple but effective solution to produce natural language utterances conveying the DM's messages was targeted. Input messages are Semantic Frames (SFs). The engine is fed with a set of templates that consist of a title (identical to an SF's goal) associated with an utterance, and whose parts may be replaced by slot names or slot name-value pairs. The result is a natural language utterance to be synthesized or displayed on a screen.

MaryTTS [23], an open-source speech synthesis framework maintained by the Cluster of Excellence MMCI and the DFKI, is used for synthesis. It offers pre-built

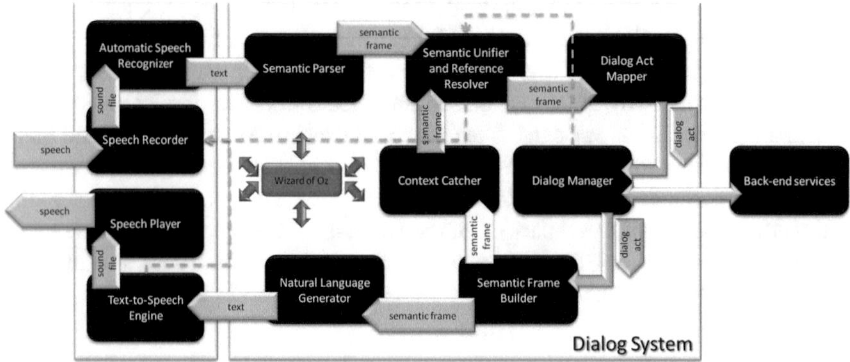

Fig. 1 Architecture of the platform

voice models for different languages as well as tools to create and manipulate them. The MaryTTS module is a client connected to a generating server (hosted local or remote). A request containing the text to be synthesized with additional prosodic information is sent to the central server, which returns the speech stream. The text-to-speech module of the present platform is a basic client program embedded into an ActiveMQ wrapper.

4.3 Semantic Parsing

The semantic parser, which gets inputs from the ASR, associates semantic labels to text utterances (or parts of them). The most commonly used parsing techniques are based on context-free grammars or probabilistic context-free grammars, which are either hand-coded, based on the analysis of collected dialog data, or designed by experts.

Our semantic parser integrates the algorithm proposed by [12], which is the application of the work from [4]. Instead of matching whole sentences with parse structures, the algorithm looks for patterns in chunks of the text-level utterance and in the temporary (i.e. currently assigned) SF. The module applies an ordered set of conditional rules, which is learned from data.

4.4 Semantic Unification and Resolution

The Semantic Unifier and Reference Resolver (SURR) holds a rather simplistic forest of nodes which is used to mine the dialog history, incorporate external information sources and add local turn context. It is the meeting point of the user's semantic frames, the external environment sensors and functions, the dialog history, and the links generated by the context catcher.

At its core the SURR embeds a forest structure. Trees consist of hierarchies of fully or partially defined SFs (some nodes are calls to external systems or services). When requested, the SURR may dynamically modify (remove/add) branches of the forest. The top node of a hierarchy defines the root.

The SURR algorithm tries to find a unique path from an input SF, i.e. from the parsed user input, to nodes of the forest, to a root node. Going up the trees, the algorithm applies the optional operations held on branches.

Reaching a root node equals the user input being contextualized [18]. In case the algorithm cannot find such a path, i.e. the SURR fails to produce a suitable SF (given the current context and available knowledge), a "NoMap" SF is generated to signal a 'non-understanding' to consecutive components.

4.5 Dialog Act Mapping

As a last stage of the NLU processing, the dialog act mapping is performed. Once an input has been parsed, external and local references have been resolved, and the semantic level has been unified, the ultimate step is to convert the SF into a DA. Following an input the mapper retrieves a set of available DAs. Then it looks for a unique match between the SF and the set of DAs.

4.6 Dialog Management Based on Disco

The core of the implemented DM is based on Disco [21], an open-source dialog management library, whose algorithm processes task hierarchy models. A dialog model is a constrained XML tree of tasks. The plan recognizer uses the recipes defined in the dialog models and this dialog state to select the best available plans for the tasks in the stack. Then the reasoning engine selects the most appropriate next step.

In an attempt to overcome the hurdles inherent to the specification of task models, the dialog modeling paradigm was shifted to a Linked-form-filling (LFF) one. Form-filling dialogs are based on structures containing sets of fields which the user needs to provide a value for in order to trigger a terminal action. The order in which the DM asks for the values is not predefined. The user may define multiple field values within a single utterance/turn.

The LFF language offers to combine these properties with the ability to trigger actions at any point of the dialog and the inclusion of subforms. Furthermore, fields and subforms can be optional, i.e. either be ignored when unset or proposed to the user. Here, we use the unlimited depth of a task model to circle tasks while keeping a sequencing order; i.e. the link between two task nodes is a reference, hence a node can point to its 'parent' node.

The aim of the LFF language is to offer a somehow simpler design method to a powerful standard dialog modeling specification. Since it is also an XML based language we opted for XSLT to convert an LFF document into a compliant dialog model.

A number of rules have been defined to create a well-formed LFF document. Doing this, the relative reduction in terms of code size and task hierarchy depth was 76 and 77 %, respectively.

4.7 Dialog Management Based on RavenClaw

RavenClaw (part of the CMU Communicator system [3]) is a task-independent DM. It manages dialogs using a task tree and an agenda.

The task tree is basically a plan to achieve the overall dialog task. At runtime, the tree is traversed recursively from left to right and from top to bottom. The execution of the dialog ends when the bottom-right node has been reached. During this process, loops and conditional control mechanisms may be added to the nodes in order to alter the normal exploration of the tree, allowing the definition of more complex dialog structures.

The second defining structure, the agenda, is an ordered list of agents that is used to dispatch inputs to appropriate agents in the task tree. It is recomputed for every turn and the current agent is placed on top of the stack. Inputs are matched to successive items on the agenda. When a match occurs the corresponding agent is activated with the matching concepts as inputs of the dialog. An agent may not consume all input concepts and thus remaining concepts are passed further down the agenda until agents can consume them.

In order to integrate RavenClaw in the architecture shown in Fig. 1, the original Disco-LFF DM was substituted by a module responsible for translating the message format defined by RavenClaw to the message format defined by the Disco-based component and vice versa.

5 Task and Experimental Scenarios

To empirically evaluate the operation of the developed voice-controlled application running on a smartphone under standardized condition, several scenarios were defined and implemented. In detail, the following scenarios and associated tasks were applied for the experimental study:

- The Prescription Management enables to monitor medical prescriptions and individual intake times. To evaluate this scenario, participants were asked to add a new predefined prescription to the application database and to set a reminder for it (AP). The app requests information regarding name of medication, quantity, dosage form, frequency, and time of intake.
- The Health Report (HR) provides an overview of physiological data. Participants filled in predefined glycaemia and blood pressure data.
- The Sleep Report (SR) monitors sleep quality. The following data was provided by the users: the time he/she went to bed, the time he/she fell asleep, and their wake-up times. Participants also reported awake periods at night and the total number of hours slept. Finally, users were asked to rate their well-being on a six-point scale. Furthermore, the evaluation included setting a reminder to remember completing the sleep report (SRR).
- Fitness Data Management consists of reporting daily activities (FD) and setting reminders for the reports. Within the evaluation, participants were asked to enter a new report including the duration of their fitness activity.
- The Communication Services include sending messages (SM) and initiating phone calls (PC). Participants were asked to test both functions.

6 Experimental Evaluation

Two series of experiments were carried out: We evaluated the vAssist system including the Disco-LFF engine in three languages: French, German and Spanish. Further, we compared the RavenClaw and Disco-LFF DMs built into the vAssist system with Spanish users.

Sixteen users took part in the experiments in each of the trial sites. In France, 14 male and 2 female persons between 65 and 90 years (mn = 77.0) participated in the study. In Austria, 8 male and 8 female participants between 60 and 76 (Mn = 68.0) years old took part. The Spanish trial site included 12 males and 4 females between 26 and 69 (Mn = 39.6) years.

Users were first shown the smartphone application, followed by a short demonstration and usage advices. The experimental scenarios were then carried out without any other introduction than the simple description of the goal. It was up to the user to figure out how to perform each task.

The system's performance was measured in terms of Task Completion (TC), i.e. success rate, and Average Dialog Length (ADL), i.e. efficiency. TC evaluates the success rate of the system in providing the user with the requested information, based on the total number of dialogs carried out and the number of successful dialogs achieved for a specific task. ADL is the average number of turns in a successful task.

For the subjective measures, a set of standardized questionnaires was applied. The standard Single Ease Questionnaire (SEQ) [24], the System Usability Scale (SUS) [5] and the Subjective Assessment of Speech System Interfaces (SASSI) [11] questionnaire were used to evaluate the vAssist system with the Disco-LFF DM. A custom set of questions was used to compare the Disco-LFF-based DM with the Ravenclaw-based DM. Results of the SEQ, SUS and SASSI are not given for Spanish, as for this language no localized mobile application interface was available.

6.1 System Performance

The first series of experiments was carried out in France, Austria and Spain, evaluating the vAssist system with the Disco-LFF DM. Table 1 shows the system performance evaluation in terms of TC and ADL values.

Table 1 reveals good TC rates, with the French version being the one generating the highest system performance and the Spanish version the one producing the lowest. Surprisingly, our results show that the vAssist system performance is not better for younger users (Spain: mn = 39.6 years) than for older ones (France: mn = 77 years). Language dependent modules, i.e. the ASR and, more importantly, the NLU, were more robust in French and German. Spanish results suffered from a less robust semantic parser and the missing mobile UI, leading to a higher number of turns to achieve the task goals.

Table 1 TC and ADL of the vAssist system using the Disco-LFF DM

	French		German		Spanish	
	TC (%)	ADL	TC (%)	ADL	TC (%)	ADL
AP	93.33	8.00	88.88	8.18	84.00	13.62
HR	100.00	3.15	93.33	3.78	100.00	4.41
SR	91.66	7.81	100.00	7.25	100.00	10.18
SRR	83.33	3.40	100.00	3.50	87.50	5.78
FD	100.00	3.00	66.66	3.00	93.75	4.53
SM	100.00	3.86	100.00	4.62	100.00	6.21
PC	100.00	1.92	100.00	1.82	100.00	2.00
Average	97.12	4.44	95.18	4.73	92.19	6.21

6.2 *Task Easiness and Usability*

Besides performance, the perceived task easiness is considered an important factor influencing user experiences [26]. This aspect was measured right after each task with the SEQ using a 7-point semantic differential ("very difficult"—"very easy"). The analysis revealed a sufficient ease of use for each task; i.e. mean ratings for the Prescription Management and for sending a message were 4.94. Initiating a phone call and the Health Report were rated 5.06.

To obtain insights regarding the prototype's usability, learnability, and intuitivity, the SUS was used. SUS scores fall between 0 and 100; the higher the score the better. The values for Austria and France were 68 (sd = 17.2) and 70 (sd = 11.5), respectively. Hence, even though the perceived easiness of single tasks was good, the overall system experience could still be improved.

6.3 *Speech Assessment*

The SASSI questionnaire was employed to examine the interaction quality. The analysis provides developers with an assessment of the system along several axes such as easiness, friendliness, speed, etc.

Results indicate that both "Response Accuracy" (Austria: 4.27, France: 3.99) and "Speed" (Austria: 4.64, France: 4.19) were judged neutral. The analysis of the French sample reveals that "Likeability" (4.9) and "Cognitive Demand" (5.15) were fair. In contrast, the Austrian participants rated these factors as good (Likeability: 5.28, Cognitive Demand: 5.15). Hence, we may argue that participants liked the system and were not overwhelmed by its cognitive demands.

Table 2 Comparing the Disco-LFF and RavenClaw DMs

	Disco-LFF DM		RavenClaw DM	
	TC (%)	ADL	TC (%)	ADL
AP	84.00	13.62	94.40	15.64
HR	100.00	4.41	100.00	4.90
SR	100.00	10.18	83.30	11.90
SRR	87.50	5.78	75.00	6.08
FD	93.75	4.53	92.80	4.30
SM	100.00	6.21	100.00	6.64
PC	100.00	2.00	100.00	2.42
Average	92.19	6.21	89.90	6.60

6.4 Disco-LFF and RavenClaw DM Comparison

The second series of experiments was carried out in Spanish only. Note that both DMs were integrated in the same architecture (Fig. 1), i.e. only the task planification and the agent execution differed. Each user carried out the scenarios defined in Sect. 5 with either of the DMs. Table 2 shows the system performance achieved by both systems in terms of TC and ADL, for each of the defined subscenarios. Both metrics show similar behavior for the Disco-LFF and the Ravenclaw DM. A Z-test comparing the average TC proportions and the ADL means showed no statistically significant difference between the two DMs (p-value = 0.05). A detailed scenario-based analysis showed, however, differences between TC values in the AP and the SR scenarios, which correspond to longer dialogs in terms of the ADL metric. A previous series of experiments has furthermore highlighted a certain lack of robustness exhibited by the language dependent modules of the Spanish vAssist version. This issue was more evident in longer dialogs (AP and SR).

As there was no mobile UI for the Spanish language, the user experience was evaluated trough a set of direct questions regarding the system efficiency, usability and user satisfaction. Task easiness received an average score of 3.00 for the Disco-LFF DM and 3.14 for the RavenClaw DM. The respective satisfaction scores were 3.57 and 3.43 and efficiency scored 3.28 and 3.14.

7 Conclusion

This article had two objectives. First, we reported on the results of the final lab evaluation of the vAssist system, and second we compared the system's core DM implementation with a publicly deployed one.

Despite minimal differences between languages, the vAssist SDS performances proved to be sufficient for its target users, i.e. older adults living with chronic diseases and persons living with (fine) motor skills impairments.

The DM comparison showed similar performance and subjective experience for the system with the Disco-LFF DM and the one with RavenClaw, promoting the Disco-LFF as a valid alternative to existing DM approaches.

Acknowledgements The presented research is conducted as part of the vAssist project (AAL-2010-3-106), which is partially funded by the European Ambient Assisted Living Joint Programme and the National Funding Agencies from Austria, France and Italy. It has also been partially supported by the Spanish Ministry of Science under grant TIN2014-54288-C4-4-R and by the Basque Government under grant IT685-13.

References

1. Aust, H., Oerder, M., Seide, F., Steinbiss, V.: The Philips automatic train timetable information system. Speech Commun. **17**(3–4), 249–262 (1995)
2. Bohus, D., Raux, A., Harris, T.K., Eskenazi, M., Rudnicky, A.I.: Olympus: an open-source framework for conversational spoken language interface research. In: Proceedings of the Workshop on Bridging the Gap: Academic and Industrial Research in Dialog Technologies, pp. 32–39 (2007)
3. Bohus, D., Rudnicky, A.I.: The RavenClaw dialog management framework: architecture and systems. Comput. Speech Lang. **23**(3), 332–361 (2009)
4. Brill, E.: Transformation-based error-driven learning and natural language processing: a case study in part-of-speech tagging. Comput. Linguist. **21**(4), 543–565 (1995)
5. Brooke, J.: SUS-a quick and dirty usability scale. Usability evaluation in industry **189**(194), 4–7 (1996)
6. Chollet, G., Caon, D.R., Simonnet, T., Boudy, J.: vAssist: Le majordome des personnes dépendantes. In: Proceedings of 2e Conférence Internationale sur l'Accessibilité et les Systémes de Suppléance aux personnes en Handicap (2011)
7. Crook, P.A., Roblin, B., Loidl, H.W., Lemon, O.: Parallel computing and practical constraints when applying the standard POMDP belief update formalism to spoken dialogue management. In: Delgado, R.L.C., Kobayashi, T. (eds.) Proceedings of the Paralinguistic Information and its Integration in Spoken Dialogue Systems Workshop, pp. 189–201. Springer, New York (2011)
8. Cuayáhuitl, H., Renals, S., Lemon, O., Shimodaira, H.: Evaluation of a hierarchical reinforcement learning spoken dialogue system. Comput. Speech Lang. **24**, 395–429 (2010)
9. Ghigi, F., Torres, M.I.: Decision making strategies for finite state bi-automaton in dialog management. In: Proceedings of the International Workshop Series on Spoken Dialogue Systems Technology, IWSDS, pp. 308–312 (2015)
10. Griol, D., Hurtado, L., Segarra, E., Sanchis, E.: A statistical approach to spoken dialog systems design and evaluation. Speech Commun. **50**, 666–682 (2008)
11. Hone, K.S., Graham, R.: Towards a tool for the subjective assessment of speech system interfaces (SASSI). Nat. Lang. Eng. **6**, 287–303 (2000)
12. Jurčíček, F., Mairesse, F., Gašić, M., Keizer, S., Thomson, B., Yu, K., Young, S.: Transformation-based Learning for semantic parsing. In: Proceedings of the InterSpeech, pp. 2719–2722 (2009)
13. Jurčíček, F., Thomson, B., Young, S.: Reinforcement learning for parameter estimation in statistical spoken dialogue systems. Comput. Speech Lang. **26**(3), 168–192 (2011)
14. Larsson, S., Traum, D.: Information state and dialogue management in the TRINDI Dialogue Move Engine Toolkit. Nat. Lang. Eng. **6**, 323–340 (1998)

15. Lee, C., Jung, S., Eun, J., Jeong, M., Lee, G.G.: A situation-based dialogue management using dialogue examples. In: Proceedings of the International Conference on Acoustics, Speech and Signal Processing, pp. 69–72 (2006)
16. Levin, E., Pieraccini, R., Eckert, W.: Using Markov decision process for learning dialogue strategies. In: Proceedings of the International Conference on Acoustics, Speech and Signal Processing, pp. 201–204 (1998)
17. Lison, P.: A hybrid approach to dialogue management based on probabilistic rules. Comput. Speech Lang. **34**(1), 232–255 (2015)
18. Milhorat, P.: An Open-source Framework for Supporting the Design and Implementation of Natural-language Spoken Dialog Systems. Ph.D. thesis, Télécom Paris-Tech - 46, rue Barrault - 75013 Paris (2015)
19. Milhorat, P., Schlögl, S., Chollet, G., Boudy, J.: Un Systéme de Dialogue Vocal pour les Seniors: Études et Spécifications. Journées d'étude sur la TéléSanté (2013)
20. Raux, A., Langner, B., Bohus, D.: Lets go public! taking a spoken dialog system to the real world. In: Proceedings of the InterSpeech (2005)
21. Rich, C.: Building task-based user interfaces with ANSI/CEA-2018. IEEE Comput. **8**, 20–27 (2009)
22. Schlögl, S., Milhorat, P., Chollet, G.: Designing, building and evaluating voice user interfaces for the home. In: Proceedings of the Workshop on Methods for Studying Technology in the Home at the ACM SIGCHI Conference on Human Factors in Computing Systems (CHI13) (2013)
23. Schröder, M., Trouvain, J.: The German text-to-speech synthesis system MARY: a tool for research, development and teaching. Int. J. Speech Technol. **6**(4), 365–377 (2003)
24. Tedesco, D., Tullis, T.: A comparison of methods for eliciting post-task subjective ratings in usability testing. Usability Prof. Assoc. (UPA) **2006**, 1–9 (2006)
25. Torres, M.I.: Stochastic bi-languages to model dialogs. In: Proceedings of the International Conference on Finite State Methods and Natural Language Processing, pp. 9–17 (2013)
26. Venkatesh, V., Bala, H.: Technology acceptance model 3 and a research agenda on interventions. Decis. Sci. **39**(2), 273–315 (2008)
27. Weizenbaum, J.: ELIZA—a computer program for the study of natural language communication between man and machine. Commun. ACM **9**(1), 36–45 (1966)
28. Williams, J.D., Young, S.: Partially observable Markov decision processes for spoken dialog systems. Comput. Speech Lang. **21**(2), 393–422 (2007)
29. Young, S.: Probabilistic methods in spoken dialogue systems. Philos. Trans. R. Soc. Lond. (2000)
30. Young, S., Gašić, M., Thomson, B., Williams, J.D.: POMDP-based statistical spoken dialog systems: a review. In: Proceedings of the IEEE **101**(5), 1160–1179 (2013)
31. Zue, V., Seneff, S., Glass, J.R., Polifroni, J., Pao, C., Hazen, T.J., Hetherington, L.: JUPITER: a telephone-based conversational interface for weather information. IEEE Trans. Speech Audio Process. **8**(1), 85–96 (2000)

Multimodal HALEF: An Open-Source Modular Web-Based Multimodal Dialog Framework

Zhou Yu, Vikram Ramanarayanan, Robert Mundkowsky, Patrick Lange, Alexei Ivanov, Alan W. Black and David Suendermann-Oeft

Abstract We present an open-source web-based multimodal dialog framework, "Multimodal HALEF", that integrates video conferencing and telephony abilities into the existing HALEF cloud-based dialog framework via the FreeSWITCH video telephony server. Due to its distributed and cloud-based architecture, Multimodal HALEF allows researchers to collect video and speech data from participants interacting with the dialog system outside of traditional lab settings, therefore largely reducing cost and labor incurred during the traditional audio-visual data collection process. The framework is equipped with a set of tools including a web-based user survey template, a speech transcription, an annotation and rating portal, a web visual processing server that performs head tracking, and a database that logs full-call audio

Z. Yu (✉) · V. Ramanarayanan · R. Mundkowsky · P. Lange · A. Ivanov · D. Suendermann-Oeft
Educational Testing Service (ETS) R&D, San Francisco, CA, USA
e-mail: zhouyu@cs.cmu.edu

V. Ramanarayanan
e-mail: vramanarayanan@ets.org

R. Mundkowsky
e-mail: rmundkowsky@ets.org

P. Lange
e-mail: plange@ets.org

A. Ivanov
e-mail: aivanou@ets.org

D. Suendermann-Oeft
e-mail: suendermann-oeft@ets.org

Z. Yu · V. Ramanarayanan · R. Mundkowsky · P. Lange · A. Ivanov · D. Suendermann-Oeft
Educational Testing Service (ETS) R&D, Princeton, NJ, USA

Z. Yu · A.W. Black
Carnegie Mellon University, Pittsburgh, PA, USA
e-mail: awb@cs.cmu.edu

K. Jokinen and G. Wilcock (eds.), *Dialogues with Social Robots*,
Lecture Notes in Electrical Engineering 427,
DOI 10.1007/978-981-10-2585-3_18

and video recordings as well as other call-specific information. We present observations from an initial data collection based on an job interview application. Finally we report on some future plans for development of the framework.

Keywords Dialog systems · Multimodal inputs

1 Introduction and Related Work

Previously, many end-to-end spoken dialog systems (SDSs) used close-talk microphones or handheld telephones to gather speech input [1, 2] in order to improve automatic speech recognition (ASR) performance of the system. However, this limits the accessibility of the system. Recently, the performance of ASR systems has improved drastically even in noisy conditions [3]. In turn, spoken dialog systems are now becoming increasingly deployable in open spaces [4]. They can also interact with users remotely through the web without specific microphone requirements [5], thus reducing the cost and effort involved in collecting interactive data.

Recently, multimodal sensing technologies such as face recognition, head tracking, etc. have also improved. Those technologies are now robust enough to tolerate a fair amount of noise in the visual and acoustic background [6, 7]. So it is now possible to incorporate these technologies into spoken dialog systems to make the system aware of the physical context, which in turn will result in more natural and effective conversations [8].

Multimodal information has been proven to be useful in dialog system design in driving both low level mechanics such as turn taking as well as high level adaptive strategies such as user attention regulation. Sciutti et al. [9] used gaze as an implicit signal for turn taking in a robotic teaching context. In [10], a direction-giving robot used conversational strategies such as pause and restarts to regulate user's attention. Kousidis et al. [11] used situated incremental speech synthesis that accommodates users' cognitive load in a in-car navigation task, which improved user experience but the task performance stays the same. However most multimodal systems suffer from not enough data for model training or evaluation, and they are not easy to access, since most of them require you to be physically co-present with the system. The community has been struggling with limited publicly available data for a long time. We propose a web-based multimodal dialog system, Multimodal HALEF, to tackle this issue. It integrates the video-enabled Freeswitch telephony server with an open-source distributed dialog system, HALEF. Multimodal HALEF records the remote user's video interaction and streams it to its servers by accessing the remote user's camera via a web browser. The HALEF source code is available at: https://sourceforge.net/projects/halef/.

2 Foundational Frameworks

In this section, we describe the prior framework that Multimodal HALEF extends and builds upon. Figure 1 schematically depicts the overall architecture of the Multimodal HALEF framework.

The HALEF dialog framework leverages different open-source components to form an SDS framework that is modular and industry-standard-compliant: Asterisk, a SIP (Session Initiation Protocol) and PSTN (Public Switched Telephone Network) compatible telephony server [12]; JVoiceXML, an open-source voice browser that can process SIP traffic [13] via a voice browser interface called Zanzibar [14]; Cairo, an MRCP (Media Resource Control Protocol) speech server, which allows the voice browser to initiate SIP and RTP (Real-time Transport Protocol) connections between the speech server and the telephony server [14]; the Sphinx automatic speech recognizer [15] and the Kaldi ASR system; Festival [16] and Mary [17] text to speech synthesis engines; and an Apache Tomcat-based web server that can host dynamic VoiceXML (VXML) pages and serve media files such as grammars and audio files to the voice browser. OpenVXML allows designers to specify the dialog workflow as a flowchart, including details of specific grammar files to be used by the speech recognizer and text-to-speech prompts that need to be synthesized. In addition, dialog designers can insert "script" blocks of Javascript code into the workflow that can be

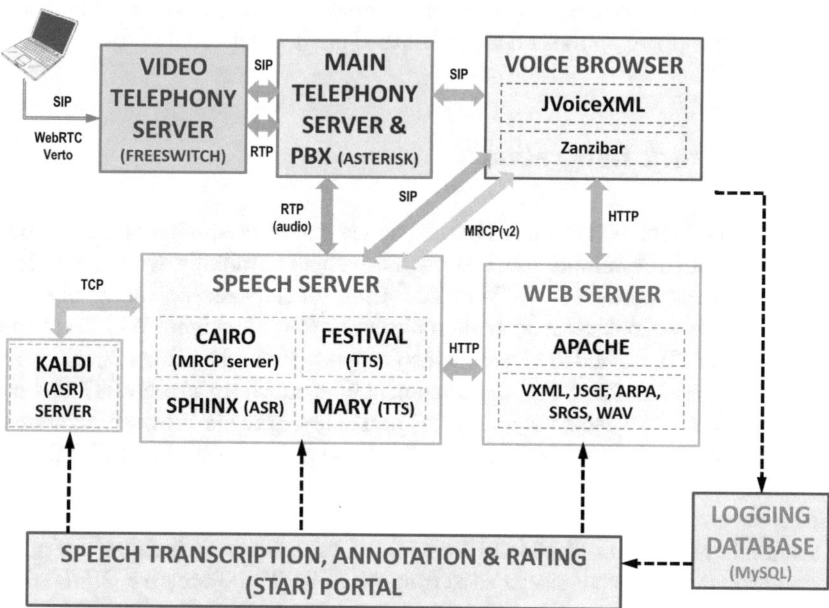

Fig. 1 System architecture of the HALEF multimodal dialog system depicting the various modular open-source components

used to perform simple processing steps, such as basic natural language understanding on the outputs of the speech recognition. The entire workflow can be exported to a Web Archive (or WAR) application, which can then be deployed on a web server running Apache Tomcat that serves Voice XML (or VXML) documents.

Note that unlike a typical SDS, which consists of sequentially-connected modules for speech recognition, language understanding, dialog management, language generation and speech synthesis, in HALEF some of these are grouped together forming independent blocks which are hosted on different virtual machines in a distributed architecture. For further details on the individual blocks as well as design choices, please refer to [18, 19].

FreeSWITCH, specifically the 1.6 Video version,[1] is a scalable open source cross-platform telephony framework designed to route and interconnect popular communication protocols using audio, video, text or any other form of media. It supports various communication technologies such as Skype, SIP and WebRTC. FreeSWITCH builds natively and runs standalone on several major operating systems, such as Linux, Windows and Mac OS. FreeSWITCH has been previously integrated with a dialog system in [20] that allows access via a web browser. However, this system can only handle audio input. FreeSWITCH is experimenter friendly. The experimenter can modify interaction settings, such as the number of people who can call in at any given time, whether to display the video of the user on the webpage, the resolution of the video, sampling rate of the audio, etc. FreeSWITCH also allows users to choose between different I/O devices for recording. They can switch between different microphones and cameras connected to their computers by selecting appropriate options on the web-based graphical user interface.

3 Framework Integration

We integrated FreeSWITCH into HALEF, so the dialog system have both video and audio as input from remote users through a webpage. Initially we planned to set up Asterisk to support video via WebRTC.[2] Unfortunately Asterisk does not support video yet. Thus we looked for other alternatives and found that FreeSWITCH released the FreeSWITCH 1.6 Video version which supports video as of May 2015.

We followed the FreeSWITCH documentation to set up FreeSWITCH 1.6 on Debian 8 (Jesse). We decide to use Verto[3] for achieving the in-browser video access capability. Verto is a FreeSWITCH module included the default FreeSWITCH configuration. Using Verto, rather than SIP/RTP over WebRTC, offered a few advantages. First it has a working webpage based conference call demo that we easily modified to allow a caller to interact with HALEF's audio dialogs. Second it is capable of recording audio and video as it is streamed to/from the caller. The alternative, SIP/RTP over

[1]https://freeswitch.org/confluence/display/FREESWITCH/FreeSWITCH+1.6+Video.

[2]http://www.webrtc.org.

[3]https://freeswitch.org/confluence/display/FREESWITCH/mod_verto.

WebRTC has more web-based clients such as sip.js,[4] sipml5,[5] and jssip,[6] but these required more work to get them to be functional on Chrome and Firefox. The problems with these clients are likely rooted in the fact that WebRTC and Web Audio are continuously evolving, experimental APIs. Therefore we choose not use them for our current framework.

The following is a description of how video and audio data flow to/from the Multimodal HALEF system. First the human user navigates to a custom webpage using their web browser and selects an extension to call. Different extensions are associated with different dialog system instances that have different task contents (for example, an interview for a job at a pizza restaurant, ...). This webpage application is written in HTML, CSS, and javascript which leverages experimental web browser APIs (WebRTC and Web Audio). The Web Audio API enable access to the audio and video devices (e.g. cameras, microphones, speakers, etc.) via the web browser. The Verto protocol, which leverages WebRTC, is then used via javascript to send video/audio to FreeSWITCH and receive audio from FreeSWITCH. Note that FreeSWITCH can also send video to the user, but HALEF currently only supports audio dialogs. The Verto teleconference webpage demo also uses the FreeSWITCH teleconference module. Overall, this is a teleconference that has two participants: HALEF and the user. When the call comes in from the user, HALEF starts the dialog with an audio prompt that flows out of HALEF system via Asterisk over SIP/RTP to FreeSWITCH. FreeSWITCH then sends the audio to the web browser over Verto. The user then gives a response to the system that is transported using Verto protocols to FreeSWITCH and then over SIP/RTP to HALEF's front end (Asterisk). During the teleconference, the users video and audio interactions are continuously streamed and recorded on the FreeSWITCH server, while HALEF audio prompts are streamed to the user.

Within the HALEF system, once the interaction starts, Asterisk makes a SIP call to the Voice Browser (JVoiceXML) to get the specific dialog started based on the extension number dialed. The Voice Browser then gets various resource files from the Web Server via HTTP that will allow it to control the call flow or call logic. For example, if the user says "yes" or "no" to a question then the Voice Browser will tell HALEF the next audio prompt to send accordingly, based on the dialog management instructions specified in the VXML application. The Voice Browser uses this information to communicate certain information to and from the Speech Servers. First, it tells the Speech Server to interact with Asterisk with regards to inbound and outbound audio (over SIP/RTP). Second, it tells the Speech Server to send the transcriptions of the audio to itself. And finally the sends instructions to receive text from the Speech Sever that will be synthesized to audio output to the user. The Voice Browser communicates with the Speech Server via MRCP and SIP.

There are other endpoints that are supported or likely can be supported with a little work by HALEF and/or Multimodal HALEF. An endpoint is defined as a device that

[4]http://sipjs.com.

[5]http://sipml5.org.

[6]http://tryit.jssip.net.

lives at the network edge (e.g. a telephone or a client application that acts like a telephone) that usually one user uses. HALEF (non-Multimodal) already supports audio-only dialogs with endpoints other than Verto. For example, we have used PSTN (public switched telephone network) endpoints, i.e., land line telephones and cell phones, to place calls to HALEF. We did this by using a PSTN/SIP proxy such as ipKall.[7] We also used SIP over WebRTC, and SIP/WebRTC clients like sipml5, jssip, etc. to connect to HALEF directly thru Asterisk as well as via webrtc2sip http://webrtc2sip.org/ to Asterisk. Note that we suggest using webrtc2sip, because it handles differences in implementations in Chrome and Firefox (in regards SRTP types and transcoding audio formats). As for Multimodal HALEF, we used Verto, but it is likely that with slightly modifications SIP/WebRTC clients could be supported.

4 Supporting Modules

We introduced four supporting modules that assist researchers in conducting scientific research on interactive dialog systems and human behavior: a relational database that stores all call-log information, a survey webpage that collects users' feedback of the interaction and pushes the information to the database, a web-based call viewing and rating portal, and a remote video processing server.

4.1 Database

We use the open-source database MySQL for our data warehousing purposes. All modules in the Multimodal HALEF connect to the database and write their log messages to it. We then post process this information with stored procedures into easily accessible views. Metadata extracted from the logs include information about the interactions such as the duration of the interaction, the audio/video recording file names, the caller IP, etc. Additionally, we store participants' survey data and expert rating information. All the modules connected to the database have been implemented such that all information will be available in the database as soon as the interaction, the survey, or the rating task is completed.

4.2 Participant Web-Survey

We created a survey webpage that participants were required to fill out upon completion of their video call. The survey was embedded along with the experimental instructions for the participant. Once the participant finishes the interaction, they

[7]http://www.ipkall.com/.

are directed to fill out the survey regarding their interaction experience as well as some demographic information. Once the participant clicks the submit button the information is pushed to the appropriate table in the MySQL database.

4.3 STAR Portal

We developed an interactive rating portal, dubbed the Speech Transcription Annotation and Rating (STAR) portal, in order to facilitate annotation, transcription and rating of calls. It is mainly implemented using PHP and the JavaScript framework jQuery. It has the advantage of accessing meta-data from the data warehouse and the audio/video data from the server as well. It provides a nice display of different information of the interaction. It also allows the experimenter to design rating questions that correspond to different components of the dialog framework or targeting the participant's performance, such as user speech input, system TTS, etc. It supports different types of questions, such as multiple choice questions, open questions, etc. Thus the rating task can not only be simple perception feedback to the interaction, but also detailed human transcription of the entire dialog. The tool also lets the experimenter manage raters by assigning different interactions for different users for the rating task. All of the information will be stored in the database for further analysis. The webpage supports playing both audio and video recordings of the collected interaction data.

4.4 Visual Processing Service

We also set up a standalone Linux server for automatic head tracking via Cambridge Head Tracker [21]. It can track a user's head movement and also estimate the head pose (e.g. 15 degrees to the left) given an input video with a clear view of the user's face. It also supports online feature extraction. When fully integrated with our system, it will take the video captured by FreeSWITCH server in real time as input and output the head tracking information to the webserver that hosts the VXML application (with dialog management instructions), thus making multimodal human behavior information available in the dialog strategy selection module for decision making. This makes the dialog system aware of the user's behaviors so it can act accordingly. This is important since previous literature suggests that "computer systems with capabilities to sense agreement and attention that are able to adapt or respond to these signals in an appropriate way will likely be perceived as more natural, efficacious and trustworthy" [8]. Visual information has also been shown to be critical in assessing the mental states of the users in other systems as well [10, 22]. So we included the visual processing service in our framework as well.

5 Example Application: Job Interview

The diagram in Fig. 2 describes the dialog workflow of one example job interview application. This conversational item was created to assess participants' English conversational skills, pragmatics and ability to comprehend stimuli and respond appropriately to questions posed during naturalistic conversational settings. Among the things we would like to study in the long run with this type of application are: (i) how users signal engagement in conversing with a task-oriented dialog system; (ii) whether the ability to show engagement has an impact on potential hiring decisions; (iii) what aspects of the system prevent or contribute to perceived naturalness of the conversations.

We conducted a preliminary data collection in order to test and evaluate the system in anticipation of a much larger planned data collection using the Amazon Mechanical Turk crowd sourcing platform. The observations from this preliminary data collection follow in the sections below.

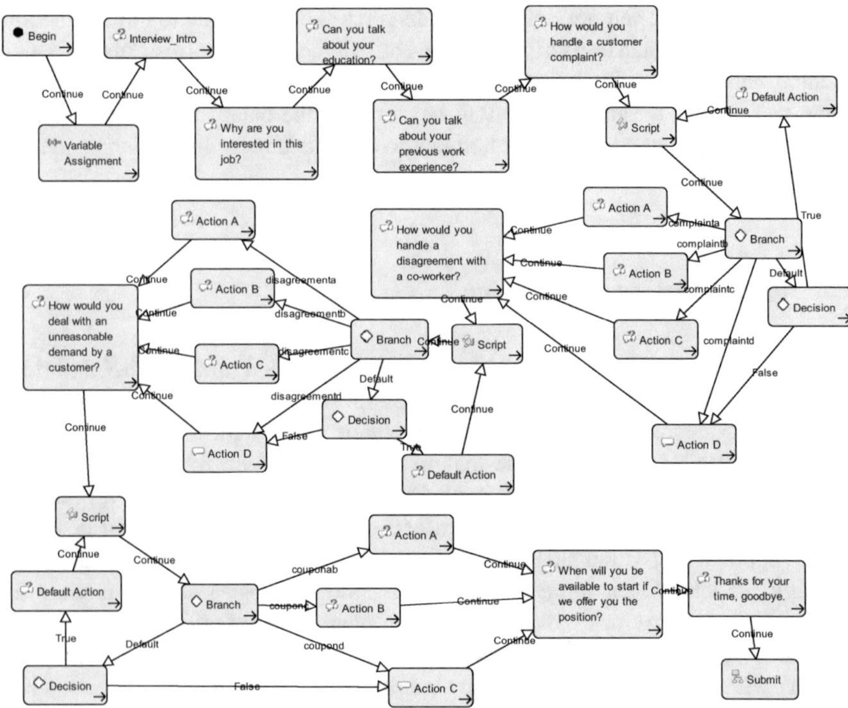

Fig. 2 Workflow of the job interview application

5.1 Preliminary Data Collection

5.1.1 Experimental Procedure

We recruited participants by sending them emails of a link to a webpage with detailed experimental instructions. All participants were given the same resume stimulus in order to control for varying personal backgrounds. The participants were instructed to access a webpage on their own computer and allow the webpage to use their local camera. Then they were asked to enter the specified extension and connected to the dialog system (every extension corresponds to a different dialog application). We instructed the user to choose a well-lit room with uniform background when interacting with the system to reduce the possible errors of the head tracker and the ASR.

5.1.2 Data Statistics

There were 13 participants in total, six males and seven females, among whom eight were native English speakers and the other five, non-native English speakers. The participants' average age was 36. The average length for all the interactions was 6 min.

In the post-interview survey, we asked users a set of questions about the system itself and the experience of interacting with it. Sample questions include, "Please rate the delay of system responses in allowing you to carry on a natural conversation?", "How engaged did you feel while interacting with the system?", etc. These ratings were on a scale from 1 to 5, 1 being least satisfactory and 5 being most satisfactory. We also had an expert rate the recordings based on similar aspects. The results of the user and expert ratings are shown in Table 1.

Table 1 User and expert ratings

User ratings	Ratings mean(SD)	Expert ratings	Ratings mean(SD)
Interaction completion	3.83(1.11)	Conversation naturalness	3.62(0.51)
System intelligibility	4.61(0.77)	User cooperation	3.92(0.64)
User reported engagement	3.23(1.01)	User engagement	3.23(0.75)
System understanding degree	2.23(0.83)	Audio/video quality	4.15(0.80)
Conversation latency	3.00(0.82)	System latency	3.92(0.28)
Overall user experience	3.00(0.82)	Overall user experience	3.62(0.50)

We used pre-recorded natural human speech for the system's voice output, and thus the intelligibility of the system is very highly rated by the users. The self-reported engagement and the expert user engagement rating is highly correlated (0.83 in Pearson correlation). We also found that the overall user experience and the conversation naturalness perceived by experts are both slightly correlated with the user's self-reported engagement during the interaction (0.50 and 0.49 respectively in Pearson correlation). This motivates us to develop systems that can improve user's engagement, which in turn may improve the overall user experience.

6 Conclusions and Future Work

We have designed and implemented an open-source web-based multimodal dialog framework, Multimodal HALEF, by integrating the existing dialog framework, HALEF, and a video conferencing framework, FreeSWITCH. The framework also includes a database, a rating portal, a survey and a video processing server. It allows an experimenter to collect video and audio data of users interacting with the dialog system outside of a controlled lab environment, and largely reduces the cost and labor in collecting audio-visual data. The framework was designed to facilitate scientific research on how humans interact with dialog systems, among other purposes. To this end, we also described and collected preliminary data on an example application that leverages the above framework and demonstrates its utility.

One limitation of the current system is that it only supports interactions with one user at time. In the future, we plan to port the system onto the Amazon Elastic Compute Cloud, thus making multiple simultaneous interactions possible. The participants of the initial data collection also provided some interesting feedback about the system which sheds light on potential future work. For example, one user suggested that having a virtual face/avatar would allow the system to signal turn taking patterns better. This is indeed along the lines of our planned future work involving multimodal synthesis and recognition (instead of simply speech recognition and synthesis, as is the case currently). In addition, we plan to incorporate and improve statistical models for automatic speech recognition, language understanding and dialog management to enhance system performance.

References

1. Eskenazi, M., Black, A.W., Raux, A., Langner, B.: Let's go lab: a platform for evaluation of spoken dialog systems with real world users. In: Proceedings of the Ninth Annual Conference of the International Speech Communication Association (2008)
2. Zue, V., Seneff, S., Glass, J.R., Polifroni, J., Pao, C., Hazen, T.J., Hetherington, L.: Jupiter: a telephone-based conversational interface for weather information. IEEE Trans. Speech Audio Process. **8**(1), 85–96 (2000)

3. Hinton, G., Deng, L., Yu, D., Dahl, G.E., Mohamed, A.R., Jaitly, N., Senior, A., Vanhoucke, V., Nguyen, P., Sainath, T.N., et al.: Deep neural networks for acoustic modeling in speech recognition: the shared views of four research groups. IEEE Signal Process. Mag. **29**(6), 82–97 (2012)
4. Bohus, D., Saw, C.W., Horvitz, E.: Directions robot: in-the-wild experiences and lessons learned. In: Proceedings of the International Conference on Autonomous Agents and Multi-Agent Systems (AAMAS), pp. 637–644. International Foundation for Autonomous Agents and Multiagent Systems (2014)
5. McGraw, I., Lee, C.Y., Hetherington, I.L., Seneff, S., Glass, J.: Collecting voices from the cloud. In: LREC (2010)
6. He, X., Yan, S., Hu, Y., Niyogi, P., Zhang, H.J.: Face recognition using laplacianfaces. IEEE Trans. Pattern Anal. Mach. Intell. **27**(3), 328–340 (2005)
7. Morency, L.P., Sidner, C., Lee, C., Darrell, T.: Contextual recognition of head gestures. In: Proceedings of the 7th international conference on Multimodal interfaces, pp. 18–24. ACM (2005)
8. Vinciarelli, A., Pantic, M., Bourlard, H.: Social signal processing: survey of an emerging domain. Image Vision Comput. J. **27**(12), 1743–1759 (2009)
9. Sciutti, A., Schillingmann, L., Palinko, O., Nagai, Y., Sandini, G.: A gaze-contingent dictating robot to study turn-taking. In: Proceedings of the Tenth Annual ACM/IEEE International Conference on Human-Robot Interaction Extended Abstracts, pp. 137–138. ACM (2015)
10. Yu, Z., Bohus, D., Horvitz, E.: Incremental coordination: attention-centric speech production in a physically situated conversational agent. In: Proceedings of the 16th Annual Meeting of the Special Interest Group on Discourse and Dialogue, p. 402 (2015)
11. Kousidis, S., Kennington, C., Baumann, T., Buschmeier, H., Kopp, S., Schlangen, D.: A multimodal in-car dialogue system that tracks the driver's attention. In: Proceedings of the 16th International Conference on Multimodal Interaction, pp. 26–33. ACM (2014)
12. Van Meggelen, J., Madsen, L., Smith, J.: Asterisk: The Future of Telephony. O'Reilly Media, Inc. (2007)
13. Schnelle-Walka, D., Radomski, S., Mühlhäuser, M.: Jvoicexml as a modality component in the w3c multimodal architecture. J. Multimodal User Interfaces **7**(3), 183–194 (2013)
14. Prylipko, D., Schnelle-Walka, D., Lord, S., Wendemuth, A.: Zanzibar openivr: an open-source framework for development of spoken dialog systems. In: Proceedings of the Text, Speech and Dialogue, pp. 372–379. Springer (2011)
15. Lamere, P., Kwok, P., Gouvea, E., Raj, B., Singh, R., Walker, W., Warmuth, M., Wolf, P.: The CMU Sphinx-4 speech recognition system. In: Proceedings of the IEEE International Conference on Acoustics, Speech and Signal Processing (ICASSP 2003), Hong Kong, vol. 1, pp. 2–5. Citeseer (2003)
16. Taylor, P., Black, A.W., Caley, R.: The architecture of the Festival speech synthesis system In: the Third ESCA Workshop in Speech Synthesis, pp. 147–151 (1998)
17. Schröder, M., Trouvain, J.: The German text-to-speech synthesis system Mary: a tool for research, development and teaching. Int. J. Speech Technol. **6**(4), 365–377 (2003)
18. Ramanarayanan, V., Suendermann-Oeft, D., Ivanov, A.V., Evanini, K.: A distributed cloud-based dialog system for conversational application development. In: Proceedings of the 16th Annual Meeting of the Special Interest Group on Discourse and Dialogue, p. 432 (2015)
19. Mehrez, T., Abdelkawy, A., Heikal, Y., Lange, P., Nabil, H., Suendermann-Oeft, D.: Who discovered the electron neutrino? A telephony-based distributed open-source standard-compliant spoken dialog system for question answering. In: Proceedings of the GSCL, Darmstadt, Germany (2013)
20. Pappu, A., Rudnicky, A.: Deploying speech interfaces to the masses. In: Proceedings of the Companion Publication of the 2013 International Conference on Intelligent User Interfaces Companion, pp. 41–42. ACM (2013)
21. Baltrusaitis, T., Robinson, P., Morency, L.P.: 3D constrained local model for rigid and non-rigid facial tracking. In: Proceedings of International Conference on Computer Vision and Pattern Recognition (CVPR), pp. 2610–2617. IEEE (2012)

22. Yu, Z., Gerritsen, D., Ogan, A., Black, A.W., Cassell, J.: Automatic prediction of friendship via multi-model dyadic features. In: Proceedings of SIGDIAL, pp. 51–60 (2013)

Data Collection and Synchronisation: Towards a Multiperspective Multimodal Dialogue System with Metacognitive Abilities

Fasih Haider, Saturnino Luz and Nick Campbell

Abstract This article describes the data collection system and methods adopted in the METALOGUE (Multiperspective Multimodal Dialogue System with Metacognitive Abilities) project. The ultimate goal of the METALOGUE project is to develop a multimodal dialogue system with abilities to deliver instructional advice by interacting with humans in a natural way. The data we are collecting will facilitate the development of a dialogue system which will exploit metacognitive reasoning in order to deliver feedback on the user's performance in debates and negotiations. The initial data collection scenario consists of debates where two students are exchanging views and arguments on a social issue, such as a proposed ban on smoking in public areas, and delivering their presentations in front of an audience. Approximately 3 hours of data has been recorded to date, and all recorded streams have been precisely synchronized and pre-processed for statistical learning. The data consists of audio, video and 3-dimensional skeletal movement information of the participants. This data will be used in the development of cognitive dialogue and discourse models to underpin educational interventions in public speaking training.

Keywords Dialogue systems · Metacognition · Multimodal data · Instructional advice · Presentation quality

F. Haider (✉) · N. Campbell
Speech Communication Lab, Trinity College Dublin, Dublin, Ireland
e-mail: haiderf@tcd.ie

N. Campbell
e-mail: nick@tcd.ie

S. Luz
Usher Institute of Population Health Sciences & Informatics, University of Edinburgh, Edinburgh, UK
e-mail: S.Luz@ed.ac.uk

K. Jokinen and G. Wilcock (eds.), *Dialogues with Social Robots*,
Lecture Notes in Electrical Engineering 427,
DOI 10.1007/978-981-10-2585-3_19

1 Introduction

In the framework of METALOGUE[1] project, we are developing a multimodal dialogue system with abilities to interact with humans in a natural way and helps them in improving their presentation and negotiation skills through instructional advice. In order to accomplish this, we have divided the development process into different pilot studies. The goal of the initial pilot system (pre-pilot system) is to primarily observe users and give feedback on their interaction, regardless of whether the interaction is successful or not. The system uses audio-visual features and limits its interventions to inaction feedback [1] in the form of a green or red light, so as to minimise participant distraction during data collection. This limited form of feedback is meant to reflect the system's view on whether a participant is interacting in a "successful" way or not based on its analysis of audio and visual input features. Prosodic features, facial expressions and body gestures will be used by the system to make a judgement about the participant's metacognitive skills. The system overview is shown in Fig. 1.

In the context of METALOGUE, a metacognitive skill is defined as the ability of a participant in an interaction to understand, control and modify his own cognitive process. Such skills are believed to be useful in real life learning and training processes, and in debating skills in particular [2]. Since our research focuses on educational and coaching situations where negotiation skills play a key role in the decision-making processes, two different real life situation scenarios have been targeted for data collection. These are customer care service training sessions and formal debates. This article describes the data collection process in the latter, with participants drawn from the Hellenic youth parliament student cohort. We focus initially on a data collection scenario with minimal feedback. Several studies use avatars in dialogue systems to make the conversation more interactive and natural [3–5]. So, in the later versions, our system will gradually incorporate the ability to interact with users through animated agents and speech synthesis technologies. Moreover it will be able to simulate a successful interaction between two animated avatars for training purposes.

2 Related Corpora

In this section we discuss the possibility of using other available corpora for the purposes of the research outlined above, and their limitations.

[1] http://www.metalogue.eu/.

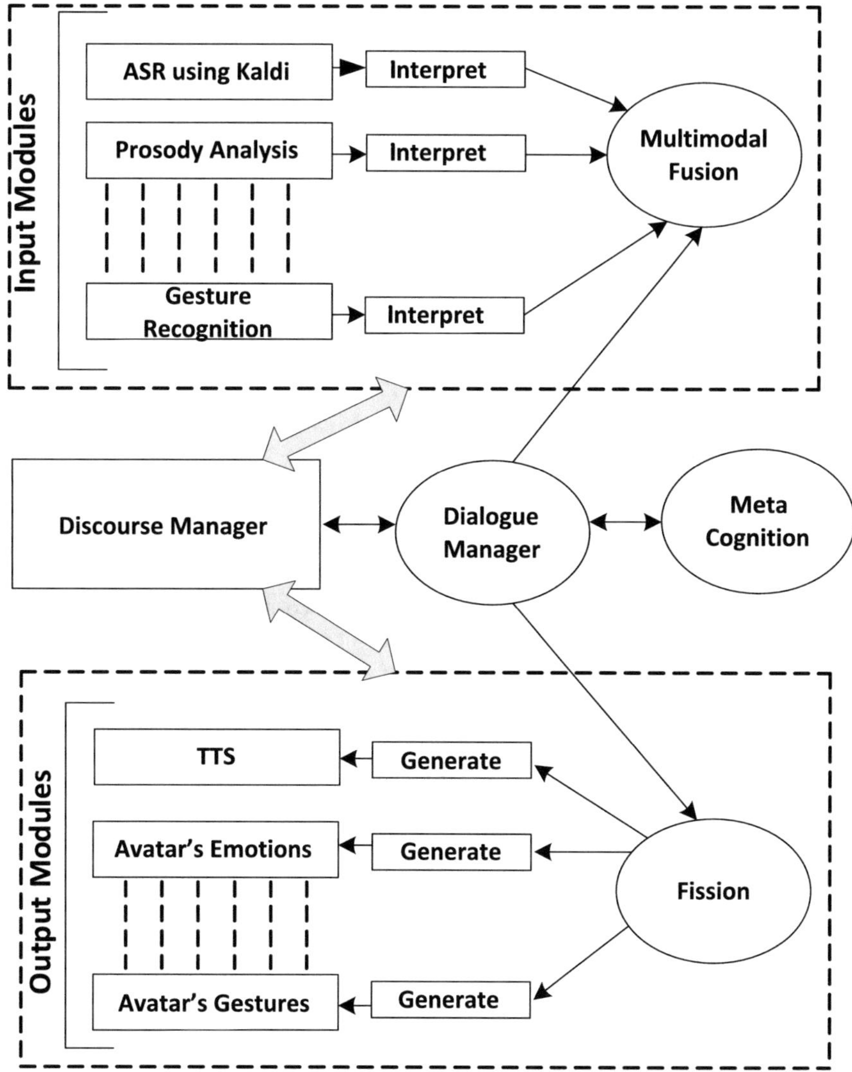

Fig. 1 METALOGUE system overview adapted from [1]

2.1 *IFA Dialogue Corpus*

The IFA (Institute of Phonetic Sciences) dialogue corpus contains a collection of face-to-face dialogue videos with annotated labels. Even though the language is Dutch the corpus gives examples of informal and friendly dialogue. This corpus could be useful to model the friendly behaviour which can be later used especially for call centre scenario of METALOGUE after annotating it for metacognitive skills.

There are in total 20 dialogue conversation videos and the duration of each video is around 15 min. There is no topic restriction imposed on the participants of the dialogue. Two video cameras are used to record the dialogue sessions. The overall duration of the corpus is around 5 h. To make the dialogue more useful and friendly, selection of participants is based on a subset of the following factors:

- Good friends
- Relatives
- Long- time colleagues

The corpus can be used and distributed under the GNU General Public Licence (an open source license) [6].

2.2 AMI Meeting Corpus

The AMI (Augmented Multi-party Interaction) meeting corpus [7] consists of 100 h of recordings. The corpus is multimodal since it includes several inputs: voice, video and writing. This corpus is annotated at different levels (dialogue acts, topic segmentation, individual actions, person location, focus of attention and AmiEmotion) and could be helpful to model the formal behaviour of a person in an interactive communicative situation. The language of the corpus is English and most of the participants are non native speakers. However the recording is performed in different rooms with different acoustic properties. The AMI meeting corpus is released under a creative common attribution shareAlike license.

2.3 Youth Parliament Corpus

The Hellenic Youth Parliament (HeP) corpus is available on line (1996 till 2013).[2] This corpus includes sessions in which members of the Hellenic youth parliament (aged between 17 and 20) discuss culture, social and economic affairs in a formal style. The sessions were audio-visually recorded and minutes of all meetings along with reports on each issue are available in PDF format. The duration of the parliament session for each year is around 16–17 h. This could be a useful corpus for analysis of real world data in the HeP scenario but we still need additional data recorded by the latest sensors (Kinect,[3] Myo gesture tracking armband[4] and Intel's RealSense depth camera[5]) to analyse body gestures and facial expressions in a more detailed way.

[2] http://www.efivoi.gr/?CMD=psifiako_iliko_intro.

[3] https://developer.microsoft.com/en-us/windows/kinect.

[4] https://www.myo.com/.

[5] https://software.intel.com/en-us/realsense/home.

2.4 MLA-14 Data

The MLA-14 (Multimodal Learning Analytics 2014) data contains students presentations [8]. In total there are 441 oral presentations delivered by Spanish-speaking students presenting projects about entrepreneurship ideas, literature reviews, research designs, software design etc. Recordings were placed in regular classroom settings and include multimodal data: speech, video, skeletal data gathered from Kinect for each individual, and presentations slides. In total, approximately 19 h of multimodal data is recorded. In addition individual ratings for each presentation is included as well as a group grade related to the quality of the slides used when doing each presentation. Each presentation has a rating based on the following performance factors:

1. Structure and connection of ideas.
2. Presents relevant information with good pronunciation.
3. Maintains an adequate voice volume for the audience.
4. Language used in presentation according to audience.
5. Grammar of presentation slides.
6. Readability of presentation slides.
7. Impact of the visual design of the presentation slides.
8. Posture and body language.
9. Eye contact.
10. Self confidence and enthusiasm.

2.5 Limitations

While the above described corpora provide useful resources in terms of training and testing general models of dialogue interaction, none of them directly fits in the scenario targeted by METALOGUE. The first two corpora (IFA and AMI) are relevant in terms of interactivity and multimodality, but lack the instructional element. The HeP corpus fits in the METALOGUE scenario but lacks the detailed multimodal input needed for system development purposes. The MLA-14 corpus is also situated in an educational context, contains rich multimodal data, but lacks the dialogue and interactivity elements. These limitations motivated the METALOGUE data collection activity described below.

3 The METALOGUE Data Collection Process

Several studies indicate that a presenter should speak with lively voice and make an eye contact with the audience. Furthermore, the presenter should stand straight, avoid crossing his legs, and use his hands, body and face to do gestures at the appropriate time [9–12]. However the face position information is also helpful in increasing speech recognition performance [13].

As METALOGUE focuses on presentation and negotiations skills, it needs to be able to gather the full skeleton/face data of participants along with audio and video recordings, specially for the HeP scenario. Since this setting requires the use of several independent cameras (Kinect and conventional video cameras), synchronisation issues need to be addressed. An example of such issues is the dropping of frames by the Kinect sensors during recording, which was addressed by duplicating some of the neighbouring frames. The motivation behind adding conventional cameras for recording is to obtain high quality frames for image analysis (emotion and effect recognition etc.) in addition to Kinect's built-in skeleton tracking functionality.

This section describes the complete data collection process including recording settings, environment, room and equipments specifications. We have recorded approximately 3 h of audio-visual training material simulating the Hellenic youth parliament scenario. In total 14 sessions have been recorded and each session lasts around 10–15 min.

3.1 Recording Settings

The recording takes place in a controlled settings and it includes a quiet room, and appropriate light conditions (no windows behind participants and adequate illumination of faces). The participants are allowed to face their opponent and audience, but are instructed to remain in the field of view of the Kinect camera during the debate. In a recording session, there are two students standing in-front of an audience debating on a social issue (whether smoking should be banned in public places). One student is in favour of such ban and the other student is against it. Any of the participants can start the debate, and after outlining their views (2–4 min), they listen to their opponent's intervention, taking turns as the debate proceeds. A schematic representation of this recording set up along with recording hardware is shown in Fig. 2.

3.2 Wizard of Oz Software

A simple WOZ (Wizard of Oz) system prototype has been implemented in Python[6] which consists of two programs:

RedGreenUser.py the user's interface, that is, a frame displaying two panels (a red and a green one) which light up according to the feedback sent to the participants by the wizard. The program starts as a server and listens for feedback from up to 10 concurrent connections, which makes it possible for multiple wizards to control the interface collaboratively.

[6]http://www.python.org/.

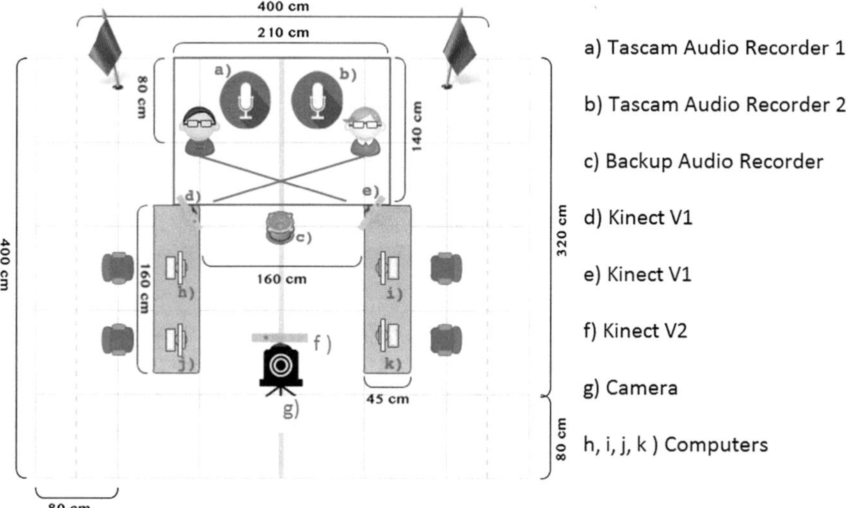

Fig. 2 Recording settings

RedGreenWOZ.py the wizard's control panel, through which the wizard chooses different categories of feedback to send to the participants' screen.

Although the feedback is categorised, the participants only sees unspecified red or green feedback on their screen. The wizard's and participant's user interfaces are shown in Fig. 3. The types of "red" (negative) feedback the wizard can choose from are:

- a participant is speaking too fast,
- a participant is speaking too softly,
- a participant has inappropriately interrupted the other participant,
- a participant speaks too much,

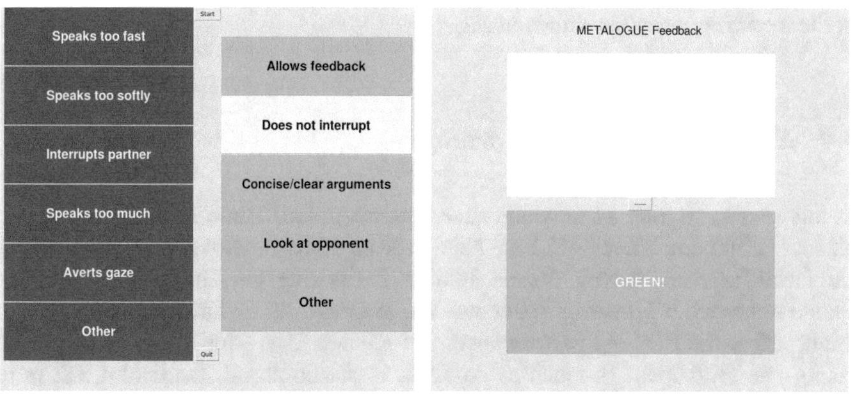

Fig. 3 Wizard (*left*) and participant (*right*) user interfaces

- a participant averts gaze, and
- other (a "catch all" category)

The types of "green" (positive) feedback the wizard can choose from are:

- a participant allows the other to provide feedback,
- a participant does not interrupt the other's speech,
- a participant presents clear and concise arguments,
- a participant looks at the other while debating, and
- other

When any of these items is chosen by the wizard, RedGreenWOZ.py timestamps it and writes a log of this feedback to an XML file. In these trials two such files are recorded per session, one containing the actual WOZ feedback (as displayed on the participant's screen) and an additional file generated by a "silent wizard" who interacts with an instance of RedGreenWOZ which simply records the feedback without actually presenting it to the participants. This file can be useful, for instance, in assessing the level of agreement between the two wizards through comparison of the feedback given by each of the participants on the time line. For instance, a wizard might decide to give a "red" feedback when participant A interrupts participant B, while the other wizard might decide to give "green" feedback to participant B for allowing A to provide feedback. The wizard log files are synchronised with the other media streams gathered during the sessions.

3.3 First Sitting for Recording

In this sitting, three sessions have been recorded using a real time feedback light (red or green) which is simulated by a wizard (as described in Sect. 3.2). Two Kinect sensors are used to track body skeleton and facial landmarks, and the facial and skeleton information (tracked by Kinect) are saved in XML files. A video camera is also used to record the whole recording session besides Kinect sensors. A snapshot of the recording set-up is shown in Fig. 4.

3.4 Second Sitting for Recording

In this sitting, in total 11 sessions have been recorded. Three kinect sensors (two kinect V1 and one kinect V2) have been used to track the movements of skeletons and facial features. To avoid the inaccuracies in tracking, the Kinects' fields of view should not overlap. However in our case the proposed set-up have an overlap which might affect the tracking performance of a Kinect. Two Kinect V1 sensors, each facing one participant as much as possible, is placed at a distance of 1.5–2 m to the participants. Participants are facing each other and/or audience, and markers are

Fig. 4 A snap shop of recording settings of the 'first sitting' using WOZ software

placed on the floor (movement space: 50 cm × 50 cm). For Kinect V1, the skeleton and face information are tracked and time stamped in real time and saved in xml files. Beside that Kinect V2 is also used to record the raw data using Kinect Studio 2.0 for off line processing. The motivation behind using the Kinect V2 is that to create backup in case any Kinect V1 crashes or vice versa. In future, Kinect V2 raw data can also be used to extract the new feature (hand open or close etc.) The difference between previous sitting and this one is the removal of feedback tool (described in Sect. 3.2), introduction of Kinect V2 and placing marker on the floor instead of just telling the participants to be remained in the range of the Kinect.

3.5 Data Synchronisation

Synchronisation of video and audio streams is performed using Final Cut pro X and a snapshot of a synchronised session is shown in Fig. 5. The details of each synchronised stream (e.g. time offsets) are imported in fcpxml files. The Kinect devices were started manually so the time offsets help us to synchronise the data (tracked by Kinect) with other streams. To accomplish this objective, we have parsed the tracked xml files using python and update their time stamps (by adding or subtracting the offset). Automatic speaker diarization is performed using the LIUM toolkit [14] and speech chunks are extracted using speaker diarization information.

Fig. 5 A snapshot showing synchronised video streams and kinects tracking of the 'second sitting' (WOZ software is not used.)

3.6 Speaker Characteristics

The subjects have been recruited by the HeP (Hellenic Parliament). The speakers are young students from two different schools, and their ages range between 17 and 20 years. They are non native speakers of English, know each other, have participated in HeP annual debating sessions, and are taking part in the METALOGUE data collection activities as volunteers. There are six participants in total: three females (S2, S3 and S5) and 3 males. The role of every speaker in the debate (pro or against a proposed smoking ban) and the duration of each session are depicted in Table 1.

4 Conclusion and Future Work

The data recorded with the setup and according to the procedures described in this article have been made available to all project partners. For future data collection, new sensor technologies will be tested including the Myo gesture tracking armband and Intel's RealSense depth camera. The synchronisation technique described above is used only for this pilot. In future versions all modules will time-stamp their output and transmit it to the METALOGUE server through sockets. The costumer care service scenario of METALOGUE, which we will address in the next phase of the project, poses additional difficulties for data collection. Due to privacy and confidentiality concerns, call centre companies are reluctant to provide real-world call data.

Table 1 Speaker ID, their role (pro and against) and the location of each speaker in a video (*left* or *right*)

Setting	Session no.	Speaker ID (left)	Role	Speaker ID (right)	Role	Duration (mm:ss)
Pre Pilot 1st sitting	1	S5	pro	S6	against	11:10
	2	S6	against	S4	pro	09:50
	3	S4	against	S5	pro	07:07
Pre Pilot 2nd sitting	1	S0	pro	S1	against	10:25
	2	S2	pro	S3	against	11:39
	3	S0	pro	S2	against	09:43
	4	S4	pro	S1	against	12:05
	5	S5	pro	S0	against	13:26
	6	S3	pro	S5	against	13:09
	7	S1	pro	S4	against	13:11
	8	S4	pro	S0	against	19:39
	9	S1	pro	S3	against	16:31
	10	S3	pro	S4	against	11:49
	11	S5	pro	S0	against	13:18

Therefore we are exploring the possibility of obtaining data used by companies to train their employees. We also consider recording training sessions between trainer and trainees.

Acknowledgements This research is supported by EU FP7-METALOGUE project under Grant No. 611073 at School of Computer Science and Statistics, Trinity College Dublin.

References

1. Helvert, J.V., Rosmalen, P.V., Börner, D., Petukhova, V., Alexandersson, J.: Observing, coaching and reflecting: a multi-modal natural language-based dialogue system in a learning context. In: Workshop Proceedings of the 11th International Conference on Intelligent Environments, vol. 19, pp. 220–227. IOS Press (2015)
2. Tumposky, N.R.: The debate debate. Clearing House: J. Educ. Strateg. Issues Ideas **78**(2), 52–56 (2004)
3. Cassell, J., Pelachaud, C., Badler, N., Steedman, M., Achorn, B., Becket, T., Douville, B., Prevost, S., Stone, M.: Animated conversation: rule-based generation of facial expression, gesture & spoken intonation for multiple conversational agents. In: Proceedings of the 21st Annual Conference on Computer Graphics and Interactive Techniques, pp. 413–420. ACM (1994)
4. Huang, L., Morency, L.P., Gratch, J.: Virtual rapport 2.0. In: Proceedings of the International Workshop on Intelligent Virtual Agents, pp. 68–79. Springer (2011)

5. Zhao, R., Papangelis, A., Cassell, J.: Towards a dyadic computational model of rapport management for human-virtual agent interaction. In: Proceedings of the International Conference on Intelligent Virtual Agents, pp. 514–527. Springer (2014)
6. van Son, R., Wesseling, W., Sanders, E., van den Heuvel, H.: The IFADV corpus: a free dialog video corpus. In: Proceedings of the Sixth International Conference on Language Resources and Evaluation (LREC'08), pp. 501–508. European Language Resources Association (ELRA), Marrakech, Morocco (May 2008). http://www.lrec-conf.org/proceedings/lrec2008/
7. McCowan, I., Lathoud, G., Lincoln, M., Lisowska, A., Post, W., Reidsma, D., Wellner, P.: The AMI meeting corpus. In: Noldus, L.P.J.J., Grieco, F., Loijens, L.W.S., Zimmerman, P.H. (eds.) Proceedings Measuring Behavior 2005, 5th International Conference on Methods and Techniques in Behavioral Research. Noldus Information Technology, Wageningen (2005)
8. Ochoa, X., Worsley, M., Chiluiza, K., Luz, S.: Mla'14: Third multimodal learning analytics workshop and grand challenges. In: Proceedings of the 16th International Conference on Multimodal Interaction, pp. 531–532. ICMI '14, ACM, New York, NY, USA (2014). http://doi.acm.org/10.1145/2663204.2668318
9. Stassen, H., et al.: Speaking behavior and voice sound characteristics in depressive patients during recovery. J. Psychiatr. Res. **27**(3), 289–307 (1993)
10. Lamerton, J.: Public Speaking. Everything You Need to Know. Harpercollins Publishers Ltd. (2001)
11. Grandstaff, D.: Speaking as a Professional: Enhance Your Therapy or Coaching Practice Through Presentations, Workshops, and Seminars. A Norton Professional Book. W.W. Norton & Company (2004). https://books.google.ie/books?id=UvmrZdAmNcYC
12. DeCoske, M.A., White, S.J.: Public speaking revisited: delivery, structure, and style. Am. J. Health-Syst. Pharm. **67**(15), 1225–1227 (2010). http://www.ajhp.org/cgi/content/full/67/15/1225
13. Slaney, M., Stolcke, A., Hakkani-Tür, D.: The relation of eye gaze and face pose: potential impact on speech recognition. In: Proceedings of the 16th International Conference on Multimodal Interaction, pp. 144–147. ACM (2014)
14. Rouvier, M., Dupuy, G., Gay, P., Khoury, E., Merlin, T., Meignier, S.: An open-source state-of-the-art toolbox for broadcast news diarization. Techniacl report, IDIAP (2013)

HELPR: A Framework to Break the Barrier Across Domains in Spoken Dialog Systems

Ming Sun, Yun-Nung Chen and Alexander I. Rudnicky

Abstract People usually interact with intelligent agents (IAs) when they have certain goals to be accomplished. Sometimes these goals are complex and may require interacting with multiple applications, which may focus on different domains. Current IAs may be of limited use in such cases and the user needs to directly manage the task at hand. An ideal personal agent would be able to learn, over time, these tasks spanning different resources. In this article, we address the problem of cross-domain task assistance in the context of spoken dialog systems, and describe our approach about discovering such tasks and how IAs learn to talk to users about the task being carried out. Specifically we investigate how to learn user activity patterns in a smartphone environment that span multiple apps and how to incorporate users' descriptions about their high-level intents into human-agent interaction.

Keywords Complex user intention · Multi-domain · Spoken dialog systems

1 Introduction

Smart devices, such as smartphones or TVs, allow users to achieve their intentions through verbal and non-verbal communication. The intention sometimes can be fulfilled in one single domain (i.e., an app). However, the user's intention is possible to span multiple domains and hence requires information coordination among these domains. A human user needs to keep track of the global context in order to organize the functionality provided by apps and coordinate information. On the other hand, although current intelligent agents can be configured by developers to passively

M. Sun (✉) · Y.-N. Chen · A.I. Rudnicky
School of Computer Science, Carnegie Mellon University, Pittsburgh, USA
e-mail: mings@cs.cmu.edu

Y.-N. Chen
e-mail: yvchen@cs.cmu.edu

A.I. Rudnicky
e-mail: air@cs.cmu.edu

K. Jokinen and G. Wilcock (eds.), *Dialogues with Social Robots*,
Lecture Notes in Electrical Engineering 427,
DOI 10.1007/978-981-10-2585-3_20

support (limited) types of cross-domain interactions, they are not capable of actively managing apps to satisfy potentially complex user intentions. As a result, multi-domain interactions with current agents may not be efficient.

Nowadays, most human-machine interactions are carried out via touch-screen. Although the vocabularies of recognizable gestures have been expanded during the past decade [1], interactive expressions are still restricted due to the limit of gestures and displays. This limit may affect usability, especially for certain populations, such as older users or users with visual disabilities. By contrast, spoken language can effectively convey high-level and complex user intentions to a device. However, the challenges are: (1) understanding both at the level of individual apps and at the level of activities that span apps; and (2) communicating a task-level functionality between user and agent. Our previous work focused on predicting follow-up action at app-level [2] or understanding the current app-level intention [3]. This article mainly addresses the high-level intention-embedded language understanding. For example, our proposed model understands that "*plan a dinner with Alex*" is composed of several domains such as YELP, OPENTABLE and MESSENGER. We also enable the system to verbally communicate its understanding of user intentions, in order to maintain a transparent communication channel.

Compared with conventional single-domain dialog systems [4–6], multi-domain systems have been studied in the past [7, 8], where a classic architecture contains multiple models developed independently for different domains and allows corresponding apps to handle user requests [3, 9–11]. Given a user utterance, a domain detector selects (1) a single domain [2, 3, 10, 12] or (2) several domains based on the functionality in the user request [13, 14]. However, neither of the two approaches considered the ultimate user intention behind the multi-domain interaction (i.e., why the user needs this set of domains). Our method bridges the low-level surface forms in cross-domain interactions and the high-level intention in the user's mind to enable systems to support intention realization. Moreover, from a personal assistant's perspective, we compare personalized models with generic ones based on personal data availability.

The rest of the article is organized as follows: we first briefly describe a data collection process to gather user's real-life multi-domain tasks. Then we discuss the methodology to discover, recognized and realize user intentions. Two user studies are described later as end-to-end and component-wise evaluation.

2 Data Collection

We undertook a data collection to investigate how participants interact with several domains (i.e., apps) to achieve their complex intentions. We first recorded the interactions with conventional touch-screen modality and then let the participants achieve the same goals but by using spoken language instead. More details of the collection are described in [15].

2.1 Smartphone Data Collection and Annotation

We continuously recorded participants' smartphone use—events such as app invocations together with its date/time and the phone's location (if GPS is enabled), over an extended period of time. Initial analysis of the data indicated that phone usage could be segmented into episodes consisting of interaction events closely spaced in time. In our pilot data, we found 3 mins of inactivity could be used to group events. Although this parameter appeared to vary across users, we used a single value for simplicity. Participants were asked to upload their logs on a daily basis. A privacy step allowed them to delete episodes that they might not wish to share.

Due to multi-tasking, an episode might consist of multiple single-app or multi-app tasks. For example one might be communicating with a friend but at the same time playing a game or surfing the web. We invited participants to our lab on a regular basis (about once a week) to annotate their submitted logs to decouple multiple tasks in the same episodes and also describe the nature (intent) of the tasks. Note that some tasks might also span episodes (for example making plans with others); we did not examine these.

Participants were presented with episodes from their log and asked to group events into sequences corresponding to individual tasks [16] (which we will also refer to as activities). Meta-information such as date, time, and street location, was shown to aid recall. Participants were asked to produce two types of annotation, using the Brat server-based tool [17]: (1) **Task Structure**: link apps that served a common intention; (2) **Task Description**: type in a brief description of the intention of the task. For example, in Fig. 1, the participant first linked four apps together and wrote that the intention was to schedule a session with our lab. Some of the task descriptions were quite detailed and provided the actual app sequence executed. However, others were quite abstract, such as "look up math problems" or "schedule a study session". In this paper, we took **task descriptions** as transcribed intent-embedded user utterances since these descriptions are usually abstract. We used these descriptions to train intention understanding models.

2.2 Interactive Dialog Task

We asked participants to talk to a Wizard-of-Oz dialog system to reenact their own multi-domain tasks using speech, instead of the GUI, in a controlled laboratory environment. Participants were shown (1) apps used; (2) task description they provided earlier; (3) meta data such as time, location to help them recall the task (see left part

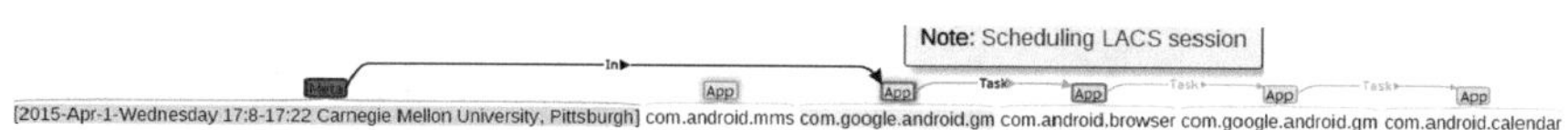

Fig. 1 Example of smartphone data annotation

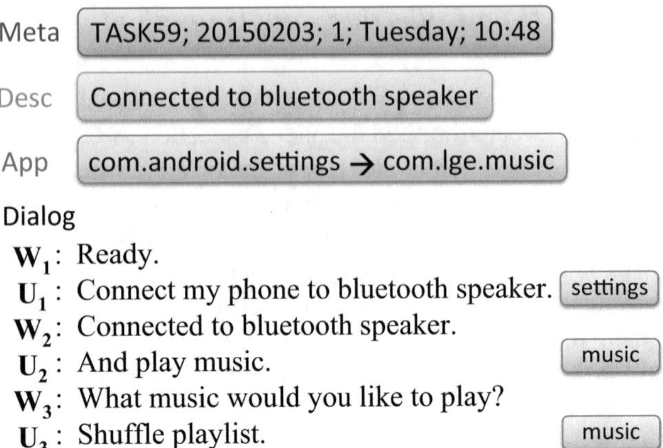

Fig. 2 Multi-app task dialog example. Meta, Desc, App were shown to the participant. Utterances were transcribed manually or via Google ASR. Apps were manually assigned to utterances

Table 1 Corpus characteristics

Category	#Participants	Age	#Apps	#Tasks	#Multi
Male	4	23.0	19.3	170	133
Female	10	34.6	19.1	363	322
Age < 25	6	21.2	19.7	418	345
Age ≥ 25	8	38.9	18.8	115	110
Native	12	31.8	19.3	269	218
Non-native	2	28.5	18.0	264	237
Overall	14	31.3	19.1	533	455

Age informally indicates young and old. A native Korean and Spanish speaker participated; both were fluent in English. #Apps is the average number of unique apps. #Multi is the number of tasks which involves multiple user turns

in Fig. 2). Participants were not required to follow the order of the applications used in their smartphone logs. Other than for remaining on-task, we did not constrain expression. The wizard (21-year-old male native English speaker) was instructed to respond directly to a participant's goal-directed requests and to not accept out-of-domain inputs. An example of a transcribed dialog is shown in Fig. 2.

This allowed us to create parallel corpora[1] of how people would use multiple apps to achieve a goal via both smartphone (touch screen) and language. We recruited 14 participants and collected 533 parallel interactions, of which 455 involve multiple user turns (see Table 1).

[1] Dataset: http://appdialogue.com.

3 Methodology

For an agent to interact with users at the level of intention, it should (1) **understand** an intention expressed by speech; and (2) be able to **convey** its understanding of the intention via natural language. For example, once the user says "I'd like to plan a farewell party for my lab-mate", the agent needs to know the intention behind this spoken input as well as be able to assist user to find a restaurant (YELP) and schedule time with other lab-mates (MESSENGER). On the other hand, the agent may reveal its inner state of understanding to the user, especially in clarification process. For instance, it may say "I think we are going to plan an evening event, right?" Channel-maintenance with such verbal cues (either implicit or explicit) is helpful in conversation [18]. We first describe modeling intention understanding, then describe the process by which the agent can verbally convey its inner state.

3.1 Models for Intention Understanding

What is user intention? We consider two possibilities. Observed interactions in the intention semantic space may be clustered into K_C groups, each representing a specific intention. We refer to this cluster-based definition as the **static** intention. On the other hand, we can also define **dynamic** intention, which is a collection of local neighbors (seen interactions) of the input user request. See Fig. 3 as an example. In the static intention setting, the agent is aware of the existence of K_C intentions and their semantics prior to invocation. However, in the dynamic setting, intention is implicitly defined by the K_N nearest neighbors during execution. In both cases, a realization process using the members of the recognized intention set maps the user utterance into a sequence/set of apps to support the user activity.

We anticipate two major differences between statically and dynamically based intentions. First, the static approach can use potentially richer information than just intention-embedded utterances when discovering basic intentions—it could use post-

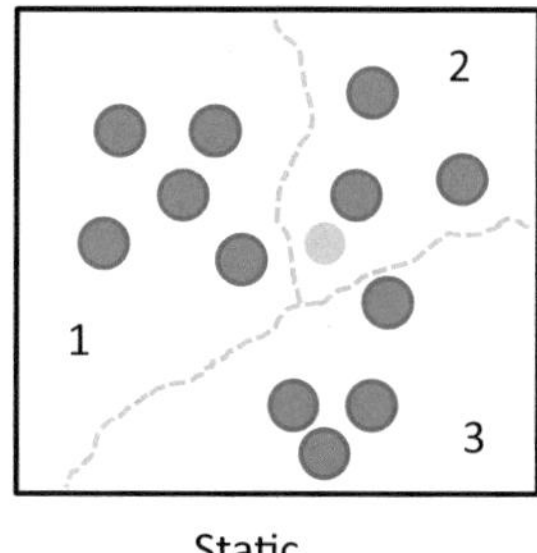

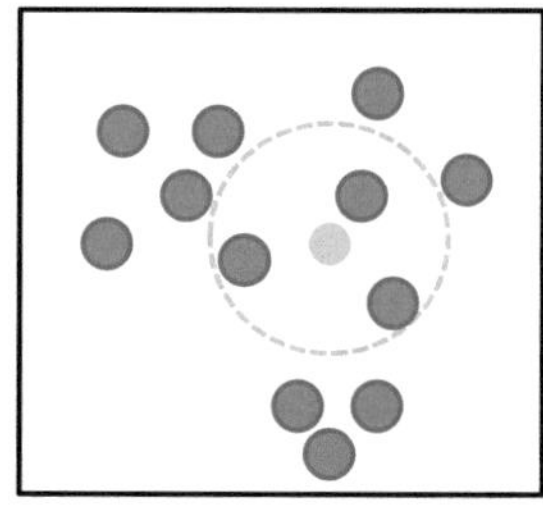

Fig. 3 Illustration of static intention versus dynamic intention. *Blue circles* denote training examples and the *yellow circle* is a testing example

initiate features such as apps launched or user utterances in the spoken dialog. Ideally, this may yield a better semantic space to categorize seen interactions. However, during execution, the input feature is the same as in the dynamic approach, i.e., task description. Second, the static approach has hard boundaries between intentions. Instances close to the boundaries may not be well characterized by their cluster members.

In both cases the agent will need to map an intention-embedded utterance into steps (i.e., sequence of apps/domains). Several techniques are available. We can combine the individual app sequences of the set members into a single app sequence that represents a common way of surfacing the intention (denoted as REPSEQ). Alternately, we can use a classifier that assigns multiple labels (apps ids) to the input (denoted as MULTLAB). Compared with the MULTLAB strategy, the advantage of REPSEQ is that it can preserve the order of the app sequence. However, once the intention is classified, the representative app sequence will always be the same, regardless of variations in the input. This could be a potential problem for statically based intentions. Arguably, during this process, we could weight each set member by its closeness to the input; we did not investigate this possibility. To evaluate, we compare the set of apps predicted by our realization model with the actual apps launched by the user and compute an F_1 score.[2]

There are two types of users—ones for which historical data are available, and the others. New users or users with privacy concerns will not have sufficient data. Thus, a *generic* model trained from large user community can be used instead of *personalized* model. We expect that a sufficiently well-trained generic model can provide reasonable performance; as history is accumulated performance will improve.

The building of intention understanding models may be impacted by intra- and inter-user inconsistency in the language/apps. We may encounter the problem of *vocabulary-mismatch* [16, 19], where interactions related with the same intention have non-overlapping (1) spoken terms (words), even caused by minor differences such as misspellings, morphologies, etc.; (2) apps, e.g., people may use different apps—MESSENGER or EMAIL with essentially similar functionality. Below we describe two techniques to overcome potential language- and app-mismatch.

3.1.1 Language Mismatch

We can consider a user's input utterances (e.g., "schedule a meeting") as a query to the intention model. To manage language inconsistency, we used a two-phase process—(1) text normalization where only verbs and nouns in the query are preserved and further lemmatized (e.g., "took" → "take"); (2) query enrichment (QryEnr) which expands the query by incorporating words related to it semantically. QryEnr can reduce the likelihood of seeing sparse input feature vector due to out-of-vocabulary [20] words. In this work, we used word2vec [21] with gensim[3] toolkit

[2] $F_1 = 2 \times Precision \times Recall/(Precision + Recall)$.

[3] Toolkit: https://radimrehurek.com/gensim/models/word2vec.html.

on the pre-trained GoogleNews word2vec[4] model. The proposed QryEnr algorithm is described in Algorithm 1. In short, each word w_i in the lemmatized query Q yields mass increases for N semantically close words in the feature vector f.

Algorithm 1 Query Enrichment

Require: lemmatized words of the query $q = \{w_1, ..., w_{|q|}\}$ and their counts $C = \{c_1, ..., c_{|q|}\}$; training vocabulary V; bag-of-words feature vector $\boldsymbol{f}_q = \{f_1, ..., f_{|V|}\}$ constructed on q; the word semantic relatedness matrix M; the number of semantically similar words N allowed to be extracted for each word in q;

Ensure: an enriched bag-of-words feature vector $\mathbf{f}_q^* = \{f_1^*, ..., f_{|V|}^*\}$

for each $w_i \in Q$ **do**
    Use M to find N words closest to w_i: $V_N = \{v_1, ..., v_N\} \in V$;
    for each $v_j \in V_N$ **do**
        $f_j^* = f_j + M_{i,j} \times c_i$
    end for
end for
return $\mathbf{f}_q^*$;

3.1.2 App Mismatch

When a generic model is used, recommended apps may not match the apps available on a specific user's device. For example, the recommended app, BROWSER should be converted to CHROME if that is the only (or preferred) app in this user's phone that can browse the Internet. Therefore, similarity metrics among apps are needed.

There are several ways to compute **app similarity** (AppSim). First, based on the edit distance between app (package) names, for example com.lge.music is similar to com.sec.android.app.music since both contains the string "music". Second, we can project an app to a vector space. Ideally, apps with similar functionalities will appear close to each other. Possible resources to use are (1) app descriptions in app stores; (2) language associated with each app when users verbally command the app (see example in Fig. 2). Third, app-store category may indicate functionality-wise similarity. However, we found Google Play category too coarse. In this work, we used the first method with 16 fillers (e.g., "android", "com", "htc") removed from package names. Examples are shown in Table 2. We found this simple method significantly improved system performance (described later).

[4]Model: https://drive.google.com/file/d/0B7XkCwpI5KDYNlNUTTlSS21pQmM/edit?usp=sharing.

Table 2 Most similar apps for Accuweather and Music among 132 apps in our data collection

Order	com.accuweather.android	com.lge.music
1	com.sec.android.widgetapp.ap.hero.accuweather	com.google.android.music
2	com.jrdcom.weather	com.sec.android.app.music
3	com.weather.Weather	com.spotify.music

3.2 Conveying Intention Understanding

IAs may need to communicate with the user in language cast at the level of intention, especially as part of a clarification process. For example, the IA may launch a short sub-dialog by saying "are you trying to *share a picture*?" This involves a template ("are you trying to ___?") and some content ("share a picture"). Instead of echoing content directly extracted from the user's current input, we abstract the semantics of similar previous interactions to provide language material which indicates (though a paraphrase) that the agent indeed understands the user's intention.

4 Study

4.1 Intention Interpretation and Realization

To evaluate intention modeling, we focus on three comparisons: (1) **intention**: static versus dynamic models; (2) **source**: personalized versus generic setups; (3) **method**: RepSeq versus MultLab realization strategies. We used the chronologically first 70 % of each user's data for training the personalized model, in principle mirroring actual data accumulation. The remaining 13 users' first 70 % data was combined to train the generic model. The number of intentions K_C for the static intention model and the number of nearest neighbors K_N for the dynamic model can be varied. We adapted K_C using gap statistics [22] to unsupervisely select the optimal K_C from 1 to 10 before KMeans. K_N was set to the square root of the number of training examples [23]. For RepSeq we used ROVER to collapse multiple app sequences into one [24]. For MultLab, we used SVM with linear kernel.

We show system performance in Table 3. This prediction task is difficult since on average each user has 19 unique apps and 25 different sequences of apps in our data collection. The upper part corresponds to static intention model and the lower part to dynamic intention. Within either approach, different intention realization strategies (QryEnr and AppSim) and their combination are also shown. We performed a balanced ANOVA test of F_1 score on the factors mentioned above: **intention, source** and **method**. The test indicates that the performance differs significantly ($p < 0.05$).

As noted earlier, the static model has the flexibility to incorporate richer information (post-initiate features) to discover the basic K_C intentions. As shown in Table 3, adding more post-initiate information (denoted with ⋆ and †) improves personalized models, but not necessarily generic models, mainly due to the inter-user difference in language and apps. But we do not observe superior performance for the static model over the dynamic one, even when richer information incorporated (⋆ and †). For REPSEQ strategy, the dynamic model is much better than the static one. REPSEQ is completely based on the selection of similar interactions. Arguably, an input may fall close to the intention boundary in the static setting, which indeed is closer to some interactions on the other side of the boundary as opposed to its own cluster members. On the other hand, the MULTLAB approach shows relatively consistent performance in both static and dynamic settings, indicating robustness and self-adaptability with respect to the choice of interactions of similar intention.

In Table 3, the fact that QryEnr improves the F_1 score in all conditions indicates that semantic similarity among words can effectively address the *language-mismatch* problem. On the other hand, although AppSim has no effect on the personalized model, it addresses the *app-mismatch* issue in generic models intuitively ($p < 0.05$ when comparing with the baseline in a balanced ANOVA on two additional factors: **intention, method**). Combining QryEnr and AppSim methods together (denoted as "+QryEnr+AppSim") consistently achieves the highest F_1 score. As we expected, the generic intention model is consistently inferior to the personalized model.

Table 3 Weighted average F_1 score (%) on test set across 14 participants, using **bag-of-words**

	REQSEQ		MULTLAB	
	Personalized	Generic	Personalized	Generic
Static (baseline)	42.8	10.1	55.7	23.8
+QryEnr	44.6	11.2	56.3	27.9
+AppSim	–	15.1	–	27.8
+QryEnr +AppSim	–	16.1	–	36.1
+QryEnr⋆ +AppSim	44.9	18.0	57.5	**37.1**
+QryEnr† +AppSim	**45.8**	**18.1**	**57.6**	35.9
Dynamic (baseline)	50.8	23.8	51.3	19.1
+QryEnr	54.9	26.2	57.0	22.9
+AppSim	–	30.1	–	22.7
+QryEnr +AppSim	**54.9**	**32.5**	**57.0**	**28.0**

Average K_C in static condition is 7.0 ± 1.0 for generic model, and 7.1 ± 1.6 for personalized model. The static condition was run 10 times and the average is reported. K_N in the dynamic condition is 18.5 ± 0.4 for the generic model and 4.9 ± 1.4 for the personalized model. ⋆ indicates both descriptions and user utterances are used in clustering and † indicates apps are used as well. +AppSim is not meaningful for Personalized models

Table 4 Mean number of phrases generated using different resources

MANUAL	ASR	DESC	DESC+ASR	DESC+MANUAL
20.0	20.3	11.3	**29.6**	29.1

4.2 *Intention Representation in Natural Language*

It should be possible to automatically abstract the semantics of the recognized intention cluster (or neighbors): Text summarization may be used to generate high-level description of the intention cluster [25, 26]. Keyphrase extraction provides another alternative [27–29]. Note that, even if the automatic generation of semantic summarization is not precise, it may still be sufficiently meaningful in context.

In this study, we used the Rapid Automatic Keyword Extraction (RAKE[5]) algorithm [29], an unsupervised, language-independent and domain-independent extraction method. This method has been reported to outperform other unsupervised methods such as TextRank [30, 31] in both precision and F score. In RAKE, we required that (1) each word have 3 or more characters; (2) each phrase have at most 3 words; and that (3) each key word appear in the text at least once. We did not investigate tuning these parameters. We use 3 individual resources and 2 combinations, reflecting constraints on the availability of different contexts in real-life. The three individual resources are manual transcription of user utterances in their dialogs (MANUAL) and their ASR transcriptions (ASR) and high-level task descriptions (DESC). The average number of key phrases generated by each resource (or their combination) is shown in Table 4.

We selected 6 users to first review their own clusters, by showing them all cluster members with (1) apps used in the member interaction; (2) dialog reproduced; (3) meta-data such as time, date, address, etc. We let them judge whether each individual phrase generated by the system summarized all the activities in the cluster (binary judgement). We used three Information Retrieval (IR) metrics to evaluate performance among different resources—(1) Precision at position K ($P@K$); (2) Mean Average Precision[6] at position K ($MAP@K$); (3) Mean Reciprocal Rank (MRR). The first two metrics emphasize on the quality of the top K phrases, MRR focuses on a practical goal—"how deep the user has to go down a ranked list to find one useful phrase?". Average MRR is 0.64, meaning that the user will find an acceptable descriptive phrase in the top 2 items shown; an ANOVA did not show significant differences between resources. With more sensitive $MAP@K$ and $P@K$ metrics, DESC+ASR and DESC+MANUAL do best. The improvement becomes significant as K increases: having a user-generated task description is very useful.

Participants were also asked to suggest carrier phrases that the agent could use to refer to activities; we found these to be unremarkable. Among the 23 phrases collected, "do you want to ___" and "would you like to ___" were the most popular.

[5]Toolkit: https://www.airpair.com/nlp/keyword-extraction-tutorial.

[6]$MAP@K$ computed as: $\sum_{k=1}^{K} precision(k) * relevance(k)/K$.

To conclude, if the IA can observe a user's speech commands or elicit descriptions from the user (ideally both), it can generate understandable activity references and might be able to avoid recourse to less efficient strategies (e.g. lists).

5 Conclusion and Future Work

We present a framework, HELPR, that is used to learn to understand a user's intention from a high-level description of goals (e.g., "go out with friends") and to link these to specific functionality available on a smart device. The proposed agent solicits descriptions from the user. We found that the language used to describe activities is sufficient to group together similar activities. Query enrichment and app similarity help with language- and domain-mismatch problems, especially when a generic model is used. We demonstrated that an agent could use data from large user community while also learning user-specific models.

The long-term goal of our work is to create agents that observe recurring human activities, understand the underlying intentions and support the task through spoken language interaction. The agent must communicate on the level of intentions instead of, or in addition to, individual apps. And it needs to manage the context of the activity so that its state can be shared between different apps.

The value of such an agent is that it would operate on a level higher than provided by app-specific interfaces. It would moreover allow the user to effectively build their own applications by composing the functionality in existing apps. We have shown that it is possible to infer user intentions; the next challenge is to capture meaningful context and actively apply it across different apps.

Acknowledgements This work was supported in part by YAHOO! InMind, and by the General Motors Advanced Technical Center.

References

1. Harrison, C., Xiao, R., Schwarz, J., Hudson, S.E.: Touchtools: leveraging familiarity and skill with physical tools to augment touch interaction. In: Proceedings of the SIGCHI Conference on Human Factors in Computing Systems, pp. 2913–2916 (2014)
2. Sun, M., Chen, Y.N., Rudnicky, A.I.: Understanding user's cross-domain intentions in spoken dialog systems. In: Proceedings of NIPS workshop on Machine Learning for SLU and Interaction (2015)
3. Chen, Y.N., Sun, M., Rudnicky, A.I.: Leveraging behavioral patterns of mobile applications for personalized spoken language understanding. In: Proceedings of the 18th ACM International Conference on Multimodal Interaction (ICMI), pp. 83–86 (2015)
4. Young, S.: Using POMDPs for dialog management. In: Proceedings of the IEEE Spoken Language Technology Workshop (SLT), pp. 8–13 (2006)
5. Sun, M., Rudnicky, A.I., Winter, U.: User adaptation in automotive environments. In: Proceedings of the 4th International Conference on Advances in Human-Factors and Ergonomics (AHFE), pp. 156–165 (2012)

6. Pappu, A., Sun, M., Sridharan, S., Rudnicky, A.I.: Situated multiparty interactions between humans and agents. In: Proceedings of the 15th International Conference on Human-Computer Interaction: Interaction Modalities and Techniques, pp. 107–116 (2013)
7. Lunati, J.M., Rudnicky, A.I.: Spoken language interfaces: the OM system. In: Proceedings of the Conference on Human Factors in Computing Systems, pp. 453–454 (1991)
8. Rudnicky, A.I., Lunati, J.M., Franz, A.M.: Spoken language recognition in an office management domain. In: Proceedings of the International Conference on Acoustics, Speech, and Signal Processing (ICASSP), pp. 829–832. IEEE (1991)
9. Lin, B.S., Wang, H.M., Lee, L.S.: A distributed architecture for cooperative spoken dialogue agents with coherent dialogue state and history. In: Proceedings of the Automatic Speech Recognition and Understanding Workshop (ASRU), vol. 99, p. 4 (1999)
10. Nakano, M., Sato, S., Komatani, K., Matsuyama, K., Funakoshi, K., Okuno, H.G.: A two-stage domain selection framework for extensible multi-domain spoken dialogue systems. In: Proceedings of the 12th Annual Meeting of the Special Interest Group on Discourse and Dialogue (SIGDIAL), pp. 18–29. ACL (2011)
11. Chen, Y.N., Rudnicky, A.I.: Dynamically supporting unexplored domains in conversational interactions by enriching semantics with neural word embeddings. In: Proceedings of the IEEE Spoken Language Technology Workshop (SLT), pp. 590–595. IEEE (2014)
12. Li, Q., Tür, G., Hakkani-Tür, D., Li, X., Paek, T., Gunawardana, A., Quirk, C.: Distributed open-domain conversational understanding framework with domain independent extractors. In: Proceedings of the IEEE Spoken Language Technology Workshop (SLT), pp. 566–571. IEEE (2014)
13. Ryu, S., Lee, D., Lee, I., Han, S., Lee, G.G., Kim, M., Kim, K.: A hierarchical domain model-based multi-domain selection framework for multi-domain dialog systems. In: Proceedings of the 24th International Conference on Computational Linguistics (ACL) (2012)
14. Ryu, S., Song, J., Koo, S., Kwon, S., Lee, G.G.: Detecting multiple domains from users utterance in spoken dialog system. In: Proceedings of the 6th International Workshop on Spoken Dialogue Systems (IWSDS), pp. 101–111 (2015)
15. Sun, M., Chen, Y.N., Hua, Z., Tamres-Rudnicky, Y., Dash, A., Rudnicky, A.I.: AppDialogue: Multi-app dialogues for intelligent assistants. In: Proceedings of the 10th International Conference on Language Resources and Evaluation (LREC) (2016)
16. Lucchese, C., Orlando, S., Perego, R., Silvestri, F., Tolomei., G.: Identifying task-based sessions in search engine query logs. In: Proceedings of the 4th ACM International Conference on Web Search and Data Mining, pp. 277–286. ACM (2011)
17. Stenetorp, P., Pyysalo, S., Topić, G., Ohta, T., Ananiadou, S., Tsujii, J.: BRAT: a web-based tool for NLP-assisted text annotation. In: Proceedings of the Demonstrations at the 13th Conference of the European Chapter of the Association for Computational Linguistics (EACL), pp. 102–107. Association for Computational Linguistics (2012)
18. Bohus, D., Rudnicky, A.I.: Sorry, I didn't catch that!-an investigation of non-understanding errors and recovery strategies. In: Proceedings of the 6th Annual Meeting of the Special Interest Group on Discourse and Dialogue (SIGDIAL) (2005)
19. Shen, X., Tan, B., Zhai., C.: Implicit user modeling for personalized search. In: Proceedings of the 14th ACM International Conference on Information and Knowledge Management, pp. 824–831 (2005)
20. Sun, M., Chen, Y.N., Rudnicky., A.I.: Learning OOV through semantic relatedness in spoken dialog systems. In: Proceedings of the 16th Annual Conference of the International Speech Communication Association (Interspeech), pp. 1453–1457 (2015)
21. Mikolov, T., Chen, K., Corrado, G., Dean, J.: Efficient estimation of word representations in vector space. In: Proceedings of the Workshop at International Conference on Learning Representations (ICLR) (2013)
22. Tibshirani, R., Walther, G., Hastie., T.: Estimating the number of clusters in a data set via the gap statistic. J. R. Stat. Soc.: Series B (Stat. Method.) **63**, 411–423 (2001)
23. Duda, R., Hart, P., Stork, D.: Pattern Classification. Wiley (2012)

24. Fiscus, J.G.: A post-processing system to yield reduced word error rates: recognizer output voting error reduction (ROVER). In: Proceedings of the Automatic Speech Recognition and Understanding Workshop (ASRU), pp. 347–352 (1997)
25. Ganesan, K., Zhai, C., Han, J.: Opinosis: a graph-based approach to abstractive summarization of highly redundant opinions. In: Proceedings of the 23rd International Conference on Computational Linguistics (COLING), pp. 340–348. ACL (2010)
26. Liu, F., Flanigan, J., Thomson, S., Sadeh, N., Smith, N.A.: Toward abstractive summarization using semantic representations. In: Proceedings of the 14th Annual Conference of the North American Chapter of the Association for Computational Linguistics: Human Language Technologie (NAACL-HLT), pp. 1077–1086 (2015)
27. Witten, I.H., Paynter, G.W., Frank, E., Gutwin, C., Nevill-Manning, C.G.: KEA: practical automatic keyphrase extraction. In: Proceedings of the 4th ACM Conference on Digital Libraries, pp. 254–255 (1999)
28. Medelyan, O.: Human-competitive automatic topic indexing. Ph.D. thesis, University of Waikato (2009)
29. Berry, M.W., Kogan., J.: Text Mining: Applications and Theory. Wiley (2010)
30. Mihalcea, R., Tarau, P.: TextRank: bringing order into texts. In: Proceedings of the Conference on Empirical Methods in Natural Language Processing (EMNLP), pp. 404–411. ACL (2004)
31. Hulth, A.: Improved automatic keyword extraction given more linguistic knowledge. In: Proceedings of the Conference on Empirical Methods in Natural Language Processing (EMNLP), pp. 216–223. ACL (2003)

Towards an Open-Domain Social Dialog System

Maria Schmidt, Jan Niehues and Alex Waibel

Abstract This article describes a text-based, open-domain dialog system developed especially for social, smalltalk-like conversations. While much research is focused on goal-oriented dialog currently, in human-to-human communication many dialogs do not have a predefined goal. In order to achieve similar communication with a computer, we propose a framework which is easily extensible by combining different response patterns. The individual components are trained on web-crawled data. Using a data-driven approach, we are able to generate a large variety of answers to diverse user inputs.

Keywords Social dialog · Open domain · Wikipedia

1 Introduction

Currently, goal-oriented dialog systems are still in the primary focus of dialog system research. In these systems, a clear goal of the dialog is defined and the user should be primarily provided with their needed information in an efficient way.

In contrast to this, in small talk-like dialogs as modeled in the approach presented in this work, no clear dialog goal is given. The main target when developing such a social system is to enable diversified dialogs that keep the user interested in the conversation. Motivated by human response patterns, our initial small talk system will use two different response patterns to show its interest in the conversation with the user. One possibility to show interest is to ask questions about the user's input.

M. Schmidt (✉) · J. Niehues · A. Waibel
Karlsruhe Institute of Technology, Institute for Anthropomatics and Robotics,
Karlsruhe, Germany
e-mail: Maria.Schmidt@kit.edu

J. Niehues
e-mail: Jan.Niehues@kit.edu

A. Waibel
e-mail: Alexander.Waibel@kit.edu

K. Jokinen and G. Wilcock (eds.), *Dialogues with Social Robots*,
Lecture Notes in Electrical Engineering 427,
DOI 10.1007/978-981-10-2585-3_21

Like it is done by humans, we can respond to the user by asking for more details or explanations. A second template to respond to the user is to report about a topic that is related to input. If the user tells us some fact, he might be interested in related information.

2 Background

A similar situation to the one described in this paper was targeted in [1]. They developed a chat-like conversation system, but in contrast to this work, their work focuses on the development of a dialog strategy based on a Markov Decision Processes (MDP) framework.

Furthermore, Jokinen and Wilcock published their work on the so called *WikiTalk* system, which is a spoken Wikipedia-based open domain knowledge access system [2]. Similar to the aim in this article, they used the WikiTalk system for constructive interaction for talking about interesting topics [3]. In another work, Jokinen and Wilcock used their system for multimodal open-domain conversations with the NAO robot [4].

3 System Description

The presented dialog system consists of different modules m_i, which analyze the input and generate a response r annotated with a confidence $c_i(r)$. These modules are developed to generate different types of answers. For example, we develop a module for questions and one to present additional information to the user. Furthermore, these answers belong to different topics, such as entertainment, sports, latest news or business.

Afterwards, a response combination is used to evaluate the different responses and select the most appropriate one. Figure 1 depicts the system and its modules.

Furthermore, we keep track of the dialog flow and its continuity with session objects, in which we store dialog states, current answers and a history. Thereby, we can for example check whether the now proposed answer has already been mentioned in the last few dialog turns.

3.1 Response Combination

The different modules m_i of the system generate possible responses r and their confidence $c_i(r)$. The response combination selects the most probable answer and presents it to the user. In conventional settings in order to train a classifier, we need user input I from many users along with the best response $\hat{r}$ given a set of possible responses R and their respective confidences $\{(r, c_i(r))\}$.

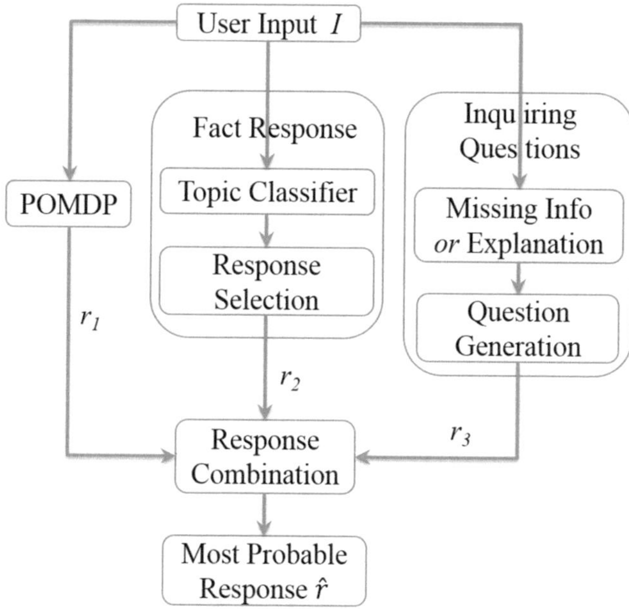

Fig. 1 System overview

3.2 *Fact Response*

Our Fact Response component is responsible for giving facts back to the user. One way of responding to the user is to provide additional information about the current conversation topic. The challenges here are to find answers that are interesting to the user and simultaneously matching the current topic. The topic classifier module is responsible for analyzing the user input and finding the most probable article a given the user input I which is called Natural Language Understanding (NLU) in most Dialog Systems (DSs). We use the response selection to select the best fitting response r given the information from the topic classifier and assigning a confidence in the selection. It is one of the Natural Language Generation (NLG) modules in our social DS.

We collected possible responses from articles crawled in the web. In the experiment we use Wikipedia and news articles. In order to efficiently retrieve the best response, we define a set of topics T. Each topic $t \in T$ consists of a set of articles A, where $a \in A$. Thereby, an article can be part of several topics. Finally, each article is associated with a set of responses R, where $r \in R$.

3.2.1 Topic Classifier

The Topic Classifier serves as an NLU component which determines the topic the user wants to talk about.

Given the user's input $I = w_0, \ldots, w_N$ with an arbitrary number of words, the topic classifier assigns a probability to the triple (t, a, k) where t is a topic, a is an article and $k \subseteq I$ is a set of keywords in I.

The article probability is defined by

$$P(a|t,k) = \frac{1_{a \in t} \sum_{k_i \in k} e^{tfidf(a,k)}}{Z(t,k)} \tag{1}$$

where $Z(t, k)$ is a normalization factor defined as:

$$Z(t,k) = \sum_a 1_{a \in t} \sum_{k_i \in k} e^{tfidf(a,k)} \tag{2}$$

For training the language models (LMs), we applied the KenLM toolkit[1] [5].

3.2.2 Response Selection

After assigning probabilities to the different articles in the database, the response selection picks an answer and assigns a confidence to it. In order to generate different answers for a classified topic t, the responses are drawn randomly given a probability distribution. The probability of a response r is given by:

$$P(r|a,t,k) = \frac{1_{|k \cap r|>0}}{Z(a,t,k)} \tag{3}$$

In order to limit the computation, we first calculate $(a, \hat{t}, k) = argmax(P((t, a, k)|I)$ and then randomly choose a response from $P(r|a, \hat{t}, k)$. As confidence for the response r we use $P((t, a, k)|I)$.

3.3 *Inquiring Questions*

In this system, we use two different templates for inquiring questions. First, we search for information that are not given in the input and ask for these information. A second possibility is to ask for an explanation, since the user has given us information we do not understand.

[1]http://kheafield.com/code/kenlm/.

3.3.1 Missing Information

We want to analyze the sentence in order to find information the user did not give to the computer. An example input would be a sentence like *I had many meetings today*. If we have this sentence, we could ask things like *What were the meetings about?* or *With whom did you meet?*.

3.3.2 Explanation

A different word template is to ask for explanation about unknown terms. Like in a human-human conversation, if there appears a word or phrase the system does not know, it should ask the user for an explanation of the term. Since we use training data covering many topics, we assume that a word, which is unknown to the system, should be a quite specific term, e.g., a name only known by some people. Therefore, it is reasonable to assume that a person in the conversation might also not know this term and would ask who or what it is.

Therefore, we use the training vocabulary to find unknown words in the input sentence and mark them as possible question items. Similar to the missing information, the actual questions for explanation are generated by the Question Generation module which is described in the upcoming section below.

This module is similar to the function of the OOV component in [6].

3.3.3 Question Generation

Given the two previously mentioned modules, we have annotated the input with possible missing information like "*meetings with*" and out-of-vocabulary words. In order to ask for this information, we need an NLG module to create questions based on the given input words.

3.4 POMDP Goal-Oriented Dialog

Our system contains a POMDP component in order to be able to handle goal-oriented dialogs in addition. Our implementation is based on the state-of-the-art technique shown in Young [7]. Currently, this subsystem is trained to work on a restaurant information scenario.

4 Results

For training the fact response, we used Wikipedia articles from the simple English edition and we crawled news articles from RSS feeds. We used the simple English Wikipedia edition, since in dialogs usually simpler and shorter sentences are used than in written text. For the news, we use articles published within the last three weeks. We use the articles from 5 different topics of Wikipedia and 12 RSS feeds. The resulting number of articles and facts are displayed in Table 1.

We trained the possible questions for missing information from the OpenSubtitle corpus and three additional English corpora. In total, these corpora contain 73M words. We applied a stop word list to the relation extracted from the dependencies generated by the Stanford Dependency Parser [8]. After filtering out all relations with a probability lower than 0.1, we generated a list of relations with 14,696 entries.

Table 1 Fact response data

	Wikipedia	News
Topics	5	12
Articles	90,514	2,896
Facts	317,732	32,993

U: FC Bayern lost this week. Barca was too strong for them.
S: Pep Guardiola will return to former club Barcelona for the first time with Bayern Munich after the two clubs were drawn against each other in the European Champions League semifinal.

U: I will go now
S: Go to?
U: I will walk to the hotel
S: Walk with?
⋮
U: Nobody. I am alone at the IWSDS
S: Iwsds?
U: It is a conference on spoken dialog systems

Fig. 2 Example of a dialog: RSS feed info, missing information, explanation

Figure 2 illustrates example dialog turns of our social DS for RSS feed information, missing information and an explanation dialog. First, we show an example of a user input triggering a current RSS news feed on soccer. According to the keywords giving back RSS feed information was more likely than responding with Wikipedia information. The following two dialog pairs show the system's behavior for missing information. Based on our corpus, the employed dependency parser detects "go *to*" and "walk *with*" as the most likely phrases for "go" and "walk", respectively. Therefore, the system assumes that the user input sentences are underspecified and asks for where the user is going and with whom they are walking. Finally, we give an example of explanation. If there is an OOV word occurring such as "IWSDS", the system will rephrase the OOV term as a question.

5 Demo Setup

The demo is accessible via a web interface. It consists of two text boxes, where the upper, bigger one is capturing the previous dialog with user and system turns indicated by `U` and `S`, respectively. The lower, smaller text box displays the user input as it is typed in. By hitting `Enter` the user submits their input which will be displayed in the upper text box followed by the corresponding system response. The UI was implemented by using the Django web framework[2] while the dialog system as such is implemented in Java.

6 Conclusion

In this system we have implemented a data-driven approach towards modeling open-domain social dialogs which is accessible through a website. We have shown a system architecture that is easily extensible and can generate different types of responses. Motivated by human small talk, we develop different modules to generate diverse responses to the user input.

A first module is able to provide additional information to the user. This information is gathered from Wikipedia and current news feeds. Using language model and information retrieval based techniques, we find the most appropriate information in a large collection of potential answers.

Secondly, we developed a module to determine which information a dialog partner might be interested in and ask for this information. Also for this module, we used a data-driven approach in order to cover a variety of topics.

---

[2]https://www.djangoproject.com/.

References

1. Shibata, T., Egashira, Y., Kurohashi, S.: Chat-like conversational system based on selection of reply generating module with reinforcement learning. In: Proceedings of the 5th International Workshop on Spoken Dialog Systems (2014)
2. Wilcock, G.: WikiTalk: a spoken Wikipedia-based open-domain knowledge access system. In: Proceedings of the COLING 2012 Workshop on Question Answering for Complex Domains, pp. 57–69 (2012)
3. Jokinen, K., Wilcock, G.: Constructive interaction for talking about interesting topics. In: Proceedings of Eighth International Conference on Language Resources and Evaluation (LREC 2012), pp. 404–410 (2012)
4. Jokinen, K., Wilcock, G.: Multimodal open-domain conversations with the Nao robot. In: Natural Interaction with Robots, Knowbots and Smartphones, pp. 213–224. Springer (2014)
5. Heafield, K., Pouzyrevsky, I., Clark, J.H., Koehn, P.: Scalable modified Kneser-Ney language model estimation. In: Proceedings of the 51st Annual Meeting of the Association for Computational Linguistics, pp. 690–696, Sofia, Bulgaria (August 2013). http://kheafield.com/professional/edinburgh/estimate_paper.pdf
6. Banchs, R.E., Li, H.: Iris: a chat-oriented dialogue system based on the vector space model. In: Proceedings of the ACL 2012 System Demonstrations, pp. 37–42, ACL '12. Association for Computational Linguistics, Stroudsburg, PA, USA (2012). http://dl.acm.org/citation.cfm?id=2390470.2390477
7. Young, S., Gašić, M., Keizer, S., Mairesse, F., Schatzmann, J., Thomson, B., Yu, K.: The hidden information state model: a practical framework for POMDP-based spoken dialogue management. Comput. Speech Lang. **24**(2), 150–174 (2010)
8. Chen, D., Manning, C.D.: A fast and accurate dependency parser using neural networks. In: Proceedings of the 2014 Conference on Empirical Methods in Natural Language Processing (EMNLP), pp. 740–750 (2014)

Part V
Evaluation of Human–Robot Dialogue in Social Robotics

Is Spoken Language All-or-Nothing? Implications for Future Speech-Based Human-Machine Interaction

Roger K. Moore

Abstract Recent years have seen significant market penetration for voice-based personal assistants such as Apple's *Siri*. However, despite this success, user take-up is frustratingly low. This article argues that there is a *habitability gap* caused by the inevitable mismatch between the capabilities and expectations of human users and the features and benefits provided by contemporary technology. Suggestions are made as to how such problems might be mitigated, but a more worrisome question emerges: "*is spoken language all-or-nothing*"? The answer, based on contemporary views on the special nature of (spoken) language, is that there may indeed be a fundamental limit to the interaction that can take place between mismatched interlocutors (such as humans and machines). However, it is concluded that interactions between native and non-native speakers, or between adults and children, or even between humans and dogs, might provide critical inspiration for the design of future speech-based human-machine interaction.

Keywords Spoken language · Habitability gap · Human-machine interaction

1 Introduction

The release in 2011 of *Siri*, Apple's voice-based personal assistant for the iPhone, signalled a step change in the public perception of spoken language technology. For the first time, a significant number of everyday users were exposed to the possibility of using their voice to enter information, navigate applications or pose questions—all by speaking to their mobile device. Of course, voice dictation software had been publicly available since the release of *Dragon Naturally Speaking* in 1997, but such technology only found success in niche market areas for document creation (by users who could

R.K. Moore (✉)
Speech and Hearing Research Group, Department of Computer Science,
University of Sheffield, Regent Court, 211 Portobello, Sheffield S1 4DP, UK
e-mail: r.k.moore@sheffield.ac.uk
URL: http://www.dcs.shef.ac.uk/~roger/

K. Jokinen and G. Wilcock (eds.), *Dialogues with Social Robots*,
Lecture Notes in Electrical Engineering 427,
DOI 10.1007/978-981-10-2585-3_22

not or would not type). In contrast, *Siri* appeared to offer a more general-purpose interface that thrust the potential benefits of automated speech-based interaction into the forefront of the public's imagination. By combining automatic speech recognition and speech synthesis with natural language processing and dialogue management, *Siri* promoted the possibility of a more *conversational* interaction between users and smart devices. As a result, competitors such as *Google Now* and Microsoft's *Cortana* soon followed.[1]

Of course, it is well established that, while voice-based personal assistants such as *Siri* are now very familiar to the majority of mobile device users, their practical value is still in doubt. This is evidenced by the preponderance of videos on *YouTube*™ that depict humorous rather than practical uses; it seems that people give such systems a try, play around with them for a short while and then go back to their more familiar ways of doing things. Indeed, this has been confirmed by a recent survey of users from around the world which showed that only 13 % of the respondents used a facility such as *Siri* every day, whereas 46 % had tried it once and then given up (citing inaccuracy and a lack of privacy as key reasons for abandoning it) [2].

This lack of serious take-up of voice-based personal assistants could be seen as the inevitable teething problems of a new(ish) technology, or it could be evidence of something more deep-seated. This article addresses these issues, and attempts to tease out some of the overlooked features of spoken language that might have a bearing on the success or failure of voice-based human-machine interaction. In particular, attention is drawn to the inevitable *mismatch* between the capabilities and expectations of human users and the features and benefits provided by contemporary technical solutions. Some suggestions are made as to how such problems might be mitigated, but a more worrisome question emerges: "*is spoken language all-or-nothing*"?

2 The Nature of the Problem

There are many challenges facing the development of effective voice-based human-machine interaction [3, 4]. As the technology has matured, so the applications that are able to be supported have grown in depth and complexity (see Fig. 1). From the earliest military *Command and Control Systems* to contemporary commercial *Interactive Voice Response (IVR) Systems* and the latest *Voice-Enabled Personal Assistants* (such as *Siri*), the variety of human accents, competing signals in the acoustic environment and the complexity of the application scenario have always presented significant barriers to practical usage. Considerable progress has been made in all of the core technologies, particularly following the emergence of the data-driven stochastic modelling paradigm [5] (now supplemented by *deep learning* [6]) as a key driver in pushing regularly benchmarked performance in a positive

[1] See [1] for a comprehensive review of the history of speech technology R&D up to, and including, the release of *Siri*.

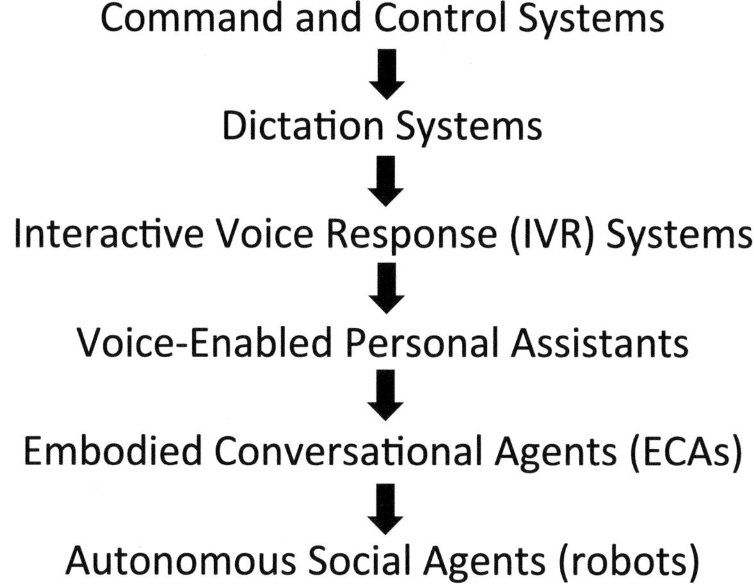

Fig. 1 The evolution of spoken language technology applications from early military *Command and Control Systems* to future *Autonomous Social Agents* (robots)

direction. Yet, as we have seen, usage remains a serious issue; not only does a speech interface compete with very effective non-speech GUIs [7], but people have a natural aversion to talking to machines in public spaces [2]. As Nass and Brave stated in their seminal book *Wired for Speech* [8]: "*voice interfaces have become notorious for fostering frustration and failure*" (p. 6).

These problems become magnified as the field moves forward to developing voice-based interaction with *Embodied Conversational Agents (ECAs)* and *Autonomous Social Agents* (robots). In these futuristic scenarios, it is assumed that spoken language will provide a "natural" conversational interface between human beings and so-called *intelligent* systems. However, there many additional challenges which need to be overcome in order to address such a requirement …

> We need to move from developing robots that simply talk and listen to evolving intelligent communicative machines that are capable of truly understanding human behaviour, and this means that we need to look beyond speech, beyond words, beyond meaning, beyond communication, beyond dialogue and beyond one-off interactions [9] (p. 321).

Of these, a perennial problem seems to be how to evolve the complexity of voice-based interfaces from simple structured dialogues to more flexible conversational designs without confusing the user [10–12]. Indeed, it has been known for some time that there appears to be a non-linear relationship between *flexibility* and *usability* [13]—see Fig. 2. As flexibility increases with advancing technology, so usability increases until users no longer know what they can and cannot say, at which point usability tumbles and interaction falls apart.

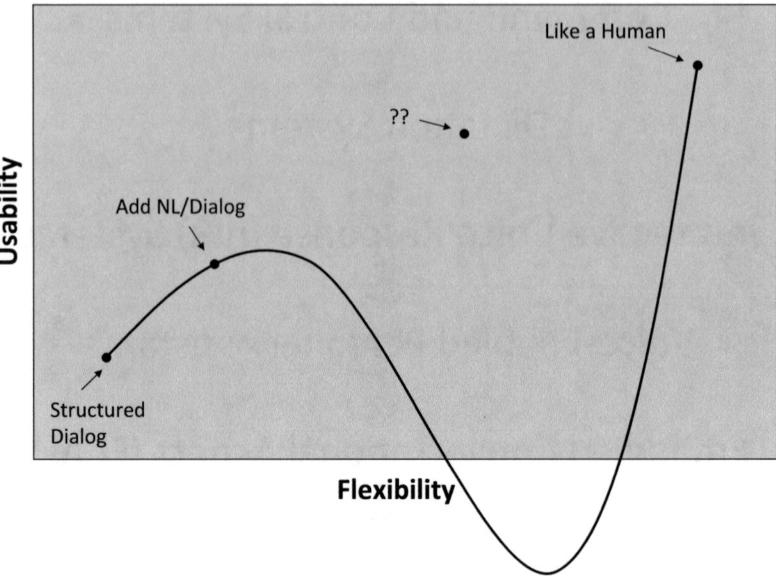

Fig. 2 Illustration of the consequences of increasing the flexibility of spoken language dialogue systems; increasing flexibility can lead to a *habitability gap* where usability drops catastrophically (reproduced, with permission, from Phillips [13]). This means that it is surprisingly difficult to deliver a technology corresponding to the point marked '??'. *Siri* corresponds to the point marked 'Add NL/Dialog'

2.1 The "Habitability Gap"

Progress is being made in this area: for example, by providing targeted help to users [14–16] and by replacing the traditional notion of turn-taking with a more fluid interaction based on *incremental processing* [17, 18]. Likewise, simple slot-filling approaches to language understanding and generation are being replaced by sophisticated statistical methods for estimating dialogue states and optimal next moves [19, 20]. Nevertheless, it is still the case that there is a *habitability gap* of the form illustrated in Fig. 2.

In fact, the shape of the curve illustrated in Fig. 2 is virtually identical to the famous *Uncanny Valley effect* [21] in which a near human-looking artefact (such as a humanoid robot) can trigger feelings of eeriness and repulsion in an observer; as *human likeness* increases, so *affinity* increases until a point where artefacts start to appear creepy and affinity goes negative. A wide variety of explanations have been suggested for this non-linear relationship but, to date, there is only one quantitative model [22], and this is founded on the combined effect of categorical perception and mismatched perceptual cues giving rise to a form of *perceptual tension*. The implication of this model is that uncanniness—and hence, habitability—can be avoided if care is taken to align how an autonomous agent looks, sounds and behaves [9, 23].

In other words, if a speech-enabled agent is to converse successfully with a human being, it should make clear its interactional *affordances* [24, 25].

This analysis leads to an important implication—since a spoken language system consists of a number of different components, each of which possesses a certain level of technical capability, then in order to be coherent (and hence *usable*), the design of the overall system needs to be aligned to the component with the *lowest* level of performance. For example, giving an automated personal assistant a natural human voice is a recipe for user confusion in the (normal) situation where the other speech technology components are limited in their abilities. In other words, in order to maximise the effectiveness of the interaction, **a speech-enabled robot should have a robot voice**. As Balentine succinctly puts it [26]: "*It's better to be a good machine than a bad person*"! This is an unpopular result,[2] but there is evidence of its effectiveness [27], and it clearly has implications for contemporary voice-based personal assistants such as *Siri*, *Google Now* and *Cortana* which employ very humanlike voices.[3]

Of course, some might claim that the habitability problem only manifests itself in applications where task-completion is a critical measure of success. The suggestion would be that the situation might be different for applications in domains such as social robots, education or games in which the emphasis would be more on the spoken interaction itself. However, the argument presented in this paper is not concerned with the nature of the interaction, rather it questions whether such speech-based interaction can be sustained without access to the notion of *full* language.

2.2 *Half a Language?*

So far, so good—as component technologies improve, so the flexibility of the overall system would increase, and as long as the capabilities of the individual components are aligned, it should be possible to avoid falling into the habitability gap.

However, sending mixed messages about the capabilities of a spoken language system is only one part of the story; even if a speech-based autonomous social agent looks, sounds and behaves in a coherent way, will users actually be able to engage in conversational interaction if the overall capability is *less* than that normally enjoyed by a human being? What does it mean for a language-based system to be compromised in some way? How can users know what they may and may not say [15, 29], or even if this is the right question? Is there such a thing as *half* a language and, if so, is it habitable? Indeed, what is language anyway?

---

[2]It is often argued that such an approach is unimportant as users will habituate. However, habituation only occurs after sustained exposure, and a key issue here is how to increase the effectiveness of first encounters (since that has a direct impact on the likelihood of further usage).

[3]Interestingly, these ideas do appear to be having some impact on the design of contemporary autonomous social agents such as *Jibo* (which has a childlike and mildly robotic voice) [28].

3 What Is Language?

Unfortunately there is no space here to review the extensive and, at times, controversial history of the scientific study of language, or of the richness and variety of its spoken (and gestural) forms. Suffice to say that human beings have evolved a prolific system of (primary vocal) interactive behaviours that is vastly superior to that enjoyed by any other animal [30–34]. As has been said a number of times …

> Spoken language is the most sophisticated behaviour of the most complex organism in the known universe [35].

The complexity and sophistication of (spoken) language tends to be masked by the apparent ease with which we, as human beings, use it. As a consequence, engineered solutions are often dominated by a somewhat naïve perspective involving the coding and decoding of messages passing from one brain (the sender) to another brain (the receiver). In reality, *languaging* is better viewed as an emergent property of the dynamic coupling between *cognitive unities* that serves to facilitate distributed sense-making through cooperative behaviours and, thereby, social structure [36–40]. Furthermore, the contemporary view is that language is based on the co-evolution of two key traits—*ostensive-inferential* communication and *recursive mind-reading* (including 'Theory-of-Mind') [41–43]—and that abstract (mental) meaning is grounded in the concrete (physical) world through *metaphor* [44, 45].

These modern perspectives on language not only place strong emphasis on *pragmatics* [46], but they are also founded on an implicit assumption that interlocutors are conspecifics[4] and hence share significant priors. Indeed, evidence suggests that some animals draw on representations of their own abilities (expressed as predictive models [47]) in order to interpret the behaviours of others [48, 49]. For human beings, this is thought to be a key enabler for efficient recursive mind-reading and hence for language [50, 51].

Several of these advanced concepts may be usefully expressed in pictographic form [52]—see Fig. 3.

So now we arrive at an interesting position; if (spoken) language interaction between human beings is grounded through shared experiences, representations and priors, to what extent is it possible to construct a technology that is intended to replace one of the participants? For example, if one of the interlocutors illustrated in Fig. 3 is replaced by a cognitive robot (as in Fig. 4), then there will be an inevitable mismatch between the capabilities of the two partners, and coupled ostensive recursive mind-reading (i.e. full language) cannot emerge.

Could it be that there is a fundamental limit to the language-based interaction that can take place between *unequal* partners—between humans and machines? Indeed, returning to the question posed in Sect. 2.2 "*Is there such a thing as half a language?*", the answer seems to be "*no*"; spoken language does appear to be all-or-nothing …

> The assumption of continuity between a fully coded communication system at one end, and language at the other, is simply not justified [41] (p. 46).

---

[4]Members of the same species.

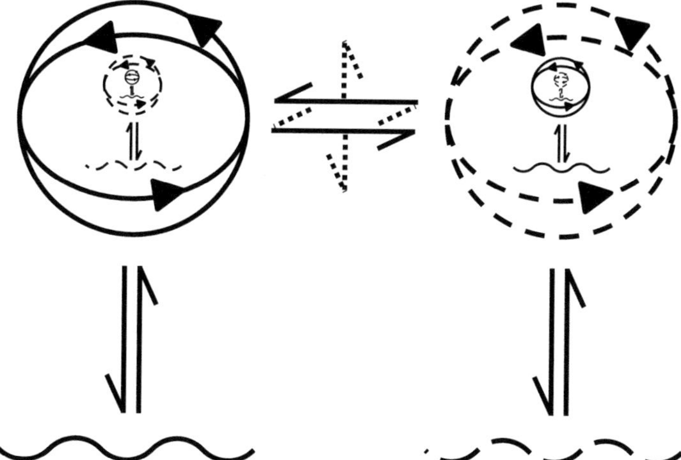

Fig. 3 Pictographic representation of language-based coupling (dialogue) between two human interlocutors [52]. One interlocutor (and its environment) is depicted using solid lines and the other interlocutor (and its environment) is depicted using broken lines. As can be seen, communicative interaction is founded on two-way ostensive recursive mind-reading (including mutual Theory-of-Mind)

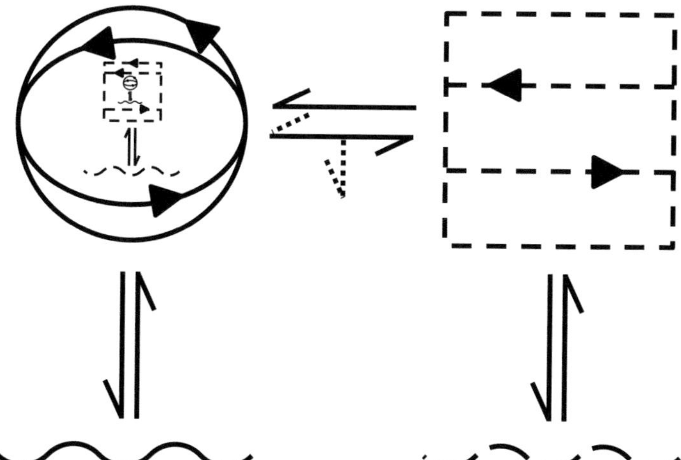

Fig. 4 Pictographic representation of coupling between a human being (on the *left*) and a cognitive robot (on the *right*). The robot lacks the capability of ostensive recursive mind-reading (it has no Theory-of-Mind), so the interaction is inevitably constrained

4 The Way Forward?

The story thus far provides a compelling explanation of the less-than-satisfactory experiences enjoyed by existing users of speech-enabled systems and identifies the source of the *habitability gap* outlined in Sect. 2.1. It would appear that, due to the gross mismatch between their respective priors, it might be impossible to create an automated system that would be capable of a sustained and productive language-based interaction with a human being (except in narrow specialised domains involving experienced users). The vision of constructing a general-purpose voice-enabled autonomous social agent may be fundamentally flawed—the equivalent of trying to build a vehicle that travels faster than light!

However, before we give up all hope, it is important to note that there are situations where voice-based interaction between mismatched partners *is* successful—but these are very different from the scenarios that are usually considered when designing current speech-based systems. For example, human beings regularly engage in vocal interaction with members of a different cultural and/or linguistic and/or generational background.[5] In such cases, all participants dynamically adjust many aspects of their behaviour—the clarity of their pronunciation, their choice of words and syntax, their style of delivery, etc.—all of which may be controlled by the perceived effectiveness of the interaction (that is, using *feedback* in a coupled system). Indeed, a particularly good example of such accommodation between mismatched interlocutors is the different way in which caregivers talk to young children (termed "*parentese*") [53]. Maybe these same principles should be applied to speech-based human-machine interaction? Indeed, perhaps we should be explicitly studying the particular adaptations that human beings make when attempting to converse with autonomous social agents—a new variety of spoken language that could be appropriately termed "*robotese*".[6]

Of course, these scenarios all involve spoken interaction between one human being and another, hence in reality there is a huge overlap of priors in terms of bodily morphology, environmental context and cognitive structure, as well as learnt social and cultural norms. Arguably the largest mismatch arises between an adult and a very young child, yet this is still interaction between members of the same species. A more extreme mismatch exists between non-conspecifics; for example, between humans and animals. However, it is interesting to note that our nearest relatives—the apes—do not have language, and this seems to be because they do not have the key precursor to language: ostensive communication (apes do not seem to understand pointing gestures) [41].

Interestingly, one animal—the domestic dog—appears to excel in ostensive communication and, as a consequence, dogs are able to engage in very productive spoken language interaction with human partners (albeit one-sided and somewhat limited in scope) [41, 55]. Spoken human-dog interaction may thus be a potentially important example of a heavily mismatched yet highly effective cooperative configuration that might usefully inform spoken human-robot interaction in hitherto unanticipated ways.

[5]Interestingly, Nass and Brave [8] noted that people speak to poor automatic speech recognition systems as if they were non-native listeners.

[6]Unfortunately, this term has already been coined to refer to a robot's natural language abilities in robot-robot and robot-human communication [54].

5 Final Remarks

This article has argued that there is a fundamental *habitability* problem facing contemporary spoken language systems, particularly as they penetrate the mass market and attempt to provide a general-purpose voice-based interface between human users and (so-called) intelligent systems. It has been suggested that the source of the difficulty in configuring genuinely usable systems is twofold: first, the need to align the visual, vocal and behavioural affordances of the system, and second, the need to overcome the huge mismatch between the capabilities and expectations of a human being and the features and benefits offered by even the most advanced autonomous social agent. This led to the preliminary conclusion that spoken language may indeed be all-or-nothing.

Finally, and on a positive note, it was observed that there are situations where successful spoken language interaction can take place between mismatched interlocutors (such as between native and non-native speakers, or between an adult and a child, or even between a human being and a dog). It is thus concluded that these scenarios might provide critical inspiration for the design of future speech-based human-machine interaction.

Acknowledgements This work was supported by the European Commission [grant numbers EU-FP6-507422, EU-FP6-034434, EU-FP7-231868 and EU-FP7-611971], and the UK Engineering and Physical Sciences Research Council [grant number EP/I013512/1].

References

1. Pieraccini, R.: The Voice in the Machine. MIT Press, Cambridge (2012)
2. Liao, S.-H.: Awareness and Usage of Speech Technology. Masters thesis, Dept. Computer Science, University of Sheffield (2015)
3. Deng, L., Huang, X.: Challenges in adopting speech recognition. Commun. ACM **47**(1), 69–75 (2004)
4. Minker, W., Pittermann, J., Pittermann, A., Strauß, P.-M., Bühler, D.: Challenges in speech-based human-computer interfaces. Int. J. Speech Technol. **10**(2–3), 109–119 (2007)
5. Gales, M., Young, S.J.: The application of hidden Markov models in speech recognition. Found. Trends Signal Process. **1**(3), 195–304 (2007)
6. Hinton, G., Deng, L., Yu, D., Dahl, G.E., Mohamed, A., Jaitly, N., Senior, A., Vanhoucke, V., Nguyen, P., Sainath, T.N., Kingsbury, B.: Deep neural networks for acoustic modeling in speech recognition: the shared views of four research groups. IEEE Signal Process. Mag. (2012)
7. Moore, R.K.: Modelling data entry rates for ASR and alternative input methods. In: Proceedings of the INTERSPEECH-ICSLP, Jeju, Korea (2004)
8. Nass, C., Brave, S.: Wired for Speech: How Voice Activates and Advances the Human-computer Relationship. MIT Press, Cambridge (2005)
9. Moore, R.K.: From talking and listening robots to intelligent communicative machines. In: Markowitz, J. (ed.) Robots That Talk and Listen, pp. 317–335. De Gruyter, Boston (2015)
10. Bernsen, N.O., Dybkjaer, H., Dybkjaer, L.: Designing Interactive Speech Systems: From First Ideas to User Testing. Springer, London (1998)
11. McTear, M.F.: Spoken Dialogue Technology: Towards the Conversational User Interface. Springer, London (2004)

12. Lopez Cozar Delgado, R.: Spoken, Multilingual and Multimodal Dialogue Systems: Development and Assessment. Wiley (2005)
13. Philips, M.: Applications of spoken language technology and systems. In: Gilbert, M., Ney, H. (eds.) IEEE/ACL Workshop on Spoken Language Technology (SLT) (2006)
14. Tomko, S., Harris, T.K., Toth, A., Sanders, J., Rudnicky, A., Rosenfeld, R.: Towards efficient human machine speech communication. ACM Trans. Speech Lang. Process. **2**(1), 1–27 (2005)
15. Tomko, S.L.: Improving User Interaction with Spoken Dialog Systems via Shaping. Ph.D. Thesis, Carnegie Mellon University (2006)
16. Komatani, K., Fukubayashi, Y., Ogata, T., Okuno, H.G.: Introducing utterance verification in spoken dialogue system to improve dynamic Help generation for novice users. In: Proceedings of the 8th SIGdial Workshop on Discourse and Dialogue, pp. 202–205 (2007)
17. Schlangen, D., Skantze, G.: A general, abstract model of incremental dialogue processing. In: Proceedings of the 12th Conference of the European Chapter of the Association for Computational Linguistics (EACL-09), Athens, Greece (2009)
18. Hastie, H., Lemon, O., Dethlefs, N.: Incremental spoken dialogue systems: tools and data. In: Proceedings of NAACL-HLT Workshop on Future Directions and Needs in the Spoken Dialog Community, pp. 15–16, Montreal, Canada (2012)
19. Williams, J.D., Young, S.J.: Partially observable Markov decision processes for spoken dialog systems. Comput. Speech Lang. **21**(2), 231–422 (2007)
20. Gašić, M., Breslin, C., Henderson, M., Kim, D., Szummer, M., Thomson, B., Tsiakoulis, P., Young, S.J.: POMDP-based dialogue manager adaptation to extended domains. In: Proceedings of 14th SIGdial Meeting on Discourse and Dialogue, pp. 214–222, Metz, France (2013)
21. Mori, M.: Bukimi no tani (the uncanny valley). Energy **7**, 33–35 (1970)
22. Moore, R.K.: A Bayesian explanation of the "Uncanny Valley" effect and related psychological phenomena. Nat. Sci. Rep. **2**(864) (2012)
23. Moore, R.K., Maier, V.: Visual, vocal and behavioural affordances: some effects of consistency. In: Proceedings of the 5th International Conference on Cognitive Systems (CogSys 2012), Vienna (2012)
24. Gibson, J.J.: The theory of affordances. In: Shaw, R., Bransford, J. (eds.) Perceiving, Acting, and Knowing: Toward an Ecological Psychology, pp. 67–82. Lawrence Erlbaum, Hillsdale (1977)
25. Worgan, S., Moore, R.K.: Speech as the perception of affordances. Ecolog. Psychol. **22**(4), 327–343 (2010)
26. Balentine, B.: It's Better to Be a Good Machine Than a Bad Person: Speech Recognition and Other Exotic User Interfaces at the Twilight of the Jetsonian Age. ICMI Press, Annapolis (2007)
27. Moore, R.K., Morris, A.: Experiences collecting genuine spoken enquiries using WOZ techniques. In: Proceedings of the 5th DARPA Workshop on Speech and Natural Language, New York (1992)
28. Jibo: The World's First Social Robot for the Home. https://www.jibo.com
29. Jokinen, K., Hurtig, T.: User expectations and real experience on a multimodal interactive system. In: Proceedings of the INTERSPEECH-ICSLP Ninth International Conference on Spoken Language Processing, Pittsburgh, PA, USA (2006)
30. Gardiner, A.H.: The Theory of Speech and Language. Oxford University Press, Oxford (1932)
31. Bickerton, D.: Language and Human Behavior. University of Washington Press, Seattle (1995)
32. Hauser, M.D.: The Evolution of Communication. The MIT Press (1997)
33. Hauser, M.D., Chomsky, N., Fitch, W.T.: The faculty of language: what is it, who has it, and how did it evolve? Science **298**, 1569–1579 (2002)
34. Everett, D.: Language: The Cultural Tool. Profile Books, London (2012)
35. Moore, R.K.: Spoken language processing: piecing together the puzzle. Speech Commun. **49**(5), 418–435 (2007)
36. Maturana, H.R., Varela, F.J.: The Tree of Knowledge: The Biological Roots of Human Understanding. New Science Library/Shambhala Publications, Boston (1987)

37. Cummins, F.: Voice, (inter-)subjectivity, and real time recurrent interaction. Front. Psychol. **5**, 760 (2014)
38. Bickhard, M.H.: Language as an interaction system. New Ideas Psychol. **25**(2), 171–187 (2007)
39. Cowley, S.J. (ed.): Distributed Language. John Benjamins Publishing Company (2011)
40. Fusaroli, R., Raczaszek-Leonardi, J., Tylén, K.: Dialog as interpersonal synergy. New Ideas Psychol. **32**, 147–157 (2014)
41. Scott-Phillips, T.: Speaking Our Minds: Why Human Communication Is Different, and How Language Evolved to Make It Special. Palgrave MacMillan (2015)
42. Baron-Cohen, S.: Evolution of a theory of mind? In: Corballis, M., Lea, S. (eds.) The Descent of Mind: Psychological Perspectives on Hominid Evolution. Oxford University Press (1999)
43. Malle, B.F.: The relation between language and theory of mind in development and evolution. In: Givón, T., Malle, B.F. (eds.) The Evolution of Language out of Pre-Language, pp. 265–284. Benjamins, Amsterdam (2002)
44. Lakoff, G., Johnson, M.: Metaphors We Live By. University of Chicago Press, Chicago (1980)
45. Feldman, J.A.: From Molecules to Metaphor: A Neural Theory of Language. Bradford Books (2008)
46. Levinson, S.C.: Pragmatics. Cambridge University Press, Cambridge (1983)
47. Friston, K., Kiebel, S.: Predictive coding under the free-energy principle. Phil. Trans. R. Soc. B **364**(1521), 1211–1221 (2009)
48. Rizzolatti, G., Craighero, L.: The mirror-neuron system. Annu. Rev. Neurosci. **27**, 169–192 (2004)
49. Wilson, M., Knoblich, G.: The case for motor involvement in perceiving conspecifics. Psychol. Bull. **131**(3), 460–473 (2005)
50. Pickering, M.J., Garrod, S.: Do people use language production to make predictions during comprehension? Trends Cogn. Sci. **11**(3), 105–110 (2007)
51. Garrod, S., Gambi, C., Pickering, M.J.: Prediction at all levels: forward model predictions can enhance comprehension. Lang. Cogn. Neurosci. **29**(1), 46–48 (2013)
52. Moore, R.K.: Introducing a pictographic language for envisioning a rich variety of enactive systems with different degrees of complexity. Int. J. Adv. Robot. Syst. **13**(74) (2016)
53. Fernald, A.: Four-month-old infants prefer to listen to Motherese. Infant Behav. Dev. **8**, 181–195 (1985)
54. Matson, E.T., Taylor, J., Raskin, V., Min, B.-C., Wilson, E.C.: A natural language exchange model for enabling human, agent, robot and machine interaction. In: Proceedings of the 5th International Conference on Automation, Robotics and Applications, pp. 340–345. IEEE (2011)
55. Serpell, J.: The Domestic Dog: Its Evolution, Behaviour and Interactions with People. Cambridge University Press (1995)

Toward a Context-Based Approach to Assess Engagement in Human-Robot Social Interaction

Laurence Devillers and Guillaume Dubuisson Duplessis

Abstract This article addresses the issue of evaluating Human-Robot spoken interactions in a social context by considering the engagement of the human participant. Our work regards the Communication Accommodation Theory (CAT) as a promising paradigm to consider for engagement, by its study of macro- and micro-contexts influencing the behaviour of dialogue participants (DPs), and by its effort in depicting the accommodation process underlying the behaviour of DPs. We draw links between the accommodation process described in this theory and human engagement that could be fruitfully used to evaluate Human-Robot social interactions. We present our goal which is to take into account a model of dialogue activities providing a convenient local interpretation context to assess human contributions (involving verbal and nonverbal channels) along with CAT to assess Human-Robot social interaction.

Keywords Human-robot interaction · Social dialogue · Communication accommodation theory · Engagement

1 Introduction

This article addresses the issue of evaluating Human-Robot spoken interactions in a social context. In order to evaluate dialogue systems, subjective and objective metrics have been proposed. For example, the PARADISE framework allows designers to

L. Devillers (✉) · G. Dubuisson Duplessis
LIMSI, CNRS, Université Paris-Saclay, 91405 Orsay, France
e-mail: devil@limsi.fr

G. Dubuisson Duplessis
e-mail: gdubuisson@limsi.fr

L. Devillers
Sorbonne Universités, Université Paris-Sorbonne, 75005 Paris, France

K. Jokinen and G. Wilcock (eds.), *Dialogues with Social Robots*,
Lecture Notes in Electrical Engineering 427,
DOI 10.1007/978-981-10-2585-3_23

predict user satisfaction from a linear combination of objective metrics such as mean recognition score and task completion [1]. On the other hand, developers of spoken dialog systems tend to use performance measures in terms of the "ability of a system to provide the function it has been designed for" [2] to assess their systems. Subjective experiments can also be carried out but they are expensive, time-consuming and not replicable. However, the big picture has been missing: there has been no clear view of how these methods relate to each other, and how they might include task and social interaction. Improvements in the quality, usability and acceptability of spoken dialogue systems used for example in social robotics can be facilitated by better evaluation methods.

Talks during social interactions naturally involve the exchange of spontaneous propositional content but also and perhaps more importantly the expression of interpersonal relationships, as well as displays of emotion, affect, interest, etc. Such social interaction with a robot requires it to possess the ability to detect and interpret social language as well as the ability to represent parts of the complex human social behaviour. Cognitive decisions have to be used for reasoning on the strategy of the dialogue and deciding social behaviours (humour, compassion, white lies, etc.) taking into account the user profile and contextual information. Research challenges include the evaluation of such systems and the definition of various metrics that could be used, like the measure of social engagement with the user. From our perspective, engagement in dialogue with a machine should not be limited to error rates. Engagement is a complex notion linked to many social interaction levels. For instance, the impact of errors made by the machine can differ depending upon the human partner (e.g., elderly people or children), or as a function of the role of the robot (e.g., companion or supervisor). In this article, we argue that the communication accommodation theory (CAT) is a promising paradigm to understand how the communicative verbal and nonverbal behaviour of dialogue participants could contribute to engagement measures in a social context. We draw links between the accommodation process described in this theory and human engagement that could be fruitfully used to evaluate Human-Robot social interactions.

Section 2 presents work related to the notion of "engagement" in HRI and describes the "Communication Accommodation Theory (CAT)". Section 3 discusses the contribution of CAT to the notion of engagement for HRI. We take advantage of the CAT to identify multi-level expectations in the human communicative behaviour that could be fruitfully used to assess the engagement of the human. Section 4 describes our current work which is directed towards a context-based approach to quantify engagement in HRI based on CAT. Section 5 concludes this article and presents future work.

2 Related Work

2.1 *Engagement in HRI*

The term "engagement" admits several definitions in the HRI literature and is recognised as a complex concept [3]. Corrigan et al. [4] propose to discern three definitions relevant to HRI: "task engagement" (involving a human and an explicit task), "social engagement" (involving a human and a robot) and "social-task engagement" (involving a human, a robot and an explicit task). Social engagement has been defined as "the process by which two (or more) participants establish, maintain and end their perceived connection during interactions they jointly undertake" [5, 6]. Engagement process involves nonverbal and verbal behaviours [5], as well as low-level processes (such as behaviour synchrony, mimetics) and high-level cognitive processes (such as answering a riddle).

Based on the previous definition of social engagement, Bohus et al. [7] have proposed a computational model for managing engagement decisions in situated, multi-party, spoken dialog systems. This latter involves the sensing of a binary engagement state (engaged/not-engaged), engagement intentions [8], a set of four engagement actions (e.g., engage, disengage), as well as high-level engagement decisions from the system. Rich et al. [6] have identified four types of connection event (*directed gaze*, *mutual facial gaze*, *delay in adjacency pair*, and *backchannel*) involved in the computation of statistics on the overall engagement process. Salam et al. [9] relate engagement to both the context (e.g., competitive, informative, educative, collaborative) and emotional and mental states from the human (e.g., listening, thinking, concentrating). They emphasise the importance of context in stating the hypothesis that "the definition of engagement varies in function of the context of the interaction".

2.2 *Communication Accommodation Theory (CAT)*

CAT is a theory of communication outlining the importance of both intergroup and interpersonal factors in predicting and understanding intergroup interactions [10]. It is chiefly interested in the links between language, context and identity.

This theory emphasises the relational function of communication, i.e. the fact that communication involves the management of interpersonal and intergroup relationships as well as the negotiation of personal and social identities. It argues that dialogue participants (DPs) adapt their communicative verbal and nonverbal behaviours via an accommodation process in which they can seek to be more alike (accommodation) or to be more distinct (nonaccommodation). CAT describes 3 main accommodative strategies involved in this process: (i) convergence, in which DPs adapt their communicative behaviour so as to become more similar to their interlocutor's behaviour (e.g., pause, utterance length, vocal intensity, gesture, gaze); (ii) divergence, in which DPs seek to accentuate differences with their interlocutor's behaviour;

(iii) maintenance, in which a DP persists in his original communicative behaviour independently of the communicative behaviour of his interlocutor. Convergence usually signals liking and closeness between DPs while divergence and maintenance are often involved in the emphasis of social group distinctiveness (e.g., marking a degree of social distance). Although convergence is clearly linked to the notion of accommodation, it can be involved in nonaccommodation. Indeed, this latter takes several forms: (i) counter-accommodation, in which DPs try to maximise the differences with their interlocutor's behaviour (ostensible negative and hostile behaviour), (ii) under-accommodation, in which DPs maintain their behaviour despite the usefulness to change it for their interlocutor (e.g., continuing speaking in a low voice volume despite the hearing difficulties of the interlocutor), and (iii) over-accommodation, in which DPs use convergence but in an inappropriate way (e.g., speaking with a voice volume beyond the optimal level to someone with hearing impairment, patronising with elderly people). DPs have expectations regarding optimal levels of accommodation based on stereotypes about their interlocutor, and by taking into consideration the social and situational norms of their interaction.

Communicative behaviours of DPs happen in a dynamic and opportunistic environment. CAT outlines 3 components of the environment in which the interaction takes place: (i) the sociohistorical context (including intergroup history, norms specifying with whom, when and how members should interact, social equality of inequality), (ii) the initial orientation of the DPs (including the interpersonal history ranging from no previous interaction to long-term relationship, stereotypes), and (iii) the immediate interaction situation (including norms, values). CAT theory informs that accommodative or non-accommodative strategies adopted by DPs are influenced (among other things) by these elements from the context, their motives (e.g., facilitating a clear communication, maintaining their identities) and the perception and evaluation of other participant's behaviour.

3 CAT and Engagement for HRI

3.1 Contributions of CAT

CAT presents a theoretical framework to study aspects of the "perceived connection" evoked in the definition of engagement. In particular, CAT offers a clear study of macro- and micro-contexts (interpersonal, intergroup, situational, societal, sociohistorical, etc.) that influence the behaviour of DPs in social interaction. In addition, CAT describes a sound accommodation process and accommodative strategies underlying the communicative behaviour of DPs. The accommodation process has tight links with social engagement in HRI. On the one hand, accommodation of the robot contributions by the human participant is a strong indicator of engagement. Ideally, accommodation from the human could be compared to the expected optimal accommodation level to discern various degrees of engagement from the human. On the other hand, nonaccommodation from the human is a clear evidence of disengagement. Studying the communicative behaviour of the human in terms of counter-accommodation, under-accommodation and over-accommodation could offer a clear frame to discern various disengagement levels.

From our perspective, the consideration of the accommodation process described by CAT in the assessment of engagement in HRI highlights two under-looked but important aspects, namely: (i) the necessity to take into account the behaviour of the robot when evaluating the communicative behaviour of the human, and (ii) the necessity to take into account the dynamic of the interaction, and the changes in the communicative behaviour of both DPs.

3.2 Multi-level Expectations

In HRI, the communicative behaviour of the human can be evaluated through expectations at different levels. The main type of contribution expected from the human in a spoken interaction with a robot is a dialogue act [11] specified by a communicative function, a semantic content and an addressee (e.g., an "answer" communicative function is expected after a question). In light of the CAT, we complete these high-level expectations with other ones related to the communicative behaviour of the human participant that lead to the production of a dialogue act (or more). They could include: (i) linguistic cues (e.g., specific lexical items), (ii) paralinguistic cues [12] (e.g., dominant expressed emotion, filled pause, utterance length, voice quality, speaking style, vocal intensity, presence/absence of bursts such as laughter), (iii) interactional cues (e.g., speech reaction time corresponding to the delay between the end of the turn of the robot and the start of the first following speech segment of the human), (iv) vision and other modalities (e.g., facial expressions such as smile, shared gaze, gestures such as nodding, postures, touch, proxemics), and (v) specific emotional states (e.g., stress, uncertainty) and mental states (e.g., "concentrating", "thinking").

3.3 Influencing Factors

It should be kept in mind that these expectations are influenced by many factors as pointed out by the CAT (cf. Sect. 2.2). First, the communicative behaviour of the human can be influenced by the relationship between the human and the system in terms of: (i) past interactions (shared interaction history involving familiarity, affinity, habits), and (ii) likelihood of future interactions (a once in a lifetime meeting may not involve the same need for accommodation). Next, the interaction occurs in a social environment which involves its own social norms (e.g., regarding how we should address people depending on their age) and stereotypes (e.g., robot and the "Frankeinstein complex") that may affect the communicative behaviour of participants. Then, the behaviour of the human is influenced by its long-term profile, including: (i) his personality (e.g., age, gender, elements from the Five-Factor Model [13]), (ii) interpersonal variability and idiosyncrasies (e.g., hearing impairments, stammer), and (iii) social statuses. Eventually, behaviour of the human is influenced by the immediate interaction, involving: (i) more or less temporary states (e.g., health state, mood), (ii) interaction scenario and the roles of DPs (e.g., teacher/student), (iii) topic (e.g., type of riddle such as absurd, excessive), and (iv) the communicative behaviour of the robot (e.g., a laughter from the robot may be expected to be catching).

4 Towards a Context-Based Approach to Quantify Engagement

In our work, we are considering useful local contexts for spoken and social HRI dedicated to (i) the interpretation and the evaluation of the human behaviour, and (ii) the management and generation of the communicative behaviour of the robot. In that direction, we are investigating a dialogue activity model providing a local interpretation context of the human communicative behaviour which can be fruitfully used to fuse verbal and non-verbal channels [14]. We view H-R social dialogue as a combination of joint activities that are activated and completed by DPs [14]. These activities can be viewed as *joint projects* [15], i.e. a bounded joint activity which can be broken down into an entry, a body, and an exit. In our model (see Fig. 1), a dialogue activity is defined by a *type* (e.g., the "riddle" activity), a *conversational topic* (e.g., a specific riddle), an *initiator* and a *partner*. The initiator may either be the robot or the human. A dialogue activity specifies *expectations* from dialogue participants in terms of *moves* (described Sect. 3.2). Moves are produced by the DPs to progress in the activity. The main type of contribution expected from the human in an activity is a move or dialogue act [11] specified by a communicative function, a semantic content and an addressee (e.g., the utterance "Émile Zola!" in a "riddle" activity realises a dialogue act with a "answer" function and the content "Émile Zola"). In the simplest case, dialogue activities specify expected sequences of moves such as an adjacency pairs with preferred and dispreferred pair parts [16]. In other words, one aspect of this model is dedicated to the interpretation of the

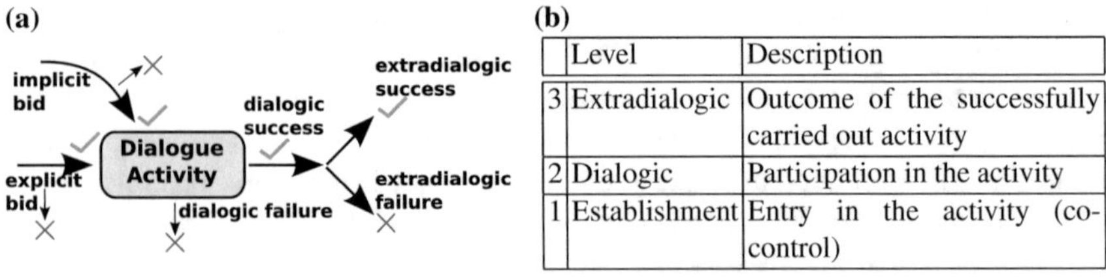

	Level	Description
3	Extradialogic	Outcome of the successfully carried out activity
2	Dialogic	Participation in the activity
1	Establishment	Entry in the activity (co-control)

Fig. 1 Dialogue activity model (from [14]) (**a**), **b** scale of completion levels

Table 1 Excerpt of a dialogue from our corpus (translated from French to English). H = Human, Nao is the robot

	Activity Type	Loc.	Transcription	Audio	Video	Activity Status
1	Greetings	Nao	Hi, I'm Nao. I like to joke and I know riddles.			
2		H		*silence*	*continued attention*	Dialogic failure
3	Dialogic failure recovery	Nao	Well, you do not want to tell me anything,			
4	Riddle		but let me tell you a riddle.			Explicit bid
5		H		*silence*	*smile*	
6		Nao	Who wrote the article "J'accuse?"			
7		H	Ah... I do not know.		*head movement*	
8		Nao	The answer was Émile Zola. (laugh)	laughter		
9		H	*(laugh)*		*laughter*	Dialogic success Extra-dialogic failure
10	Riddle	Nao	I love riddles, let me tell you another one!			Explicit bid
			(...)			

human communicative behaviour as being appropriate or inappropriate considering the current activity. From our perspective, dialogue emerges opportunistically from dialogue activity combinations such as (but not limited to) sequence, embedding and parallel execution.

This model discerns various success and failure statuses relatively to the entry, the body and the exit of the activity. Our model conceptually distinguishes three completion levels forming a scale in which a success at a level allows to pass to the next one, while a failure at a given level prevents from moving on to the next (see Table 1b). The first one is the *establishment of the activity* via an implicit or an explicit mechanism. It captures a part of the effort of the DPs to co-control the dialogue, seen as an opportunistic joint activity [15]. The second one is the *progress of the activity* at the dialogic level, concerned with the participation of DPs in the dialogue activity. The dialogic status reflects whether the DPs have performed their moves in order to get to the end of the activity (success) or not (failure). While previous levels are dedicated to the dialogue activity in itself, the last level takes a "task-oriented perspective" by dealing with the outcome of the activity. Thus, the extradialogic status is the result of a dialogue activity that has been successfully carried out by dialogue participants at the dialogic level.

Table 1 presents an example of dialogue that we collected and analysed according to our dialogue activity model. It illustrates the chaining of three activities ("greetings" followed by two "riddle" ones). It starts with a "greeting" activity, implicitly established by the robot (turn 1). The human participant shows continued attention whereas he is expected to return the greetings or to explicitly reject the activity. Thus, the activity turns out to be a dialogic failure, notified by the robot (turn 3). Next, Nao explicitly introduces a "riddle" activity, accepted by a positive sign from the human (turn 5). This activity progresses as expected (turns 6–9), and terminates on a dialogic success and an extradialogic failure (the answer to the riddle has not be found). Then, Nao introduces another riddle (turn 10).

We believe that our dialogue activity model provides a convenient local context to take into account the multi-level expectations described in Sect. 3.2. The implementation of these expectations obviously depends on the perceptual and computational capabilities of the system, which could exploit only a subset of these cues. Depending on the interaction context, we imagine that the system could select the most salient cues to assess whether the expectations are met or not (e.g., in terms of robustness). For instance, an interaction in a very noisy environment could lead the robot to rely more on visual cues than audio ones.

A first prototype based on this model has been implemented in the context of the Joker project [17]. It features a system-directed entertaining interaction dialogue that includes the telling of riddles and other humorous contributions, synthesised through the Nao robot via speech, laughter, movement and eye colour variations. Currently, our system is based on the paralinguistic analysis of the audio channel (involving emotion detection, laughter detection and speaker identification [18]). In [14], we have reported preliminary results using this system, showing the ability of our model to discern three groups of human participants in terms of a participation score derived from dialogue activities (namely, the low, medium and high participation groups). This score quantifies the participation of the human in the dialogue activities initiated by Nao during the interaction. It is worth noting that this score is not a task completion one. As such, it is concerned with the participation in the activity rather than the activity outcome (e.g., what matters in a "riddle" activity is to participate by providing an answer, regardless of whether it is the right one or not).

5 Conclusion and Future Work

In this article, we have addressed the issue of evaluating H-R spoken social interaction. To that purpose, we have presented the Communication Accommodation Theory (CAT) which argues that dialogue participants (DPs) adapt their communicative verbal and nonverbal behaviours via an accommodation process in order to be more similar or more distinct depending on the context. Interesting lessons from CAT with regard to engagement measures include the importance of considering low-level and high-level processes involved in communication, the necessity to evaluate the communicative behaviour of the human w.r.t. the behaviour of the robot, and the necessity to take into account the dynamic of the interaction (i.e. the changes in the communicative behaviour of both DPs). In light of the CAT, we have identified some types of communicative behaviour cues (e.g., linguistic, paralinguistic, visual) and contexts (e.g., H-R relationship, social, situational, human profile) that could enter into consideration to specify multi-level expectations from the behaviour of the human participant. These expectations could be advantageously taken into account to quantify human engagement. In future work, we aim to exploit CAT along with a dialogue activity model for H-R social interaction, which provides a convenient local interpretation context to assess human contributions [14].

Acknowledgements This work has been funded by the JOKER project and supported by ERA-Net CHIST-ERA, and the "Agence Nationale pour la Recherche" (ANR, France).

References

1. Walker, M.A., Litman, D.J., Kamm, C.A., Abella, A.: PARADISE: a framework for evaluating spoken dialogue agents. In: Proceedings of the Eighth Conference on European Chapter of the Association for Computational Linguistics, pp. 271–280. Association for Computational Linguistics (1997)
2. Möller, S., Ward, N.G.: A framework for model-based evaluation of spoken dialog systems. In: Proceedings of the 9th SIGdial Workshop on Discourse and Dialogue, pp. 182–189. Association for Computational Linguistics (2008)
3. Glas, N., Pelachaud, C.: Definitions of engagement in human-agent interaction. In: Proceedings of the International Workshop on Engagement in Human Computer Interaction (ENHANCE), pp. 944–949 (2015)
4. Corrigan, L.J., Peters, C., Castellano, G., Papadopoulos, F., Jones, A., Bhargava, S., Janarthanam, S., Hastie, H., Deshmukh, A., Aylett, R.: Social-task engagement: striking a balance between the robot and the task. In: Proceedings of the Embodied Communication of Goals Intentions Workshop ICSR, vol. 13, pp. 1–7 (2013)
5. Sidner, C.L., Lee, C., Kidd, C.D., Lesh, N., Rich, C.: Explorations in engagement for humans and robots. Artif. Intell. **166**(1), 140–164 (2005)
6. Rich, C., Sidner, C.L.: Collaborative discourse, engagement and always-on relational agents. In: Proceedings of the AAAI Fall Symposium: Dialog with Robots (2010)
7. Bohus, D., Horvitz, E.: Models for multiparty engagement in open-world dialog. In: Proceedings of the SIGDIAL 2009 Conference: The 10th Annual Meeting of the Special Interest Group on Discourse and Dialogue, pp. 225–234. Association for Computational Linguistics (2009)
8. Bohus, D., Horvitz, E.: Learning to predict engagement with a spoken dialog system in open-world settings. In: Proceedings of the SIGDIAL 2009 Conference: The 10th Annual Meeting of the Special Interest Group on Discourse and Dialogue, pp. 244–252. Association for Computational Linguistics (2009)
9. Salam, H., Chetouani, M.: A multi-level context-based modeling of engagement in human-robot interaction. In: Proceedings of the International Workshop on Context Based Affect Recognition (2015)
10. Gallois, C., Ogay, T., Giles, Howard, H.: Communication accommodation theory: a look back and a look ahead. In: Gudykunst, W. (ed.) Theorizing About Intercultural Communication, pp. 121–148. Sage, Thousand Oaks (2005)
11. Bunt, H.: The DIT++ taxonomy for functional dialogue markup. In: Proceedings of the AAMAS Workshop, Towards a Standard Markup Language for Embodied Dialogue Acts, pp. 13–24 (2009)
12. Schuller, B., Steidl, S., Batliner, A., Burkhardt, F., Devillers, L., Müller, C., Narayanan, S.: Paralinguistics in speech and language–state-of-the-art and the challenge. Comput. Speech Lang. **27**(1), 4–39 (2013)
13. McCrae, R.R., John, O.: An introduction to the five-factor model and its applications. J. Pers. **60**, 175–215 (1992)
14. Dubuisson Duplessis, G., Devillers, L.: Towards the consideration of dialogue activities in engagement measures for human-robot social interaction. In: Proceedings of the workshop Designing & Evaluating Social Robots for Public Settings, International Conference on Intelligent Robots and Systems, pp. 19–24, Hambourg, Germany (Sep 2015)
15. Clark, H.: Using language, vol. 4. Cambridge University Press (1996)
16. Schegloff, E.A., Sacks, H.: Opening up closings. Semiotica **8**(4), 289–327 (1973)
17. Devillers, L., Rosset, S., Dubuisson Duplessis, G., Sehili, M., Béchade, L., Delaborde, A., Gossart, C., Letard, V., Yang, F., Yemez, Y., Türker, B., Sezgin, M., El Haddad, K., Dupont, S., Luzzati, D., Estève, Y., Gilmartin, E., Campbell, N.: Multimodal data collection of human-robot humorous interactions in the Joker project. In: Proceedings of the 6th International Conference on Affective Computing and Intelligent Interaction (ACII) (2015)
18. Devillers, L., Tahon, M., Sehili, M.A., Delaborde, A.: Inference of human beings' emotional states from speech in human–robot interactions. Int. J. Soc. Robot. 1–13 (2015)

Extrinsic Versus Intrinsic Evaluation of Natural Language Generation for Spoken Dialogue Systems and Social Robotics

Helen Hastie, Heriberto Cuayáhuitl, Nina Dethlefs, Simon Keizer and Xingkun Liu

Abstract In the past 10 years, very few published studies include some kind of extrinsic evaluation of an NLG component in an end-to-end-system, be it for phone or mobile-based dialogues or social robotic interaction. This may be attributed to the fact that these types of evaluations are very costly to set-up and run for a single component. The question therefore arises whether there is anything to be gained over and above intrinsic quality measures obtained in off-line experiments? In this article, we describe a case study of evaluating two variants of an NLG surface realiser and show that there are significant differences in both extrinsic measures and intrinsic measures. These differences can be used to inform further iterations of component and system development.

Keywords Natural language generation · Evaluation · Spoken dialogue systems

H. Hastie (✉) · S. Keizer · X. Liu
School of Mathematical and Computer Sciences, Heriot-Watt University, Edinburgh, UK
e-mail: h.hastie@hw.ac.uk

S. Keizer
e-mail: s.keizer@hw.ac.uk

X. Liu
e-mail: x.liu@hw.ac.uk

H. Cuayáhuitl
School of Computer Science, University of Lincoln, Lincoln, UK
e-mail: hcuayahuitl@lincoln.ac.uk

N. Dethlefs
School of Engineering and Computer Science, University of Hull, Hull, UK
e-mail: n.dethlefs@hull.ac.uk

K. Jokinen and G. Wilcock (eds.), *Dialogues with Social Robots*, Lecture Notes in Electrical Engineering 427,
DOI 10.1007/978-981-10-2585-3_24

1 Introduction

The ultimate goal of the design of socially intelligent agents, including robots, is the development of systems capable of generating natural language in order to interact with humans at a human-like level of proficiency [1, 2]. This includes the use of naturally sounding Natural Language Generation (NLG) to portray emotional feelings. Extrinsic evaluations are key to improving the performance of social agents. However, it can be difficult to isolate one aspect of interaction such as NLG in extrinsic evaluations of social agents due to the complex interaction of multiple factors such as NLG, voice, personality, gestures as well as task completion if the agent is assisting in a task.

Extrinsic evaluations of output components (both NLG and Text-to-Speech Synthesis (TTS)) are relatively rare with only around 15 % of publications reporting on NLG evaluations referring to extrinsic evaluations over intrinsic [3]. Similar published TTS studies are even rarer. Extrinsic evaluations that include testing end-to-end spoken dialogue systems are highly labour-intensive to set up even more so for robotic platforms. In addition, these types of evaluations cost on average more per data-point to collect due to the time taken to complete a whole dialogue rather than, for example, an off-line rating of written output. Input components (both Automatic Speech Recognition (ASR) and Spoken Language Understanding (SLU)) rely less on subjective measures of quality than output components and one is easily able to obtain intrinsic quality measures, such as WER, as well as extrinsic measures during end-to-end system evaluations.

The question arises, therefore, whether it is worth running extrinsic evaluations for output components and whether the user can perceive any differences in quality whilst performing a task through dialogue with an agent. One could hypothesise that with all that is going on in a social robot-human interaction such as the study reported in [4], that subtle nuances in the style and naturalness of the output will not be perceivable by the user.

The study described in this article compares two surface realisers and we show that significant differences can be found between them in terms of both *intrinsic measures*, such as *user like measures*, and *extrinsic measures*, such as *task success*. These differences are significant despite the added complication and potential increase in cognitive load of conducting a task-based interaction with a spoken dialogue system. We conclude that assessing the performance of output components in an end-to-end evaluation is indeed useful and important in that it highlights areas for improvement that would not be detected in a purely intrinsic evaluation. Here, we discuss the context of a spoken dialogue system for restaurant recommendations for which one calls up or uses an app to talk to the system. Given the relevance of natural language generation in conversational scenarios, we argue that these findings are transferable to other types of human-computer interaction, as well as social robotics and human-robot interaction in general.

2 Background Literature

Previous end-to-end evaluations reported in the literature have had mixed results with regards evaluating NLG components embedded in a spoken dialogue system. One study does show significant differences in *extrinsic user-task-success* measures and one *user like measure* on voice quality when testing an adaptive and non-adaptive version for Information Presentation in an end-to-end dialogue system [5]. For the ILEX project [6], two versions of NLG for museum artefact text generation were evaluated and found significant differences in *user like measures* but not for *extrinsic measures* of user task or system purpose success measures. For the M-PIRO system [7], a follow-on from ILEX, the authors were interestingly able to show significant improvement on *extrinsic user-task-success measures* of comprehension accuracy and *extrinsic user-assessed learning gain* but not *intrinsic user like measures* such as 'interestingness' and enjoyment.

SkillSum is an NLG system that generates a personalised feedback report for low-skilled readers who have just completed a screening assessment of their basic literacy and numeracy skills [8]. The evaluation included several automatically computed *intrinsic, automatic output-quality measures* (mean sentence length, mean word length, Flesch reading ease, Flesch-Kincaid grade); several *extrinsic user-task-success measures* (reading-aloud speed, reading-aloud errors, silent-reading speed, comprehension accuracy); and some *intrinsic user like measures* (preference judgements between alternative versions of reports). The team found no statistically significant difference for any of these measures. As is the case with a number of these NLG systems, differences between report versions were fairly subtle. The authors conclude that the lack of statistical results is perhaps due to the great variation between the needs of students with low literacy skills and that modelling multiple different user types, and evaluating these different models, might have worked better [8]. This was confirmed in work reported in [9], showing that while human judges assign subjective ratings to utterances that are inherently consistent, cross-user assessment can vary widely for the same utterance.

Results are presented here from an end-to-end evaluation comparing a realiser trained from labelled data with one trained from unlabelled data. This work follows on from [10], where we show through *intrinsic* measures that no differences are perceivable between an NLG trained on automatically labelled data and an NLG trained on hand-labelled data. There was, however, a significant difference between these two systems and hand-crafted output.

These two variants of the NLG component were integrated into an end-to-end system and an evaluation of comparable size, in terms of generated utterances, was conducted. Our contribution is in the fact that, where *intrinsic measures* in a comparable component-only study showed no significant differences, there were significant differences to be found in the end-to-end system study in terms of similar *intrinsic measures* as well as other *extrinsic measures*.

In this article, we firstly discuss the NLG variants briefly, referring the reader to previous work [10, 11]; we present the end-to-end system; describe the evaluation and results and analysis; and finally provide a discussion and ideas for future work.

3 Natural Language Generation Module

Our domain is an interactive system that provides restaurant recommendations to users with varying preferences and constraints. From the surface realisation perspective, we assume that a semantic form is provided by a preceding module, such as the Interaction Manager, and the task of the surface realiser is to find an appropriate linguistic realisation. An example input is *inform(restaurant, food=Italian, venueName=Roma)*, which could be expressed as *The restaurant Roma serves Italian food*, or a number of alternative candidates.

Ultimately, the research aim behind developing the NLG module is to move towards open-domain surface realisation, where a realiser is retrained during deployment from unlabelled data, e.g. retrieved from the web. An example scenario is as follows: at some point in time a (spoken) dialogue system may know a set of semantic slots in an ontology (O1), and at some other point it becomes aware or learns of new additional semantic slots, which are added to O1 to create a larger ontology O2. Here, O1 represents the original ontology that the interactive system is deployed with, including its corresponding surface realisation resources, which may be trained from labelled data. It contains semantic concepts for *venueName, foodType, Area* and *priceRange*. The assumption is that over time, through domain adaptivity, the system will acquire new ontological concepts, representing an extensible, growing ontology O2. In our system, these new concepts are *kidsAllowed, goodForMeal* and *nearLocation*. In order to produce meaningful outputs from these new semantic concepts (also referred to as 'semantic slots'), the surface realiser needs to be retrained during deployment and without human intervention. In other words, we might suddenly encounter the semantic form *inform(foodType=Italian, venueName=Roma, kidsAllowed=true)*, where we assume that we know the semantic concept *kidsAllowed*, because it is in the ontology, but we do not have labelled training data to produce a realisation.

Our method of porting a statistical surface realiser to these extended domains has two stages. In the first stage, automatic semantic labelling is applied to unlabelled utterances [10]. In the second stage, the automatically labelled data is then used to train an existing statistical surface realiser [11], which uses Conditional Random Fields (CRFs) to generate outputs from input dialogue acts in the restaurant domain. Our semantic labelling method is based on unsupervised clustering of clauses found in unlabelled input data according to their lexical and semantic similarity. The underlying hypothesis is that the more similar clauses are—in terms of their lexical and semantic properties—the more likely they are to represent the same semantic slot, e.g. *kidsAllowed*, or *goodForMeal*. Research reported earlier [10], confirmed that this automatic labelling technique can attain good clustering accuracy results (*automatic intrinsic quality measures*) as well as in a human rating study (*intrinsic user like measures*) as discussed above.

This study is extended here by comparing a surface realiser in an end-to-end system trained from human-labelled data (**TopBound**) against one trained from the automatically labelled data using the unsupervised learning technique (**Proposed**). Example outputs from the TopBound system and the Proposed system are given here below:

TopBound: "Right in the heart of central Richmond lies Kirin Chinese restaurant, a well-established neighbourhood favourite."; and

Proposed: "Right in the heart of central Richmond lies the well-established neighbourhood favourite of Kirin Chinese restaurant."

4 The End-to-End Spoken Dialogue System

While the problem of domain extensibility is more general and requires growing resources for all system modules, such as language understanding and interaction management, this paper focuses on the evaluation of two surface realisers where everything else in the end-to-end system is the same. The two versions of the surface realiser were swapped in and out of an end-to-end system whereby the other components were all trained to understand and process the larger O2 ontology, i.e. with the assumption that they are able to cover the new ontological concepts.

The statistical surface realiser was integrated as part of the PARLANCE dialogue system[1]. An earlier version of the PARLANCE system is described in [12], which was part of a wider project whose aim it was to automatically adapt to ever expanding domains. The system architecture includes the following components: the ATK speech recogniser [13]; an SLU dependency parser with unsupervised word embeddings [14]; an Interaction Manager (IM) that uses Gaussian Process reinforcement learning with a policy trained on the top-bound ontology [15]; a micro-turn Interaction Manager that enables rapid system responses and generation of backchannels and user barge-ins and finally, the generated outputs are given as input to the TTS engine described in [16].

5 Evaluating the NLG component in an End-to-End System

A task-based evaluation was conducted with workers recruited via Crowdflower[2]. The workers were asked to call the system and find restaurants in certain areas of San Francisco (U.S.A.) according to certain predefined scenarios. 664 dialogues were collected from 72 participants. Participants were paid $2.00 on completion of

[1] http://www.parlance-project.eu

[2] http://crowdflower.com

four dialogues. After the participants have completed a dialogue, they were given 5 subjective questions where Q1 was a binary Yes/No for perceived *user-task-success* and Q2-5 were on a 6-point rating scale and cover a variety of aspects of dialogue.

- Q1 (**TaskSuccess**): Did you manage to find a restaurant?
- Q2 (**InfoFound**): I found all the information I was looking for.
- Q3 (**Understanding**): The system appeared to understand what was saying.
- Q4 (**InfoPresentation**): The information on restaurants was well-presented.
- Q5 (**Repetitiveness**): The system's utterances were not repetitive.

5.1 The Tasks

There were 6 main semantic slots for O1 listed here with the number of possible values in brackets: *foodType(59)*, *pricerange(3)*, *area(155)*, *phoneNumber*, *address* and *postcode*. As mentioned above, for O2 these were expanded to include: *nearLocation(39)*, *kidsAllowed(2)* and *goodForMeal(4)*. A set of 3,000 tasks were automatically generated with randomly sampled constraints from this larger domain ontology O2. Participants were asked to find restaurants that have particular features, e.g. "Find a restaurant in the center and it should serve African food. You want to know the address and whether it is good for lunch." To elicit more complex dialogues, the users were sometimes asked to find a restaurant type that did not exist and asked to find an alternative, e.g. "You want to find a restaurant in the center and it should serve Indian food. If there is no such venue how about African type food? You want to know the address and whether it is good for lunch."

6 Results

As seen in Table 1, the TopBound system is *perceived* as significantly more successful than the Proposed system in retrieving a relevant restaurant (SubjTS). However, in terms of actual hand-annotated task success[3] (TS), there is no significant difference. The difference in TS and SubjTS, we believe, is due to participants overestimating their success, i.e. if they got information on *any* restaurant they marked it as a success. The Proposed system has significantly shorter turns in terms of average words per turn (an *intrinsic automatic output quality measure*), which may be due to missing or confused slot information.

As seen in Table 2, the only significant difference between the *intrinsic user like measures* was for Information Presentation. Therefore, even though the outputs are relatively similar, the user is still able to perceive a difference in the Information

[3]*Extrinsic user-task-success* was hand-annotated by a single annotator, being set to 1 if the caller received information on a restaurant that matched their request and if other information (e.g. *address, name, phoneNumber*) was asked for and correctly received.

Table 1 *Extrinsic user-task-success (TS), subjective user-task-success (SubjTS)* and various *intrinsic output quality measures* capturing dialogue length

Evaluation mode		Intrinsic output quality			Extrinsic	
System	Num of dialogues	Num of turns	Length (s)	Avg wds per turn	TS (%)	SubjTS (%)
TopBound	365	15.23	10.51	12.32	60.82	94.79*
Proposed	299	14.71	10.23	11.89*	61.20	89.63

*Indicates $p < 0.05$ using a χ^2 test for TS/SubjTS and 1-way unpaired t-test for length metrics

Table 2 *Intrinsic user like measures* from the post-questionnaire on a 6-point rating scale for the mean (mode)

Evaluation mode	Intrinsic user like			
System	InfoFound	Understanding	InfoPresentation	Repetitiveness
TopBound	4.70(6)	4.54(6)	4.56(6)*	4.36(6)
Proposed	4.60(6)	4.31(6)	4.31(6)	4.22(6)

*Indicates $p < 0.05$ for a Mann-Whitney U test

Presentation category, thus highlighting areas for improvement. In [11], we show through *user like measures* that, by their very nature, using CRFs for surface realisation results in utterances that are *less repetitive* than baseline systems. This still holds when trained on automatically labelled data, as there is little perceived difference in terms of repetitiveness between the TopBound and Proposed systems.

Error analysis reveals that in approximately 4 % of the cases, the CRF trained on automatically labelled data realises somewhat anomalous utterances such as "The Kirin restaurant is a perfect place for children out". This may contribute to the decrease in subjective evaluation scores. A further aspect revealed in our error analysis is that generated outputs can occasionally contain segments of information that are not part of the original semantic input form. An example is the realisation "Right in the heart of central Richmond lies the well-established neighbourhood favourite of Kirin Chinese restaurant." for semantic input form *inform(restaurant, area="central Richmond", venueName=Kirin, foodType=Chinese)*. The fact that the restaurant is "a well-established neighbourhood favourite" is not derived from the knowledge base but rather constitutes an artefact of the training data. End-to-end evaluations, particularly those "in the wild" with users actually visiting the recommended restaurants, may show this to be a false statement; again this would not be evident in isolated utterances evaluated off-line.

7 Discussion and Future Work

Where no significant differences have been found for *intrinsic measures* in previous component studies [10], the evaluation described here not only discovers differences in *intrinsic* measures but also *extrinsic measures*. Without an end-to-end study, the comparative differences between the perceived and the objective *user-task-success* would not have been discovered.

Future work will transfer what we have learned in this evaluation, develop our approach to new domains and go beyond restaurant recommendations. Encouraging results from a component *intrinsic user like* study [17] have already shown that similar output quality can be achieved for domains such as film reviews. Moreover, we aim to move coverage towards a larger amount of unknown slots and investigate methods to learn surface realisations for new semantic concepts on the fly rather than from corpora. As [5] also suggests, this work supports the argument for joint optimisation of NLG with other components such as the Interaction Manager, so that NLG optimisation techniques take into consideration factors falling out of the interaction with the system as evidenced here. Finally, future work would include performing such an extrinsic evaluation in multimodal settings, such as a social-robotics setting, as reported in [18].

Acknowledgements This research was funded by the European Commission FP7 programme FP7/2011-14 under grant agreement no. 287615 (PARLANCE). We thank all members of the PARLANCE consortium for their help in designing, building and testing the Parlance end-to-end spoken dialogue system. We would also like to acknowledge other members of the Heriot-Watt Parlance team in particular Prof. Oliver Lemon and Dr. Verena Rieser.

References

1. Kelleher, J.D., Kruijff, G.J.M.: Incremental generation of spatial referring expressions in situated dialog. In: Proceedings of ACL, Sydney, Australia (2006)
2. Giuliani, M., Foster, M.E., Isard, A., Matheson, C., Oberlander, J., Knoll, A.: Situated reference in a hybrid human-robot interaction system. In: Proceedings of the INLG, Trim, Ireland (2010)
3. Gkatzia, D., Mahamood, S.: A snapshot of NLG evaluation practices 2005 to 2014. In: Proceedings of ENLG (2015)
4. Deshmukh, A., Janarthanam, S., Hastie, H., Lim, M.Y., Aylett, R., Castellano, G.: How expressiveness of a robotic tutor is perceived by children in a learning environment. In: Proceedings of HRI (2016)
5. Rieser, V., Lemon, O., Keizer, S.: Natural language generation as incremental planning under uncertainty: adaptive information presentation for statistical dialogue systems. IEEE/ACM Trans. Audio Speech Lang. Process. **22**(5) (2014)
6. Cox, R., O'Donnell, M., Oberlander, J.: Dynamic versus static hypermedia in museum education: an evaluation of ILEX, the intelligent labelling explorer. In: Proceedings of AIED (1999)
7. Karasimos, A., Isard, A.: Multi-lingual evaluation of a natural language generation systems. In: Proceedings of LREC (2004)
8. Williams, S., Reiter, E.: Generating basic skills reports for low-skilled readers. Nat. Lang. Eng. **14**(4), 495–525 (2008)
9. Dethlefs, N., Cuayáhuitl, H., Hastie, H., Rieser, V., Lemon, O.: Cluster-based prediction of user ratings for stylistic surface realisation. In: Proceedings of the European Chapter of the Annual Meeting of the Association for Computational Linguistics (EACL), Gothenburg, Sweden (2014)
10. Cuayáhuitl, H., Dethlefs, N., Hastie, H., Liu, X.: Training a statistical surface realiser from automatic slot labelling. In: Proceedings of SLT, South Lake Tahoe, CA, USA (2014)

11. Dethlefs, N., Hastie, H., Cuayáhuitl, H., Lemon, O.: Conditional random fields for responsive surface realisation using global features. In: Proceedings of ACL (2013)
12. Hastie, H., Aufaure, M.A., Alexopoulos, P., Cuayáhuitl, H., Dethlefs, N., Gašić, M., Henderson, J., Lemon, O., Liu, X., Mika, P., Ben Mustapha, N., Rieser, V., Thomson, B., Tsiakoulis, P., Vanrompay, Y., Villazon-Terrazas, B.: Demonstration of the PARLANCE system: a data-driven incremental, spoken dialogue system for interactive search. In: Proceedings of SIGDIAL (2013)
13. Young, S., Kershaw, D., Odell, J., Ollason, D., Valtchev, V., Woodland, P.: The HTK Book Version 3.0. Cambridge University, UK (2000)
14. Yazdani, M., Breslin, C., Tsiakoulis, P., Young, S., Henderson, J.: Domain adaptation in ASR and SLU. Technical report, PARLANCE FP7 Project (2014)
15. Gašić, M., Breslin, C., Henderson, M., Kim, D., Szummer, M., Thomson, B., Tsiakoulis, P., Young, S.: POMDP-based dialogue manager adaptation to extended domains. In: Proceedings of SIGDIAL (2013)
16. Tsiakoulis, P., Breslin, C., Gašić, M., Henderson, M., Kim, D., Young, S.J.: Dialogue context sensitive speech synthesis using factorized decision trees. In: Proceedings of INTERSPEECH (2014)
17. Cuayáhuitl, H., Dethlefs, N., Hastie, H.: A semi-supervised clustering approach for semantic slot labelling. In: Proceedings of ICMLA, Detroit, MI, USA (2014)
18. Castellano, G., Paiva, A., Kappas, A., Aylett, R., Hastie, H., Barendregt, W., Nabais, F., Bull, S.: Towards empathic virtual and robotic tutors. In: Artificial Intelligence in Education, pp. 733–736. Springer, Berlin (2013)

Engagement in Dialogue with Social Robots

Loredana Cerrato and Nick Campbell

Abstract It is becoming increasingly clear that social and interactive skills are necessary requirements in many application areas where robots interact, communicate and collaborate with humans or other connected devices. The social aspects of human-computer interaction and the connection between humans and robots has recently received considerable attention in the fields of artificial intelligence, cognitive science, healthcare, companion technologies, and industrial/commercial robotics. This article addresses some dimensions of near-term future HRI, with focus on engagement detection for conversational efficiency. We present some findings from HRI research conducted at the Speech Communication Lab at Trinity College Dublin, report our experiences with a publicly exhibited conversational robot, and discuss some future research trends.

Keywords Human-robot-interaction · Social engagement · Functional dialogue processing · Interactive speech-synthesis

1 Introduction

Researchers in the field of human-robot interaction (HRI) are typically addressing the question of how to make robots more 'attractive' to people interacting with them and are concerned with the motivation and engagement of users in their interactions with robots [1, 2]. There are several attempts to make robot speech sound more human and even to include laughter, turn-taking signals [3, 4] and extra-linguistic sounds to mimic natural human conversational interaction. However, given the 'uncanny-valley effect' [5] wherein too close a mimic can appear disturbing, our view is that

L. Cerrato (✉) · N. Campbell
ADAPT Centre, Trinity College Dublin, Dublin, Ireland
e-mail: cerratol@tcd.ie

N. Campbell
Speech Communication Lab, Trinity College Dublin, Dublin, Ireland
e-mail: nick@tcd.ie

K. Jokinen and G. Wilcock (eds.), *Dialogues with Social Robots*,
Lecture Notes in Electrical Engineering 427,
DOI 10.1007/978-981-10-2585-3_25

while robots should perhaps continue to sound somewhat robotic, they should at the same time be able to parse and interpret natural social human speech phenomena. We anticipate many situations where robots should be able to interpret a given utterance as a joke, for example, and even to make jokey or humorous responses to human utterances.

It might be necessary to first define how we use the term *robot* in this context; and here we include many devices that might not normally come under the typical understanding of 'robotics'. The fundamental defining characteristic of a robot as we see it is that it has multiple sensors and can autonomously effect changes in the real world as a result of different sensor inputs. Of course androids and humanoid devices come under this definition, but so too, we argue, does the iPhone, a Google car, and several present-day domestic appliances. We foresee a near-term future where, in some specific contexts, ubiquitous devices are highly interconnected and able to communicate with humans through speech, as well as being able to 'understand' (i.e., parse and appropriately process) subtle human social signals including speech. The new challenge in this case lies in learning to cope with massive streams of real-world sensor data in the wild, and in finding robust significant meaningful trends emerging from these noisy signals. It may be particularly advantageous if the devices can do this with a minimal knowledge of the actual spoken content, since the ethical issues of eavesdropping machines are of considerable concern.

2 Engagement

The social connection between people and robots is already of fundamental importance when it comes to service robotics applications or robots as assistants. Looking further ahead, with intelligent homes, full interconnectivity, and ubiquitous sensing, we anticipate a future where a broad range of devices will need a social-awareness component. They will need to understand and interpret human social behavioural patterns, including but not limited to speech, and to adapt their interventions accordingly. They will need to be particularly aware of cognitive engagement among humans and to sense and measure the attentional states of people they routinely interact with. The literature proposes numerous methods to measure engagement [6, 7], however when it comes to validate and relate these measures or to provide a firm basis for assessing the quality of the user experience, little has been done.

Engagement is a fundamental aspect of the user experience with socially intelligent robots. It emphasises both positive and negative aspects of the interaction. Many objective and subjective measures have been used to design experiments aiming at assessing engagement with respect to, for instance, perceived user feedback and automatic classification of multimodal behavioural features. However it is still an open question as to how user engagement can be studied, measured, and explained, in order to facilitate the development of robots able to engage people in effective social communication.

Siri is by now well known, and provides a robust service with very little social interaction (except for the built-in joke routines: such as *will you marry me?*, that have proved so popular). In recent times robots have become involved in increasingly more complex and less structured tasks and activities, now including social interaction with people [8, 9]. This new complexity has encouraged the study of how humans interact with robots, and introduced challenging new research fields dedicated to finding out how to design and implement robot systems capable of accomplishing social interactive tasks in human environments.

Crucial components of human communication include body posture, gestures and facial expression since they represent an important source of information for social interaction.

Just as humans engage in social interactions through different means of communications, so robots should appreciate and respond to social signals in user language and behaviour, to create and maintain social bonds which are the roots of our human society's cohesion. Kendon and others (e.g., [10]) have shown that people are well able to read bodily signs and gestures in addition to linguistic components as a functional part of normal social interaction.

We are working to provide a similar rich information stream that can be processed alongside (or perhaps even instead of) the speech recognition results that provide an indication of the function of linguistic content of a spoken interaction. There are many situations where the actual speech content is secondary in importance to the social function of an interpersonal interaction; in greetings, for example, or jokes and banter, and these can occur in noisy situations where the spoken content per se might be difficult to hear or parse correctly. If we can provide a parallel stream of functional inference relating to the interpersonal rather than the linguistic exchange, then we might be able to relieve the speech recogniser for much of the time; or at least to provide a complementary stream of backup information that a higher component might call upon.

Some of this behavioural information can be inferred directly from the increasingly available number of sensors equipped in the portable devices now used in daily life, enabling context-awareness for an advanced semantic/pragmatic analysis of the human speech and gestures, as well as combining social signal processing and computer vision techniques to capture and interpret social behaviour from non-verbal, posture, gesture and facial signals [11]. Particularly in terms of social interaction, there are a great number of opportunities to improve the 'intelligent' aspect of a robot by developing response strategies to adapt both its language content and the way it speaks and behaves according to the input factors captured from the surrounding environment and the human-computer interaction itself [12].

3 Research into Interactive Speech Synthesis

In the Speech Communication Lab at the Trinity College Dublin we address several research issues related to how humans engage in social interactions with machines. We build on corpus-based studies of spontaneous human-human social conversations, including multimedia data streams aligned with biometric feedback data and human annotations [13, 14]. We consider this work a natural extension of previous speech synthesis research and are developing interactive forms of digital content delivery for use in human-machine and human-human interaction. If a speech synthesiser can be made aware of the responses of its interlocutor, then it can be better able to produce a more intelligible rendering of the intended content. With recent developments in speech delivery technology, including smart-phones, and multimedia-equipped interfaces, the listener need no longer be a distant and abstract entity, and the conversation can thereby become a more naturally interactive process.

There are two forms of engagement being simultaneously processed in our speech-based content delivery; the first is active, the second more passive. The former can employ speaking style and content manipulations to ensure that the attention of the listener is maintained, while the latter attempts to measure engagement, or more accurately, attempts to infer the cognitive/attentional state of the listener through processing of dynamic real-time signals elicited proactively through the interactive speech process itself.

For a device (robot, 'fridge' or speech synthesiser) with a message to deliver, it should make an important difference whether or not there is a listener present, and whether that person (as is usually the case) has a) ears to listen, and b) (not so often) a mind directed towards listening. To most humans, these elementary forms of interaction processing are so fundamentally obvious as to be unthought of, but to a robot or similar device employing a speech synthesiser, they are necessary procedural steps.

We therefore consider a series of five (5) preliminary tests sufficient for efficient content delivery in first encounters. The last is the most difficult and forms the core of our current research. The first four can be solved through the clever use of sensors in conjunction with the delivery of the speech signal itself:

- first: is there a listener present?
- second: does the person have functioning ears?
- third: is he/she actively listening?
- fourth: is she/he following (able to comprehend) the speech? and only then,
- fifth: do they understand/agree? (are they understanding as we speak?)

The first can be answered by image processing, backed up by e.g., depth-sensing of physical movement in the proximity of the listener. The second by making a sound (e.g., saying *hello*) and watching for a response. It may not be a verbal answer (a direct *hello* in reply) but is typically just a synchronous movement of the head in relation to the sound. The third can be tackled by repeating that sound (or making another) to see if a similar synchronous response occurs, so that systematicity can be inferred. The fourth by asking a question and receiving an answer (*Hello, Hi?* → *Hi.*) or an informed response.

The fourth takes more time and has to be inferred from a series of systematic synchronised responses, but again is reducible to stimulus-response pairs co-occurring within a certain time window or period. The fifth, however, requires intelligent inference and remains a hard problem. We believe that to some extent, it can be tackled by analysing biometric signals without resource to semantics or sophisticated text processing, much like a human performing in a foreign language or culture. Of course if there is more than one interlocutor present, then the problem is confounded by deciding who should take priority. But this is a matter for another paper.

4 Natural Human-Machine Conversational Interaction

We use various devices equipped with a number of sensors to capture a variety of visual, physiological and gestural information which can be used to estimate the user's social and physical contexts for optimal adaptation of content delivery strategies. We also carry out experiments to investigate the degree to which we can classify engagement in existing corpora of multi-party conversations by using a combination of voice and visual features. Initial results show that it is possible to obtain high accuracy rates in automatic engagement detection from social signals extracted from speech prosody and visual features in human-human conversation [15].

Similar experiments have been carried out to develop and test an engagement classifier for human-computer interaction systems. We recently tested a conversational robot for three months in real-world live experiments with unpaid (and uninstructed) off-the-street volunteers, to analyse unstructured and unprompted human-robot interaction (first encounters) in order to observe the natural levels of engagement of users. The recorded interactions constitute the Herme Data [16]: a collection of human-robot conversations gathered using the Wizard of Oz technique. The collection took place at the Science Gallery in Dublin, where as part of a larger exhibition ('*Human+*') visitors were invited to engage in a conversation with a small talking robot, Herme, part of which included a request to sign a consent form that allowed their speech to be used for research purposes.

One of the main interests in this data collection is the fact that the interactions are not task-based, but rather a sort of bond-forming social interaction typical of early speech among human participants who are initially strangers to each other. The data were collected with the aim of observing the way people interact with a speaking robot, and although we were initially sceptical, we soon found that people fall easily into well-established and predictable patterns of behaviour that can be reliably prompted with simple conversational strategies.

We tested several different conversation strategies ('engagement techniques') in both fully-automatic and Wizard-mediated form. The automatic system reliably failed to produce engagement, but the human-mediated conversations (where a person was responsible for initiating each utterance but only in control of its timing, not the utterance sequence per se) were consistently more successful, and some human operators proved capable of reliably maintaining a conversation to its full conclusion.

5 Findings and Observations

From the Herme data we can spot where the interaction breaks down and why, and this gives an indication of the user's engagement with the Herme robot. Apparently the main difference between successful and unsuccessful interactions can be reduced to a simple timing measurement: the length of the gap between the end of the user's utterance and the start of the subsequent utterance by the robot. This gap is variable and depends on when the wizard triggers the robot's utterance.

If the timing is appropriate then the interactions proceeds smoothly. Knowing where the interaction stops is the first step in defining where our human-robot social interaction model needs to be improved. From preliminary observations the principal reasons for the interruption of the interactions and the consequent loss of engagement broken were:

- The users realize that the robot was not listening to what was being said, for instance when users asked questions that could not get an answer. When instead users were more passive and led along, following the predetermined dialogue sequence, the interaction proceeded in a smoother way.
- Users had difficulty to understand Herme's voice—partly due to the environmental noise in the open public environment, partly perhaps to inattention. The robot was not able to repeat the same utterance, so this might have been a problem for comprehension in some cases.
- The timing between the utterances was inappropriate, resulting in a rapid loss of engagement from the users. This third reason offers us considerable room for improvement of the system. We believe that once the timing structure and its features are better understood and appropriately modelled, the interactions between users and the robot can proceed smoothly thus supporting users engagement in the conversation.

6 Conclusions

In order to achieve a strong social connection between humans and robots it is necessary that the linguistic, verbal and visual animation is in synchrony and that the robot or device (delivery platform) is able to process and respond to equivalent social signals in user language and behaviour. These can be inferred from the simple output of an increasingly available number of sensors provided in portable devices used in our daily life. In many cases these simple biometrics can provide rich sources of inference particularly when used in synchronous relationships. This may alleviate the need for sophisticated semantic analysis of the human speech and gestures by the application of social signal processing and computer vision techniques to capture and interpret social behaviour. While state-of-the-art robots are now capable of processing some of the linguistic aspects of human communication, they are not yet fully capable of processing the complex dynamics involved in social interaction and often fail to capture the coordination and adaptation on the part of interlocutors. Such processing and capture as well as the effective and accurate interpretation of such signals in dialogue are crucial parts of achieving truly social interactions between the humans and robots.

Acknowledgements This research is supported by Science Foundation Ireland under Grant No. 13/RC/2016, through the ADAPT Centre for Digital Content Technology at Trinity College, Dublin (www.adaptcentre.ie). We are grateful to the School of Computer Science and Statistics at Trinity College Dublin for their continuing support of the Speech Communication Lab.

References

1. Dautenhahn, K.: Socially intelligent robots: dimensions of human-robot interaction. Philos. Trans. R. Soc. Lond. B Biol. Sci. **362**(1480), 679–704 (2007)
2. Lungarella, M., Metta, G., Pfeifer, R., Sandini, G.: Developmental robotics: a survey. Connect. Sci. **15**, 151190 (2004). doi:10.1080/09540090310001655110
3. Becker-Asano, C., Kanda, C.T., Ishiguro H.: How about laughter? Perceived naturalness of two laughing humanoid robots. In: Proceedings of the 3rd International Conference on Affective Computing and Intelligent Interaction and Workshops, Amsterdam, Nederlands (2009)
4. Skantze, G., Johansson, M., Beskow, J.: Exploring turn-taking cues in multi-party human-robot discussions about objects. In: Proceedings of ICMI, Seattle, Washington, USA (2015)
5. Tinwell, A.: The Uncanny Valley in Games and Animation. CRC Press (2014). ISBN: 9781466586956
6. Peters, C., Castellano, G., de Freitas, S.: An exploration of user engagement in HCI. In: Proceedings of the International Workshop on Affective-Aware Virtual Agents and Social Robots Article (AFFINE 09), No. 9 ACM, New York, NY, USA (2009)
7. Campbell, N.: Machine Processing of Dialogue States; Speculations on Conversational Entropy. Invited keynote in Proceeding of the Specom 2016. Budapest, Hungary (2016)
8. Shibata, T.: An overview of human interactive robots for psychological enrichment. In: Proceedings of IEEE (2004)
9. Pepper: https://www.aldebaran.com/en/a-robots/pepper/more-about-pepper
10. Kendon, A.: Gesture and speech: how they interact. In: Wiemann, Harrison, R. (eds.) Nonverbal Interaction (Sage Annual Reviews of Communication, vol. 11), Sage Publications, Beverly Hills, California. pp.13–46 (1983)
11. Huang, Y., Gilmartin, E., Campbell, N.: Conversational engagement recognition using auditory and visual cues. In: Proceedings of Interspeech 2016, San Francisco, USA (2016)
12. Sidner, L.C., Dzikovska, M.: Robot Interaction: Engagement between Humans and Robots for Hosting Activities Pittsburgh, PA, USA, pp. 123–130 (2002). ISBN: 0-7695-1834-6
13. Gilmartin, E., Bonin, F., Vogel C., Campbell, N.: Exploring the role of laughter in multiparty conversation, DialDam. In: Proceedings of 17th Workshop on the Semantics and Pragmatics of Dialogue, (SEMDIAL) Amsterdam (2013)
14. Oertel, C., Cummins, F., Edlund, J., Wagner, P., Campbell, N.: D64: a corpus of richly recorded conversational interaction. J. Multimodal User Interfaces (2010)
15. Hayakawa, A., Haider, F., Cerrato, L., Campbell, N., Luz, S.: Detection of cognitive states and their correlation to speech recognition performance in speech-to-speech machine translation systems. In: Proceedings of Interspeech 2015. Dresden, Germany (2015)
16. Han, J.G. et al.: Speech & multimodal resources: the herme database of spontaneous multimodal human-robot dialogues. In: Proceedings of the 8th LREC, Istanbul, Turkey, 23–25 May 2012

Lend a Hand to Service Robots: Overcoming System Limitations by Asking Humans

Felix Schüssel, Marcel Walch, Katja Rogers, Frank Honold and Michael Weber

Abstract Service robots such as vacuum-cleaning robots have already entered our homes. But in the future there will also be robots in public spaces. These robots may often reach their system limitations while performing their day-to-day work. To address this issue we suggest asking passersby for help. We enhanced an iRobot Roomba vacuum cleaning robot to set up a low-budget Wizard-of-Oz (WOZ) evaluation platform designed to investigate human-robot interaction (HRI). Furthermore, we suggest how HRI can be investigated in public spaces with a robot in need. An early evaluation shows that our prototype is a promising approach to explore how robots can cope with their limitations by asking somebody for help.

Keywords Human robot interaction · Public spaces · Service robots · System limitations · Dialog strategies · WOZ-study

1 Introduction

Service and domestic robots have already become part of our homes. For instance, vacuum-cleaning robots are very popular. In contrast, we are not yet confronted with service robots in public spaces. Before service robots can be introduced in public

F. Schüssel (✉) · M. Walch · K. Rogers · F. Honold · M. Weber
Institute of Media Informatics, Ulm University, James-Franck-Ring, 89081 Ulm, Germany
e-mail: felix.schuessel@uni-ulm.de

M. Walch
e-mail: marcel.walch@uni-ulm.de

K. Rogers
e-mail: katja.rogers@uni-ulm.de

F. Honold
e-mail: frank.honold@uni-ulm.de

M. Weber
e-mail: michael.weber@uni-ulm.de

K. Jokinen and G. Wilcock (eds.), *Dialogues with Social Robots*,
Lecture Notes in Electrical Engineering 427,
DOI 10.1007/978-981-10-2585-3_26

spaces, several HRI issues have to be solved. Users of current service robots working in housekeeping or healthcare are familiar with *their* robots. Owners know about the robots' limitations. In contrast, robots that will be introduced in public spaces will be unfamiliar to people in their immediate surrounding. A robot is almost on its own when it is confronted with a problem. If such robots reach a system limitation, they could call a technician for maintenance, however this would cause an absence for a longer period of time. Many issues may not need a professional technician to be solved, they can rather be solved by any person near the robot in need. For instance, a vacuum-cleaning robot in a public space may get stuck below furniture—a situation the robot cannot solve on its own, but any human could "rescue" it. The main challenge for the robot in this situation is to gain the attention of somebody and to convince this person to help. In the article, after a discussion of related work, the WOZ platform based on a vacuum cleaning robot is described. Section 4 then presents a first user study proving the viability of the platform and identifying differences between two dialog strategies for gaining attention and help from passersby.

2 Related Work

Human-robot interaction is a growing, interdisciplinary field [1] and there has been substantial work on service robots as sociable agents. There are many paradigms for human-robot relationships; as for example suggested by Breazeal et al. [2] these raise questions of whether robots can or should take the role of a tool, a pet, or a person. In fact, recent studies by Dautenhahn et al. [3] and Ezer et al. [4] reveal that most people can think of robots as an assistant or servant with a clear purpose and usage, rather than being purely social with human-like behavior and appearance.

Service robots have been developed for several kinds of domestic tasks, especially for the elderly (e.g. [5, 6]), differing wildly in their behavior and design, depending on their intended functionality [7]. A survey of such is given by Smarr et al. [8]. A task that is applicable for a broader range of people is housecleaning. With the rise of iRobot's Roomba [9, 10] and models from other manufacturers, vacuum cleaning is one of the fields in which robots are already entering people's homes. Other cleaning robots for tasks like floor wiping or pool and window cleaning have already gained in popularity. Although there have been several studies with cleaning robots, they focus on the private use and acceptance of the robot (e.g. [11–13]), or even personalization as investigated by Sung et al. [14–16].

While domestic robots have become popular in recent years, their application as service robots in public spaces is still scarce. This may be due to the obvious limitations of available domestic systems, as they have a high rate of failure, lack the ability to move between floors (due to stairs or elevators), easily get stuck and cannot overcome even the smallest obstacles. While sometimes expensive robots in the private domain are shepherded by their owners who help the robot when it reaches a limitation, they are almost on their own when confronted with a problem in public spaces. Calling a technician for help is not always a feasible option. Many problems could be solved by passersby as well, but the robot needs to attract somebody's attention and convince this person to help which raises issues of HRI. The famous example of hitchBOT[1] shows that people are mostly helpful (until its sudden "death" on August 5, 2015). But hitchBOT was not a service robot, its sole purpose was to make people help it travel around; it was especially designed for this very reason. As service robots are not yet capable of prevailing in public spaces, research on their potential to enter public environments requires a WOZ setting (for an overview of WOZ human-robot studies, see [17]). So the task at hand is to get passersby to help the robot without having a personal relation. To the best of our knowledge, this has not yet been the focus of scientific research.

3 Wizard of Oz Platform

As robots become a larger part of our lives, there will not only be personal robots in our homes; we will also be confronted with unfamiliar robots in public environments. It is important to examine how people react to such robots at first contact.

Everyday experience with an office-bound roomba has showed the authors the boundaries of current domestic robots quite clearly—the roomba often got stuck in cables or under radiators, and could not extract itself from these situations without human assistance. Similarly, future robots may reach their system boundaries during their daily work in public spaces. We propose that they will be able solve many issues by asking bystanders or passersby for help. Thus, the roomba was extended to a Wizard of Oz Platform to investigate HRI. Figure 1 shows the assembled platform in a public waiting room. In the following, the underlying hardware and software components are described.

Fig. 1 The service robot platform performing its vacuum cleaning task in a public waiting room

[1] http://m.hitchbot.me/ (accessed 11/2015).

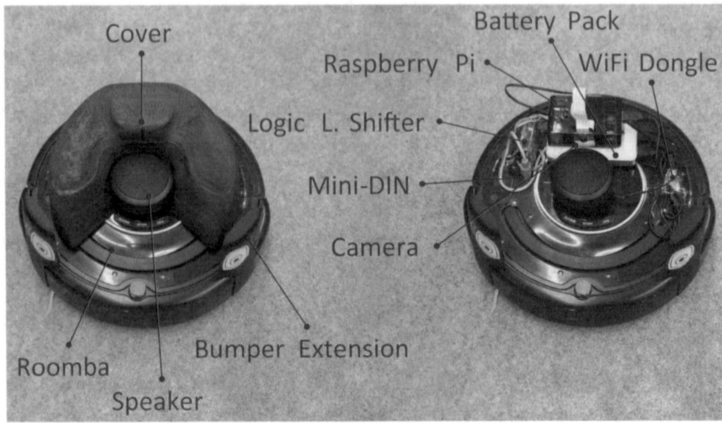

Fig. 2 The service robot platform based on an iRobot Roomba 581. The platform is shown completely covered (*left*) and with all internal components (*right*)

3.1 Hardware

The hardware base is an iRobot Roomba 581 vacuum-cleaning robot controlled by a Raspberry Pi 2 Model B.[2] It is connected to the Open Interface serial port of the Roomba via a logic level shifter that converts the 3.3 V logic levels of the Pi's GPIO pins to the 5 V of the Roomba's 7-pin mini-DIN socket. In addition, it is equipped with a Raspberry Pi Camera, a USB-powered speaker and a WiFi USB dongle. A 12,000 mAh battery pack serves as a long-lasting power supply, providing more than enough power to withstand a complete cleaning cycle of the robot.

All these components are fitted on top of the robot. A 3D-printed cover and two extensions of the robot's front bumper serve protective purposes and allow a safe performance of the robot's autonomous cleaning without risking damage to the platform hardware. Two comic-like eyes on the bumper extensions give the robot a more pleasant and less daunting appearance. Figure 2 shows the platform with all described components. In order to avoid manually plugging in the battery pack for charging, it is co-charged with the roomba itself on its docking station. This is achieved by directing the battery's charging cable to the bottom of the robot and equipping the docking station with matching contactors (see Fig. 3).

3.2 Software

The platform's software mainly consists of two separate web interfaces. One provides the live video from the Raspberry Pi Camera using the RPi-Cam-Web-Interface,[3]

---

[2]https://www.raspberrypi.org/products/raspberry-pi-2-model-b/ (accessed 11/2015).

[3]http://elinux.org/RPi-Cam-Web-Interface (accessed 11/2015).

Fig. 3 Additional copper contactors on the robot's bottom (*left*) and the docking station (*right*) allow simultaneous charging of the roomba's internal battery and the battery pack for the additional hardware components

while the other is a self-developed HTTP server written in Python, which provides the control interface. Embedding both into a single HTML page (using Iframes) on a client within the same network allows remote operation of the robot from an on-board perspective, as shown in Fig. 4. Switching between manual control and autonomous cleaning is possible at any time. Moreover, the interface lists all previously stored audio files (e.g. speech samples).

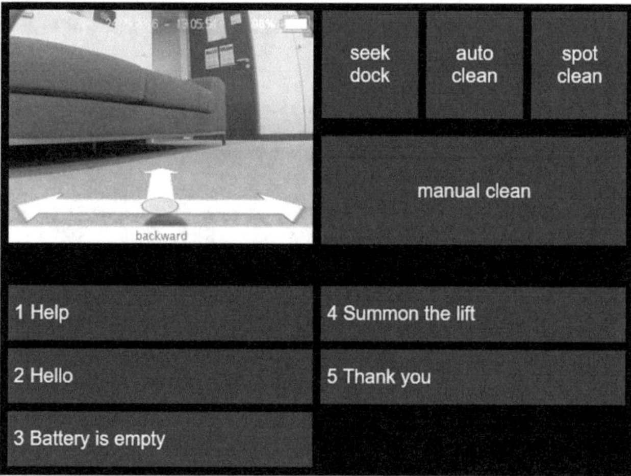

Fig. 4 The web interface for the operator allows complete remote control of the robot's movement and cleaning functions as well as triggering pre-recorded voice outputs

4 First User Study

To test the system's applicability for WOZ scenarios in general and to gather first insights for getting human assistance, we conducted a study on the university campus. The robot was placed near an elevator, which it obviously could not use without additional help, waiting for a helping hand in the form of any passersby (see Fig. 5). As soon as a passerby was approaching the scene, the robot started to attract attention using a pre-recorded male speech of an Ivona Text-to-Speech system. This happened in one of two different flavors:

Peer Condition: On the one hand, the robot acted like a peer by simply uttering "Hello" and stating its need, and once attention was gained: "Please summon the elevator and select floor 4".

Help Condition: On the other hand, it could clearly state its need for help by uttering "Help" and expressing its dependency: "My battery is almost empty. I need to get to my charging station on floor 4. I can't reach the button on my own. Can you help me please?" This was then followed by the specific need as in the first condition ("Please summon ...").

Independent of the applied dialog strategy, once the passerby summoned the elevator, the robot entered it on arrival and thanked the helper: "Thank you for your assistance." Both strategies are illustrated in Fig. 6. At this point of interaction, the experiment was stopped and the passerby was informed of the WOZ study, before they were asked to fill in basic demographic questionnaires. In total, 19 trials were conducted (11 in the peer condition, 8 in the help condition). Regarding the overall number of attempts to get help, only 10 were actually successful (52.6 %). The successful attempts yielded 13 active participants (3 attempts were conducted on passersby

Fig. 5 Study scenario: A passerby is summoning the elevator for the cleaning robot right after the robot asked for assistance

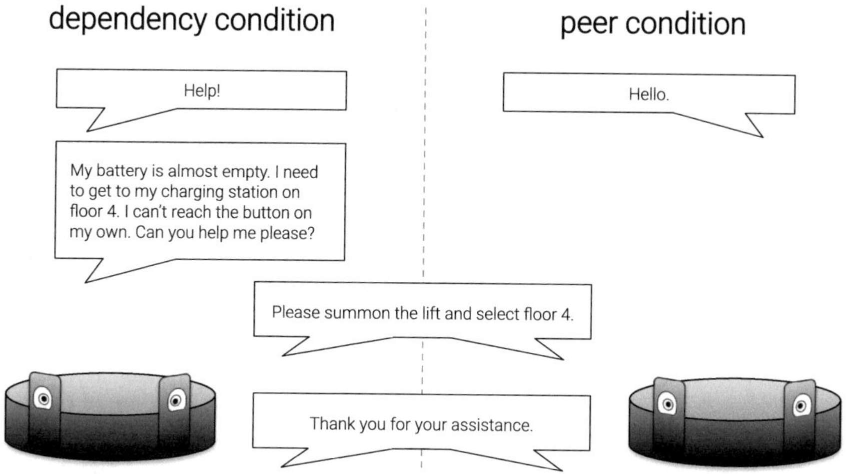

Fig. 6 The dialogues of the dependency and peer conditions

pairs) with a mean age of 32.2 years ($SD = 13.1$). Regarding the difference of the two variants, the peer condition had a success rate of 36.4 % (4 trials), while the help condition performed much better with 75 % success rate (6 trials).

Independent of the two dialog strategies, our results show that gaining attention is rather difficult. This may be due to the relatively poor visibility of the almost black robot on the dark floor in front of the elevator. Although most people seemed to hear the robot's voice, they sometimes could not locate it and so finally ignored it. Other passersby seemed in such a hurry that they simply did not care about the robot. A more striking design with a more easily noticeable color could prevent such oversights.

Once the robot successfully gathered the subject's attention, the difference in the applied dialog strategy is rather remarkable. The help condition's success rate more than doubled that of the peer condition. This shows that a clear statement of the helplessness and explanation for the robot's situation is essential. This differs a lot from the error messages given by the commercially available systems for the home use today, as they usually purely state their needs as in the peer condition, e.g. "clear the path".

5 Conclusion and Future Work

While domestic cleaning robots have become popular in recent years, they are not (yet) found in public spaces although this application seems very reasonable. One handicap for public use are the system's limitations. This means that robots will need help from unknown passersby, e.g. when they get stuck or cannot reach their destination on their own. We presented a low-cost platform, based on an iRobot Roomba, which allows WOZ studies in the wild to investigate promising spontaneous social interactions. The first user study presented here resulted in two key findings. First, gathering attention is rather difficult for small, easily overlooked robots. Thus, the design of the robot should optimize recognizability in the intended surroundings (Fig. 7 shows a new version of the robot with LED-lit eyes). Second, the robot should clearly state the reason for its need of assistance. Simply stating the need (like a peer) is often not enough to convince a stranger to provide help.

For the future, there are a number of technical enhancements that facilitate the system's applicability for further studies in the wild. Currently, the system lacks the ability to record sound. Verbal utterances from the subjects interacting with the robot could not been recorded yet. So we are currently investigating the options to use a standard USB webcam instead of the video-only Raspberry Pi camera. In addition, simultaneous recording of the video and audio data would be helpful, as long as it is not prohibited by ethical regulations. Another subject that needs improvement concerns the sound reproduction capabilities. Currently, the system is limited to previously stored audio files. In the future, we will investigate text-to-speech production possibilities on the Raspberry Pi itself. Another option would be a live stream of audio data to the robot from the operator's machine. This would allow any audio data (including text-to-speech audio) to be played in real-time, without being limited to previously stored data and independent of the Raspberry Pi's processing limits.

From a scientific perspective, we will continue investigating spontaneous social interactions with robots. This includes measurement of the passersby's attitude and empathy towards the robot in need, in combination with more affectively designed dialog strategies. Besides the mere willingness to help the robot, the subjects' behavior can also be analyzed in a more systematic and differentiated manner. This includes factors like the mean distance towards the robot, manual support of the robot's movement, as well as the observed tendency to test the robot's capabilities or even impeding it on purpose.

Fig. 7 New version of the robot using LED-lit eyes for easier recognizability (*left*). The back-mounted touch display allows direct control of the robot (*right*)

Acknowledgements The authors would like to thank all passersby who participated in our study. Moreover, the authors would like to thank the Carl-Zeiss-Foundation for the partially funding of this work. This work was also supported by the Transregional Collaborative Research Center SFB/TRR 62 "Companion-Technology for Cognitive Technical Systems", which is funded by the German Research Foundation (DFG).

References

1. Goodrich, M.A., Schultz, A.C.: Human-robot interaction: a survey. Found. Trends Human-Comput. Interact. **1**(3), 203–275 (2007)
2. Breazeal, C.: Social interactions in HRI: the robot view. IEEE Trans. Syst. Man Cybern. C Appl. Rev. **34**(2), 181–186 (2004)
3. Dautenhahn, K., Woods, S., Kaouri, C., Walters, M.L., Koay, K.L., Werry, I.: What is a robot companion—friend, assistant or butler? In: Proceedings of the IEEE/RSJ International Conference on Intelligent Robots and Systems, 2005 (IROS 2005), pp. 1192–1197. IEEE (2005)
4. Ezer, N., Fisk, A.D., Rogers, W.A.: More than a servant: self-reported willingness of younger and older adults to having a robot perform interactive and critical tasks in the home. In: Proceedings of the Human Factors and Ergonomics Society Annual Meeting, vol. 53, pp. 136–140. SAGE Publications (2009)
5. Kidd, C.D., Taggart, W., Turkle, S.: A Sociable robot to encourage social interaction among the elderly. In: Proceedings of the IEEE International Conference on Robotics and Automation, 2006, ICRA 2006, pp. 3972–3976. IEEE (2006)
6. Hutson, S., Lim, S.L., Bentley, P.J., Bianchi-Berthouze, N., Bowling, A.: Investigating the suitability of social robots for the wellbeing of the elderly. In: Affective Computing and Intelligent Interaction, pp. 578–587. Springer (2011)
7. Dario, P., Guglielmelli, E., Laschi, C.: Humanoids and personal robots: design and experiments. J. Rob. Syst. **18**(12), 673–690 (2001)
8. Smarr, C., Fausset, C.B., Rogers, W.A.: Understanding the potential for robot assistance for older adults in the home environment. Technical report, Georgia Institute of Technology (2011)
9. Jones, J.L.: Robots at the tipping point: the road to iRobot Roomba. IEEE Robot. Autom. Mag. **13**(1), 76–78 (2006)
10. Tribelhorn, B., Dodds, Z.: Evaluating the Roomba: A low-cost, ubiquitous platform for robotics research and education. In: Proceedings of the IEEE International Conference on Robotics and Automation, pp. 1393–1399. IEEE (2007)
11. Fink, J.: Dynamics of Human-Robot Interaction in Domestic Environments. Ph.D. thesis, École Polytechnique Fédérale de Lausanne (EPFL) (2014)
12. Forlizzi, J.: How robotic products become social products: an ethnographic study of cleaning in the home. In: Proceedings of the ACM/IEEE International Conference on Human-Robot Interaction, pp. 129–136. ACM (2007)
13. Broadbent, E., Stafford, R., MacDonald, B.: Acceptance of healthcare robots for the older population: review and future directions. Int. J. Soc. Robot. **1**(4), 319–330 (2009)
14. Sung, J.Y., Guo, L., Grinter, R.E., Christensen, H.I.: My Roomba Is Rambo: Intimate Home Appliances. Springer (2007)
15. Sung, J.Y., Grinter, R.E., Christensen, H., Guo, L. et al.: Housewives or technophiles?: understanding domestic robot owners. In: Proceedings of the 3rd ACM/IEEE International Conference on Human-Robot Interaction (HRI), pp. 129–136. IEEE (2008)
16. Sung, J.Y., Grinter, R.E., Christensen, H.I.: Pimp my Roomba: designing for personalization. In: Proceedings of the SIGCHI Conference on Human Factors in Computing Systems, pp. 193–196. ACM (2009)
17. Riek, L.D.: Wizard of oz studies in HRI: a systematic review and new reporting guidelines. J. Human-Robot Interact. **1**(1) (2012)

Dialogue with Robots to Support Symbiotic Autonomy

Andrea Vanzo, Danilo Croce, Emanuele Bastianelli, Guglielmo Gemignani, Roberto Basili and Daniele Nardi

Abstract Service Robotics is finding solutions to enable effective interaction with users. Among the several issues, the need of adapting robots to the way humans usually communicate is becoming a key and challenging task. In this context the design of robots that understand and reply in Natural Language plays a central role, especially when interactions involve untrained users. In particular, this is even more stressed in the framework of Symbiotic Autonomy, where an interaction is always required for the robot to accomplish a given task. In this article, we propose a framework to model dialogues with robotic platforms, enabling effective and natural dialogic interactions. The framework relies on well-known theories as well as on perceptually informed spoken language understanding processors, giving rise to interactions that are tightly bound to the operating scenario.

Keywords Dialogue modeling · Symbiotic autonomy · Mobile service robots

A. Vanzo (✉) · G. Gemignani · D. Nardi
Department of Computer, Control and Management Engineering "Antonio Ruberti", Sapienza University of Rome, Rome, Italy
e-mail: vanzo@dis.uniroma1.it

G. Gemignani
e-mail: gemignani@dis.uniroma1.it

D. Nardi
e-mail: nardi@dis.uniroma1.it

D. Croce · R. Basili
Department of Enterprise Engineering, University of Roma Tor Vergata, Rome, Italy
e-mail: croce@info.uniroma2.it

R. Basili
e-mail: basili@info.uniroma2.it

E. Bastianelli
Department of Civil Engineering and Computer Science Engineering, University of Roma Tor Vergata, Rome, Italy
e-mail: bastianelli@ing.uniroma2.it

K. Jokinen and G. Wilcock (eds.), *Dialogues with Social Robots*,
Lecture Notes in Electrical Engineering 427,
DOI 10.1007/978-981-10-2585-3_27

1 Introduction

Robots are expected to support human activities in everyday environments, interacting with different kinds of users. In particular, domestic robots (i.e. robots operating in our homes) have already entered the market, e.g. cleaning robots or telepresence robots for elderly care. In these contexts, the interaction with the user plays a key role.

The current development of robotics technology is facing several difficulties in providing general solutions to this problem. The major causes that withhold the realization of a robust Natural Language interface consist of the enormous variety of environments, involved users and tasks to be executed, aspects that need to be understood by a robot. On the one hand, the perception capabilities of the robots make it difficult to build rich and reliable representations of the operational environment; on the other hand, combining motion and manipulation capabilities on a single platform is still very expensive and makes the size of the robot not well suited for operation in homes. While these difficulties may require some time before satisfactory solutions became available, a number of researchers are proposing to exploit Human-Robot Interaction (HRI) to enable the robot to understand the environment and accomplish tasks that would be otherwise unachievable. This line of research has been termed *Symbiotic Autonomy* [1] and it substantially relies on spoken dialogue between robots and users.

In fact, given the recent advancements in *Spoken Language Recognition and Understanding*, dialogue in Natural Language will be a major component of robotic interfaces, also considering that it will certainly be coupled with other multi-modal communication channels. In this respect, "Dialogue with Robots" has been the focus of recent research, as confirmed by special issues dedicated by several journals to this topic [2].

In the context of HRI, the focus is usually on situated dialogues. In situated dialogues, robots and humans have different representations of the shared environment, because of their mismatched perceptual capabilities. Accordingly, understanding dialogue is about more than just understanding the speech signal, words, or the utterance. Hence, for a robot that is expected to understand dialogue when talking with a human, it is essential to interpret how that dialogue relates and refers to the surrounding world.

The aim of the present work is twofold. First, we identify different HRI scenarios and situations where dialogue can be beneficial and plays a key role. As an example, dialogic interactions allow to fulfill missing information when a command has not been completely understood by the robot as well as when the resulting interpretation involves manifold ambiguities. Second, we provide pragmatic solutions to deal with this problem, along with possible frameworks that enable an effective interaction between humans and robots.

The rest of the article is structured as follows. Section 2 reports related works, while Sect. 3 describes backgrounds and proposed solutions. In Sect. 4, we identify some use cases in the context of HRI. Finally, Sect. 5 provides final remarks.

2 Related Work

In the context of Human-Robot Interaction (HRI), Natural Language Understanding has been studied starting from [3], where the focus was on a system able to process NL instructions to perform actions in a virtual environment. In the Robotic field, speech-based approaches have been applied to deploy robotic platforms in a wide plethora of environments. These techniques have been used in manipulation tasks [4], and for wheeled platforms [5, 6]. Dialogue has also been employed to instruct robots to accomplish a given unknown task, such as giving a tour [7], delivering objects [8], or manipulating them [9]. Other related works have combined speech-based approaches with other types of interactions [6, 10].

More recently, several domain-specific systems that allow users to instruct robots through Natural Language have been presented in literature. For example, in [11, 12], the authors present different methods for following Natural Language route instructions, by decoupling the semantic parsing problem from the grounding problem. In these works, the input sentences are first translated to intermediate representations, which are then grounded into the available knowledge. In [13], the authors present a preliminary version of a cascade of reusable Natural Language Processing (NLP) modules, that can be adapted to changing operational scenarios, through trainable statistical models for which HRI specific learning algorithms. These modules range from ASR re-ranking functions (e.g. [14, 15]) up to techniques to ground entities according to lexical references [14, 16]. A further refinement of such a cascade has been proposed in [17], where a standard pipeline for semantic parsing is extended toward a form of perceptually informed NLP, by combining discriminative learning, distributional semantics and perceptual knowledge. In [18], the authors show how to enable Natural Language interactions in a scenario of collaborative human-robot tasks, by mining past interactions between humans in online multiplayer games.

In [19], the authors present a probabilistic approach able to learn referring expressions for robot primitives and physical locations in a map, by exploiting the dialogue with the user. The problem of Referring Expressions Generation (REG) has also been taken into account by in [20]. They propose a hypergraph-based approach to account for group-based spatial relations and uncertainties in perceiving the environment, in the context of situated dialogues. A further refinement of their approach is introduced in [21]. Here, they develop two collaborative models for REG. Both models, instead of generating a single referring expression to describe a target object as in the previous work, generate multiple small expressions that lead to the target object with the goal of minimizing the collaborative effort. A study examining the generation of noun phrases within a spoken dialogue agent for a navigation domain is presented in [22]. Here the noun phrase generation is driven by both the dialogue history and spatial context features, e.g. view angle of the agent, distance from the target referent and the number of similar items in view. In [23], the authors present a Natural Language generation approach which models, exploits, and manipulates the non-linguistic context in situated communication. The proposed method for the generation of referring expressions is tightly integrated with the syntactic realization of the sentence.

The problem of tackling the vocabulary in conversational systems has been addressed in [24]. They propose approaches that incorporate user language behavior, domain knowledge, and conversation context in word acquisition, evaluating such methods in the context of situated dialogue in a virtual world. In [25], the authors present four case studies of implementing a typical HRI scenario with different state-of-the-art dialogue frameworks with the goal to identify pitfalls and potential remedies for dialogue modeling on robots. They show that none of the investigated frameworks overcomes all problems in one solution. In [26], the authors focused on recovery from *situated grounding problems*, a type of miscommunication that occurs when an agent fails to uniquely map a person's instructions to its surroundings.

NLP in Robotics can be coupled with other communication channels. In [27], a flexible dialogue-based robotic system for humanlike interaction is proposed. In particular, they focus on task-based dialogues, where the robot behavior is changed based on a tight integration between Natural Language and action execution. An algorithmic framework, Continual Collaborative Planning (CCP), for modeling the integration of the different channels in situated dialogues has been proposed in [28]. This framework allows to integrate planning, acting and perception with communication. Similarly, in [29] the authors propose information-state dialogue management models for the situated domain. Here, the dialogue management model fuses information-state update theory, with a light-weight rational agency model.

Nevertheless, all the presented works are not able to recover when manifold ambiguities and missing information are found and to incrementally enhance their Natural Language Understanding from the continuous interaction with the user. Moreover, the state of the robot is often neglected, giving rise to additional ambiguities and misunderstandings. In the next section all these aspects are addressed in detail.

3 A Pragmatic Approach for Dialogue Modeling

According to the Symbiotic Autonomy paradigm, we investigated several realms, where dialogic interactions between a user and a robot are beneficial: from the robot perspective, they allow to better understand the user needs while from the user perspective, it is the most natural way to support the robot in a better comprehension of the user's requests. Section 3.1 provides some of the motivations of this work. A possible dialogue-based framework for Human-Robot Interaction is presented in Sect. 3.2.

3.1 Background and Motivation

In our earlier research on Human-Robot Interaction, we addressed the **Spoken Language Understanding** (SLU) task for the automatic interpretation of commands. Given a spoken command, this process aims at automatically analyzing the user's utterances to derive computational structures that (i) reflect the meaning of the commands and (ii) activate the robot plans. Nevertheless, the correct interpretation of a command does not merely depend only on the linguistic information that is derivable from the utterance. As suggested in [17] the SLU process does depend also on other factors, e.g., the environment surrounding the robot. As an example, the command *Take the book on the table* requires the robot to *Take* the book from the table only if there is actually a book on it; on the contrary, the same command requires the robot to *Bring* the book over the table.[1] Dialogue is crucial in order to support a proper comprehension of a command, e.g., when some information is missing. A command such as *Take the book* cannot be executed if the robot is unaware of the position of the book. In these cases, the robot could require some additional information to complete the task and fulfill the user needs.

Secondly, we analyzed the process called **Human Augmented Mapping** (HAM) that corresponds to a specific approach to support a robot in acquiring representations of the environment, in order to associate symbols to objects and locations perceived by the robot. These representations enable the robot to actually execute commands like "*go to my bedroom*", without being tele-operated by the user or requiring him to specify a target position in terms of coordinates. This process provides a general framework that does not depend from the underlying platform, also improving the adoption of a map in different robots. Moreover, it enables an incremental construction of the representation, as well as its revision in accordance with the changes in the environment [6]. In the HAM process, dialogue is crucial for a natural interaction between the user and the robot, especially when some properties of the entities or the environment itself cannot be directly derived from the sensory apparatus (e.g., whether an object is fragile or not).

Finally, we considered the **Task Teaching** process, that involves the interaction between the user and the robot to teach complex commands, which can be composed by primitive actions. In this respect, dialogue can support the extension of previous approaches by enabling the robot to learn parametric commands, as well as exploiting the knowledge about tasks to simplify the learning process [8, 31].

Hereafter, we will discuss a possible dialogue-based framework for Human-Robot Interaction to support the above tasks.

[1] A video describing this example is available at https://goo.gl/bpXmln and the underlying system is presented in [30].

3.2 A Framework for Flexible Pragmatic Task-Based Dialogues

We propose the adoption of an approach that we consider, to some extent, to be *pragmatic*. In fact, the final aim of the dialogue is to fulfill the information required to accomplish a given task, regardless it is an activity required to the robot or a step in the overall interpretation/mapping process.

We will adhere to the theory of *Information State* [32] for the management of the dialogues between the user and the robot. Such theory contemplates **informational components** (i.e. description of the context shared by the participants), **formal representation** of the aforementioned components, **dialogue moves** that trigger the update of such information, the **update rules** to be applied and the **update strategy** that is supposed to trigger the proper update rule.

The proposed framework will thus rely on the above (general) definitions to allow an easy and cost-effective design of dialogic interactions, specialized for a targeted task. These ideas are reflected by our framework, that is sketched in Fig. 1 and described hereafter.

The first module to be invoked during the processing of a user's utterance is the *Dialogue Act Classifier*: it extracts the intent of the user, expressed as a subset of Dialogue Acts (DA) proposed in [33]. This module gets the transcription of spoken utterances from Automatic Speech Recognition (ASR). Once the intent of a sentence

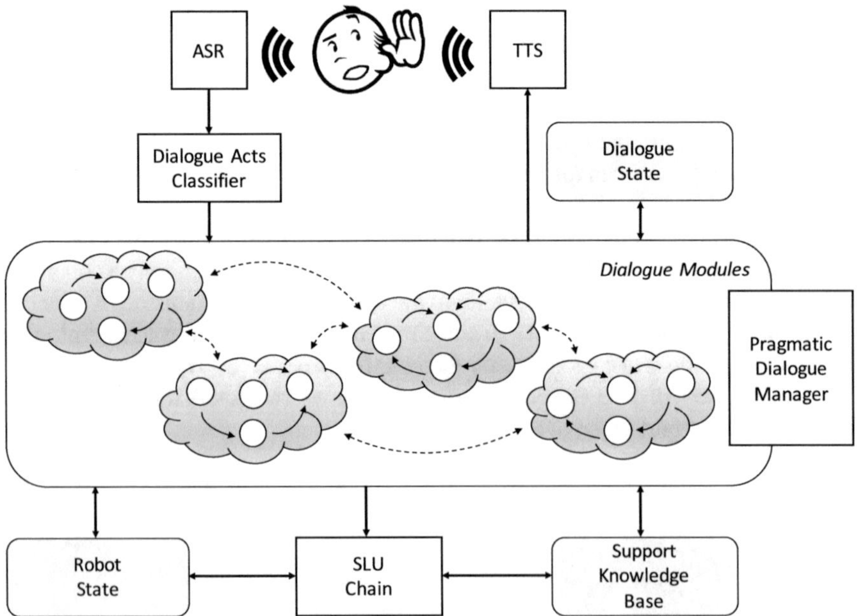

Fig. 1 The proposed framework for pragmatic task-based dialogue modeling

has been extracted (i.e. the user needs), the control of the dialogue is delivered to the *Pragmatic Dialogue Manager*, that controls the dialogue flow. It activates task-based representation structures (*Dialogue Modules*) that are in charge of fulfilling the missing information to accomplish the task. Such modules can be realized as (Partially Observable) Markov Decision Processes (POMDPs) or relying on Petri Net Plans (PNPs). The second solution allows to take into account the robot behavior and to harmonize the dialogue flow with the actions performed by the platform. Regardless of the implementation, possible interrupts are considered in the dialogue flow, to allow the user to control the overall dialogue and to facilitate timely reactions of the robotic platform to the user needs.

The status of the overall dialogue is traced by the Pragmatic Dialogue Manager by updating the *Dialogue State*, which stores aspects of the dialogue, such as the shared context and the parameters required by the robot to accomplish a given task. We decoupled such information with other aspects that are strictly related to the robot, namely the *Robot State* and the *Support Knowledge Base*. While the former collects physical and abstract aspects of the robot (e.g. manipulator availability, inability to perform a task, ...), the latter maintains a structured representation of the environment, formalized through semantic maps and domain models. These resources are employed by the *Spoken Language Understanding Chain* (SLU Chain), that produces an interpretation of a user's utterance. The adopted SLU Chain [30] makes the interpretation process dependent also on the robot capabilities and the environmental settings, such as existence of entities referred in a user utterance, as well as spatial relations among them.

4 Use Cases

We identified several situations where such a framework can be used. Such scenarios are summarized hereafter and a more detailed use-case (related to the Human Augmented Mapping task) is reported at the end of this section.

Reasoning about the environment. In order to enable a semantic-aware navigation of the environment, the robot needs a structured representation of the world in which it operates. This representation is often built by relying on the interaction with the user, that instruct the robot for the operating environment. Often, this representation presents some mismatches with the real world, e.g. a book that is not into the semantic map but it is present the real environment. When the user asks to *Take the book*, the robot is supposed to start a dialogue to detect the book position, add it into the semantic map and complete the task.

Management of robot's self-awareness. Another situation where the dialogue is able to recover from undesired situations is when the robot is aware of its state (e.g. busy tray or manipulator, capability to perform some actions, ...) and use a dialogic interaction to solve possible issues. In this case, when the robot detects a mismatch between the user needs and its state, it should be able to leverage an interaction to solve potential hindrances.

Dealing with persisting ambiguities. In [13], a lexical grounding function has been proposed. Such function is able to link lexical references to real object, by observing linguistic aspects of the referring objects. We propose to use the dialogue to solve uncertainties about objects in the environment. A common scenario is when the user asks for an object and two entities of the same type are present. A proper interaction is supposed to solve such ambiguities.

Dialogue-based Human Augmented Mapping and Task Teaching. Each of these two tasks (Human Augmented Mapping and Task Teaching) required an ad-hoc interaction to accomplish the desiderata. In these cases, the dialogue can be employed for two purposes: (i) to acquire all the information needed in order to add a new entity to its semantic representation of the environment and (ii) to instruct the robot to perform new and unknown actions.

Interaction with user for re-training. The last scenario we are dealing with is when the sentence is misinterpreted. Assuming that our interpretation chain is based on Machine Learning techniques, correcting of a wrong interpretation represents a further step toward a system that is able to continuously learn by mistakes and improve its accuracy as interactions occurs. In this case, a suitable dialogue can lead the user to provide the correct interpretation of a sentence and this new observation can be employed to re-train the models the chain relies on.

4.1 Dialogue for Human Augmented Mapping

In Human Augmented Mapping, the user instructs the robot in constructing a structured representation of the environment. In this representation symbols are associated to the objects and locations in the environment. This operation is performed through interactions in Natural Language, where the robot learns the entities to be included in the semantic map. To this end, the sentences uttered by the user provide a mechanism to acquire the symbolic representation of the targeted object and location that populates the knowledge base.

When the robot is idle, the user can give commands to the robot. Once the user's intent and a structured representation of the sentence meaning are extracted by the SLU Chain, the robot attempts to ground each entity within the command. If an argument of the command denoting an object/location can not be grounded, the robot asks the user to help him in acquiring the position and features of the object/location. During this process, the user drives the robot with spoken commands such as *Turn right*, *Follow me* or *Go to the laboratory*. When the robot stands in front of the object/location that needs to be grounded, the user points to the object[2] and, once the

[2]This is currently done by a laser pen.

dot is recognized by a dot detection system, tells the robot the referring expression for the object, e.g., *Robot, this is the book.*

In the example below, the user refers to an object (a *book*) that is not present in the map. In other words, the robot does not know where the *book* is. Hence, the command *Bring me the book* can not be executed without the information about the existence of a book. The Human Augmented Mapping allows to complete such lack of knowledge and accomplish the task. To this end, once the request is given to the robot, the Dialogue Manager activates the dialogue module related to the Human Augmented Mapping.

User: `Hi robot, I'm Andrea.`
Robot: `Hi Andrea. How can I help you?`
User: `Yes. Could you please bring me the book on the table?`
Robot: `I'm sorry, but I don't see any book. Could you help me?`
User: `Yes, sure. Follow me!`
Robot: `Okay, I'm following you!`
`[The robot starts following the user]`
User: `Okay, here we are.`
`[The robot stops following]`
User: `Now turn left.`
`[The robot turns left and the user points the book]`
User: `This is the book.`
Robot: `I suppose that I am able to grasp the book. Is it correct?`
User: `Yes, it is.`
`[The robot label the Graspable property of the book as true]`
Robot: `Thank you, now I know what to bring you.`
...

The Dialogue Manager is also in charge of handling interrupts of the dialogue flow when commands that are not exclusively related to the Human Augmented Mapping, such as *Follow me* or *Turn left*, are provided. Finally, once the user points the object and assigns a lexical reference to it (i.e., *book*), the Support Knowledge Base is updated with this information (i.e., position and affordances), also enabling the interpretation of future commands.

Within the same scenario, the next example shows how a dialogue can be used to solve ambiguous spatial references in the environment. In fact, we consider the same request of the previous example to a robot that does not have any reference of the book in the map, but it has references of two different tables:

User: `Hi robot.`
Robot: `Hi. Can I help you?`
User: `Yes. Bring me the book on the table.`
Robot: `I'm sorry, but I don't see any book. Could you help me?`
User: `Yes. The book is on the table.`
Robot:`I see two tables in my map. Which one are you referring to?`
User: `The table of the laboratory.`
Robot: `Okay. I'm going to take the book in the laboratory. Please, wait.`
User: `Okay, thank you.`
...

Notice that the Dialogue Manager deals with persisting ambiguities of the interpretation. In fact, the sentence *The book is on the table* is inherently ambiguous, whenever two tables co-exist in the same environment.

In conclusion, the above examples show different pragmatic interactions with different objectives. At the very beginning, the greeting dialogue is always activated whenever the user introduces himself. Then, once the intent of the user is understood, the request is delivered to the specific dialogue module, until the request is accomplished. Additionally, during the interactions the Dialogue Manager activates the flows to (i) fill missing knowledge, e.g., the robot does not know where is the book, (ii) drive it through specific commands, e.g., following and turning, (iii) characterize the entity, e.g., affordances, and (iv) resolve persisting ambiguities, e.g. the robot attempts to get the referred table. These features are essential to enable a natural interaction when teaching robots as in a Symbiotic Autonomy approach.

5 Conclusion

In this article, we propose a pragmatic framework aiming at effectively modeling dialogues within robotics platforms. The proposed approach aims at providing a natural interaction between a user and a robot by jointly exploiting (i) contextual information acquired by the robot, i.e., from the semantic map reflecting a semantically enriched representation of robot perceptions, (ii) knowledge related to the task to be accomplished, and (iii) other knowledge essential when dealing with robotic platforms, i.e., the robot state. Additionally, the framework allows to incrementally expand such resources, resulting in a more accurate and natural interaction with a robot that adapts itself to the user's profile.

The framework is based on the theory of Information State and the resulting architecture is decoupled in several task-based modules that are designed to support the robot in accomplishing the user's requests. The resulting architecture is thus biased towards the information required by the robot to determine the objectives of each interaction. The framework relies on a perceptually informed Spoken Language Understanding Chain to extract a structured representation of the meaning of user's utterances. In fact, such a chain exploits contextual information, e.g., existence of entities within the environment and spatial relations among them, to provide unambiguous interpretations and groundings that indeed depend on the environment where the interactions arise, as discussed in [17].

In order to support the potential contribution of the proposed approach, we identified some scenarios, in the context of Symbiotic Autonomy, that can benefit from the adoption of this framework, ranging from solving possible ambiguities of a command up to dialogue-based interactions for Human Augmented Mapping.

References

1. Rosenthal, S., Biswas, J., Veloso, M.: An effective personal mobile robot agent through symbiotic human-robot interaction. In: Proceedings of the 9th International Conference on Autonomous Agents and Multiagent Systems, vol. 1, pp. 915–922. International Foundation for Autonomous Agents and Multiagent Systems (2010)
2. Bohus, D., Horvitz, E., Kanda, T., Mutlu, B., Raux, A.: Introduction to the special issue on dialog with robots. AI Mag. **32**(4), 15–16 (2011)
3. Winograd, T.: Procedures as a representation for data in a computer program for understanding natural language. Cogn. Psychol. **3**(1), 1–191 (1972)
4. Zuo, X., Iwahashi, N., Funakoshi, K., Nakano, M., Taguchi, R., Matsuda, S., Sugiura, K., Oka, N.: Detecting robot-directed speech by situated understanding in physical interaction. Trans. Jpn. Soc. Artif. Intell. **25**(6), 670–682 (2010)
5. Kruijff, G., Zender, H., Jensfelt, P., Christensen, H.: Situated dialogue and understanding spatial organization: knowing what is where and what you can do there. In: Proceedings of the 15th IEEE International Symposium on Robot and Human Interactive Communication, 2006, ROMAN 2006, pp. 328–333 (2006)
6. Bastianelli, E., Bloisi, D., Capobianco, R., Cossu, F., Gemignani, G., Iocchi, L., Nardi, D.: On-line semantic mapping. In: Proceedings of the 16th International Conference on Advanced Robotics (ICAR). pp. 1–6 (2013)
7. Rybski, P.E., Yoon, K., Stolarz, J., Veloso, M.M.: Interactive robot task training through dialog and demonstration. In: Proceedings of the ACM/IEEE International Conference on Human-robot Interaction, pp. 49–56. HRI '07, ACM, New York, NY, USA (2007)
8. Gemignani, G., Bastianelli, E., Nardi, D.: Teaching robots parametrized executable plans through spoken interaction. In: Proceedings of the 2015 International Conference on Autonomous Agents and Multiagent Systems, pp. 851–859. AAMAS '15, International Foundation for Autonomous Agents and Multiagent Systems, Richland, SC (2015)
9. Gemignani, G., Klee, S.D., Veloso, M., Nardi, D.: On task recognition and generalization in long-term robot teaching. In: Proceedings of the 2015 International Conference on Autonomous Agents and Multiagent Systems. pp. 1879–1880. AAMAS '15, International Foundation for Autonomous Agents and Multiagent Systems, Richland, SC (2015)
10. Lemaignan, S., Alami, R.: "talking to my robot": From knowledge grounding to dialogue processing. In: Proceedings of the 8th ACM/IEEE International Conference on Human-robot Interaction. pp. 409–410. HRI '13, IEEE Press, Piscataway, NJ, USA (2013)
11. Kollar, T., Tellex, S., Roy, D., Roy, N.: Toward understanding natural language directions. In: Proceedings of the 5th ACM/IEEE International Conference on Human-robot Interaction, pp. 259–266. HRI '10, IEEE Press, Piscataway, NJ, USA (2010)
12. MacMahon, M., Stankiewicz, B., Kuipers, B.: Walk the talk: Connecting language, knowledge, and action in route instructions. In: Proceedings of the 21st National Conference on Artificial Intelligence, vol. 2, pp. 1475–1482. AAAI'06, AAAI Press (2006)
13. Bastianelli, E., Castellucci, G., Croce, D., Basili, R., Nardi, D.: Effective and robust natural language understanding for human-robot interaction. In: Proceedings of 21st European Conference on Artificial Intelligence, pp. 57–62. IOS Press (2014)
14. Bastianelli, E., Croce, D., Basili, R., Nardi, D.: Using semantic maps for robust natural language interaction with robots. In: Sixteenth Annual Conference of the International Speech Communication Association (2015)
15. Vanzo, A., Croce, D., Bastianelli, E., Basili, R., Nardi, D.: Robust spoken language understanding for house service robots. In: Proceedings of the 17th International Conference on Intelligent Text Processing and Computational Linguistics CICLing 2016. Konya, Turkey, April 3–9, 2016
16. Bastianelli, E., Croce, D., Basili, R., Nardi, D.: Using semantic models for robust natural language human robot interaction. In: Gavanelli, M., Lamma, E., Riguzzi, F. (eds.) AI*IA 2015 Advances in Artificial Intelligence: XIVth International Conference of the Italian Association for Artificial Intelligence, pp. 343–356. Springer International Publishing (2015)

17. Bastianelli, E., Croce, D., Vanzo, A., Basili, R., Nardi, D.: A discriminative approach to grounded spoken language understanding in interactive robotics. In: Proceedings of the Twenty-Fifth International Joint Conference on Artificial Intelligence, IJCAI 2016. New York, USA, July 9–15, 2016
18. Chernova, S., DePalma, N., Breazeal, C.: Crowdsourcing real world human-robot dialog and teamwork through online multiplayer games. AI Mag. **32**(4), 100–111 (2011)
19. Kollar, T., Perera, V., Nardi, D., Veloso, M.M.: Learning environmental knowledge from task-based human-robot dialog. In: ICRA, pp. 4304–4309. IEEE (2013)
20. Fang, R., Liu, C., She, L., Chai, J.Y.: Towards situated dialogue: revisiting referring expression generation. In: EMNLP, pp. 392–402 (2013)
21. Fang, R., Doering, M., Chai, J.Y.: Collaborative models for referring expression generation in situated dialogue. In: Proceedings of the Twenty-Eighth AAAI Conference on Artificial Intelligence, pp. 1544–1550 (2014)
22. Stoia, L., Shockley, D.M., Byron, D.K., Fosler-Lussier, E.: Noun phrase generation for situated dialogs. In: Proceedings of the Fourth International Natural Language Generation Conference, pp. 81–88. INLG '06, Association for Computational Linguistics, Stroudsburg, PA, USA (2006)
23. Garoufi, K., Koller, A.: Automated planning for situated natural language generation. In: Proceedings of the 48th Annual Meeting of the Association for Computational Linguistics, pp. 1573–1582. ACL '10, Association for Computational Linguistics, Stroudsburg, PA, USA (2010)
24. Qu, S., Chai, J.Y.: Context-based word acquisition for situated dialogue in a virtual world. J. Artif. Intell. Res. 247–277 (2010)
25. Peltason, J., Wrede, B.: The curious robot as a case-study for comparing dialog systems. AI Mag. **32**(4), 85–99 (2011)
26. Marge, M., Rudnicky, A.I.: Miscommunication recovery in physically situated dialogue. In: Proceedings of the 16th Annual Meeting of the Special Interest Group on Discourse and Dialogue, p. 22 (2015)
27. Scheutz, M., Cantrell, R., Schermerhorn, P.W.: Toward humanlike task-based dialogue processing for human robot interaction. AI Mag. **32**(4), 77–84 (2011)
28. Brenner, M., Kruijff-Korbayová, I.: A continual multiagent planning approach to situated dialogue. In: Proceedings of the Semantics and Pragmatics of Dialogue (LONDIAL), p. 61 (2008)
29. Ross, R.J., Bateman, J.: Agency & information state in situated dialogues: analysis & computational modelling. In: Proceedings of the DiaHolmia, p. 113 (2009)
30. Bastianelli, E., Croce, D., Vanzo, A., Basili, R., Nardi, D.: Perceptually informed spoken language understanding for service robotics. In: Proceedings of the IJCAI2016 Workshop on Autonomous Mobile Service Robots. To appear (2016)
31. Klee, S.D., Gemignani, G., Nardi, D., Veloso, M.: Graph-based task libraries for robots: generalization and autocompletion. In: Proceedings of the 14th Italian Conference on Artificial Intelligence (2015)
32. Traum, D., Larsson, S.: The information state approach to dialogue management. In: Current and New Directions in Discourse and Dialogue, Text, Speech and Language Technology, vol. 22, pp. 325–353. Springer, Netherlands (2003)
33. Stolcke, A., Coccaro, N., Bates, R., Taylor, P., Van Ess-Dykema, C., Ries, K., Shriberg, E., Jurafsky, D., Martin, R., Meteer, M.: Dialogue act modeling for automatic tagging and recognition of conversational speech. Comput. Linguist. **26**(3), 339–373 (2000)

Towards SamiTalk: A Sami-Speaking Robot Linked to Sami Wikipedia

Graham Wilcock, Niklas Laxström, Juho Leinonen, Peter Smit, Mikko Kurimo and Kristiina Jokinen

Abstract We describe our work towards developing SamiTalk, a robot application for the North Sami language. With SamiTalk, users will hold spoken dialogues with a humanoid robot that speaks and recognizes North Sami. The robot will access information from the Sami Wikipedia, talk about requested topics using the Wikipedia texts, and make smooth topic shifts to related topics using the Wikipedia links. SamiTalk will be based on the existing WikiTalk system for Wikipedia-based spoken dialogues, with newly developed speech components for North Sami.

Keywords Language revitalisation · Speech technology · Humanoid robots · Spoken dialogue systems

1 Introduction

In this article, we describe our work towards developing SamiTalk, a robot application for the North Sami language. This robot application is chosen because it is an interface to collaboratively edited Wikipedia information, and as a novel application, it is

G. Wilcock (✉) · N. Laxström · K. Jokinen
University of Helsinki, Helsinki, Finland
e-mail: graham.wilcock@helsinki.fi

N. Laxström
e-mail: niklas.laxstrom@helsinki.fi

K. Jokinen
e-mail: kristiina.jokinen@helsinki.fi

J. Leinonen · P. Smit · M. Kurimo
Aalto University, Espoo, Finland
e-mail: juho.leinonen@aalto.fi

P. Smit
e-mail: peter.smit@aalto.fi

M. Kurimo
e-mail: mikko.kurimo@aalto.fi

K. Jokinen and G. Wilcock (eds.), *Dialogues with Social Robots*,
Lecture Notes in Electrical Engineering 427,
DOI 10.1007/978-981-10-2585-3_28

expected to increase the visibility of the language as well as interest in it. In particular, it is expected that young people may become more interested in using the language, which is regarded as an important and effective strategy for language revival in general. The motivation for creating robot applications to support revitalisation of endangered languages is discussed in more detail in [1].

SamiTalk will be the first robot application in the North Sami language, enabling users to hold spoken dialogues with a humanoid robot that speaks and recognizes North Sami. The robot will access information from the Sami Wikipedia, will talk about requested topics using the Wikipedia texts, and will make smooth topic shifts to related topics using the Wikipedia links. SamiTalk will be based on the existing WikiTalk system for Wikipedia-based spoken dialogues (see Sect. 2), with newly developed speech components for North Sami.

The article is structured as follows. Section 2 summarizes the existing WikiTalk system and explains the differences between a localisation of WikiTalk for North Sami and previous localisations. Section 3 describes the development of the speech synthesizer and speech recognizer for North Sami that will be essential components of SamiTalk. Section 4 gives an example of the style of interaction for getting Wikipedia information with Samitalk. Section 5 indicates plans for future work.

2 WikiTalk and SamiTalk

The WikiTalk system [2, 3] accesses Wikipedia directly online. Using paragraphs and sentences from Wikipedia, the system can talk about whatever topics are of interest to the user. The overall architecture of the WikiTalk system on the Nao robot is shown in Fig. 1.

There are other applications that can read out Wikipedia articles, but simply reading out an article is a monologue rather than a dialogue. The key feature that enables WikiTalk to manage dialogues rather than monologues is its ability to handle smooth topic shifts, as described by [2]. Hyperlinks in the Wikipedia text are extracted, to be used as potential topic shifts. The system predicts that the user will often want to shift the topic to one of the extracted links.

The main challenge for dialogue modelling in WikiTalk is to present information in a way that makes the structure of the articles clear, and to distinguish between two conditions: the user shows interest and wants the system to continue on the current topic, or the user is not interested in the topic and the system should stop or find some other topic to talk about [3].

WikiTalk has been internationalised, and localised versions are now available for English, Finnish and Japanese. A number of major issues in internationalisation and localisation of spoken dialogue systems are discussed by [5], using WikiTalk as an example. In the case of WikiTalk, each new language version requires speech recognition and speech synthesis components for the language, an available Wikipedia in the language, and a localisation of the WikiTalk multimodal user interface.

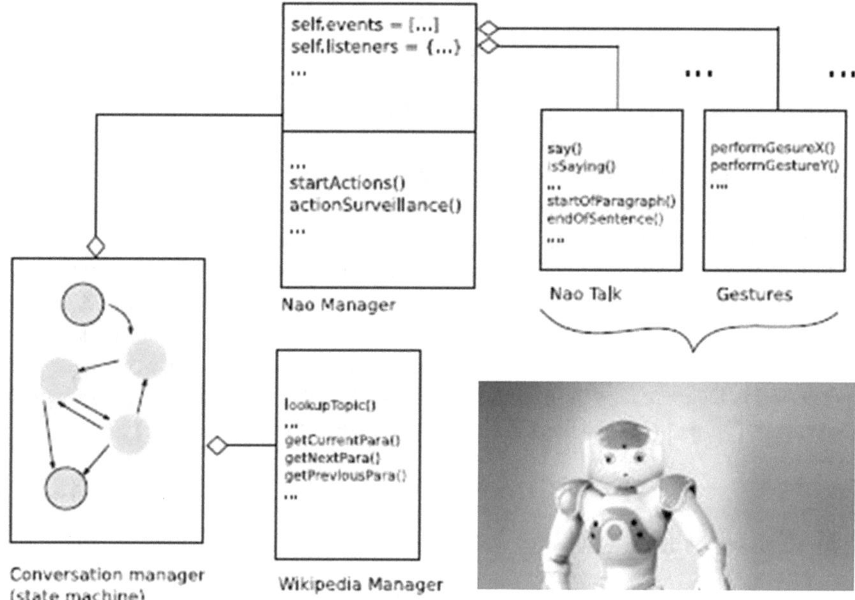

Fig. 1 System architecture of WikiTalk on the Nao robot. From [4]

Due to the internationalisation of WikiTalk, new language localisations can generally be produced relatively rapidly. However, a localisation of WikiTalk for North Sami involves more than previous localisations of WikiTalk. Up to now, the speech recognition and speech synthesis components for the languages supported by WikiTalk have been the components provided by the Nao robot. In order to use these languages, the WikiTalk system simply needs to check that the required language is installed on the robot. All Nao robots can speak English, but other languages need to be installed. Currently 19 languages (including Finnish and Japanese) are available for purchase. If the required language is installed on the robot, it can be selected and set as the current language until another language is selected.

The situation for North Sami is quite different, as it is not one of the 19 available languages. Moreover, there are no off-the-shelf speech technology tools available for North Sami, so it is necessary to develop them from scratch. The required speech recognition and speech synthesis components are being newly developed for the DigiSami project using the new DigiSami corpus.

3 Speech Technology for SamiTalk

Both a speech recognizer and speech synthesizer are essential parts for a Sami-speaking robot. Ideally the robot would be able to speak any free-form North Sami sentence and be able to recognize any spoken sentence from any North Sami speaker.

Given the amount of available resources however, this recognizer is not attainable at the moment and the current recognizer can either understand large vocabulary continuous speech from a single speaker, or only a small list of words, e.g. a list of common Wikipedia titles, from a wide range of speakers.

The biggest obstacle to creating these systems is the limited availability of high quality data. Especially for large-vocabulary continuous speech recognition, which normally requires more then 100 different speakers for a good model, the amount of available data limits both functionality and performance.

In this research we used audio data available from only two speakers; one male and one female. The speech is clean, read speech and amounts to 4.6 and 3.3 hours respectively. Two text corpora are available. The first is a download of the North Sami Wikipedia which contains 10 K sentences with 20 K different word types and the second is *Den samiske tekstbanken* provided by the University of Tromsø, which has almost 1 M sentences with 475 K different word types.

3.1 Speech Synthesis

A statistical Text-To-Speech (TTS) system was built using the Ossian toolkit,[1] which is suitable for building synthetic voices from small amounts of data with a minimal requirement of linguistic expertise.

A TTS system commonly has two parts. The first part predicts a sequence of labels from a piece of plain text. These labels contain a multitude of information, such as the phoneme, the phoneme context, stress and prosody information. Normally these labels are generated using rule-based systems and lexicons that were purposely designed for the language. The Ossian toolkit however, utilizes vector space models to predict the pronunciation, stress and other relevant factors needed to create a voice [6]. Hence, the process does not require any linguistic expertise or resources.

The second part of the TTS systems uses the generated labels to create an audio file. There are two common approaches, Statistical Parametric Speech Synthesis (SPSS) and Unit Selection Systems (USS). SPSS systems require in general less data than USS systems, but generate a less natural result. As the data is limited, we created an SPSS model, which uses the HMM-based Speech Synthesis System (HTS) and the GlottHMM vocoder [7].

Given the small amount of data used (approximately 3 hours), the system performs reasonably well and informal listening tests show that the results are approaching those of a commercial solution by Acapela. The main difference is that the commercial system has a hand-crafted preprocessor to transcribe numbers and abbreviations and the use of punctuation to determine prosodic structures.

[1]Open source, available from http://simple4all.org/product/ossian/.

3.2 Speech Recognition

An Automatic Speech Recognition (ASR) system has two main components, the acoustic model and the language model, trained from the audio data and the text data respectively. In order to make an ASR system that can recognize a large vocabulary continuous speech of any person, both the acoustic model and the language model need to be big enough and of high quality. Unfortunately a high quality acoustic model that is speaker-independent—i.e. it can perform recognition with any speaker of a language, not only speakers present in the training data—normally requires data from at least 100 different speakers. As this data is not available, we designed two other recognizers. The first recognizes a large vocabulary but is speaker-dependent—i.e. it can perform recognition with only one specific speaker. The second recognizes with any speaker, but with a limited vocabulary.

The acoustic model of the first recognizer, the speaker-dependent system, is trained using conventional methods using the data of only one speaker. The amount of audio data available is enough to build a well-performing system [8]. The language model is trained in a similar way as is done normally for Finnish, using a sub-word language model [9] to combat the high number of word forms in the North Sami languages that are caused by its agglutinative nature.

The second, speaker-independent, model is trained in a more unconventional way. Instead of using North Sami speech data, a big database of Finnish speech with over 200 different speakers is used. After that, the sounds (or phonemes) of North Sami are mapped to the closest sounding Finnish phoneme. This gives a speaker-independent system which has limited accuracy because of the initial language mismatch. To overcome the limited accuracy of the acoustic model, the language model is dramatically reduced in complexity, to a small vocabulary system. The exact vocabulary can be varied based upon the needs of the dialogue task, but it should stay small to keep an acceptable performance. This system requires also a significant amount of linguistic knowledge as the mapping between Finnish and North Sami words has to be done by hand.

Both systems are tri-state tri-phone hidden Markov models with Gaussian mixture model emission distributions, trained with the AaltoASR[2] toolkit [10, 11]. In the case of the large vocabulary system, Morfessor[3] [12] is used to split words into segments to reduce the number of types in the lexicon. For language modelling a varigram model created with VariKN toolkit[4] [13] is used. The language model is trained on the *Den samiske tekstbanken* corpus.

In [8] the performance of the large vocabulary speaker-dependent system is compared with similar systems in Finnish and Estonian. The reported accuracy is similar to the systems in other languages and gives on average a 23 % Word Error Rate

---

[2]Open source, available from https://github.com/aalto-speech/AaltoASR.

[3]Open source, available from http://www.cis.hut.fi/projects/morpho/.

[4]Open source, available from https://github.com/vsiivola/variKN.

(WER) for a speaker-dependent system trained with 2.5 h of data. In the other language systems the error rate can be decreased to approximately 18 % WER if 8 h of data per speaker is available. In [8] it is also shown that if a full speech database with more than 100 different speakers were available, it would be possible to build a large vocabulary speaker-independent system with an expected WER around 20 %.

Overall we have succeeded in making a synthesizer and recognizer which are suitable for the SamiTalk application. Further development requires more data, or alternatively, new techniques such as deep learning, to create high-quality speaker-independent models, possibly using the more widely available data of related languages such as Finnish.

4 Example Interaction

This section shows an example of how users will interact with SamiTalk. Figure 2 shows a screenshot of the interaction and the video can be viewed at http://www.helsinki.fi/digisami/#samitalk.

The robot first suggests a few topics that it can talk about using information from Sami Wikipedia. Among them, *Tiina Sanila-Aikio* is the current president of the Sami Parliament of Finland, while Japanese *Manga* shows that "local" topics such

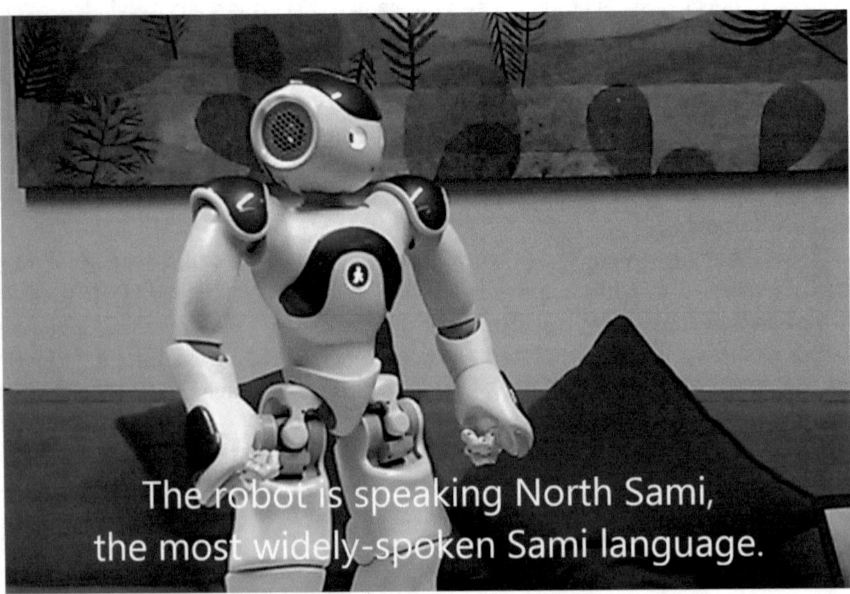

Fig. 2 Screenshot from an example interaction with SamiTalk on a Nao robot

as *Snowmobile* are not the only topics in Sami Wikipedia. The user asks the robot to talk about the situation of the Sami language in Finland.

Robot: *Man fáttá birra don háliidat gullat? Soames mu favorihta fáttát leat Sámegiella, Roavvenjárga, Mohtorgielká, Tiina Sanila-Aikio, Manga.* (What topic do you want to hear about? Some of my favourites are: Sami language, Rovaniemi, Snowmobile, Tiina Sanila-Aikio, Manga.)
User: *Sámegiella* (Sami language)
Robot: *Sámegiela dilli Norgga, Ruoŧa, Suoma ja Ruošša bealde lea earálágan.* (The situation of Sami language is different in Norway, Sweden, Finland and Russia.)
User: *Dilli Suomas* (The situation in Finland.)
Robot: *Bealli Suoma 9 350 sámis máhtet sámegiela. Suomas hállojuvvo davvisámegiella, anárašgiella ja nuortalašgiella, main anárašgiella dušše Suomas. Buoremus dilli lea davvisámegielain, mii ii leat seamma áitatvuloš go eará sámegielat. Anárašgiela dilli lea vehá buoret go nuortalašgiela, dan dihte go anárašgielat giellabeassi doaibma lea joatkašuvvan guhkit áiggi. Suomas sámiid ruoktoguovllus ássi sámiin lea leamaš riekti geavahit sámegiela virgeolbmuiguin jagi 1992 rájes, muhto virgeolbmuin ii gáibiduvvo sámegielmáhttu baicce geavahuvvojit jorgaleaddjit ja dulkat.* (Half of the 9,350 Sami in Finland speak Sami language. Three Sami languages are spoken in Finland: North Sami, Inari Sami and Skolt Sami. Inari Sami is spoken only in Finland. North Sami has the best situation, and it is not as endangered as other Sami languages. The situation of Inari Sami is a bit better than of Skolt Sami, because the language nest of Inari Sami has existed longer than the language nest of Skolt Sami. In Finland the Sami people who live in the Sami region have had the right to use Sami language with officials since 1992, but the language knowledge is not required of officials, they use translators and interpreters.)
Robot: *Leage buorre, vállje čuovvovaččain: Joatkke, Ovddit, dahje daja Bisán nu ahte bisánuvvo.* (Please select from: Continue, Previous, or say Stop if you want to stop.)
User: *Joatkke.* (Continue.)
Robot: *Suoma sámiid ruovttoguovllus mánáin lea riekti lohkat vuođđoskuvllas ja logahagas sámegiela eatnigiellan, válljaávnnasin dahje eaktodáhtolaččat. Dasa lassin giella oahppahuvvo maid ruovttoguovllu olggobealde Oulus, Roavvenjárggas ja Soađegili girkogilis. Sámegiela lea vejolaš čállit studeantadutkosis eatnigiellan. Vuosttas háve dát dáhpáhuvai jagis 1994.* (In the Sami region in Finland the children have the right to study Sami as mother tongue, optional subject, or voluntarily in basic school and senior secondary school. In addition, the Sami language is taught outside the Sami region, in Oulu, Rovaniemi and the center of Sodankylä. It is possible to take the mother tongue part of the matriculation examination in Sami. This was done for the first time in 1994.)
User: *Joatkke.* (Continue.)
Robot: *Sámit leat ain juo guovttegiellagat, mii dárkkuha dan, ahte sii hállet sámegiela lassin riikkaid váldogiela. Suomas skuvllain oahpahuvvo*

davvisámegiela lassin maid anáraš- ja nuortalašgillii, vaikke eanas oahpahusas lea davvisámegillii. Sámegielat oahpahusa birrasis leat Suomas sullii 500 oahppi, geain 150 ož̌žot oahpahusa sámegillii. Váldooassi oahpahusas dáhpáhuvvá 1-6-luohkáin. (The Sami are now always bilingual, which means that in addition to Sami they speak the main language of the state they live in. In Finland the schools teach in North Sami, Inari Sami and Skolt Sami, although mostly the teaching language is North Sami. Under the teaching in Sami language there are around 500 pupils in Finland, and 150 of them gain teaching in Sami. The main part of the teaching in Sami language is given in classes 1–6.)
User: *Joatkke.* (Continue.)
Robot: *Suomas golbma universitehta fállet sámegiela oahpahusa. Oulu universitehta Giellagas-instituhtas lea vejolaš studeret sámegiela váldoávnnasin. Oalgeávnnasin sámegiela lea vejolaš studeret Helssega universitehtas ja Lappi universitehtas.* (In Finland three universities offer education in Sami language. The Giellagas Institute of Oulu University offers Sami language as main subject. It is possible to study Sami language as minor subject at University of Helsinki and University of Lapland.)
User: *Bisán.* (Stop.)
Robot: *Háliidatgo duođ at bisánit?* (Do you really want to stop?)
User: *Juo.* (Yes.)
Robot: *Ortnegis. Giitu. Báze dearvan.* (OK. Thank you. Goodbye.)

5 Conclusions and Future Work

The speech components for North Sami are under further development and have not yet been integrated and tested with the Nao robot software and the WikiTalk software. For demo purposes we hope to use a speaker-independent, but closed vocabulary, speech recognizer. The dialogue has been restricted in order to show the general concept even with this limitation.

In future work, we will explore options for creating a large-vocabulary continuous speech recognizer that is speaker-independent with an open vocabulary. The options include not only collecting more data, but also techniques for utilizing available North Sami resources such as news broadcasts, and resources from related Finno-Ugric languages such as Finnish.

In future the robot will be able to speak and understand at least English and Finnish as well as North Sami. This brings the risk that users in Finland, who are normally bilingual in North Sami and Finnish, will switch the robot to Finnish and continue using Finnish because there are many more articles in Finnish Wikipedia than in Sami Wikipedia. As that would defeat the aim of using the robot to support revitalization of North Sami, it is proposed that the robot will take the initiative and keep switching to North Sami whenever there is an equivalent article in Sami Wikipedia.

Acknowledgements We thank our DigiSami project colleagues for Sami and English translations and the University of Tromsø for access to their North Sami datasets, and acknowledge the computational resources provided by the Aalto Science-IT project. The work has received funding from the EC's 7th Framework Programme (FP7/2007–2013) under grant agreement n°287678 and from the Academy of Finland under the Finnish Centre of Excellence Program 2012–2017 (grant n°251170, COIN) and through the project Fenno-Ugric Digital Citizens (grant n°270082).

References

1. Jokinen, K., Hiovain, K., Laxström, N., Rauhala, I., Wilcock, G.: DigiSami and digital natives: interaction technology for the North Sami language. In: Jokinen, K. and Wilcock, G. (Eds.) Dialogues with Social Robots, Springer (2016) pp. 3–19 (this volume)
2. Wilcock, G.: WikiTalk: a spoken Wikipedia-based open-domain knowledge access system. In: Proceedings of the COLING 2012 Workshop on Question Answering for Complex Domains, pp. 57–69. Mumbai (2012)
3. Jokinen, K., Wilcock, G.: Multimodal open-domain conversations with the Nao robot. In: Mariani, J., Rosset, S., Garnier-Rizet, M., Devillers, L. (eds.) Natural Interaction with Robots, Knowbots and Smartphones: Putting Spoken Dialogue Systems into Practice, pp. 213–224. Springer (2014)
4. Csapo, A., Gilmartin, E., Grizou, J., Han, J., Meena, R., Anastasiou, D., Jokinen, K., Wilcock, G.: Multimodal conversational interaction with a humanoid robot. In: Proceedings of 3rd IEEE International Conference on Cognitive Infocommunications (CogInfoCom 2012), pp. 667–672. Kosice (2012)
5. Laxström, N., Wilcock, G., Jokinen, K.: Internationalisation and localisation of spoken dialogue systems. In: Jokinen, K. and Wilcock, G. (Eds.) Dialogues with Social Robots, Springer (2016) pp. 207–219 (this volume)
6. Watts, O.: Unsupervised Learning for Text-to-Speech Synthesis. Ph.D. thesis, University of Edinburgh (2012)
7. Raitio, T., Suni, A., Yamagishi, J., Pulakka, H., Nurminen, J., Vainio, M., Alku, P.: HMM-based speech synthesis utilizing glottal inverse filtering. IEEE Trans. Audio Speech Lang. Process. **19**(1), 153–165 (2011)
8. Smit, P., Leinonen, J., Jokinen, K., Kurimo, M.: Automatic speech recognition for Northern Sami with comparison to other Uralic languages. In: Presentation at Second International Workshop on Computational Linguistics for Uralic Languages, Szeged, Hungary (2016)
9. Hirsimäki, T., Creutz, M., Siivola, V., Kurimo, M., Virpioja, S., Pylkkönen, J.: Unlimited vocabulary speech recognition with morph language models applied to Finnish. Comput. Speech Lang. **20**(4), 515–541 (2006)
10. Hirsimäki, T., Pylkkönen, J., Kurimo, M.: Importance of high-order n-gram models in morph-based speech recognition. IEEE Trans. Audio Speech Lang. Process. **17**(4), 724–732 (2009)
11. Pylkkönen, J.: An efficient one-pass decoder for Finnish large vocabulary continuous speech recognition. In: Proceedings of the 2nd Baltic Conference on Human Language Technologies, pp. 167–172 (2005)
12. Virpioja, S., Smit, P., Grönroos, S.A., Kurimo, M.: Morfessor 2.0: Python implementation and extensions for Morfessor baseline. Technical report (2013)
13. Siivola, V., Creutz, M., Kurimo, M.: Morfessor and variKN machine learning tools for speech and language technology. In: INTERSPEECH 2007, 8th Annual Conference of the International Speech Communication Association, pp. 1549–1552 (2007)

Part VI
Dialogue Quality Assessment

Utterance Selection Using Discourse Relation Filter for Chat-oriented Dialogue Systems

Atsushi Otsuka, Toru Hirano, Chiaki Miyazaki, Ryuichiro Higashinaka, Toshiro Makino and Yoshihiro Matsuo

Abstract We propose a novel utterance selection method for chat-oriented dialogue systems. Many chat-oriented dialogue systems have huge databases of candidate utterances for utterance generation. However, many of these systems have a critical issue in that they select utterances that are inappropriate to the past conversation due to a limitation in contextual understanding. We solve this problem with our proposed method, which uses a discourse relation to the last utterance when selecting an utterance from candidate utterances. We aim to improve the performance of system utterance selection by preferentially selecting an utterance that has a discourse relation to the last utterance. Experimental results with human subjects showed that our proposed method was more effective than previous utterance selection methods.

Keywords Chat-oriented dialogue system · Discourse relation · Utterance selection · Utterance filtering

1 Introduction

Chat-oriented dialogue systems are beginning to be actively investigated from their social and entertainment aspects [1, 2]. The basic flow of a chat-oriented dialogue system consists of three processes as shown in Fig. 1. First, the system analyzes an input user utterance using such processes as dialogue-act estimation and topic-word extraction in the utterance understanding unit. Next, in the utterance generation unit, the system generates utterance candidates from various kinds of stored knowledge such as Web data and past conversations. Finally, the system selects only one utterance from the candidate utterances by ranking them using the context of conversation in the utterance selection unit. Because the utterance candidates are created from various kinds of knowledge as mentioned above, the candidates are likely to consist

---

A. Otsuka (✉) · T. Hirano · C. Miyazaki · R. Higashinaka · T. Makino
NTT Media Intelligence Laboratories NTT Corporation, 1-1 Hikarinooka, Yokosuka-shi, Kanagawa 239-0847, Japan
e-mail: otsuka.atsushi@lab.ntt.co.jp

K. Jokinen and G. Wilcock (eds.), *Dialogues with Social Robots*, Lecture Notes in Electrical Engineering 427,
DOI 10.1007/978-981-10-2585-3_29

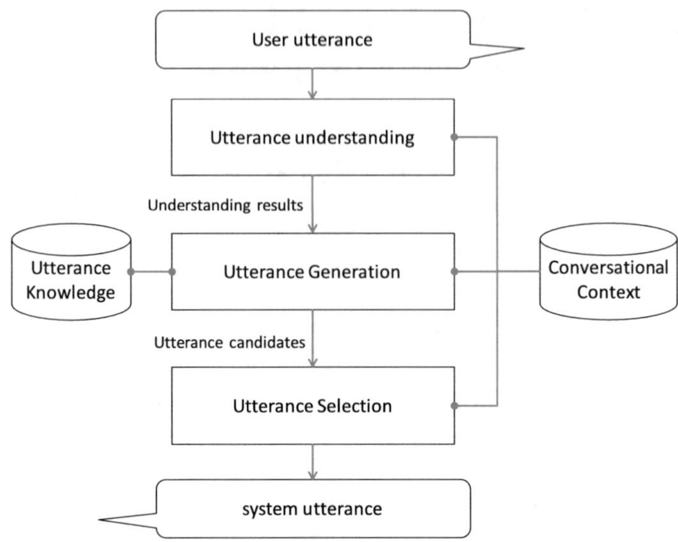

Fig. 1 Architecture of chat-oriented dialogue system

of both appropriate and inappropriate utterances. Thus, utterance selection is one of the most important elements of chat-oriented dialogue systems to obtain appropriate utterances.

We tackle the problem of selecting appropriate utterances by using discourse relations for utterance selection. Discourse relations define the relationship between two sentences in a discourse structure. Two utterances with a discourse relation have a semantic connection. Therefore, the system utterance is likely to be an appropriate response for a user utterance if the system is able to select an utterance that has a discourse relation to the last user utterance. Our approach checks the discourse relation between a candidate utterance and the last user utterance for all candidates and then selects the utterance that has a discourse relation with the last user utterance. In experiments, our proposed method was able to select a more appropriate utterance than a previous method that does not use discourse relations.

In this article, we first introduce related work in Sect. 2. In Sect. 3, we describe the utterance selection method using discourse relations. We explain the utterance selection experiment in Sect. 4. Finally, we conclude the article in Sect. 5.

2 Related Work

Chat-oriented dialogue systems have attracted a lot of attention recently in spoken dialogue system studies. Higashinaka et al. implemented an open domain conversational system based on a natural language processing approach [3]. The system consists of several utterance understanding modules and dialogue control modules.

The dialogue control modules include utterance ranking modules based on cohesion. Banchs and Li proposed a chat-oriented dialogue system using a vector space model [4]. This system selects an utterance by searching for the most similar utterance in the vector space from a large dialogue database. Bang et al. developed a personalized chat-oriented dialogue system by extracting user-related facts and storing the facts in long short-term memory [5].

Earlier studies applied discourse relations and structures approach to dialogues. Tonelli et al. created a dialogue corpus annotated with Penn Discourse Tree Bank (PDTB)-styled discourse relations [6]. Afantenos et al. proposed a discourse parser for multi-party chat dialogues [7]. They studied the attachment problem in detail for multi-party chat dialogues from the aspect of discourse structures and implemented a discourse parser to capitalize on peculiarities of chat dialogues. Stoyanchev et al. implemented a Text-to-Dialogue generation system with discourse relations [8]. When a discourse relation is input into this system, the system outputs a set of dialogue act sequences. Our work differs from previous studies in that we utilize discourse relations for utterance selection in chat-oriented dialogue systems.

3 Utterance Selection with Discourse Relations

In this section, we describe the utterance selection method using discourse relations. The architecture of our utterance selection method is shown in Fig. 2. The method consists of three blocks: utterance ranking, discourse relation identification, and utterance filtering.

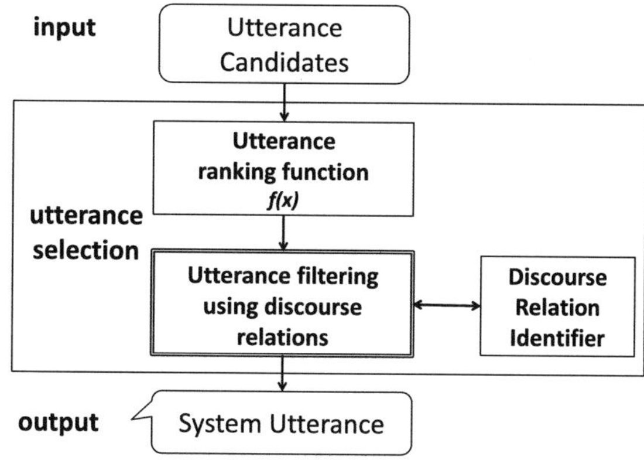

Fig. 2 Flow of utterance selection method in chat-oriented dialogue system

3.1 Utterance Ranking Function

The utterance ranking function $f(x)$ computes the relevance between each candidate utterance and the past conversation. Here, we assume that the relevance is computed in consideration of surface cohesion and shallow semantic coherence as in previous studies [3, 9].

On top of this utterance ranking function, our proposed method applies an utterance filter using discourse relations to the utterance candidates. The utterance filter checks the discourse relation between the last user utterance and the candidate utterances. Finally, the dialogue system outputs the utterance that has a discourse relation and the highest score given by the ranking function.

3.2 Discourse Relation Identification

Discourse relations are relationships defined in discourse structures. The Rhetorical Structure Theory (RST) [10] and PDTB [11] are two well-known discourse structure theories. The PDTB is based on the connective relations between two arbitrary text spans, which means that it is excellent for our purposes because we want to detect discourse relations between the context (or the previous user utterance) and one of the utterance candidates. In fact, an earlier study used PDTB-styled discourse relations for dialogue annotation [6].

The discourse relation of the PDTB is defined by connectives (e.g., *but, and, because*). Connectives are grammatical representations of logical relationships. Table 1 gives examples of PDTB-styled discourse relations between pairs of utterances. In the PDTB, discourse relations are categorized into explicit and implicit relations. We use as an example the first two utterances in Table 1:

$utt1$: *Do you drink alcoholic beverages in your daily life?*

$utt2$: *I often drink Smirnoff-Ice.*

Table 1 Examples of discourse relations between utterance pairs

Utterance 1	Utterance 2	Category	Relation	Connective
Do you drink alcoholic beverages in your daily life?	*I often drink Smirnoff-Ice*	Implicit	Instantiation	(For example)
I often drink Smirnoff-Ice	*It has a refreshing taste*	Implicit	Cause	(Because)
I often drink Smirnoff-Ice	*I rarely drink Japanese sake*	Implicit	Contrast	(But)
I rarely drink Japanese sake	*Because I think its taste is so unique*	Explicit	Cause	Because

In *utt*2, an example is given of an alcoholic beverage referred to in *utt*1. Thus, *utt*1 and *utt*2 have an instantiation relation, and the most reasonable connective is "*For example*" in this case. However, *utt*2 does not contain the connective "*For example*". Therefore, this sentence pairis categorized as having an implicit relation. In contrast, a discourse relation where a discourse marker is explicitly stated in either sentence is categorized as an explicit relation.

The task of identifying a discourse relation between a user utterance and each candidate utterance is similar to the task of implicit discourse relation identification because the candidates are taken from some other context and are unlikely to have connectives.

One of the most typical approaches for implicit discourse identification is the word-pair based method. This method extracts words in each input utterance. For example, the word pair "*alcoholic-Smirnoff-Ice*" is extracted from *utt*1 and *utt*2 in the first example of Table 1. Previous studies have proposed many methods of implicit discourse recognition, and many of these earlier studies [12–17] focused on using word pairs or their derivative features. The recursive neural network based implicit discourse identification method has also been proposed recently [18].

3.3 Utterance Filtering Using Discourse Relations

The algorithm used in our proposed method is indicated as Algorithm 1. We first select the utterance with the highest score from the results of the ranking function and then check the discourse relation between the selected utterance and the last utterance. If they have a discourse relation, the dialogue system utters the selected utterance. If they have no discourse relation, the algorithm rejects the first selected utterance and selects the runner-up utterance. The algorithm checks whether there is a discourse relation with the reselected utterance just as it did with the first selected utterance. These processes are repeated until the algorithm finds an utterance that has a discourse relation or the score of the decision function exceeds the lower limit of the threshold; low score utterances have little relevance to past conversations even if they have a discourse relation to the last utterance. If the algorithm does not find an utterance that has a discourse relation, the dialogue system utters the highest scoring utterance.

4 Experiment

We conducted an experiment using data from a Japanese chat-oriented dialogue system created by Higashinaka et al. [3]. We had human subjects evaluate the quality of system utterances using our proposed method and comparative methods. We first describe the dataset used in the experiment. Then, we explain the comparative methods, and finally, we describe the results of the experiment.

Algorithm 1 Utterance filtering algorithm with discourse relation

utterance candidates sorted by relevance: $U = [utt_1, utt_2, ..., utt_n]$
ranking function score: $Score(U(i))$
discourse relation identification function: $F_{DR}(utt1, utt2)$
user utterance: U_{user}, discourse relation: DR
ranking position: i, filtering threshold: θ
$i \leftarrow 1$
while $Score(U(i)) > \theta$ **do**
  $DR \leftarrow F_{dr}(U_{user}, U(i))$
  if $DR \neq NoRel$ **then**
    return $U(i)$
  else
    $i \leftarrow i + 1$
  end if
end while
return $U(1)$

4.1 Dataset

We created an evaluation dataset for the experiment. The dataset is the collection of utterance candidates created by our system [3]. We annotated the utterance candidates for 147 utterances in 23 conversations to create the dataset. The dialogue system typically outputs 10–30 candidates for each input user utterance. We annotated all candidates with the utterance evaluation “OK” or “NG”.

An example of utterance candidate evaluation is shown in Fig. 3. The dataset consists of the conversational context and the system utterance candidates. The bottommost system utterance is blank in the conversational context, and the system utterance candidates are the candidates for the blank utterance. Each utterance candidate is annotated with a label indicating the appropriateness for the conversational context. The utterance with the label “OK” is an appropriate utterance for the conversational context. In contrast, the utterance with the label “NG” is an inappropriate utterance. In this paper, we evaluate the utterance selection according to the ratio of appropriate utterances in the evaluation dataset.

4.2 Comparative Methods

Our proposed method uses an utterance filter based on discourse relation identification. Thus, in this experiment, we compared the utterance selections by focusing on utterance filtering. The baseline method without filtering and the methods using filtering are described below.

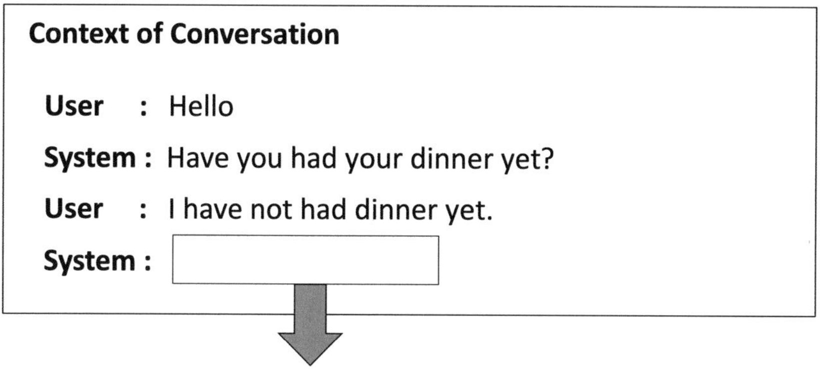

Context of Conversation

User : Hello
System : Have you had your dinner yet?
User : I have not had dinner yet.
System :

Candidates of system utterance

No	Utterance	Appropriate?
1	Are you full?	OK
2	What do you want to eat in dinner?	OK
3	I can't not take a dinner.	NG
4	You must start preparations for dinner.	NG

Fig. 3 Example of evaluation dataset for utterance selection

No filtering (baseline)
This method employs only an utterance ranking function; thus, it uses no filtering at all. The utterance ranking function is based on cohesion of utterance pairs [3]. Note that in the experiment, we used the same utterance ranking function with all of the comparative methods.

Filtering with human based discourse relation identification (upper)
We manually give the discourse relation for the last user utterance to each utterance candidate. Examples of discourse relation annotated utterance candidates are shown in Fig. 4. The discourse relation label for each candidate represents the relation to the user utterance. For example, in Fig. 4, candidate 1 "*Are you full?*" has a causal relation to the user utterance "*I have not had dinner yet*". Note that these discourse labels were annotated by several different annotators from the evaluation data described in Sect. 4.1.

Filtering with word-pair based discourse relation identification
The word-pair based method uses word-pair features to identify discourse relations. The word-pair feature is one of the most typical features in discourse relation identification [12–17]. Thus, this method is useful for evaluating the basic performance of discourse relation filtering in comparison to no filtering. Word pairs are extracted from the user and candidate utterances. For example, the word pair "*dinner-full*" is extracted from the user utterance "*I have not had dinner yet*" and the candidate utterance "*Are you full?*" in Fig. 4.

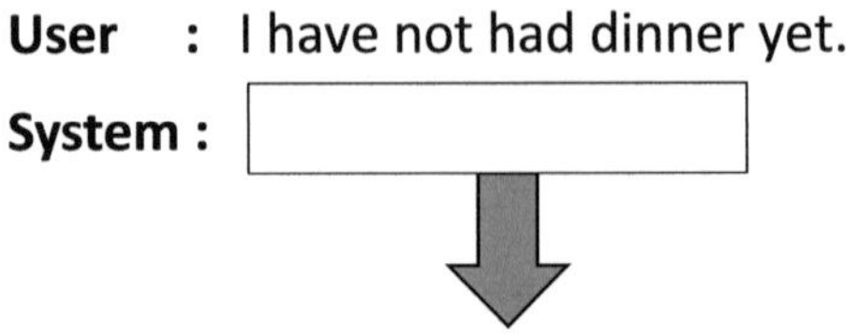

User : I have not had dinner yet.

System :

Candidates of system utterance

No	Utterance	Discourse relation
1	Are you full?	Cause
2	What do you want to eat in dinner?	Instantiation
3	I can't not take a dinner.	NoRel
4	You must start preparations for dinner.	Condition

Fig. 4 Examples of utterance candidates with discourse relations annotated manually

Our discourse relation classifier implemented by using an SVM has three discourse relation classes ("COMPARISION", "CONTINGENCY", and "EXPANSION") based on the PDTB tag-set, and it was trained by 15,000 utterance pairs that have a discourse relation.

Filtering with RNN based discourse relation identification

In our previous work, we approached discourse relation identification using intermediate expressions of sentences with a recursive neural network (RNN) [18]. Our proposed discourse relation identification method significantly improved the accuracy of identifying discourse relations compared with the word-pair method. Therefore, here, we use an RNN-based discourse relation identification method. The discourse relation classes and the training data are the same as those used for the word-pair based classifier.

4.3 Experimental Results

We evaluated the appropriate utterance ratio for all comparative methods in 147 utterance selections from 23 dialogues. The results of the experiment are shown in Fig. 5. When we used no discourse relation filter, the ratio of appropriate utterances was 0.558. All the methods using a discourse relation filter improved the appropriate utterance ratio from that of the method without a discourse filter. The discourse relation filter with gold annotation improved the utterance selection accuracy by 22.1 % from that of the baseline method. When we compare machine-learning based discourse relation filters, the RNN based discourse relation filter outperformed the word-pair based filter. The RNN based filter improved the accuracy by 12 % from the baseline and achieved 86 % of the upper bound.

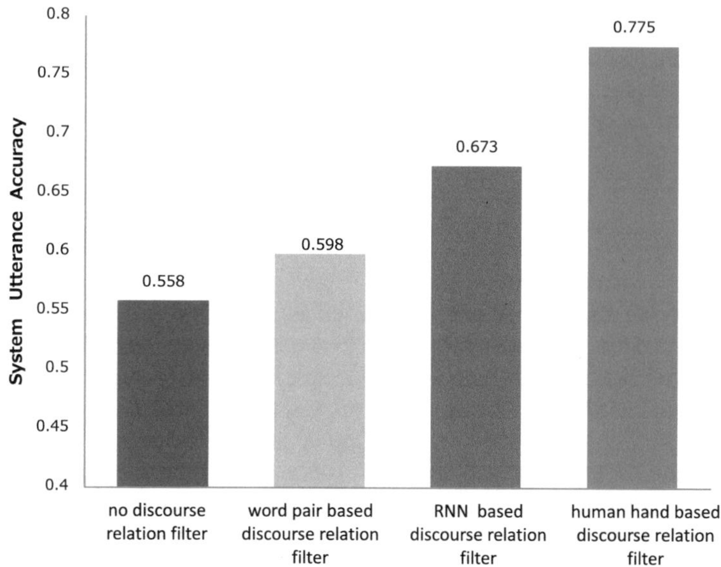

Fig. 5 Accuracy of output utterances

Examples of changes in utterance selection by the discourse relation filter are shown in Fig. 6. Both dialogues are examples of how the system utterance was changed from an inappropriate utterance, indicated by "NG", to an appropriate one, indicated by "OK". In example 1, the system utterance was changed from "*What is this food?*" to "*Do you know any good foods to eat at work?*". The first utterance

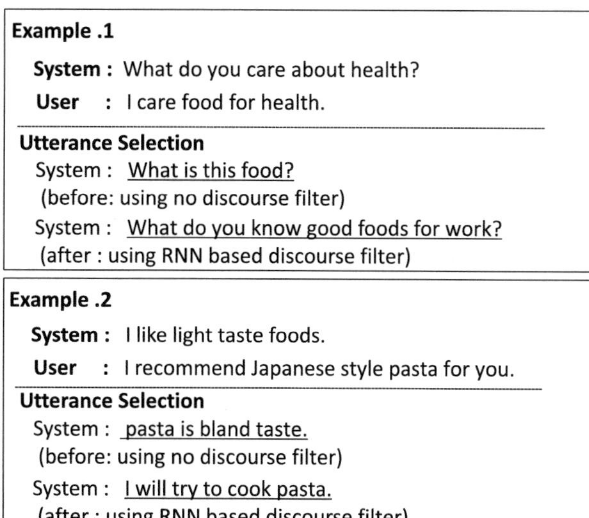

Fig. 6 Examples of changes in system utterances by RNN based discourse relation filter

had no discourse relation, whereas the second one had an instantiated relation to the user utterance. Just as with example 1, the changed system utterance in example 2 "*I will try to cook pasta.*" has a conditional relation to the user utterance "*I recommend Japanese style pasta for you*".

5 Conclusion

In this article, we proposed an utterance selection method using discourse relations. We incorporated the discourse relation filter into utterance selection. The filter checks whether candidate utterances have a discourse relation to the last user utterance. The contribution of this article is that we empirically demonstrated that discourse relations are efficient for improving utterance selection in dialogue systems.

In the future, we plan to investigate ways to increase the sophistication of utterance selection using discourse relation labels. For example, we want to consider the use of longer context for utterance selection. The experimental results demonstrated that a discourse relation filter improves the quality of utterance selection in dialogue systems. However, even when the dialogue system used a discourse relation filter, approximately 20 % of the utterances were not improved. We think that the systems must consider not only the last user utterance but also the whole dialogue. A previous study reported that context-level errors were numerous, second only to the number of response-level errors [19].

Our proposed method uses discourse relation as a Boolean value in this paper. Namely, the discourse relation filter only checks whether or not the candidate has a discourse relation. We surmise that the performance of a dialogue system improves by checking the past direction of conversation with discourse relation labels and selecting the most appropriate utterance from candidates that have discourse relation labels.

We consider that the dialogue systems are able to control dialogues by using discourse relation labels. For example, an "Instantiation" utterance can deepen the conversation. In contrast, a "Contrast" utterance can make the systems seem more distant. We aim to achieve a chat-oriented dialog system based on planning using discourse relations.

References

1. Bickmore, T.W., Picard, R.W.: Establishing and maintaining long-term human-computer relationships. ACM Trans. Comput. Hum. Interact. 293–327 (2005)
2. Ritter, A., Cherry, C., Dolan, W.B.: Data-driven response generation in social media. In: Proceedings of the 2011 Conference on Empirical Methods in Natural Language Processing (EMNLP2011), pp. 583–593 (2011)
3. Higashinaka, R., Imamura, K., Meguro, T., Miyazaki, C., Kobayashi, N., Sugiyama, H., Hirano, T., Makino, T., Matsuo, Y.: Towards an open domain conversational system fully based on

natural language processing. In: Proceedings of the 25th International Conference on Computational Linguistics (COLING 2014), pp. 928–939 (2014)
4. Banchs, R.E., Li, H.: Iris: a chat-oriented dialogue system based on the vector space model. In: Proceedings of the ACL 2012 System Demonstrations (ACL2012), pp. 37–42 (2012)
5. Bang, J., Noh, H., Kim, Y., Lee, G.: Example-based chat-oriented dialogue system with personalized long-term memory. In: Proceedings of 2015 International Conference on Big Data and Smart Computing (Big-Comp2015), pp. 238–243 (2015)
6. Tonelli, S., Riccardi, G., Prasad, R., Joshi, A.: Annotation of discourse relations for conversational spoken dialogs. In: Proceedings of the Seventh International Conference on Language Resources and Evaluation (LREC 2010), pp. 19–21 (2010)
7. Afantenos, S., Kow, E., Asher, N., Perret, J.: Discourse parsing for multi-party chat dialogues. In: Proceedings of the 2015 Conference on Empirical Methods in Natural Language Processing (EMNLP2015), pp. 928–937 (2015)
8. Stoyanchev, S., Piwek, P.: The coda system for monologue-to-dialogue generation. In: Proceedings of the 12th Annual Meeting of the Special Interest Group on Discourse and Dialogue (SIGDIAL2011), pp. 335–337 (2011)
9. Shibata, M., Nishiguchi, T., Tomiura, Y.: Dialog system for open-ended conversation using web documents. Informatica **33**, 277–284 (2009)
10. Mann, W.C., Thompson, S.A.: Rhetorical Structure Theory: A Framework for the Analysis of Texts. Information Sciences Institute (ISI/RS- 87-185) (1987)
11. Prasad, R., Dinesh, N., Lee, A., Miltsakaki, E., Robaldo, L., Joshi, A., Webber, B.: The penn discourse treebank 2.0. In: Proceedings of the Sixth International Conference on Language Resources and Evaluation (LREC 2008) (2008)
12. Marcu, D., Echihabi, A.: An unsupervised approach to recognizing discourse relations. In: Proceedings of the 40th Annual Meeting on Association for Computational Linguistics (ACL 2002) pp. 368–375 (2002)
13. Lin, Z., Kan, M.Y., Ng, H.T.: Recognizing implicit discourse rela- tions in the penn discourse treebank. In: Proceedings of the 2009 Conference on Empirical Methods in Natural Language Processing (EMNLP 2009), pp. 343–351 (2009)
14. Pitler, E., Louis, A., Nenkova, A.: Automatic sense prediction for implicit discourse relations in text. In: Proceedings of the Joint Conference of the 47th Annual Meeting of the ACL and the 4th International Joint Conference on Natural Language Asian Federation of Natural Language Processing (AFNLP 2009), pp. 683–691 (2009)
15. Wang, X., Li, S., Li, J., Li, W.: Implicit discourse relation recognition by selecting typical training examples. In: Proceedings of the 24th International Conference on Computational Linguistics (COLING 2012), pp. 2757–2772 (2012)
16. Biran, O., McKeown, K.: Aggregated word pair features for implicit discourse relation disambiguation. In: Proceedings of the 51st Annual Meeting of the Association for Computational Linguistics (ACL 2013), pp. 69–73 (2013)
17. Rutherford, A., Xue, N.: Discovering implicit discourse relations through brown cluster pair representation and coreference patterns. In: Proceedings of the 14th Conference of the European Chapter of the Association for Computational Linguistics (EACL 2014), pp. 645–654 (2014)
18. Otsuka, A., Hirano, T., Miyazaki, C., Masumura, R., Higashinaka, R., Makino, T., Matsuo, Y.: Discourse relation recognition by comparing various units of sentence expression with recursive neural network. In: Proceedings of the 28th Pacific Asia Conference on Language, Information, and Computation (PACLIC2015), pp. 63–72 (2015)
19. Higashinaka, R., Funakoshi, K., Araki, M., Tsukahara, H., Kobayashi, Y., Mizukami, M.: Towards taxonomy of errors in chat-oriented dialogue systems. In: Proceedings of the 16th Annual Meeting of the Special Interest Group on Discourse and Dialogue (SIGDIAL2015), pp. 87–95 (2015)

Analysis of Temporal Features for Interaction Quality Estimation

Stefan Ultes, Alexander Schmitt and Wolfgang Minker

Abstract Many different approaches for Interaction Quality (IQ) estimating of Spoken Dialogue Systems have been investigated. While dialogues clearly have a sequential nature, statistical classification approaches designed for sequential problems do not seem to work better on automatic IQ estimation than static approaches, i.e., regarding each turn as being independent of the corresponding dialogue. Hence, we analyse this effect by investigating the subset of temporal features used as input for statistical classification of IQ. We extend the set of temporal features to contain the system and the user view. We determine the contribution of each feature sub-group showing that temporal features contribute most to the classification performance. Furthermore, for the feature sub-group modeling the temporal effects with a window, we modify the window size increasing the overall performance significantly by +15.69 % achieving an Unweighted Average Recall of 0.562.

Keywords Spoken dialogue system evaluation · User satisfaction · Support vector machine classification

1 Introduction

Due to recent advances in Speech Recognition Technology, technical systems with speech interfaces are becoming more and more prevalent in our everyday lives.

Research conducted while working at Ulm University.

S. Ultes (✉)
Engineering Department, University of Cambridge, Cambridge, UK
e-mail: su259@cam.ac.uk

A. Schmitt · W. Minker
Institute of Communications Engineerig, Ulm University, Ulm, Germany
e-mail: alexander.schmitt@ulm.de

W. Minker
e-mail: wolfgang.minker@ulm.de

K. Jokinen and G. Wilcock (eds.), *Dialogues with Social Robots*,
Lecture Notes in Electrical Engineering 427,
DOI 10.1007/978-981-10-2585-3_30

Comparing the performance of such SDSs is a non-trivial task which has not yet been solved. While many (mostly statistical) approaches on spoken dialogue assessment take into account objective criteria like dialogue length or task success rate, the focus has shifted to more user-centered criteria, such as the satisfaction level measured while or after users have interacted with an SDS.

To achieve automatic evaluation, recent approaches focus on User Satisfaction (US) recognition employing state-of-the-art statistical classification systems. While some of these approaches deal with US on the dialogue level, i.e., providing a satisfaction score for the complete interaction, recent work focuses on US on the system-user-exchange level. Having this information, the system may react adaptively depending on the current US score [1–4].

In this work, we focus on a US-based exchange-level measure called Interaction Quality (IQ) [5]. Naturally, having a dialogue consisting of several turns (or system-user-exchanges) in a temporal order may evidently be regarded as a sequence of exchanges. Furthermore, the IQ score on the exchange-level measuring the quality up to a single interaction step highly depends on the IQ scores of the previous dialogue steps [6]. For such a sequential problem, many classification approaches specifically designed for the needs inherent in sequential problems exist, e.g., Hidden Markov Models (HMMs) [7]. However, applying these approaches to IQ estimation has not resulted in an increase in performance compared static approaches,[1] e.g., applying a support vector machine (SVM) [8]. First experiments applying a Conditioned Random Field [9] or Recurrent Neural Networks [10], which have both shown to achieve good performance for other tasks, also resulted in low performance.

Thus, while the problem of estimating the Interaction Quality clearly seems as if it would benefit from sequential classification, it is not a straight forward problem to outperform static approaches. Here, we assume that the reason why static approaches perform that well lies in the modelling of the interaction parameters. They consist of a high number of temporal features thus encoding temporal information about the dialogue. However, simply adding the previous IQ value to the feature set is not sufficient and does result in worse performance [6]. To prove our assumption, these temporal features are analyzed.

Additionally, the set of temporal features is extended to contain both the system and the user view and some temporal features are modified. Thus, in contrast to previous work where different classification approaches have been investigated, we aim at increasing the estimation performance by extending and optimizing the feature set.

The remainder of this article is organized as follows: significant related work on user satisfaction recognition in general as well as on Interaction Quality estimation specifically is stated in Sect. 2. Our main contribution of analyzing the temporal features for IQ estimation is presented in Sect. 3 including a thorough description of the Interaction Quality paradigm. Furthermore, in the same section, we will argue for a recalculation of some features and describe the new extended feature set. We compare the newly created feature set with the original features for different feature

[1] Regarding each exchange being independent of all other exchanges, and not as part of a sequence.

sub-groups as well as analyse the performance analysis of different temporal contexts. The results are discussed in Sect. 4. Finally, the results are interpreted in Sect. 5 and future work is outlined.

2 Significant Related Work

Estimating User Satisfaction (US) for SDSs has been in the focus of research for many years. In this section, we present approaches on US recognition in general and on IQ estimation specifically.

Groundbreaking work on automatic SDS evaluation has been presented by Walker et al. [11] with the PARADISE framework. The authors assume a linear dependency between quantitative parameters derived from the dialogue and User Satisfaction on the dialogue level, modeling this dependency using linear regression. Unfortunately, for generating the regression model, weighting factors have to be computed for each system anew. This generates high costs as dialogues have to be performed with real users where each user further has to complete a questionnaire after completing the dialogue. Moreover, in the PARADISE framework, only quality measurement for whole dialogues (or system) is allowed. However, this is not suitable for online adaptation of the dialogue [12]. Furthermore, PARADISE relies on questionnaires while we focus on work using single-valued ratings.

Numerous work on predicting User Satisfaction as a single-valued rating task for each system-user-exchange has been performed using both static and sequential approaches. Hara et al. [13] derived turn level ratings from an overall score applied by the users after the dialogue. Using n-gram models reflecting the dialogue history, the estimation results for US on a 5 point scale showed to be hardly above chance.

Higashinaka et al. [14] proposed a model to predict turn-wise ratings for human-human dialogues (transcribed conversation) and human-machine dialogues (text from chat system). Ratings ranging from 1–7 were applied by two expert raters labeling “Smoothness”, “Closeness”, and “Willingness” not achieving a Match Rate per Rating (MR/R) of more than 0.2–0.24 applying Hidden Markov Modes as well as Conditioned Random Fields. These results are only slightly above the random baseline of 0.14. Further work by Higashinaka et al. [15] uses ratings for overall dialogues to predict ratings for each system-user-exchange using HMMs. Again, evaluating in three user satisfaction categories “Smoothness”, “Closeness”, and “Willingness” with ratings ranging from 1–7 achieved best performance of 0.19 MR/R.

An approach presented by Engelbrecht et al. [16] uses Hidden Markov Models (HMMs) to model the SDS as a process evolving over time. User Satisfaction was predicted at any point within the dialogue on a 5 point scale. Evaluation was performed based on labels the users applied themselves during the dialogue.

Work by Schmitt et al. [17] deals with determining User Satisfaction from ratings applied by the users themselves during the dialogues. A Support Vector Machine (SVM) was trained using automatically derived interaction parameter to predict User

Satisfaction for each system-user-exchange on a 5-point scale achieving an MR/R of 0.49.

Interaction Quality (IQ) has been proposed by Schmitt et al. [5] as an alternative performance measure to US. In their terminology, US ratings are only applied by users. As their presented measure uses ratings applied by expert raters, a different term is used. Using the same approach as for estimating US, they achieve an MR/R of 0.58 for estimating IQ on the turn level on a 5-point scale. The IQ paradigm will be described in more detail in Sect. 3.1.

To improve the performance of static classifiers for IQ recognition, Ultes et al. [18] proposed a hierarchical approach: first, IQ is predicted using a static classifier. Then, the prediction error is calculated and a second classifier is trained targeting the error value. In a final step, the initial hypothesis may then be corrected by the estimated error. This approach has been successfully applied improving the recognition performance relatively by up to +4.1 % Unweighted Average Recall (UAR, the average of all class-wise recalls).

Work on rendering IQ prediction as a sequential task analyzing HMMs and Conditioned Hidden Markov Models has been performed by Ultes et al. [19]. They achieved an UAR of 0.39 for CHMMs. This was outperformed by regular HMMs (0.44 UAR) using Gaussian mixture models for modeling the observation probability for both approaches. Replacing the observation probability model with the confidence scores of static classification methods, Ultes et al. [6] achieved a significant improvement of the baseline with an UAR of 0.51.

Unfortunately, applying classification approaches which render the task of IQ prediction as a sequential problem do not seem to increase the estimation performance. Therefore, as the feature set used for classification also models the temporal effects inherent in IQ estimation to a certain degree, it will be analysed more closely in the following section.

3 Temporal Feature Analysis

The main goal of this paper is to analyze why the set of interaction parameters used for IQ estimation is so powerful that most static approaches on IQ estimation outperform sequential ones. Moreover, in this paper, the emphasis lies on the temporal features, i.e., parameters on the window and dialogue levels. First, the Interaction Quality Paradigm including the definition of IQ and a description of the Interaction Parameters is presented followed by a brief description of the used evaluation methods. Then, the temporal subset of those parameters is manually analyzed to find peculiarities. Furthermore, the performance of each level separately as well as the absence of each level is analyzed. Finally, the effect of the window-size on the estimation performance in investigated.

3.1 The Interaction Quality Paradigm

The Interaction Quality paradigm has been originally introduced by Schmitt et al. [20]. It represents a scheme for bridging the gap between the subjective nature of user satisfaction and objective ratings. The general idea is to utilize statistical models to predict the Interaction Quality (IQ) of system-user-exchanges based on interaction parameters which are both described in the following. The annotated exchanges along with the interaction parameters have been combined within the LEGO corpus [21] forming the base of this work.

The Interaction Quality as a measure for dialogue performance intends to overcome the problems inherit with the purely subjective measure user satisfaction. Hence, it is defined similarly to user satisfaction: while the latter represents the true disposition of the user, IQ is the disposition of the user assumed by expert annotators.[2] Here, expert annotators listen to recorded dialogues after the interactions and rate them by assuming the point of view of the actual person performing the dialogue. These experts are supposed to have some experience with dialogue systems. In this work, expert annotators were "advanced students of computer science and engineering" [20], i.e., grad students.

The employed corpus ("LEGO corpus") is based on 200 calls to the "Let's Go Bus Information System" of the Carnegie Mellon University in Pittsburgh [23] recorded in 2006. Labels for IQ have been assigned by three expert annotators, achieving a total of 4,885 system-user-exchanges[3] with an inter-annotator agreement of $\kappa = 0.54$. This may be considered as a moderate agreement (cf. Landis and Koch's Kappa Benchmark Scale [24]) which is quite good considering the difficulty of the task that required to rate each exchange. The final label was assigned to each exchange by using the median of all three individual ratings.

The experts applied ratings at a scale from 1 (extremely unsatisfied) to 5 (satisfied) considering the complete dialogue up to the current exchange. Thus, each exchange has been rated without regarding any upcoming user utterance. Each dialogue starts with a rating of 5 since the user is expected to be satisfied in the beginning. To compensate the subjective nature and to ensure consistent labeling, the expert annotators had to follow labeling guidelines [21].

The set of *interaction parameters* used as input variables for the IQ model consists of a total of 53 parameters automatically derived from three SDS modules: Automatic Speech Recognition, Spoken Language Understanding, and Dialogue Management. Furthermore, to account for the temporal nature of the system, the parameters are modeled on three different levels (Fig. 1): *Exchange level* parameters are derived directly from the respective dialogue modules, e.g., `ASRConfidence`. *Dialogue level* parameters consist of counts (#), means (Mean), etc. of the exchange level parameters calculated from all exchanges of the whole dialogue up to the current exchange, e.g., `MeanASRConfidence`. *Window level* parameters consist of counts ({#}), means ({Mean}), etc. of the exchange level parameters calculated from the

[2] IQ is strongly related to user satisfaction [22] with a Spearman's ρ of 0.66 ($\alpha < 0.01$).

[3] A system turn followed by a user turn.

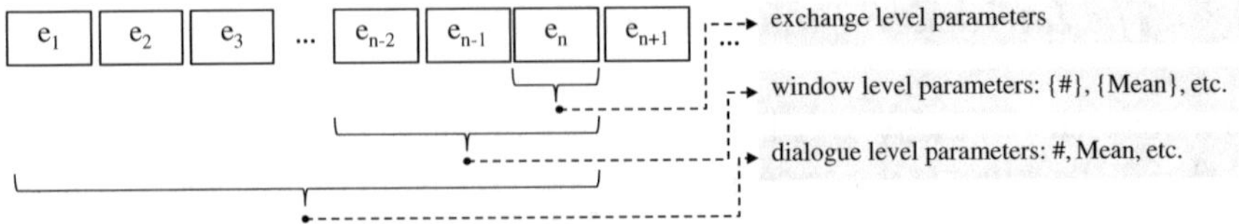

Fig. 1 The three different modeling levels [21] representing the interaction at exchange e_n: The most detailed exchange level, comprising parameters of the current exchange; the window level, capturing important parameters from the previous n dialog steps (here $n = 3$); the dialog level, measuring overall performance values from the entire previous interaction

last three exchanges, e.g., `{Mean}ASRConfidence`. A thorough description of all interaction parameters on all levels can be found in Schmitt et al. [21].

For measuring the performance of the classification algorithms, we rely on *Unweighted Average Recall (UAR)* as the average of all class-wise recalls, *Cohen's Kappa* [26] linearly weighted [27] and *Spearman's Rho* [28]. The latter two also represent a measure for similarity of paired data.

3.2 Manual Analysis and Feature Set Extension

The temporal effects of the dialogue are captured within the interaction parameters by the dialogue and window levels. To get a better understanding, the corpus is analysed more closely. The observation been made is that some system-user-exchanges contain only a system utterance without user input, e.g., the "Welcome" message of the system (Fig. 2, line 1). While the system and the user have a different view on the interaction in general, this is especially the case regarding the number of dialogue turns. However, as this information is used for computing parameters on the dialogue and window level, both views should be reflected by the interaction parameters.

				%ASR-Succ.		{#}ASR-Succ.	
	System DA	User DA	ASR-Status	User	System	User	System
1	[Welcome]	-	-	0.0	0.0	0	0
2	[Help_info]	-	-	0.0	0.00	0	0
3	[Open]	[Inform_origin]	complete	1.0	0.33	1	1
4	[Confirm_origin]	[Inform_origin]	complete	1.0	0.50	2	2
5	[Confirm_origin]	[Affirm]	incomplete	0.66	0.40	2	2
6	[Filler]	-	-	0.66	0.33	2	1
7	[Ask_destination]	[Inform_destination]	complete	0.75	0.43	2	1

Fig. 2 The computation of dialog level parameters %ASR-Success (percentage of successful ASR events) and window-level parameters {#}ASR-Success (number of successful ASR events within the window frame) from the view of the user and the system [25]

Hence, the parameters should be computed with respect to the number of system turns and with respect to the number of user turns as well. Thus, the original feature set is extended to contain both variants of parameters.

An example dialogue snippet showing parameters originating from the *ASR-Status* is illustrated in Fig. 2. It shows both calculation variants for the window parameter *{#}ASR-Success* and the dialogue parameter *%ASR-Success*. The differences are clearly visible: while *%ASR-Success* is either 0 or 1 for the user's view (only successful ASR events occur), the numbers are different for the system's view.

To reflect this system and user view for the complete corpus, a number of parameters are recalculated for both variants.[4] The window size remained the same with $n = 3$. This results in an extended feature set consisting of 65 features. For the remainder of the paper, we will refer to the original feature set as $LEGO_{orig}$ and to the extended feature set as $LEGO_{ext}$.

3.3 Analysis of Parameter Levels

To get a better understanding of the different parameter level and their contribution to the overall estimation performance, experiments have been conducted using each combination of parameter levels as a feature set, e.g., using only parameters on one level or using parameters from all but one levels. Furthermore, to get a better understanding of the extension of the feature set, the experiments are performed for $LEGO_{orig}$ and for $LEGO_{ext}$. Some interaction parameters with constant value and textual interaction parameters with a task-dependent nature have been discarded[5] leaving 38 parameters for $LEGO_{orig}$ and 50 for $LEGO_{ext}$. The results are computed using a linear Support Vector Machine (SVM) [8] in a 10-fold cross-validation setting. The results are stated in Table 1 and visualized in Fig. 3.

Best performance for both $LEGO_{orig}$ and $LEGO_{ext}$ in terms of UAR, κ, and ρ is achieved by using all parameters. However, it is highly notable that the results are very similar compared to the results of using all but the exchange level parameters (*no exchange*). In fact, applying the Wilcoxon test [29] for statistical significance proves the difference to be non-significant ($LEGO_{orig}$: $p > 0.15$, $LEGO_{ext}$: $p > 0.94$). This is underpinned by the results of only using the parameters on the exchange level (*only exchange*) being among the worst performing configurations together with the *no window* results. However, comparing the *all* results to the *no window* results ($LEGO_{orig}$: $p < 0.1$, $LEGO_{ext}$: $p < 0.001$) reveals that the window parameters play a bigger role in the overall performance.

---

[4] Recalculated parameters: %ASRSuccess, %TimeOutPrompts, %ASRRejections, %TimeOuts_ASRRej, %Barge-Ins, MeanASRConfidence, {#}ASRSuccess, {#}TimeOutPrompts, {#}ASRRejections, {#}TimeOuts_ASRRej, {#}Barge-Ins, {Mean}ASRConfidence.

[5] Discarded parameters: Activity, LoopName, Prompt, SemanticParse, SystemDialogueAct, UserDialogueAct, Utterance, parameters related to modality and help requests on all levels.

Table 1 Results in UAR, κ and ρ for including and excluding different parameter levels for $LEGO_{orig}$ and for $LEGO_{ext}$ [25]

	UAR		κ		ρ	
	$LEGO_{orig}$	$LEGO_{ext}$	$LEGO_{orig}$	$LEGO_{ext}$	$LEGO_{orig}$	$LEGO_{ext}$
Only exchange	0.328	0.328	0.310	0.310	0.456	0.456
Only window	0.338	0.363	0.333	0.380	0.479	0.558
No dialogue	0.398	0.415	0.457	0.480	0.622	0.643
Only dialogue	0.443	0.454	0.559	0.571	0.726	0.738
No window	0.460	0.471	0.578	0.589	0.737	0.747
No exchange	0.466	0.494	0.584	0.611	0.747	0.764
All	0.475	0.495	0.596	0.616	0.757	0.770

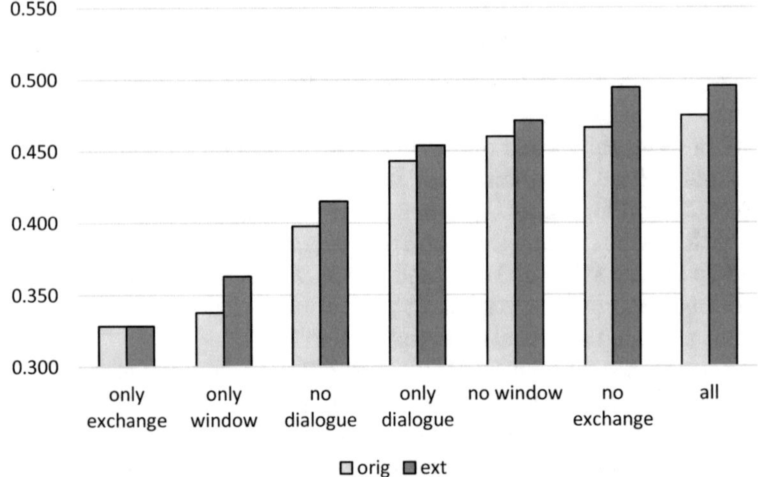

Fig. 3 SVM performance in UAR for including and excluding different parameter levels for $LEGO_{orig}$ and for $LEGO_{ext}$

While the analysis above is true for both feature sets $LEGO_{orig}$ and $LEGO_{ext}$, the results clearly show that the extension of the feature set results in an increased performance on almost all levels. The overall performance using *all* parameters has been relatively increased by 4.4 % ($p < 0.001$) in UAR and the performance of *no exchange* has been relatively increased by 6.0 % ($p < 0.001$). The results of the *only exchange* parameters are the same for both feature sets as the parameters on this level are the same, i.e., have not been computed a new.

3.4 Analysis of Window Size

While the impact of the different parameter levels on the overall estimation performance is of interest, we are also interested in how the window size influences the estimation performance. Hence, experiments have been conducted with different window sizes. As the experiments above showed that $LEGO_{ext}$ performed significantly better than $LEGO_{orig}$, only the $LEGO_{ext}$ feature set is used. Again, all experiments are conducted applying 10-fold cross-validation using a linear SVM. The results for UAR, κ, and ρ are depicted in Fig. 4. Table 2 shows also the relative improvement compared to a window size of three used as baseline for this experiment.

For calculating the window level parameters of all exchanges below the window size, the maximum possible window size of using all exchanges of the respective dialogue are used. Hence, for those exchanges, there is virtually no difference between the parameters on the dialogue and the window level. Of course, this is different for the parameters of exchanges above the window size.

A maximum performance is reached with a window size of 9 for UAR, κ and ρ alike. In fact, an UAR of 0.549 represents a relative improvement compared to a window size of 3 by +10.82 %. This clearly shows the potential hidden in these window parameters. If these results are compared to the performance of the original feature set of $LEGO_{orig}$, the performance is even relatively improved by +15.69 %. This clearly outperforms the currently best know sequential appoach to IQ estimation applying a Hybrid-HMM [6] by +8.5 %.

It is interesting though that the best window size is nine. We believe that this is system dependent and, in Let's go, related to the minimum number of system-user-exchanges necessary to perform a successful dialogue. Looking at the corpus reveals that a minimum of nine exchanges is needed.

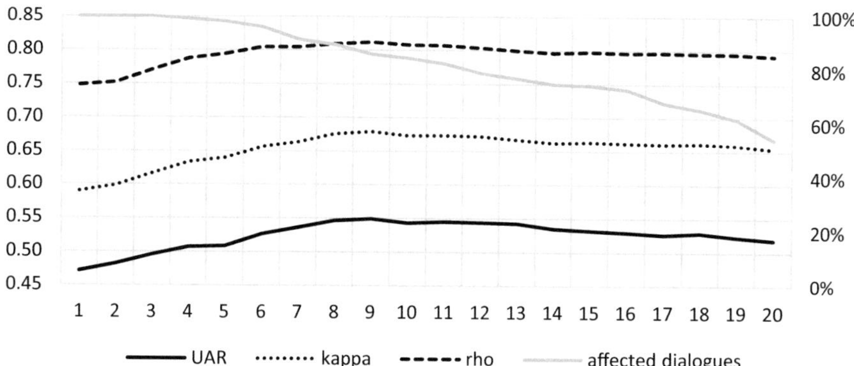

Fig. 4 SVM performance (*left ordinate*) for $LEGO_{ext}$ using different window sizes from $n = 1$ (no window) to $n = 20$ (abscissa). The percentage of affected dialogue, i.e., which have a length greater than the window size, is shown on the *right ordinate* [25]

Table 2 Results of different window sizes for IQ recognition in UAR, κ, and ρ

Window	UAR %		κ	ρ	#dial. (%)
1	0.471	−4.93**	0.589	0.747	100
2	0.482	−2.76**	0.598	0.752	100
3	0.495	–	0.616	0.770	100
4	0.507	+2.30	0.633	0.787	99
5	0.508	+2.57**	0.639	0.794	98
6	0.526	+6.16**	0.656	0.804	96
7	0.536	+8.16**	0.663	0.804	92
8	0.546	+10.22**	0.675	0.809	90
9	0.549	+10.82**	0.679	0.812	86
10	0.543	+9.61**	0.673	0.808	85
11	0.545	+9.97**	0.674	0.807	83
12	0.544	+9.76**	0.672	0.804	79
13	0.542	+9.50**	0.668	0.800	77
14	0.535	+7.99*	0.663	0.797	75
15	0.532	+7.42**	0.664	0.798	75
16	0.530	+6.90**	0.663	0.796	73
17	0.526	+6.23**	0.661	0.797	68
18	0.529	+6.75*	0.662	0.796	66
19	0.523	+5.54*	0.659	0.795	62
20	0.519	+4.66*	0.654	0.792	55

In addition, the relative improvement in UAR with respect to a window size of 3 is depicted Significance is indicated with *($\alpha < 0.05$) and **($\alpha < 0.01$) determined using the Wilcoxon test [29]. Best performance is achieved for a window size of 9 [25]

4 Discussion

When analyzing the results, clearly, the temporal information has a major effect on the Interaction Quality. In fact, the dialogue level parameters contributing most may be interpreted as the satisfaction of the user (represented by IQ) mainly depends on the complete dialogue and not on short-term events. However, putting this long-term information in the context of a shorter more recent period modelled by the window level achieves an even better performance. This increase is even more evident when further adjusting the window size. Hence, it may be concluded that IQ does not purely depend on local effects but those local effects have to be interpreted within the context of the dialogue.

5 Conclusion and Future Work

In this work, we analyzed the set of temporal parameters used as input for statistical classification approaches to estimate Interaction Quality. We showed that proper modeling of temporal aspects within the feature set may outperform sequential classification approaches like Hidden Markov Models drastically. For some temporal parameters, we introduced both the system and the user view thus extending the feature set. This results in an significant relative increase in UAR by +4.4 %. Furthermore, analysing the temporal features, i.e., features on the dialogue and window level, showed that both levels contribute most to the overall estimation performance (in contrast to the exchange level). Furthermore, we modified the window size further achieving a statistically significant relative improvement for IQ estimation by +15.69 % with an UAR of 0.549. The optimal window size of 9 for IQ recognition is attributed to the task complexity. Here, a minimum of 9 exchanges is necessary to successfully complete the task.

More generally, while the Interaction Quality is clearly influenced by local events, the complete course of the dialogue plays a major role. In other words, the user are clearly aware of the complete dialogue and do not "forget" events which may occur at the beginning of the dialogue.

For future work, previous successful experiments improving the estimation performance like applying a Hybrid-HMM or a hierarchical error-correction approach should be investigated using this extended and optimized feature set. Furthermore, repeating the experiments with more data of the *LEGOv2* corpus [4] might also give further insight into the problem.

References

1. Ultes, S., Heinroth, T., Schmitt, A., Minker, W.: A theoretical framework for a user-centered spoken dialog manager. In: López-Cózar, R., Kobayashi, T. (eds.) Proceedings of the Paralinguistic Information and its Integration in Spoken Dialogue Systems Workshop, pp. 241–246. Springer, New York, NY (2011)
2. Ultes, S., Dikme, H., Minker, W.: Dialogue management for user-centered adaptive dialogue. In: Rudnicky, A.I., Raux, A., Lane, I., Misu, T. (eds.) Situated Dialog in Speech-Based Human-Computer Interaction, pp. 51–61. Springer International Publishing, Cham (2016)
3. Ultes, S., Dikme, H., Minker, W.: First insight into quality-adaptive dialogue. In: Proceedings of the International Conference on Language Resources and Evaluation (LREC), pp. 246–251 (2014)
4. Ultes, S., Platero Sánchez, M.J., Schmitt, A., Minker, W.: Analysis of an extended interaction quality corpus. In: Lee, G.G., Kim, H.K., Jeong, M., Kim, J.H. (eds.) Natural Language Dialog Systems and Intelligent Assistants, pp. 41–52. Springer International Publishing (2015)
5. Schmitt, A., Ultes, S.: Interaction quality: assessing the quality of ongoing spoken dialog interaction by experts—and how it relates to user satisfaction. Speech Commun. **74**, 12–36 (2015). http://www.sciencedirect.com/science/article/pii/S0167639315000679
6. Ultes, S., Minker, W.: Interaction quality estimation in spoken dialogue systems using hybrid-HMMs. In: Proceedings of the 15th Annual Meeting of the Special Interest Group on Discourse

and Dialogue (SIGDIAL), pp. 208–217. Association for Computational Linguistics (2014). http://www.aclweb.org/anthology/W14-4328
7. Rabiner, L.R.: A tutorial on hidden Markov models and selected applications in speech recognition. Morgan Kaufmann Publishers Inc., San Francisco (1989)
8. Vapnik, V.N.: The Nature of Statistical Learning Theory. Springer, New York (1995)
9. Lafferty, J.D., McCallum, A., Pereira, F.C.N.: Conditional random fields: probabilistic models for segmenting and labeling sequence data. In: Proceedings of the Eighteenth International Conference on Machine Learning, pp. 282–289. Morgan Kaufmann Publishers Inc. (2001)
10. Hochreiter, S., Schmidhuber, J.: Long short-term memory. Neural Comput. **9**(8), 1735–1780 (1997)
11. Walker, M., Litman, D.J., Kamm, C.A., Abella, A.: PARADISE: a framework for evaluating spoken dialogue agents. In: Proceedings of the Eighth Conference on European Chapter of the Association for Computational Linguistics (EACL), pp. 271–280. Association for Computational Linguistics, Morristown, NJ, USA (1997)
12. Ultes, S., Schmitt, A., Minker, W.: Towards quality-adaptive spoken dialogue management. In: Proceedings of the NAACL-HLT Workshop on Future directions and needs in the Spoken Dialog Community: Tools and Data (SDCTD 2012), pp. 49–52. Association for Computational Linguistics, Montréal, Canada (2012). http://www.aclweb.org/anthology/W12-1819
13. Hara, S., Kitaoka, N., Takeda, K.: Estimation method of user satisfaction using n-gram-based dialog history model for spoken dialog system. In: Calzolari, N., Choukri, K., Maegaard, B., Mariani, J., Odijk, J., Piperidis, S., Rosner, M., Tapias, D. (eds.) Proceedings of the Seventh Conference on International Language Resources and Evaluation (LREC'10). European Language Resources Association (ELRA), Valletta, Malta (2010)
14. Higashinaka, R., Minami, Y., Dohsaka, K., Meguro, T.: Issues in predicting user satisfaction transitions in dialogues: individual differences, evaluation criteria, and prediction models. In: Lee, G., Mariani, J., Minker, W., Nakamura, S. (eds.) Spoken Dialogue Systems for Ambient Environments, Lecture Notes in Computer Science, vol. 6392, pp. 48–60. Springer, Heidelberg (2010). doi:10.1007/978-3-642-16202-2_5
15. Higashinaka, R., Minami, Y., Dohsaka, K., Meguro, T.: Modeling user satisfaction transitions in dialogues from overall ratings. In: Proceedings of the SIGDIAL 2010 Conference, pp. 18–27. Association for Computational Linguistics, Tokyo, Japan (2010)
16. Engelbrecht, K.P., Gödde, F., Hartard, F., Ketabdar, H., Möller, S.: Modeling user satisfaction with Hidden Markov Model. In: Proceedings of the 10th Annual Meeting of the Special Interest Group on Discourse and Dialogue (SIGDIAL), pp. 170–177. Association for Computational Linguistics, Morristown, NJ, USA (2009)
17. Schmitt, A., Schatz, B., Minker, W.: A statistical approach for estimating user satisfaction in spoken human-machine interaction. In: Proceedings of the IEEE Jordan Conference on Applied Electrical Engineering and Computing Technologies (AEECT), pp. 1–6. IEEE, Amman, Jordan (2011)
18. Ultes, S., Minker, W.: Improving interaction quality recognition using error correction. In: Proceedings of the 14th Annual Meeting of the Special Interest Group on Discourse and Dialogue, pp. 122–126. Association for Computational Linguistics (2013). http://www.aclweb.org/anthology/W/W13/W13-4018
19. Ultes, S., ElChabb, R., Minker, W.: Application and evaluation of a conditioned Hidden Markov Model for estimating interaction quality of spoken dialogue systems. In: Mariani, J., Devillers, L., Garnier-Rizet, M., Rosset, S. (eds.) Natural Interaction with Robots, Knowbots and Smartphones: Putting Spoken Dialog Systems into Practice, pp. 141–150. Springer, New York, NY (2012)
20. Schmitt, A., Schatz, B., Minker, W.: Modeling and predicting quality in spoken human-computer interaction. In: Proceedings of the SIGDIAL 2011 Conference, pp. 173–184. Association for Computational Linguistics, Portland, Oregon, USA (2011)
21. Schmitt, A., Ultes, S., Minker, W.: A parameterized and annotated spoken dialog corpus of the cmu let's go bus information system. In: International Conference on Language Resources and Evaluation (LREC), pp. 3369–3337 (2012)

22. Ultes, S., Schmitt, A., Minker, W.: On quality ratings for spoken dialogue systems—experts vs. users. In: Proceedings of the 2013 Conference of the North American Chapter of the Association for Computational Linguistics: Human Language Technologies, pp. 569–578. Association for Computational Linguistics (2013)
23. Raux, A., Bohus, D., Langner, B., Black, A.W., Eskenazi, M.: Doing research on a deployed spoken dialogue system: one year of let's go! experience. In: Proceedings of the International Conference on Speech and Language Processing (ICSLP) (2006)
24. Landis, J.R., Koch, G.G.: The measurement of observer agreement for categorical data. Biometrics **33**(1), 159–174 (1977)
25. Ultes, S.: User-centred Adaptive Spoken Dialogue Modelling. Dissertation, Ulm University, Ulm (2015)
26. Cohen, J.: A coefficient of agreement for nominal scales. Educ. Psychol. Meas. **20**, 37–46 (1960). Apr
27. Cohen, J.: Weighted kappa: nominal scale agreement provision for scaled disagreement or partial credit. Psychol. Bull. **70**(4), 213 (1968)
28. Spearman, C.E.: The proof and measurement of association between two things. Am. J. Psychol. **15**, 88–103 (1904)
29. Wilcoxon, F.: Individual comparisons by ranking methods. Biom. Bull. **1**(6), 80–83 (1945)

Recurrent Neural Network Interaction Quality Estimation

Louisa Pragst, Stefan Ultes and Wolfgang Minker

Abstract Getting a good estimation of the Interaction Quality (IQ) of a spoken dialogue helps to increase the user satisfaction as the dialogue strategy may be adapted accordingly. Therefore, some research has already been conducted in order to automatically estimate the Interaction Quality. This article adds to this by describing how Recurrent Neural Networks may be used to estimate the Interaction Quality for each dialogue turn and by evaluating their performance on this task. Here, we will show that RNNs may outperform non-recurrent neural networks.

Keywords RNN · Sequential data · Quality of dialogue · Recurrent neural network · Neural network · Interaction quality · User satisfaction · Spoken dialogue system

1 Introduction

While spoken human-computer interaction (HCI) is now wildly available on mobile technical devices, still, the applications are limited and the system behaves in a non-user-adaptive manner, i.e., the system does not take into account short-term properties of the user. However, having information about these short-term properties or user states like user satisfaction will help rendering this type of HCI more natural and user friendly [1–4].

Automatically deriving the user satisfaction during the ongoing interaction for each turn is a difficult task. Here, we have proposed several approaches for solving this problem before using either static methods [5–7] or sequential methods [8, 9]. For

L. Pragst (✉) · W. Minker
Institute of Communications Engineering, Ulm University, Ulm, Germany
e-mail: louisa.pragst@uni-ulm.de

W. Minker
e-mail: wolfgang.minker@uni-ulm.de

S. Ultes
Engineering Department, Cambridge University, Cambridge, UK
e-mail: su259@cam.ac.uk

K. Jokinen and G. Wilcock (eds.), *Dialogues with Social Robots*,
Lecture Notes in Electrical Engineering 427,
DOI 10.1007/978-981-10-2585-3_31

the latter, the problem is regarded as a sequence of user satisfaction values which are not independent from each other (which seems to be natural as a dialogue interaction may also be considered to be a sequence).

For such sequential problems, recurrent neural networks (RNNs) [10] have shown to perform well for modelling specific problems of Spoken Dialogue Systems (SDSs) [11–14]. An RNN extends a regular neural network by adding cyclic connections with a time delay thus making them very suitable for their application for sequential problems such as the automatic estimation of the user satisfaction which will be represented by the Interaction Quality (IQ) in this work. Hence, in this work, we will investigate and analyse the potential of RNNs for estimating the IQ level which will be presented in Sect. 2 together with other relevant related work. For applying RNNs for IQ estimation, we will propose possible approaches with different topologies (Sect. 3). Based on a data set of annotated dialogues, all approaches will be evaluated and their results presented in Sect. 4.

2 Significant Related Work

Famous work on determining the satisfaction level automatically is the PARADISE framework by Walker et al. [15]. Assuming a linear dependency between objective measures and User Satisfaction (US), a linear regression model is applied to determine US on the *dialogue level*. This is not only very costly, as dialogues must be performed with real users, but also inadequate if quality on a finer level is of interest, e.g., on the *exchange level*. In the following, we will present work on automatically determining the user satisfaction on the exchange level followed by a description of the Interaction Quality and related estimation approaches.

2.1 User Satisfaction

For predicting subjective quality measures on the exchange level, several research work has been conducted hitherto. However, most of this body of work lacks of either taking account of the sequential structure of the dialogue or resulting in insufficient performance.

Engelbrecht et al. [16] presented an approach using Hidden Markov Models (HMMs) to model the SDS as a process evolving over time. Performance ratings on a 5 point scale ("bad", "poor", "fair", "good", "excellent") have been applied by the users of the SDS during the dialogue. The interaction was halted while the user rated. A HMM was created consisting of 5 states (one for each rating) and a 6-dimensional input vector. While Engelbrecht et al. [16] relied on only 6 input variables, we will pursue an approach with 29 input variables. Moreover, we will investigate dialogues of a real world dialogue system annotated with quality labels by expert annotators.

Higashinaka et al. [17] proposed a model for predicting turn-wise ratings for human-human dialogues. Ratings ranging from 1 to 7 were applied by two expert annotators labeling for smoothness, closeness, and willingness. They achieved an UAR[1] of only 0.2–0.24 which is only slightly above the random baseline of 0.14.

Hara et al. [18] derived turn level ratings from overall ratings of the dialogue which were applied by the users *after* the interaction on a five point scale within an online questionnaire. Using n-grams to model the dialogue by calculating n-gram occurrence frequencies for each satisfaction value showed that results for distinguishing between six classes at any point in the dialogue to be hardly above chance.

2.2 The Interaction Quality Paradigm and the LEGO Corpus

As a more objective means of describing the user's satisfaction with the interaction, Schmitt and Ultes [19] proposed a measure called "Interaction Quality" (IQ) which fulfills the requirements of a quality metric for adaptive dialogue identified by Ultes et al. [20]. For the authors, the main aspect of user satisfaction is that it is assigned by real users. However, this is impractical in many real world scenarios. Therefore, the usage of expert raters is proposed. Further studies have also shown that ratings applied by experts and users have a high correlation [21].

The general idea of the Interaction Quality (IQ) paradigm—IQ being defined as user satisfaction annotated by expert raters—is to derive a number of interaction parameters from the dialogue system and use those as input variables to train a statistical classifier targeting IQ. Interaction quality is modelled on a scale from 5 to 1 representing the ratings "satisfied" (5), "slightly unsatisfied" (4), "unsatisfied" (3), "strongly unsatisfied" (2), and "extremely unsatisfied" (1).

The IQ paradigm [22] is based on automatically deriving interaction parameters from the SDS. These parameters are fed into a statistical classification module which estimates the IQ level of the ongoing interaction at the current system-user-exchange.[2] The interaction parameters are rendered on three levels (see Fig. 1): the exchange level, the window level, and the dialogue level. The exchange level comprises parameters derived from SDS modules Automatic Speech Recognition (ASR), Spoken Language Understanding (SLU), and Dialogue Management (DM) directly. Parameters on the window and the dialogue level are sums, means, frequencies or counts of exchange level parameters. While dialogue level parameters are computed out of all exchanges of the dialogue up to the current exchange, window level parameters are only computed out of the last three exchanges.

These interaction parameters are used as input variables to a statistical classification module with the goal of estimating the IQ value. The statistical model is trained based on annotated dialogues of the Lets Go Bus Information System in Pittsburgh, USA [23]. For the *LEGO* corpus [24], 200 calls from 2006 consisting of

[1]Unweighted Average Recall, see Sect. 4.1.

[2]A system-user exchange comprises a system turn followed by a user turn.

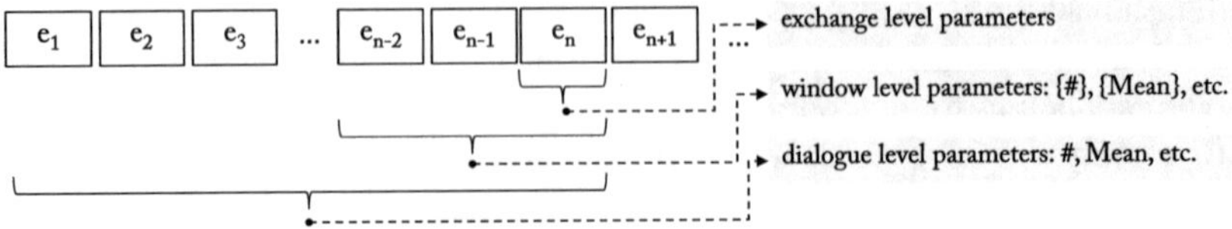

Fig. 1 This figure originally published by Schmitt et al. [24] shows the three parameter levels constituting the interaction parameters: the exchange level containing information about the current exchange, the window level, containing information about the last three exchanges, and the dialogue level containing information about the complete dialogue up to the current exchange

4,885 exchanges have been annotated by three different raters resulting in a rating agreement of $\kappa = 0.54$.[3] Furthermore, the raters had to follow labeling guidelines to enable a consistent labeling process [24]. The median of all three ratings is used to establish a ground truth for the IQ value of a given turn.

Schmitt et al. [19, 22] performed IQ recognition on the LEGO corpus using linear support vector machines (SVMs). They achieved an UAR[1] of 0.58 based on 10-fold cross-validation which is clearly above the random baseline of 0.2. Using the same data, Ultes et al. [8] put an emphasis on the sequential character of the IQ measure by applying Hidden Markov Models (HMMs) and Conditioned Hidden Markov Models (CHMMs). Both have been applied using 6-fold cross validation and a reduced feature set of the LEGO corpus achieving an UAR[1] of 0.44 for HMMs and 0.39 for CHMMs. In addition to modelling the observation probabilities of the HMM with generic Gaussian Mixture Models [8], Ultes et al. also proposed the combination of static classifiers with HMMs resulting in a hybrid approach [9]: there, the observation probabilities are modelled using the confidence scores of the static classification algorithms. By that, they were able to beat the performance of the confidence-providing SVM by 2.2 % in UAR.

3 Recurrent Neural Networks for IQ Estimation

Following the approaches presented by Ultes et al. [8], we focus on the sequential character of the IQ measure and propose recurrent neural networks [10] for IQ estimation. Recurrent neural networks enhance regular neural networks with cyclic, time delayed connections, that can be used as memory for earlier events. This makes them a reasonable choice for sequential problems such as the one presented.

The architecture of recurrent neural network can be varied by many factors, e.g. the number of layers, the number of nodes in a layer as well as the number, placement and time delay of the recurrent connections. All of these factors potentially influence the performance of the recurrent neural network.

[3]UAR, κ and ρ are defined in Sect. 4.1.

In this work, we evaluate recurrent architectures that differ in the placement of the recurrent connections, as depicted in Fig. 2, as well as the number of nodes in the hidden layer. More variations, such as the number of hidden layers and different time delays, were assessed, but yielded comparable results and are therefore not discussed further in this work.

We analyse the IQ estimation capability of three recurrent architectures for neural networks. They are introduced in the following.

An Elman network [25] is a simple recurrent network, with a recurrent connection from the hidden layer to itself. The previous results of the hidden layer may be combined with the current input values in order to achieve better a estimate of the IQ value.

A NARX network [26] utilises a recurrent connection from the output layer to the hidden layer. Thus, the estimated IQ value of the preceding exchange may be used for the current estimation. As raters were instructed to change the IQ value only by one from one exchange to the next (except in severe cases), the estimation might prove to be a valuable input.

Finally, a time delay network [27] is assessed. This architecture incorporates a recurrent connection from the input to the hidden layer. By comparing preceding input values with the current ones, differences that lead to an adjustment of the IQ value might be discovered and taken into consideration.

For all architectures, the nodes of both layers involved in recurrence are interconnected with each other by recurrent connections, and the time delay is set to one time step.

The number of nodes in the hidden layer varies from 2 to 20. The number of nodes available has an impact on the complexity of the calculation the network is able to perform. A high amount of nodes is beneficial if IQ estimation from the provided features requires an intricate mathematical formula. On the other hand, a high amount of nodes impedes the training process, and a satisfactory result might not be reached with the given amount of training data. By varying the number of nodes, the best trade-off can be determined.

The following characteristics are shared by all networks covered in this work. The networks contain only one hidden layer. All nodes of this layer are connected to all input variables as well as a bias and use a tanh transfer function. The output layer consists of one node, that gets its input from all nodes of the hidden layer and possibly a recurrent connection. It has a bias and uses a linear transfer function.

The networks perform a regression by implementing this architecture in their output layer, although the presented problem is a classification task. A different output layer implementing classification was tested and yielded worse results. Therefore, this architecture was adopted. The final estimated IQ value is obtained by rounding the result of the network.

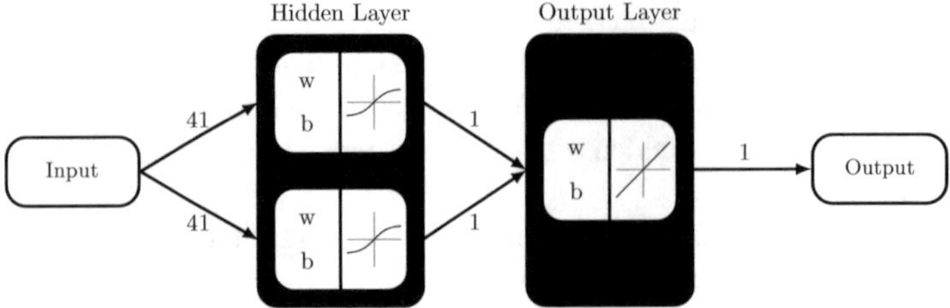

(a) Architecture of a non-recurrent neural network

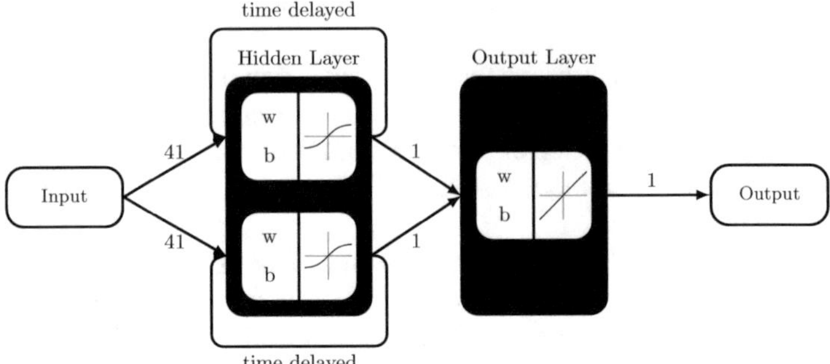

(b) Architecture of an Elman network

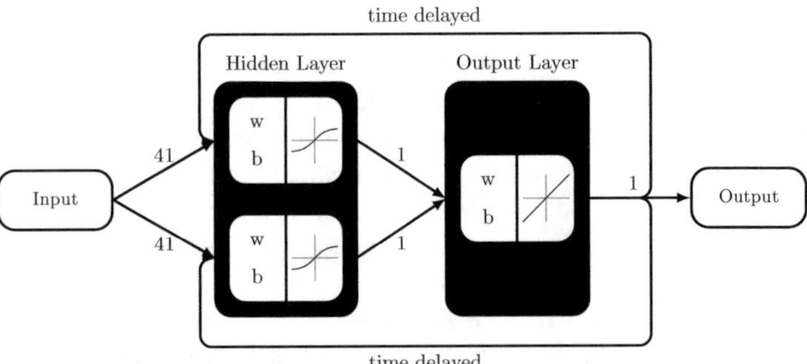

(c) Architecture of a NARX network

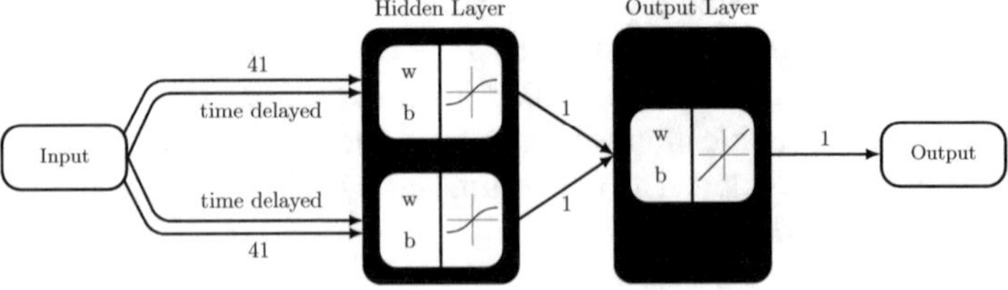

(d) Architecture of a time delay network

Fig. 2 The different architectures of recurrent neural networks we evaluate in this work, exemplary with 2 nodes in the hidden layer. The nodes of the hidden layer are connected to the input variables and a bias. They use a tanh transfer function. The output layer contains one node that gets it input from the nodes of the previous layer and a bias. It uses a linear transfer function

4 Experiments and Results

The performance of different recurrent neural networks is evaluated with the metrics described in the following.

4.1 Evaluation Metrics

Interaction Quality estimation is done by using three commonly used evaluation metrics: *Unweighted Average Recall (UAR)*, *Cohen's Kappa* [28] and *Spearman's Rho* [29]. These are also selected as the same metrics have been used in Schmitt et al. [22] as well.

Recall in general is defined as the rate of correctly classified samples belonging to one class. The recall in **UAR** for multi-class classification problems with N classes $recall_i$ is computed for each class i and then averaged over all class-wise recalls:

$$UAR = \frac{1}{N}\sum_{i=1}^{N} recall_i. \tag{1}$$

Cohen's Kappa measures the relative agreement between two corresponding sets of ratings. In our case, we compute the number of label agreements corrected by the chance level of agreement divided by the maximum proportion of times the labelers could agree. However, Cohen's weighted Kappa is applied as ordinal scores are compared [30]. A weighting factor w is introduced reducing the discount of disagreements the smaller the difference is between two ratings:

$$w = \frac{|r_1 - r_2|}{|r_{max} - r_{min}|}. \tag{2}$$

Here, r_1 and r_2 denote the rating pair and r_{max} and r_{min} the maximum and minimum ratings possible.

Correlation between two variables describes the degree by which one variable can be expressed by the other. **Spearman's Rho** is a non-parametric method assuming a monotonic function between the two variables [29].

4.2 Experimental Setup

To evaluate the performance of recurrent neural networks for IQ estimation, we use the LEGO corpus presented in Sect. 2.2. Exchanges with less than three IQ ratings are excluded from the evaluation for being too unreliable, just as dialogues with less than three exchanges, as the IQ value can not be adjusted before the third exchange.

Furthermore, only numeric features which can be automatically derived are used, leaving 41 input features.

For this corpus, we evaluate the performance of the recurrent neural network architectures described in Sect. 3 using tenfold cross validation, as well as a neural network without recurrence in order to determine the impact of sequential information. Whole dialogues are assigned to the train or test condition of the tenfold cross validation, instead of assigning each exchange separately, in order to account for the sequential character of the data. The neural networks are trained with the dialogues of the train condition using dynamic backpropagation [31]. Each dialogue is provided to the neural network sequentially, as in real conversation.

In related work [8, 9, 19, 22], the evaluation metrics are obtained using the available data in its entirety. As we feel that this approach does not adequately take into account the sequentiality of the data, we utilised a different approach: the evaluation metrics are obtained for each one of the 200 dialogues separately and then averaged over all dialogues. The performance of SVMs [19, 22] has been re-evaluated using this new approach in order to achieve comparability. This results in an UAR of 0.55, a κ of 0.30, and a ρ of 0.50 for SVMs.

4.3 Results

The average results over all 200 dialogues for all configurations and evaluation metrics can be found in Table 1.

The best results are achieved using a NARX network with ten nodes in the hidden layer: this architecture yields an UAR of 0.60, a κ of 0.37, and a ρ of 0.55. It outperforms the SVM approach (UAR: 0.55, κ: 0.30, ρ: 0.50).

When comparing the recurrent architectures with the non-recurrent architecture, statistical tests suggest that recurrence does not significantly increase the performance of neural networks for IQ estimation ($p > 0.05$ for Elman and NARX network). The time delay network even significantly decreases the performance ($p < 0.001$).

These results might be due to the fact that training recurrent neural networks is more complex than training non-recurrent neural networks [32, 33]. By utilising more elaborate training approaches, it may be possible to achieve different results.

Furthermore, the number of nodes does not appear to have a significant impact on the ability of a neural network to estimate the IQ value. While occasional significant differences between consecutive levels of the number of nodes can be found, those differences are isolated and not consistent across evaluation metrics. The assumption that the performance would become better at first as more complex calculations are possible, and then decrease as training becomes more difficult, can not be affirmed.

Table 1 Results given by the Unweighted Average Recall (UAR), Cohen's Kappa (κ), and Spearman's Rho (ρ)

#nodes	UAR	κ	ρ
(a) Performance of the baseline network			
2	0.56	0.35	0.53
3	0.55	0.34	0.52
4	0.56	0.34	0.52
5	0.56	0.35	0.51
6	0.55	0.35	0.53
7	0.57	0.36	0.52
8	0.54	0.33	0.52
9	0.56	0.35	0.52
10	0.55	0.35	0.52
11	0.53	0.31	0.49
12	0.52	0.34	**0.54**
13	0.57	0.36	0.53
14	**0.58**	**0.38**	**0.54**
15	0.56	0.35	0.52
16	0.57	0.37	0.53
17	0.57	0.35	0.53
18	0.57	0.37	0.53
19	**0.58**	0.36	0.53
20	0.56	0.35	0.53
(b) Performance of the Elman network			
2	0.54	0.31	0.48
3	0.55	0.33	0.51
4	0.56	0.33	0.52
5	0.56	0.33	0.50
6	0.55	0.32	0.52
7	0.57	0.35	0.52
8	0.55	0.34	0.51
9	0.55	0.34	0.50
10	0.56	0.34	0.52
11	0.56	0.35	0.54
12	0.57	0.35	0.53
13	0.57	0.35	0.52
14	0.58	**0.37**	0.53
15	0.58	0.35	0.51
16	0.55	0.35	0.53
17	**0.59**	**0.37**	**0.55**
18	0.58	0.35	0.53
19	0.57	0.36	0.53
20	0.58	0.36	0.53

(continued)

Table 1 (continued)

#nodes	UAR	κ	ρ
(c) Performance of the NARX network			
2	0.57	0.34	0.52
3	0.55	0.33	0.53
4	0.57	0.35	0.52
5	0.54	0.32	0.52
6	0.56	0.32	0.51
7	0.55	0.33	0.52
8	0.57	0.35	0.54
9	0.57	0.35	0.53
10	**0.60**	**0.37**	**0.55**
11	0.58	0.36	0.53
12	0.57	0.33	0.51
13	0.52	0.31	0.51
14	0.58	0.35	0.52
15	0.57	0.35	0.52
16	0.57	0.34	0.53
17	0.54	0.31	0.49
18	0.56	0.33	0.51
19	0.52	0.32	0.51
20	0.58	0.35	0.52
(d) Performance of the time delay network			
2	0.55	0.32	0.50
3	0.55	**0.33**	**0.51**
4	0.55	0.32	0.49
5	0.56	**0.33**	0.49
6	0.51	0.30	0.49
7	0.53	0.30	0.48
8	0.55	**0.33**	0.50
9	0.52	0.29	0.47
10	0.55	**0.33**	0.50
11	0.55	0.32	0.48
12	0.55	0.32	0.49
13	0.54	0.32	0.49
14	0.56	**0.33**	0.50
15	**0.57**	0.31	0.48
16	0.53	0.30	0.47
17	0.55	0.32	0.47
18	0.52	0.29	0.46
19	0.55	0.32	0.48
20	0.51	0.29	0.44

5 Conclusion

This work presented recurrent neural networks for the estimation of Interaction Quality on the exchange level. It could be shown that both non-recurrent and recurrent neural networks can outperform previous approaches such as SVMs in this task.

The sequential structure of dialogues suggests that recurrent neural networks might exceed the performance of non-recurrent neural networks in estimating the Interaction Quality on the exchange level, as they incorporate information about previous exchanges in their calculations. By evaluating the performance of different recurrent neural network architectures and comparing them to the performance of a neural network without recurrence, we found that Elman and NARX networks do not significantly outperform a corresponding non-recurrent neural network. Moreover, the time-delay network performs significantly worse than the non-recurrent network. This might be due to the fact that the training of recurrent networks is more complex than the training of non-recurrent networks. It is probable that more training data or a more refined training approach is needed.

For future research, the number of input features may be reduced in a meaningful way, or a larger corpus of training data may be used, in order to train the recurrent neural networks more efficiently. Furthermore, the training approach could be refined. Different recurrent architectures, for example a combination of the networks presented in this work, could be tested in addition.

Acknowledgements This paper is part of a project that has received funding from the European Union's Horizon 2020 research and innovation programme under grant agreement No 645012.

References

1. Ultes, S., Heinroth, T., Schmitt, A., Minker, W.: A theoretical framework for a user-centered spoken dialog manager. In: Proceedings of the Paralinguistic Information and its Integration in Spoken Dialogue Systems Workshop. pp. 241—246. Springer (2011)
2. Ultes, S., Dikme, H., Minker, W.: Dialogue management for user-centered adaptive dialogue. In: Proceedings of IWSDS (2014). http://www.uni-ulm.de/fileadmin/website_uni_ulm/allgemein/2014_iwsds/iwsds2014_lp_ultes.pdf
3. Ultes, S., Dikme, H., Minker, W.: First insight into quality-adaptive dialogue. In: Proceedings of the LREC, pp. 246–251 (2014)
4. Ultes, S., Kraus, M., Schmitt, A., Minker, W.: Quality-adaptive spoken dialogue initiative selection and implications on reward modelling. In: Proceedings of SIGDIAL, pp. 374–383. ACL (2015)
5. Ultes, S., Minker, W.: Improving interaction quality recognition using error correction. In: Proceedings of SIGDIAL, pp. 122–126. ACL (2013). http://www.aclweb.org/anthology/W/W13/W13-4018
6. Ultes, S., Minker, W.: Interaction quality: a review. Bulletin of Siberian State Aerospace University named after academician M.F. Reshetnev (4), 153–156 (2013). http://www.vestnik.sibsau.ru/images/vestnik/ves450.pdf
7. Ultes, S., Platero Sánchez, M.J., Schmitt, A., Minker, W.: Analysis of an extended interaction quality corpus. In: Proceedings of IWSDS (2015)

8. Ultes, S., ElChabb, R., Minker, W.: Application and evaluation of a conditioned hidden markov model for estimating interaction quality of spoken dialogue systems. In: Proceedings of IWSDS, pp. 141–150. Springer (2012)
9. Ultes, S., Minker, W.: Interaction quality estimation in spoken dialogue systems using hybrid-hmms. In: Proceedings of SIGDIAL, pp. 208–217. ACL (2014). http://www.aclweb.org/anthology/W14-4328
10. Hochreiter, S., Schmidhuber, J.: Long short-term memory. Neural Comput. **9**(8), 1735–1780 (1997)
11. Mesnil, G., He, X., Deng, L., Bengio, Y.: Investigation of recurrent-neural-network architectures and learning methods for spoken language understanding. In: Proceedings of the INTERSPEECH, pp. 3771–3775 (2013)
12. Henderson, M., Thomson, B., Young, S.: Robust dialog state tracking using delexicalised recurrent neural networks and unsupervised adaptation. In: Proceedings of the SLT, pp. 360–365. IEEE (2014)
13. Wen, T.H., Gačić, M., Kim, D., Mrkšic, N., Su, P.H., Vandyke, D., Young, S.: Stochastic language generation in dialogue using recurrent neural networks with convolutional sentence reranking. In: Proceedings of the ACL, pp. 275–284. (2015)
14. Su, P.H., Vandyke, D., Gasic, M., Kim, D., Mrksic, N., Wen, T.H., Young, S.: Learning from real users: rating dialogue success with neural networks for reinforcement learning in spoken dialogue systems. arXiv:1508.03386 (2015)
15. Walker, M., Litman, D.J., Kamm, C.A., Abella, A.: Paradise: a framework for evaluating spoken dialogue agents. In: Proceedings of ACL, pp. 271–280. ACL, Morristown, NJ, USA (1997)
16. Engelbrecht, K.P., Gödde, F., Hartard, F., Ketabdar, H., Möller, S.: Modeling user satisfaction with hidden markov model. In: Proceedings of SIGDIAL, pp. 170–177. ACL, Morristown, NJ, USA (2009)
17. Higashinaka, R., Minami, Y., Dohsaka, K., Meguro, T.: Issues in predicting user satisfaction transitions in dialogues: Individual differences, evaluation criteria, and prediction models. In: Spoken Dialogue Systems for Ambient Environments, Lecture Notes in Computer Science, vol. 6392, pp. 48–60. Springer, Heidelberg (2010). doi:10.1007/978-3-642-16202-2_5
18. Hara, S., Kitaoka, N., Takeda, K.: Estimation method of user satisfaction using n-gram-based dialog history model for spoken dialog system. In: Proceedings of LREC. ELRA, Valletta, Malta (2010)
19. Schmitt, A., Ultes, S.: Interaction quality: assessing the quality of ongoing spoken dialog interaction by experts—and how it relates to user satisfaction. Speech Commun. (2015) accepted for publication
20. Ultes, S., Schmitt, A., Minker, W.: Towards quality-adaptive spoken dialogue management. In: Proceedings of the NAACL-HLT: SDCTD, pp. 49–52. ACL, Montréal, Canada (2012). http://www.aclweb.org/anthology/W12-1819
21. Ultes, S., Schmitt, A., Minker, W.: On quality ratings for spoken dialogue systems—experts vs. users. In: Proceedings of NAACL-HLT, pp. 569–578. ACL (2013)
22. Schmitt, A., Schatz, B., Minker, W.: Modeling and predicting quality in spoken human-computer interaction. In: Proceedings of SIGDIAL, pp. 173–184. ACL, Portland, Oregon, USA (2011)
23. Raux, A., Bohus, D., Langner, B., Black, A.W., Eskenazi, M.: Doing research on a deployed spoken dialogue system: one year of let's go! experience. In: Proceedings of ICSLP (2006)
24. Schmitt, A., Ultes, S., Minker, W.: A parameterized and annotated spoken dialog corpus of the CMU Let's go bus information system. In: Proceedings of the LREC, pp. 3369–337 (2012)
25. Elman, J.L.: Finding structure in time. Cogn. Sci. **14**(2), 179–211 (1990)
26. Lin, T., Horne, B.G., Tiňo, P., Giles, C.L.: Learning long-term dependencies in narx recurrent neural networks. IEEE Trans. Neural Netw. **7**(6), 1329–1338 (1996)
27. Waibel, A., Hanazawa, T., Hinton, G., Shikano, K., Lang, K.J.: Phoneme recognition using time-delay neural networks. IEEE Trans. Acoust. Speech Signal Process. **37**(3), 328–339 (1989)
28. Cohen, J.: A coefficient of agreement for nominal scales. Educ. Psychol. Meas. **20**, 37–46 (1960)

29. Spearman, C.E.: The proof and measurement of association between two things. Am. J. Psychol. **15**, 88–103 (1904)
30. Cohen, J.: Weighted kappa: nominal scale agreement provision for scaled disagreement or partial credit. Psychol. Bull. **70**(4), 213 (1968)
31. De Jesús, O., Hagan, M.T.: Backpropagation algorithms for a broad class of dynamic networks. IEEE Trans. Neural Netw. **18**(1), 14–27 (2007)
32. De Jesús, O., Horn, J.M., Hagan, M.T.: Analysis of recurrent network training and suggestions for improvements. In: Proccedings of the International Joint Conference on Neural Networks (IJCNN'01) 2001. vol. 4, pp. 2632–2637. IEEE (2001)
33. Horn, J., De Jesús, O., Hagan, M.T.: Spurious valleys in the error surface of recurrent networksanalysis and avoidance. IEEE Trans. Neural Netw. **20**(4), 686–700 (2009)

An Evaluation Method for System Response in Chat-Oriented Dialogue System

Hideaki Mori, Atsushi Yasuda and Masahiro Araki

Abstract Chat functionality is currently regarded as an important factor in spoken dialogue systems. In this article, we explore the architecture of a chat-oriented dialogue system that can continue a long conversation with users and can be used for a long time. To achieve this goal, we propose a method which combines various types of response generation modules, such as a statistical model-based module, a rule-based module and a topic transition-oriented module. The core of this architecture is a method for selecting the most appropriate response based on a breakdown index and a development index. The experimental results show that the weighted sum of these indexes can be used for evaluating system utterances from the viewpoint of continuous dialogue and long-term usage.

Keywords Chat-oriented dialogue system · Response generation · SVM regression

1 Introduction

In recent years, there have been some research and development case studies on open-domain chat dialogue systems. The merit of chat functionality in a dialogue system is to encourage the daily use of the system so as to accustom the user to the speech interface. Moreover, chat dialogue functionality can give a user a sense of closeness to the system, especially for novice users of the speech interface. Considering this situation, the requirements of a chat dialogue system are (1) to maintain a longer dialogue without a breakdown of the conversation and (2) to maintain the long duration of use. We call the property of the first requirement as "continuous" and that of the second as "long-term."

H. Mori (✉) · A. Yasuda · M. Araki
Kyoto Institute of Technology, Matsugasaki Sakyo-ku, Kyoto 6068585, Japan
e-mail: mori@ii.is.kit.ac.jp
URL:http://www.ii.is.kit.ac.jp/araki/FSwiki/wiki.cgi?page=English+page

M. Araki
e-mail: araki@kit.ac.jp

K. Jokinen and G. Wilcock (eds.), *Dialogues with Social Robots*,
Lecture Notes in Electrical Engineering 427,
DOI 10.1007/978-981-10-2585-3_32

The aim of this article is to propose a framework for realizing a continuous and long-term chat-oriented dialogue system. To achieve this goal, we propose a method combining various types of response generation modules, such as a statistical model-based module, a rule-based module, and a topic transition-oriented module. The core of this architecture is a method for selecting the most appropriate response based on a breakdown index, that evaluates a continuous aspect of the system utterance, and a development index, that evaluates long-term aspect. From the results of an experiment, we found that the weighted sum of these indices could be used for evaluating system utterances from the viewpoint of continuous dialogue and long-term usage.

2 Related Works

In previous research studies on chat dialogue systems, the central theme of these studies is how to generate an appropriate and natural response to the user's utterance [1, 2]. There was little effort to realize both continuous and long-term features in chat-oriented dialogue systems.

The chat dialogue system's robustness to respond to any user utterance is a key functionality that must be implemented to make it continuous. Therefore, a statistical response generation method is used in recent chat-oriented dialogue systems [1, 3]. Moreover, appropriateness and naturalness of the response are required.

To realize these functionalities, Higashinaka et al. [1] proposed a method for evaluating the coherence of the system utterance to judge its appropriateness. They developed a classifier that distinguishes a pair of a user utterance and that generated from a system as coherent or incoherent by using various pieces of information related to dialogue exchange, such as dialogue acts, question types, predicate-argument structures, extended names entities and dependency structures. From a pair of utterances, their method creates a single tree structure containing above mentioned various types of information. Having created such trees for many utterance pairs and labeled them as coherent of incoherent, they then mine the trees to obtain patterns (subtrees) that are used in classification stage using Boosting-based Algorithm for Classifying Trees.

On the other hand, a chat system with a long-term feature should have the ability to keep the user interested and not bored (e.g. [4]). For example, it should be able to provide a new topic in a chat based on the recent news or seasonal event. It should also be able to develop a current topic for the dialogue by bringing up related topics. In general, it is difficult to realize such a topic shift in a statistical method. The rule-based method or the hybrid of the rule-based and the statistical method is appropriate for implementing such functionalities.

3 Framework of Response Generator

Because of the difference in methods in implementing a suitable functionality for a continuous and long-term chat dialogue system described in the previous section, it is difficult to realize the aforementioned functionalities in one response generation module. Such module could be complex and difficult to maintain. Therefore, it is reasonable to implement the elementary functionalities in separate modules and combine them to generate one plausible response for the purpose of the continuous and long-term chat dialogue.

To realize a continuous chat dialogue, the system needs to be robust to various user utterances. Statistical methods [1, 3] are popular in realizing the robust response generation. These methods can also generate a high-quality response in terms of appropriateness and naturalness. On the flip side of this strength, the system response tends to be confined to the expectations and, sometimes, the user considers it boring. As a result, the appropriateness and naturalness are not necessarily connected with the long-term use of the system.

Occasional and sometimes unexpected topic shift could make the chat interesting, but it requires a different response generation algorithm aiming for an appropriate and natural response.

Keeping the interest of the user in the chat system for a long-term requires changing the behavior of the system. If the system's utterance is gradually matching the user's preference, the user can feel a sense of closeness to the system. Such behavior is difficult to implement using the statistical method only. Some type of control by handwritten rule is required to begin the conversation with a new topic from the system side. In addition, the functionality of delivering the news filtered by the user's preference can encourage the daily use of the system. Such dialogue does not require robust dialogue management. The simple pattern is beneficial for both the user and the system.

As a result, the requirement of a continuous and long-term chat dialogue system is "to generate an appropriate and natural response as a majority behavior, but sometimes the system may generate an unexpected but interesting response and, sometimes, may start the chat by following the user's preference and recent news/topics." It is natural to divide the aforementioned, sometimes conflicting, functionality into individual specific modules and select the most plausible response among the candidates. Figure 1 shows our proposed architecture for realizing multiple response generation modules, and the selection method.

In the architecture, we used the following three chat dialogue systems:

- Rule-based system,
- Statistical model-based system,
- Topic transition-oriented system.

The rule-based system is based on the ELIZA-type system.[1] It manages greetings and encourages next user utterance when a typical curt reply is observed. This

[1] http://search.cpan.org/~neilb/Chatbot-Eliza-1.05/.

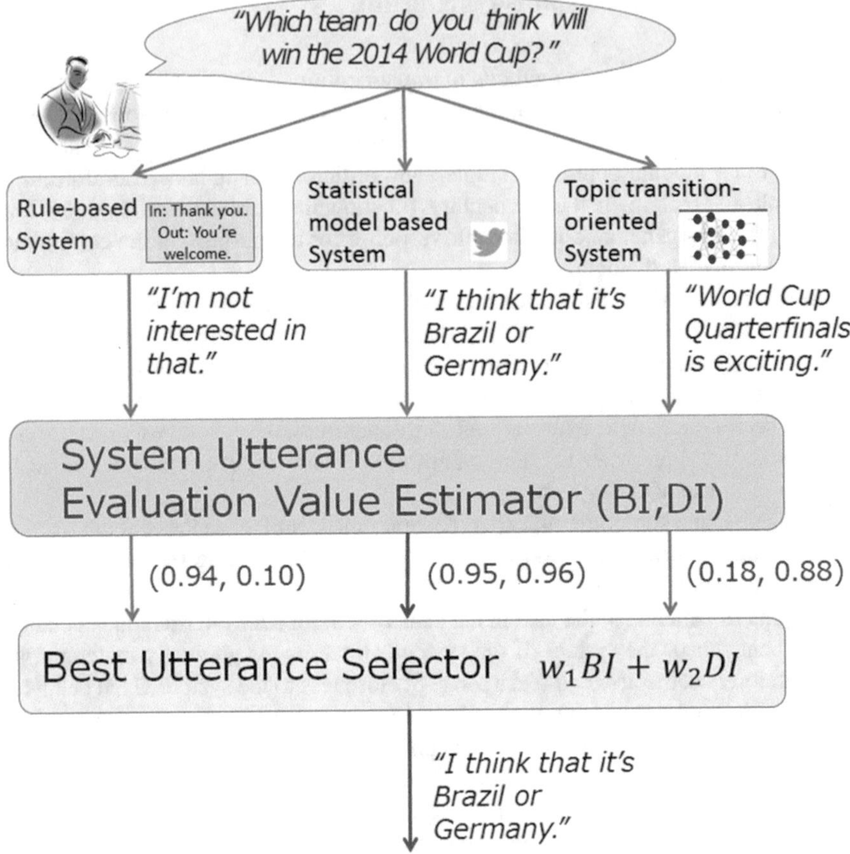

Fig. 1 Proposed architecture for realizing multiple response generation modules, and the selection method

rule-based system generates the response by the word-based similarity matching to the head of the rule. It can reply naturally when the rules match the user utterance appropriately, but it does not have a wide coverage.

The statistical model-based system uses the NTT chat dialogue API [5]. The system utterance is generated from (1) pre-stored knowledge (predicate-noun sets created by analyzing text on the Internet) to generate responses or (2) being selected from large-scale utterance data. The generated utterance is converted in pre-output conversion module in order to realize the gender-specific inflections at the end of sentences in the Japanese language and a more consistent personality. It can respond to various topics, but sometimes it replies inappropriately, from the viewpoint of topics (caused by the misunderstanding of the current topic) or of sociality (caused by picking up the utterance from dispute interaction).

The topic transition-oriented system [6] extracts topics from the preceding user utterances and select a neighboring topic in the word embedded space made by Word2Vec [7]. This topic transition-oriented system tends to generate an unnatural response, but sometimes it can generate appropriate ones and stimulate the user's willingness to chat effectively. We try to realize a continuous and long-term chat-oriented dialogue system by using the good aspects of these modules. It picks up a topic word from the user's utterance and calculates 10-best nearest words in the vector space acquired by the method of Word2Vec. From this neighboring word set, it randomly picks up one word and replace it by the user's topic word in the preceding user utterance, reform the utterance as a reply form and output as a response.

4 Evaluation Method for System Response

4.1 Evaluation Indices for Continuous and Long-Term Dialogue

As a result of the requirements discussed in Sect. 3, we created two evaluation indices, the Breakdown Index (BI) and the Development Index (DI).

The Breakdown Index (BI) determines how natural the system utterance is. For example, as a system response to Yes/No type question, "Yes." and "No" get high BI value. On the other hand, the utterance expressing system's impression to this type of question gets low BI value. This index typically reflects the validity of the adjacent pair that consists of user's utterance and system's utterance. It does not evaluate the development ability to the dialogue topic.

The Development Index (DI) determines how easily users come up with the next utterance. For example, a question about the current topic or an introduction of a related topic in the current dialogue gets high DI value. On the other hand, a simple reply or a curt answer, e.g. "I don't know." gets low DI value. It does not evaluate the appropriateness as the response of the previous user's utterance.

For example, in the context of the football World Cup, and at the dialogue state after the user utterance "Which team do you think will win the 2014 World Cup?" (in Fig. 1), the system utterance "I'm not interested in that." seems to be coherent in the dialogue context (high BI) but it will reduce the willingness to talk to the system both in short-term, i.e. this dialogue, and long-term (low DI). The other system output "World Cup Quarterfinals is exciting." seems to be incoherent (low BI) but introduce a new keyword that expands the range of the conversation (high DI). The system output "I think that it's Brazil or Germany" is the most appropriate reply from the viewpoint of both BI and DI.

4.2 Evaluation Indices Estimator

To create an estimator for the BI, we used a chat-oriented dialogue corpus collected by the dialogue task group of Project Next NLP, Japan.[2] The training data consists of 1,000 utterances (10 turns * 100 dialogues) that have breakdown annotations by 24 annotators for each utterance. We set the target value as the ratio of annotators annotated in the utterance as a breakdown. We extracted Bag-of-Words (BoW) features for each data and used a poly-kernel support vector machine (SVM) as the regressor for the target value.

To create an estimator of the DI, we used the same method used for the BI estimator. We annotated the data from the viewpoint of the development of dialogue by two annotators. The features and regression algorithm are also the same as those for the BI estimator.

The weighted sum of BI and DI are calculated for each output of the module and the system selects the highest system utterance as a final output.

5 Experimental Results

To verify the validity of the proposed architecture, we examined the accuracy of the system output. We prepared 500 utterances (10 utterances * 50 dialogues) using the three types of systems described below. In this experiment, we selected an appropriate system utterance by the majority vote of three annotators. If all annotators selected a different system utterance, we used the system utterance selected by the annotator who selected the best utterance more often. In this experiment, we used three chat dialogue systems described in Sect. 3.

The output of the proposed method explained in Sect. 3 was evaluated by regarding the majority vote of the annotators as the right answer. We examined the validity of the proposed method by varying the weights of BI and DI. Table 1 is an example of dialogue and evaluation values.

As shown in Fig. 2, the proposed method outperformed the individual estimators, especially the ratio of the negative instance in the train data are 12.0 and 89.5 %. This result indicates that weighted sum works effectively in case of the training data show a high precision rate or a high recall rate.

In the case of breakdown ratio of the training data are 12.0 and 89.5 %, the accuracy is 47.0 and 50.8 %, respectively. These values may seem to be low. However, by considering the low agreement rate of human annotation (a fair agreement by Fleiss' κ is 0.32 [8]), the accuracy of the proposed method can be regarded to be within an acceptable range.

[2]https://sites.google.com/site/dialoguebreakdowndetection/chat-dialogue-corpus.

Table 1 An example of dialogue and evaluation values

–	Utterance	BI	DI
User	It reminds me of the elementary school days	–	–
Rule-based system	You have something you want to tell me, aren't you?	0.487	0.526
Statistical-based system	I miss summer vacation	0.754	0.380
Topic transition-oriented system	It reminds me of the junior high school days too	0.637	0.528

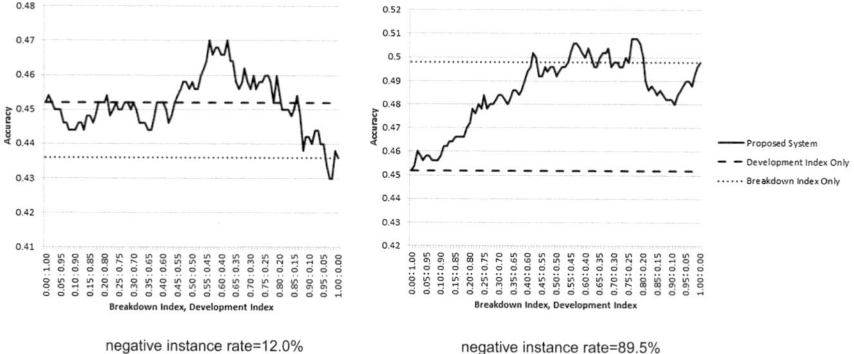

Fig. 2 Selection accuracy

6 Conclusion

In this work, we proposed a selection method for the most appropriate response using a regression technique.

Three different response generation modules are implemented: (1) The rule-based system that can reply naturally when the rules match the user utterance appropriately but does not have a wide coverage, (2) The statistical model-based system that can respond to various topics, but sometimes replies inappropriately, (3) The topic transition-oriented system tends to generate unnatural responses, but sometimes it can generate appropriate ones and stimulate the user's willingness to chat effectively.

Both a breakdown index and a development index, which are related to continuous and long-term functionality, respectively, contribute to deciding good utterances in a chat dialogue.

In future work, we plan to conduct an experiment for long-term usage of the chat dialogue system.

Acknowledgements We thank all members of the dialogue task of "Project Next NLP" for the data collection and annotation.

References

1. Higashinaka, R., Meguro, T., Imamura, K., Sugiyama, H., Makino, T., Matsuo, Y.: Evaluating coherence in open domain conversational systems. In: Proceedings of Interspeech, pp. 130–134 (2014)
2. Xiang, Y., Zhang, Y., Zhou, X., Wang, X., Qin, Y.: Problematic situation analysis and automatic recognition for chinese online conversational system. In: Proceedings of the Third CIPS-SIGHAN Joint Conference on Chinese Language Processing, pp. 43–51. Association for Computational Linguistics, Wuhan, China (2014). http://www.aclweb.org/anthology/W14-6808
3. Banchs, R.E., Li, H.: Iris: a chat-oriented dialogue system based on the vector space model. In: Proceedings of the ACL 2012 System Demonstrations, pp. 37–42. ACL '12, Association for Computational Linguistics, Stroudsburg, PA, USA (2012). http://dl.acm.org/citation.cfm?id=2390470.2390477
4. Dybala, P., Ptaszynski, M., Maciejewski, J., Takahashi, M., Rzepka, R., Araki, K.: Multiagent system for joke generation: humor and emotions combined in human-agent conversation. J. Ambient Intell. Smart Environ. **2**(1), 31–48 (2010). http://dl.acm.org/citation.cfm?id=1735862.1735867
5. Onishi, K., Yoshimura, T.: Casual conversation technology achieving natural dialog with computers. NTT DOCOMO Tech. J. **15**(4), 16–21 (2014)
6. Nakano, A., Araki, M.: Topic development method in chat dialogue system using distributed word representation (in Japanese). In: Proceedings of Annual Meeting of The Association for Natural Language Processing, pp. 269–272 (2015)
7. Mikolov, T., Chen, K., Corrado, G., Dean, J.: Efficient estimation of word representations in vector space. In: ICLR Workshop (2013)
8. Fleiss, J.L.: Measuring nominal scale agreement among many raters. Psychol. Bull. **76**(5), 378–382 (1971)

The Negotiation Dialogue Game

Romain Laroche and Aude Genevay

Abstract This article presents the design of a generic negotiation dialogue game between two or more players. The goal is to reach an agreement, each player having his own preferences over a shared set of options. Several simulated users have been implemented. An MDP policy has been optimised individually with Fitted Q-Iteration for several user instances. Then, the learnt policies have been cross evaluated with other users. Results show strong disparity of inter-user performances. This illustrates the importance of user adaptation in negotiation-based dialogue systems.

Keywords Spoken dialogue systems · Dialogue simulation · Reinforcement learning · Negotiation · Game theory

1 Introduction

Research on negotiation dialogue experiences a growth of interest. At first, Reinforcement Learning [1], the most popular framework for dialogue management in spoken dialogue systems [2–4], has been applied to negotiation with mitigated results [5, 6], because the non-stationary policy of the opposing player prevents those algorithms from converging consistently. Then, Multi-Agent Reinforcement Learning [7] was applied but still with convergence difficulties [8]. Finally, recently, Stochastic Games [9] were applied successfully [10], with convergence guarantees, but still only for zero-sum games, which is quite restrictive in a dialogue setting where noisy communication and miscommunication are the bases of the game.

In this article, the negotiation dialogue games in the literature ([5] considers sets of furniture, [8, 11] resource trading, and [12–15] appointment scheduling) have been abstracted as an agreement problem over a shared set of options. The goal for

R. Laroche (✉) · A. Genevay
Orange Labs, Châtillon, France
e-mail: romain.laroche@orange.com

A. Genevay
e-mail: aude.genevay@gmail.com

K. Jokinen and G. Wilcock (eds.), *Dialogues with Social Robots*,
Lecture Notes in Electrical Engineering 427,
DOI 10.1007/978-981-10-2585-3_33

the players is to reach an agreement and select an option. This negotiation dialogue game can be parametrised to make it zero-sum, purely cooperative, or general sum.

In addition to the study of negotiation dialogue, we claim that this game can be used for user adaptation in dialogue systems [16, 17], which is not progressing as fast as it should because of lack of data. Indeed, while one used to need only a dataset to learn from, user adaptation requires as many datasets as users in order to learn and evaluate the algorithms. The negotiation game enables to introduce several handcrafted user simulators with a set of parameters. An MDP policy has been individually optimised for five user instances. Then, these policies have been cross evaluated on all users. Results show strong disparity of inter-user performance. This illustrates the importance of user adaptation in negotiation-based dialogue.

2 The Negotiation Dialogue Game

The goal for each participant is to reach an agreement. The game involves a set of m players $\mathcal{P} = \{p^i\}_{i\in[1,m]}$. n options (in resource trading, it is an exchange proposal, in appointment scheduling, it is a time-slot) are considered, and for each option τ, players have a cost $c^i_\tau \sim \delta^i$ randomly sampled from distribution $\delta^i \in \Delta_{\mathbb{R}^+}$ to agree on it. Players also have a utility $\omega^i \in \mathbb{R}^+$ for reaching an agreement. For each player, a parameter of cooperation with other players $\alpha^i \in \mathbb{R}$ is introduced. As a result, player p^i's immediate reward at the end of the dialogue is:

$$R^i(s^i_T) = \omega^i - c^i_\tau + \alpha^i \sum_{j\neq i}(\omega^j - c^j_\tau) \tag{1}$$

where s^i_T is the last state reached by player p^i at the end of the dialogue, τ is the agreed option. If players fail to agree, the final immediate rewards $R^i(s^i_T) = 0$ for all players p^i. If at least one player p^j misunderstands and agrees on a wrong option τ^j which was not the one proposed by the other players, this is even worse, since each player p^i gets the cost of selecting option τ^i without the reward of successfully reaching an agreement:

$$R^i(s^i_T) = -c^i_{\tau^i} - \alpha^i \sum_{j\neq i} c^j_{\tau^j} \tag{2}$$

The values of α^i give a description of the nature of the players, and therefore of the game as modelled in game theory [9]. If $\alpha^i < 0$, player p^i is said to be antagonist: he has an interest in making the other players lose. In particular, if $m = 2$ and $\alpha^1 = \alpha^2 = -1$, it is a zero-sum game. If $\alpha^i = 0$, player p^i is said to be self-centred: he does not care if the other player is winning or losing. Finally, if $\alpha^i > 0$, player p^i is said to be cooperative, and in particular, if $\forall i \in [1, m], \alpha^i = 1$, the game is said to be fully cooperative because $\forall (i, j) \in [1, m]^2$, $R^i(s^i_T) = R^j(s^j_T)$.

From now on, and until the end of the article, we suppose that there are only $m = 2$ players: a system $\mathscr{p}_s$ and a user $\mathscr{p}_u$. They act each one in turn, starting randomly by one or the other. They have four possible actions. ACCEPT(τ) means that the user accepts the option τ (independently from the fact that τ has actually been proposed by the other player. If it has not, this induces the use of Eq. 1 to determine the reward). This act ends the dialogue. REFPROP(τ) means that the user refuses the proposed option and proposes instead option τ. REPEAT means that the player asks the other player to repeat his proposition. And finally, ENDDIAL denotes the fact that the player does not want to negotiate anymore, and terminates the dialogue.

Understanding through speech recognition of system $\mathscr{p}_s$ is assumed to be noisy with a sentence error rate SER_s^u after listening to a user $\mathscr{p}_u$: with probability SER_s^u, an error is made, and the system understands a random option instead of the one that was actually pronounced. In order to reflect human-machine dialogue reality, a simulated user always understands what the system says: $SER_u^s = 0$. We adopt the way [18] generates speech recognition confidence scores: $score_{reco} = \frac{1}{1+e^{-X}}$ $where X \sim \mathcal{N}(c, 0.2)$ given a user $\mathscr{p}_u$, two parameters $(c_\perp^u, c_\top^u)$ with $c_\perp^u < c_\top^u$ are defined such that if the player understood the right option, $c = c_\top^u$ otherwise $c = c_\perp^u$. The further apart the normal distribution centres are, the easier it will be for the system to know if it understood the right option, given the score.

3 The Inter-User Policy Experiment

This section intends to show that, in the negotiation game, a policy that is good or optimal against a given user might yield very poor performance against another user. First, it introduces two classes of handcrafted users. Then, it designs a linear parametrisation in order to use Fitted Q-Iteration [19, 20] for policy optimisation. And finally, it shows that policies that have been learnt and optimised on specific users are only marginally successfully reusable on other users.

3.1 User Profiles

A straightforward characteristic of a user $\mathscr{p}_u$ is its intelligibility by the system $\mathscr{p}_s$, parametrised by its average sentence error rate SER_s^u. Another understanding characteristic consists in varying centres $(c_\perp^u, c_\top^u)$ for the speech recognition score. For distant $(c_\perp^u, c_\top^u)$ values, the system will easily know if it understood well.

In order to add more variability in our simulated users, two handcrafted classes of users have been implemented:

- The *Deterministic User* (parameter x) ACCEPT(τ) if and only if $\tau \in \mathcal{T}_x$, where $\mathcal{T}_x$ is the set of its x preferred options. If $\tau \notin \mathcal{T}_x$, he REFPROP($\tau'$), $\tau' \in \mathcal{T}_x$ being his preferred options that was not proposed before. If all $\tau \in \mathcal{T}_x$ have been refused, or if the system insists by proposing the same option twice, he ENDDIAL.

- The *Random User* (parameter p) ACCEPT(τ) any option τ asked by the system, with probability p. With probability $1 - p$, he REFPROP(τ') an option τ' randomly. If he's asked to repeat, he'll make a new random proposition.

3.2 Reinforcement Learning Implementation

The system $\mathcal{p}_s$ learns the optimal policy with the Fitted Q-Iteration algorithm [19, 20], when playing against user $\mathcal{p}_u$. This subsection details the design of the Reinforcement Learning implementation.

The dialogue system is formalised as an MDP $\langle \mathcal{S}, \mathcal{A}, R, P, \gamma \rangle$ where $\mathcal{S}$ is the state space, $\mathcal{A}$ is the action space, $R : \mathcal{S} \rightarrow \mathbb{R}$ is the immediate reward function, $P : \mathcal{S} \times \mathcal{A} \rightarrow \mathcal{S}$ is the Markovian transition function and γ is the discount factor.

Least-squares Fitted Q-Iteration is used to learn the policy with a linear parametrisation of the Q-function. The optimal Q-function Q^* verifies Bellman's equation:

$$Q^*(s, a) = \mathbb{E}\left[R(s) + \gamma \max_{a'} Q^*(s', a')\right] \Leftrightarrow \quad Q^* = T^* Q^* \tag{3}$$

The optimal Q-function is thus the fixed point of Bellman's operator T^* and since it is a contraction ($\gamma < 1$), Banach's theorem ensures its uniqueness. Hence, the optimal Q-function is obtained by iteratively applying Bellman's operator.

When the state space is continuous (or very large) it is impossible to use Value-Iteration as such. The Q-function must be parametrised. A popular choice is the linear parametrisation of the Q-function [20]: $Q_a(s) = \theta_a^\top \Phi_a(s)$ where $\Phi = \{\Phi_a\}_{a \in \mathcal{A}}$ is the feature vector for linear state representation and $\theta = \{\theta_a\}_{a \in \mathcal{A}}$ are the parameters that have to be optimised. Each dimension of θ_a represents the influence of the corresponding feature in the Q_a-function.

In the experiment, the feature vector Φ_a is a 5-dimensional vector composed of the following features for each action: utility loss between the last proposed option and the next one, the square of the previous value, number of options which can still be proposed, length of the dialogue, speech recognition score. $\mathcal{A}$ is defined according to notations in Sect. 3.1 as follows: ACCEPT(τ), REFINSIST(τ_k) $\Leftrightarrow$ REFPROP(τ_k), with τ_k equal to the last proposed option by the system, REFNEWPROP(τ_{k+1}) $\Leftrightarrow$ REFPROP(τ_{k+1}), with τ_{k+1} the preferred one after τ_k, and REPEAT.

3.3 Experiment Results

The experiment includes nine different users $\mathcal{p}_u^i$ whose characteristics are described in Table 1. The systems are fully cooperative ($\alpha_s^i = 1$) with discount factor $\gamma = 0.9$

Table 1 Simulated users with the average return $\gamma^T R^i(s_T^i)$ obtained by the systems that were previously learnt with other simulated users

	User characteristics			Average return w. policy $\mathcal{P}_s^i$ learnt w. $\mathcal{P}_u^i$				
Name	Type	x/p	Centres	$\mathcal{P}_s^1$	$\mathcal{P}_s^2$	$\mathcal{P}_s^3$	$\mathcal{P}_s^4$	$\mathcal{P}_s^5$
$\mathcal{P}_u^1$	Deterministic	3	(0, 0)	**0.94**	0.38	0.55	0.33	0.35
$\mathcal{P}_u^2$	Deterministic	3	(−5, 5)	1.04	**1.23**	0.95	0.50	0.52
$\mathcal{P}_u^3$	Deterministic	6	(−5, 5)	1.06	**1.23**	**1.23**	0.61	0.65
$\mathcal{P}_u^4$	Random	0.3	(−5, 5)	0.79	0.92	0.94	**1.02**	0.98
$\mathcal{P}_u^5$	Random	0.5	(−5, 5)	0.83	0.97	1.02	1.08	**1.10**
$\mathcal{P}_u^6$	Deterministic	6	(−1, 1)	1.02	0.95	**1.08**	0.54	0.54
$\mathcal{P}_u^7$	Deterministic	6	(0, 0)	**0.91**	0.46	0.64	0.47	0.46
$\mathcal{P}_u^8$	Random	0.3	(−1, 1)	0.76	0.95	0.86	**1.02**	**1.01**

and sentence error rate $SER^i = 0.3$. The immediate reward $\omega_s^i = \omega_u^i = 1$ for reaching an agreement is the same for all players. The cost distributions are set as the uniform distribution over [0, 1]: $\delta^{\mathcal{P}_s^i} = \delta^{\mathcal{U}_s^i} = \mathcal{U}_{[0,1]}$. The costs are sampled independently at the beginning of each dialogue.

At first, learning is performed individually on the first five users $\mathcal{P}_u^i$ with Fitted Q-Iteration. The policy is updated every 500 dialogues for a total of 5000 dialogues to ensure convergence. An ϵ-greedy policy is used with $\epsilon = \frac{1}{2j}$ where j is the iteration index. Then, the policy at the end of the learning phase is saved into a player instance: system $\mathcal{P}_s^i$. Finally, systems $\mathcal{P}_s^i$ for $i \in [1, 5]$ are evaluated against all nine users $\mathcal{P}_u^j$ for $j \in [1, 8]$.

Table 1 reports all the results. Using a policy learnt with a user on another user can yield very low return if the users are too different. In particular, using a policy learnt with a random user on a deterministic user is highly inefficient, but the same statement can be made with users with more subtle differences such as $\mathcal{P}_s^2$ versus $\mathcal{P}_u^1$ with only a 0.38 average return.

4 Towards Real Users Profiling

It is planned to develop a web client enabling any human user to play the negotiation game with a simulated user or another human. For the sake of simplicity (it is easier to develop such a web client without handling the speech and natural language understanding and generation), efficiency (it is faster to generate a lot of data with a click-based navigation) and generality (the experiments and results will not be dependent on a specific implementation), the vocal interaction will remain simulated, meaning that instead of interacting naturally, the users will be asked to click on the action they want to perform. Nevertheless, their actions will be corrupted with noise later in the same way as in the simulation.

If we suppose that the human users are rational, different human user behaviours might be induced by the setting of four parameters:

- Discount factor γ: the lower γ is, the more impatient the user will be.
- Reward for reaching an agreement ω^i: the lower ω^i is, the less inclined the user will be to make efforts to find an agreement.
- Cost distribution δ^i: the higher the mean of δ^i, the more difficult it will be for the user to find a suitable option. The higher the variance of δ^i, the more stubborn the user will be.
- Cooperation parameter α^i: the lower the cooperation parameter α^i, the less empathic the user will be.

A setting of these parameters are called a role. For instance a boss should have a standard discount factor, a low reward for reaching an agreement, a high-mean and high-variance cost distribution, and a low cooperation parameter. Thus, a human can be assigned to any role in a given situation. Data will be gathered from a set of ξ humans adopting a set of ρ roles, which will allow the learning of $\xi \cdot \rho$ user models. Human models can be learnt through imitation learning or inverse reinforcement learning [21, 22], and be used for further studies.

5 Conclusion

This article presented the model of the negotiation dialogue game in order to generate artificial dialogue datasets that can be used to train and test data-driven methods later on. Several handcrafted heterogeneous users are developed and policies that are learnt with Fitted Q-Iteration individually on each of them are shown to be inefficient against other users. This game intends to be useful for experimenting data driven algorithms for negotiation and/or user adaptation.

For the near future, we plan to use the negotiation dialogue game to study Knowledge Transfer for Reinforcement Learning [23, 24] applied to dialogue systems [17, 25]. We also project to use this game to generalise the work in [10] for general-sum games. Finally, co-adaptation [16] in dialogue will be tackled.

Two improvements of the game are already considered: we will implement a web client for human data collection; we will eventually use a more accurate model for the option proposition: often, in negotiation games, options are not monolithic, they have a complex structure, which implies two things: they cannot always be expressed and understood in a single dialogue turn, and they are not necessarily proposed by a single player, but are rather co-built.

References

1. Sutton, R.S., Barto, A.G.: Reinforcement Learning: An Introduction, vol. 1. MIT Press Cambridge (1998)
2. Levin, E., Pieraccini, R.: A stochastic model of computer-human interaction for learning dialogue strategies. In: Proceedings of the 5th European Conference on Speech Communication and Technology (Eurospeech) (1997)
3. Laroche, R., Putois, G., Bretier, P., Bouchon-Meunier, B.: Hybridisation of expertise and reinforcement learning in dialogue systems. In: Proceedings of the 9th Annual Conference of the International Speech Communication Association (Interspeech), pp. 2479–2482 (2009)
4. Lemon, O., Pietquin, O.: Data-Driven Methods for Adaptive Spoken Dialogue Systems: Computational Learning for Conversational Interfaces. Springer (2012)
5. English, M.S., Heeman, P.A.: Learning mixed initiative dialogue strategies by using reinforcement learning on both conversants. In: Proceedings of the Conference on Human Language Technology (HLT) (2005)
6. Georgila, K., Traum, D.R.: Reinforcement learning of argumentation dialogue policies in negotiation. In: Proceedings of the 11th Annual Conference of the International Speech Communication Association (Interspeech), pp. 2073–2076 (2011)
7. Bowling, M., Veloso, M.: Multiagent learning using a variable learning rate. Artif. Intell. **136**(2), 215–250 (2002)
8. Georgila, K., Nelson, C., Traum, D.: Single-agent vs. multi-agent techniques for concurrent reinforcement learning of negotiation dialogue policies. In: Proceedings of the 52nd Annual Meeting of the Association for Computational Linguistics (ACL) (2014)
9. Shapley, L.S.: Stochastic games. Proc. Natl. Acad. Sci. U.S.A. **39**(10), 1095 (1953)
10. Barlier, M., Perolat, J., Laroche, R., Pietquin, O.: Human-machine dialogue as a stochastic game. In: Proceedings of the 16th Annual Meeting of the Special Interest Group on Discourse and Dialogue (Sigdial) (2015)
11. Efstathiou, I., Lemon, O.: Learning non-cooperative dialogue behaviours. In: Proceedings of the 15th Annual Meeting of the Special Interest Group on Discourse and Dialogue (Sigdial)
12. Putois, G., Laroche, R., Bretier, P.: Online reinforcement learning for spoken dialogue systems: the story of a commercial deployment success. In: Proceedings of the 11th Annual Meeting of the Special Interest Group on Discourse and Dialogue, pp. 185–192. Citeseer (2010)
13. Laroche, R., Putois, G., Bretier, P., Aranguren, M., Velkovska, J., Hastie, H., Keizer, S., Yu, K., Jurcicek, F., Lemon, O., Young, S.: D6.4: final evaluation of classic towninfo and appointment scheduling systems. Report **D6**, 4 (2011)
14. El Asri, L., Lemonnier, R., Laroche, R., Pietquin, O., Khouzaimi, H.: Nastia: negotiating appointment setting interface. In: Proceedings of the 9th Edition of Language Resources and Evaluation Conference (LREC) (2014)
15. Genevay, A., Laroche, R.: Transfer learning for user adaptation in spoken dialogue systems. In: Proceedings of the 15th International Conference on Autonomous Agents and Multi-Agent Systems (AAMAS). International Foundation for Autonomous Agents and Multiagent Systems (2016)
16. Chandramohan, S., Geist, M., Lefèvre, F., Pietquin, O.: Co-adaptation in spoken dialogue systems. In: Proceedings of the 4th International Workshop on Spoken Dialogue Systems (IWSDS), p. 1. Paris, France (Nov 2012)
17. Casanueva, I., Hain, T., Christensen, H., Marxer, R., Green, P.: Knowledge transfer between speakers for personalised dialogue management. In: Proceedings of the 16th Annual Meeting of the Special Interest Group on Discourse and Dialogue (Sigdial) (2015)
18. Khouzaimi, H., Laroche, R., Lefevre, F.: Optimising turn-taking strategies with reinforcement learning. In: Proceedings of the 16th Annual Meeting of the Special Interest Group on Discourse and Dialogue (Sigdial) (2015)
19. Gordon, G.J.: Stable function approximation in dynamic programming. In: Proceedings of the 12th International Conference on Machine Learning (ICML) (1995)

20. Chandramohan, S., Geist, M., Pietquin, O.: Optimizing spoken dialogue management with fitted value iteration. In: Proceedings of the 10th Annual Conference of the International Speech Communication Association (Interspeech) (2010)
21. Ng, A.Y., Russell, S.: Algorithms for inverse reinforcement learning. In: Proceedings of the 17th International Conference on Machine Learning (ICML), pp. 663–670. Morgan Kaufmann (2000)
22. El Asri, L., Piot, B., Geist, M., Laroche, R., Pietquin, O.: Score-based inverse reinforcement learning. In: Proceedings of the 15th International Conference on Autonomous Agents and Multi-Agent Systems (AAMAS). International Foundation for Autonomous Agents and Multiagent Systems (2016)
23. Taylor, M.E., Stone, P.: Transfer learning for reinforcement learning domains: a survey. J. Mach. Learn. Res. **10**, 1633–1685 (2009)
24. Lazaric, A.: Transfer in reinforcement learning: a framework and a survey. In: Reinforcement Learning, pp. 143–173. Springer (2012)
25. Gašic, M., Breslin, C., Henderson, M., Kim, D., Szummer, M., Thomson, B., Tsiakoulis, P., Young, S.: POMDP-based dialogue manager adaptation to extended domains. In: Proceedings of the 14th Annual Meeting of the Special Interest Group on Discourse and Dialogue (Sigdial) (2013)

Estimation of User's Willingness to Talk About the Topic: Analysis of Interviews Between Humans

Yuya Chiba and Akinori Ito

Abstract This research tried to estimate the user's willingness to talk about the topic provided by the dialog system. Dialog management based on the user's willingness is assumed to improve the satisfaction the user gets from the dialog with the system. We collected interview dialogs between humans to analyze the features for estimation, and found that significant differences of the statistics of F_0 and power of the speech, and the degree of the facial movements by a statistical test. We conducted discrimination experiments by using multi-modal features with SVM, and obtained the best result when we used the audio-visual information. We obtained 80.4 % of discrimination ratio under leave-one-out condition and 77.1 % discrimination ratio under subject-open condition.

Keywords User's willingness to talk · Spoken dialog system · Multi-modal information

1 Introduction

In a human-human conversation, a speaker talks considering the interlocutor's willingness to talk about the topic. For example, a participant in a conversation tries to continue to talk about the same topic when the interlocutor is willing to talk about that topic, and changes a subject or stops the conversation otherwise. We assume the dialog system can increase the satisfaction the user gets from the dialog by introducing dialog management based on the user's willingness to talk about the topic that the system provides. The human estimates the interlocutor's willingness by the non-verbal cues of the user, such as tone of the speech, facial expression, and gestures in addition to the verbal information. Thus, we aim to discriminate the user's willingness using multi-modal information.

Y. Chiba (✉) · A. Ito
Graduate School of Engineering, Tohoku University, Sendai, Japan
e-mail: yuya@spcom.ecei.tohoku.ac.jp

A. Ito
e-mail: aito@spcom.ecei.tohoku.ac.jp

K. Jokinen and G. Wilcock (eds.), *Dialogues with Social Robots*,
Lecture Notes in Electrical Engineering 427,
DOI 10.1007/978-981-10-2585-3_34

In this research, we deal with the dialog system which actively poses the question to collect the user's information. Dialog management based on the user's willingness is important for such system because the user can talk a lot about himself/herself (e.g. his/her favorites or pleasures) in the topic which he/she wants to talk, but the user does not talk if the user is not interested in the topic. A spoken dialog system that collects information is studied by Meena et al. [1]. Their system collects information about what can be seen from the user's position by the dialog for routing service. We assume the information obtained from the user is used for information recommendation or travel guidance. In this article, we collected dialogs of the interviews between humans by a web camera for preliminary analysis. Then, we investigate what behavior of the user was efficient to judge the user's willingness. Finally, we conduct the discrimination experiments with investigated multi-modal features, and show the highest discrimination ratio was obtained when we used audio-visual information.

2 Related Work

Several researches studied the user's state related to the willingness. Dohsaka et al. implemented a quiz-style dialog system based on natural language processing and measured the willingness of a user to make conversation with the system again [2]. In contrast, we focus on the user's willingness to talk about the topic or question provided by the system. Especially, the user's willingness to talk about the topic seems to be related to the "involvement" in conversation [3] and activation level of the dialog [4]. In these researches, the system stores the dialog history for estimating the conversation participant's state. One of the significant differences from our approach is that they used only linguistic information of the transcription of the dialog. We assume the user's non-verbal information is an important cue to recognize the user's willingness, and tried to discriminate the user's state by using multi-modal information. Using multi-modal information is actively studied for emotion recognition [5]. For example, Schuller et al. studied the method to estimate the "level of interest (LOI)," which is the user's interest in the products presented by the system [6]. We have been also investigated the method to discriminate the user's state whether the user was thinking to answer the system's prompt or the user was embarrassed by the prompt [7]. The target dialog system is similar to the chat-talk dialog system [3] and listening-oriented dialog system [8] because those systems are for the situation that the user talked to the system actively. Meguro et al. built a dialog system that gave the impression to the user that the system listens his/her talk to satisfy the user's desire to make conversation [8]. In contrast, we aim to improve the user's satisfaction by regulating the dialog based on the user's willingness.

3 Experimental Dialog Data

3.1 Experimental Setup

We collected the interview data in which one person (the questioner) asked questions to another person (the answerer). The first author played a role of the questioner and several participants played a role of answerers. The participants were 7 male undergraduate students. The questioner and each answerer met for the first time at the interview. We selected 10 topics (such as "movie" or "fashion") for interviews referring the setup of the previous researches of the chat-talk dialog system [3]. All interviews were conducted in Japanese. The questioner paid attention not to interrupt the answerer's speech to make clear a pair of the question and answer, but gave feedback at a proper part to encourage the answerer to talk. The interviews were conducted in a silent room. The questioner and answerer sat face to face, and wore pin-microphone at their chest. A web camera was placed between them to capture the upper half of the answerer's body from the front. The collected dialog data were stored at 48 kHz sampling rate, 32 bit quantized sound data and color images of 15 frames per seconds. After the dialog collection, the sound data were down-sampled to 16 kHz and requantized to 16 bit for the analysis. Sound data have two channels, and each channel includes speech recorded by the microphone of the questioner and answerer, individually. The interview was conducted after explaining the answerer that we want to analyze the conversation between humans to apply dialog management of the spoken dialog system. The conversation about one topic was stopped when the questioner judged that the answerer is not willing to talk about the topic more, and the interview was finished after talking on all the topics. Then, we asked the participants to answer the following questionnaire.

Q1 Did you want to talk about each topic?
Q2 Were you satisfied with the interview?
Q3 How did you enjoyed the interview?

The participants rated scores on a scale of one (not at all) to five (very much). We treat the scores of Q1 as the subjective rating of the willingness to talk about each topic. Hereafter, we call this value "subjective willingness."

3.2 Results of Recording

The number of collected dialog data was 70 (7 participants × 10 topics). We defined the dialog that was rated the highest subjective willingness by each answerer as the "high willingness" dialog and that with the lowest score as the "low willingness" dialog. The speech signals of the dialogs were transcribed manually to extract the linguistic information. The dialog data of "high willingness" were 12 dialogs and "low willingness" were 17 dialogs. The total number of the utterance was 441. Table 1 shows the number of dialog and the answerer's utterances.

Table 1 Number of collected dialog data

Rating	Dialog data	Utterances of answerer
High	12	98
Low	17	343
Total	29	441

Here, the average and standard deviation of ratings of Q2 and Q3 were 4.43 ± 0.49 and 4.71 ± 0.45. These scores indicate that the answerers conversed with the questioner comfortably at the experiments.

4 Extraction and Analysis of Multi-modal Information

We extracted F_0, power, and the part of speech (POS) tags of the answerer's speech in addition to the facial movements and physical movements. These features were chosen based on the introspection of third person (2 males and 1 female). We used Snack[1] to extract F_0 by using ESPS method, and translated to log scale. The F_0 value were obtained every 10 ms. We set the frame width to 25 ms and frame shift to 10 ms for calculating the power. We used the transcription of the dialog for extracting the linguistic information. The POS tags are obtained by using morphemic analyzer MeCab.[2] We quantified the facial movements of the answerer as the magnitude of the optical flow [9] detected from the facial region. The answerer's facial region is detected by Constrained Local Model (CLM) [10]. The extracted facial region is scaled to 100×100 pixels, and then optical flows of the pixels inside of the facial region are calculated. Finally, the magnitude of optical flows of all pixels is summed up and used as an the index of the facial movement of the frame. The facial movements include the movements of the facial expression and the gaze. On the other hand, the physical movements were quantified by the absolute value of the different images of the successive frames. Firstly, the brightness of the frame was transformed so that the average was 0.0 and the variance was 1.0, and the facial region was masked by setting the brightness of the facial region to 0.0. Then, difference images between the current and previous frames were calculated, and the absolute value of pixels of difference image were summed up and defined as the physical movement of the frame. Figures 1 and 2 show the examples of the facial movement and physical movement. The green lines of Fig. 1 represent the optical flows every 2 pixels in the facial region. Figure 2 is the difference image of the answerer's body. The behavior of the questioner seems to affect the answerer's willingness, but we did not consider the interaction between the speakers in this paper.

[1]http://www.speech.kth.se/snack/.

[2]http://mecab.sourceforge.net/.

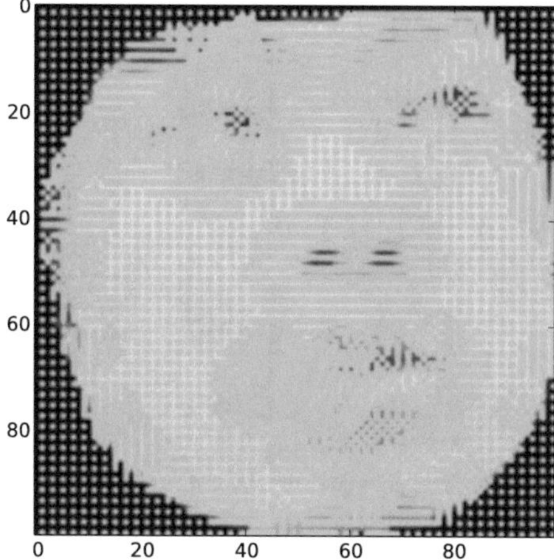

Fig. 1 Example of facial movements

Fig. 2 Example of physical movements

We investigated differences of the distributions of the multi-modal information. Figure 3 shows the distributions of the average facial and physical movements of each dialog data. Here, we excluded the data of one answerer because the reliable facial feature points could not be extracted because the answerer kept looking away from the questioner during dialog. We conducted two-way layout ANOVA with the subjective willingness and the answerer as factors, and we obtained significant difference at facial movements ($F = 19.132$, $p = 6.36 \times 10^{-4}$). Thus, considering the user's facial movements is efficient to discriminate the user's willingness to talk about the topic. On the other hand, we did not observe a significant difference at the

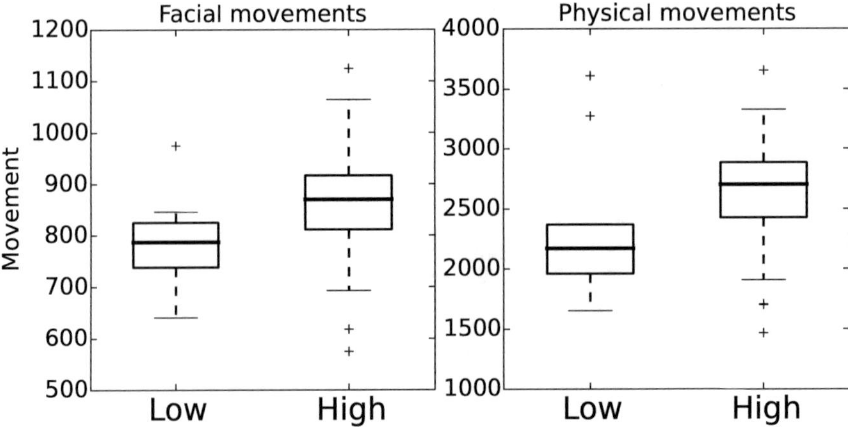

Fig. 3 Distribution of the average facial movements and physical movements

Table 2 Results of statistical test of each stochastic of prosodic features (p-value)

	Per dialog			
	Ave.	Var.	Max.	Min.
Ave.F0	$p = 0.997$	$p = 0.338$	$p = 0.786$	$p = 0.112$
Var.F0	$p = 0.129$	$p = 0.132$	$p = 0.020^a$	$p = 0.026^a$
Range F0	$p = 0.276$	$p = 0.072$	$p = 0.010^a$	$p = 0.111$
Ave.Pow	$p = 0.473$	$p = 0.625$	$p = 0.239$	$p = 0.043^a$
Var.Pow	$p = 0.317$	$p = 0.334$	$p = 0.012^b$	$p = 0.117$
Range Pow	$p = 0.499$	$p = 0.994$	$p \leq 0.001^c$	$p = 0.045^a$

$^a p \leq 0.05$, $^b p \leq 0.01$, $^c p \leq 0.001$

physical movements ($F = 1.633$, $p = 0.222$). This result reflects the expression of gesture is different from subject to subject.

To analyze the prosodic features, we calculated the average, variance, and range of F_0 and power of answerer's speech at first, and compared the average, variance, maximum, and minimum of these values per dialog. The p-values of ANOVA of each statistic are shown in Table 2. As shown in the table, the significant differences tended to be obtained at the maximum or minimum of the range or variance of F_0 and power of speech. These results suggest that the answerer's willingness is expressed at specific utterance and show the possibility of discrimination of the user's willingness by using the features obtained from a short interaction such as question and answer. As linguistic information, we also analyzed the number of morphemes per utterance to investigate the duration of the speech, but we only obtained marginally significance when the maximum value among a dialog is taken ($F = 4.114$, $p = 0.061$).

5 Discrimination of Willingness to Talk About the Topic

Finally, we conducted discrimination experiments using multi-modal features with Support Vector Machine (SVM). For the linguistic features, we included the average frequency of POS tags per utterance in addition to the average number of morpheme. The POS tags include pre-noun adjectival, prefixes (4 sub-categories), nouns (15 sub-categories), verbs (3 sub-categories), adjectives (3 sub-categories), adverbs (2 sub-categories), conjunction, interjection, filler, and other 3 categories. The post-positional particle and an auxiliary verb of Japanese were excluded as the stop words. We summarized the features used the discrimination in Table 3. We extracted the average, variance, maximum, and minimum of each feature per dialog. Therefore, the dimension of the acoustic features was 6 features $\times$ 4 statistics $=$ 24 and visual features was 2 features $\times 4$ statistics $=$ 8. We only used the average frequency for the POS tags, and the total dimension of the linguistic features was 4 statistics $+$ 34 tags $=$ 38. Each feature was scaled by z-score. We used the RBF kernel as the kernel function of SVM, and decided the hyper-parameters by grid-searching. The experiments were conducted under the leave-one-out condition and subject-open condition. Under the subject-open condition, we used the dialog of one answerer as the test data and the rest of data for training. The experiments were two-class discrimination between the high willingness and low willingness.

Table 4 shows the discrimination results. As shown in the results, we obtained the best result when we used the combination of the acoustic and visual features. In our previous experiment, the concordance of human evaluation and subjective willingness was 78.9 % under a condition equal to leave-one-out, thus these results indicate the efficacy of the multi-modal information investigated in this paper. The combination of the linguistic features gave a bad effect on the total discrimination results, which suggests the linguistic features selected in this paper was not efficient for discrimination. Thus, further investigation will be required for selecting linguistic features and efficient combination method.

Table 3 Multi-modal features used for the discrimination

Modality	Feature for the discrimination	
Audio	Average log $F0$ of speech	*Ave., Var., Max., Min.*
	Variance of log $F0$ of speech	*Ave., Var., Max., Min.*
	Range of log $F0$ of speech	*Ave., Var., Max., Min.*
	Average of log power of speech	*Ave., Var., Max., Min.*
	Variance of log power of speech	*Ave., Var., Max., Min.*
	Range of log power of speech	*Ave., Var., Max., Min.*
Visual	Facial movements	*Ave., Var., Max., Min.*
	Physical movements	*Ave., Var., Max., Min.*
Linguistic	Frequency of POS tags of speech	*Ave.* (34 tags)
	Frequency of morphemes of speech	*Ave., Var., Max., Min.*

Table 4 Discrimination results by using multi-modal information (%)

	Leave-one-out			Subject-open		
Modality	Low	High	Ave.	Low	High	Ave.
Audio	66.7	88.2	77.5	71.4	82.1	76.8
Visual	**83.3**	52.9	68.1	71.4	64.3	67.9
Linguistic	50.0	76.5	63.2	55.7	77.4	66.5
Audio, Visual	66.7	94.1	**80.4**	61.4	**92.9**	**77.1**
Audio, Linguistic	33.3	**100.0**	66.7	51.4	82.1	66.8
Visual, Linguistic	58.3	70.6	64.5	**87.1**	46.4	66.8
Audio, Visual, Linguistic	50.0	82.4	66.2	48.6	82.1	65.4

6 Conclusion

This research assumed the dialog system acquiring the user's information actively by posing questions, and investigated the effect of the multi-modal information to discriminate the user's willingness to talk about the topic by interview dialog. We analyzed the efficient multi-modal information to discriminate the user's willingness, and indicated the facial movements and some statistics of F_0 and power of the answerer's speech are efficient for the discrimination. Then, we conducted the discrimination experiments by using acoustic, visual, and linguistic information with SVM. We obtained the best result when using combination of acoustic and visual features, and the discrimination result was 80.4 % under leave-one-out condition and 77.1 % under subject-open condition. In a future work, we will build the Wizard of Oz system, and investigate the user's willingness in the human-machine dialog.

References

1. Meena, R., Boye, J., Skantze, G., Gustafson, J.: Crowdsourcing street-level geographic information using a spoken dialogue system. In: Proceedings of the 15th Annual Meeting of the Special Interest Group on Discourse and Dialogue, pp. 2–11 (2014)
2. Dohsaka, K., Higashinaka, R., Minami, Y., Maeda, E.: Effects of conversational agents on activation of communication in thought-evoking multi-party dialogues. IEICE Trans. Inf. Syst. **E97-D**(8), 2147–2156 (2014)
3. Tokuhisa, R., Terashima, R.: Relationship between utterances and enthusiasm in non-task-oriented conversational dialogue. In: Proceedings of the 7th SIGdial Workshop on Discourse and Dialogue, pp. 161–167 (2009)
4. Moriya, Y., Tanaka, T., Miyajima, T., Fujita, K.: Estimation of conversational activation level during video chat using turn-taking information. In: Proceedings of the 10th Asia Pacific Conference on Computer Human Interaction, pp. 289–298 (2012)
5. Pantic, M., Rothkrantz, L.J.: Toward an affect-sensitive multimodal human-computer interaction. In: Proceedings of the IEEE **91**, 1370–1390 (2003)

6. Schuller, B., Müeller, R., Höernler, B., Höethker, A., Konosu, H., Rigoll, G.: Audiovisual recognition of spontaneous interest within conversations. In: Proceedings of the 9th International Conference on Multimodal Interfaces, pp. 30–37 (2007)
7. Chiba, Y., Nose, T., Ito, A., Ito, M.: User modeling by using bag-of-behaviors for building a dialog system sensitive to the interlocutor's internal state. In: Proceedings of the 15th Annual Meeting of the Special Interest Group on Discourse and Dialogue, pp. 74–78 (2014)
8. Meguro, T., Higashinaka, R., Dohsaka, K., Minami, Y., Isozaki, H.: Analysis of listening-oriented dialogue for building listening agents. In: Proceedings of the 10th Annual Meeting of the Special Interest Group on Discourse and Dialogue, pp. 124–127 (2009)
9. Farnebäck, G.: Image Analysis, Chap. Two-frame Motion Estimation Based on Polynomial Expansion, pp. 363–370. Springer (2003)
10. Saragih, J.M., Lucey, S., Cohn, J.F.: Deformable model fitting by regularized landmark mean-shift. Int. J. Comput. Vis. **91**(2), 200–215 (2011)

Recognising Conversational Speech: What an Incremental ASR Should Do for a Dialogue System and How to Get There

Timo Baumann, Casey Kennington, Julian Hough and David Schlangen

Abstract Automatic speech recognition (ASR) is not only becoming increasingly accurate, but also increasingly adapted for producing timely, incremental output. However, overall accuracy and timeliness alone are insufficient when it comes to interactive dialogue systems which require stability in the output and responsivity to the utterance as it is unfolding. Furthermore, for a dialogue system to deal with phenomena such as disfluencies, to achieve deep understanding of user utterances these should be preserved or marked up for use by downstream components, such as language understanding, rather than be filtered out. Similarly, word timing can be informative for analyzing deictic expressions in a situated environment and should be available for analysis. Here we investigate the overall accuracy and incremental performance of three widely used systems and discuss their suitability for the aforementioned perspectives. From the differing performance along these measures we provide a picture of the requirements for incremental ASR in dialogue systems and describe freely available tools for using and evaluating incremental ASR.

Keywords Incremental ASR · Conversational speech · System requirements · Evaluation

T. Baumann (✉)
Natural Language Systems Group, Informatics Department, Universität Hamburg, Hamburg, Germany
e-mail: baumann@informatik.uni-hamburg.de

C. Kennington · J. Hough · D. Schlangen
Dialogue Systems Group, Faculty of Linguistics and Literature and CITEC, Bielefeld University, Bielefeld, Germany
e-mail: ckennington@citec.uni-bielefeld.de

J. Hough
e-mail: julian.hough@uni-bielefeld.de

D. Schlangen
e-mail: david.schlangen@uni-bielefeld.de

K. Jokinen and G. Wilcock (eds.), *Dialogues with Social Robots*,
Lecture Notes in Electrical Engineering 427,
DOI 10.1007/978-981-10-2585-3_35

1 Introduction

Incremental ASR is becoming increasingly popular and is available both commercially and as open-source. Given this recent development of systems, the question arises as to how they perform and compare to each other, not just in terms of utterance-final accuracy but also in terms of their *incremental* performance.

For a spoken dialogue system (SDS) consuming ASR output, incrementally receiving partial results for an on-going utterance means the system can start processing words before the utterance is complete, leading to advantages such as quicker responses, better interactive behaviour and dialogue management, more efficient database queries, and compensation for inefficient downstream processors such as slow robot actuators—see [1] for an overview. SDSs that process incrementally produce behaviour that is perceived to be more natural than systems that use the traditional turn-based approach [2–5], offer a more human-like experience for users [6], and are more satisfying to interact with than non-incremental systems [7].

Metrics have been proposed to evaluate incremental performance for ASR [8–10], which we build on in this paper. We also deal with evaluating an incremental ASR's performance on difficult phenomena from *conversational speech* such as disfluency. In this paper we investigate these challenges, firstly by outlining suitable evaluation criteria for incremental ASRs for dialogue systems, then investigating how off-the-shelf ASRs deal with speech from participants in a task-oriented dialogue domain, both with and without training on in-domain data. We present findings using our criteria to help SDS builders in their decision as to which ASR is suitable for their domain. The alternative ASR engines that are evaluated in this paper are all accessible in a uniform way with the freely available InproTK[1] [11], as is the evaluation toolbox InTELiDa[2] that we use.

2 The Challenge of Interactive, Conversational Speech

While many current SDSs claim to deal with spontaneous speech, this is often in the form of voice commands that do not require a fast verbal response, with some exceptions [3, 4]. When using voice commands, it has been established that people use more controlled, fluent and restricted speech than when in a human-only dialogue [12], with users often defaulting to what [13] calls 'Computer Talk'.

We argue ASR evaluation currently does not focus on the challenge of interactive speech as required for a highly interactive SDS. While popular dictation evaluation domains such as the spoken Wall Street Journal [14] are clearly unsuitable, even the more SLU (Spoken Language Understanding)-based benchmarks such as the ATIS

[1] http://bitbucket.org/inpro/inprotk.

[2] http://bitbucket.org/inpro/intelida.

(Airline Travel Information Systems) corpus and other genres mentioned in [15]'s ASR analysis do not meet the demands of ASR for high levels of interactivity and responsiveness.

3 Desiderata for Incremental ASR for Interactive SDSs

To address the challenge of interactive, conversational speech, here we briefly set out requirements for ASR for its suitability for interactive SDSs.

3.1 Incrementality and Timing Information

In addition to being timely and accurate in terms of the final output at the end of an utterance, we would like timeliness and accuracy on the word level from an ASR. In Fig. 1 we demonstrate the qualities needed by representing the evolution of hypotheses made by a system over time, going from bottom to top, for the reference transcription 'take the red cross': (a) is the ideal behaviour as it produces fully incremental output which is completely accurate, occasionally predicting the word before it is over, whilst the failings in (b), (c) and (d) give us the incremental desiderata of *stability of output*, *word-by-word incremental output* and *timeliness of output*. Metrics and tools for measuring these incremental qualities will be described in Sect. 4.

Another factor of situated conversational speech are deictic references that, in a fast-moving environment, can only be interpreted correctly if the timing of deictic references (and possibly co-occurring pointing gestures) is available for analysis. It is thus crucial that an ASR provide, in a timely manner, *timing of the recognized words*.

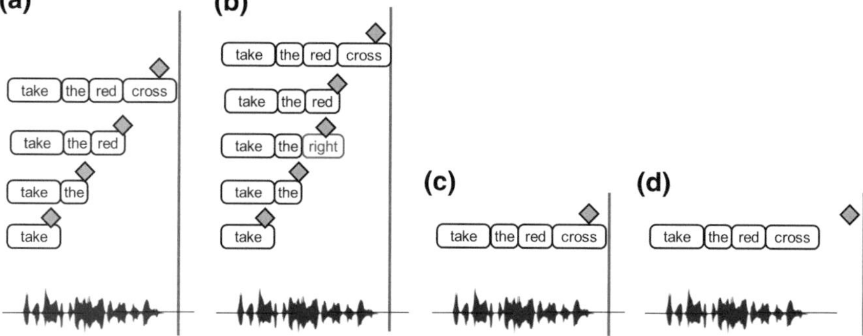

Fig. 1 Incrementality in ASR: vertical line indicates current time, diamond the time of update. **a** Perfect output, **b** unstable output, **c** non-incremental but timely, **d** non-incremental and latent

3.2 *Suitability for Disfluency*

One principal feature strikingly absent from Computer Talk but abundant in human conversational speech is disfluency. Within the larger goal of incorporating understanding of disfluent behaviour to dialogue systems [16], we require an ASR to detect all words in speech repairs, preserving the elements of the well-established structure in (1) from [17]'s mark-up.

$$\text{John}\ \underbrace{[\ \text{likes}}_{\text{reparandum}}\ +\ \underbrace{\{\text{F uh}\}}_{\text{interregnum}}\ \underbrace{\text{loves}\ ]}_{\text{repair}}\ \text{Mary} \tag{1}$$

There is evidence that repairs are reasoned with on an incredibly time-critical level in terms of understanding [18] and there are clear examples of the reparandum being needed to calculate meaning—such as in (2) and (3) where semantic processing access to "the interview" is required to resolve the anaphoric "it" and "the oranges" is required to resolve "them". If an incremental disfluency detector such as [19] is to work in a live system, all words within a disfluency must become available in the ASR output, and not be filtered out.

(2) "[the interview, was + {...} it was] all right." [20]
(3) "have the engine [take the oranges to Elmira, + { um, I mean, } take them to Corning]" [21]

Filled pauses 'um' and 'uh' can be considered English words in terms of their meaning in conversation [22] and transcribers can reliably transcribe them. While they can form interregna as in (1), isolated, non-repair filled pauses can indicate forward-looking trouble from conversation participants [23]. These should therefore not be filtered out during speech recognition if we are to build truly interactive systems.

Given this motivation, in additional to good incremental properties, we would also like an ASR to exhibit *preservation of disfluent material*, that is, we would prefer word hypotheses that are useful for disfluency detection and processing, with no filtering out of reparanda and filled pauses.

4 Evaluation Metrics

To address the desiderata we split our evaluation methods into accuracy, timing and evolution of hypotheses over time. Incremental metrics are provided by the InTELiDa toolbox [24].

4.1 Utterance-Level Accuracy and Disfluency Suitability

We use standard Word Error Rate (WER) of the final (non-incremental) hypothesis. Incremental ASR cannot reliably outperform the accuracy of non-incremental systems, hence its utterance-final quality is what matters most. To measure accuracy on disfluencies, we filter all filled pauses and all reparanda from the transcripts (leaving only the repair phases), so the standard reference 'John likes uh loves Mary' becomes 'John loves Mary' and compare WER before and after filtering. This is in order to find how much disfluent material is recovered (which would result in worse performance on the filtered reference) or whether the ASR itself filters disfluencies accurately (in which case the performance would improve on the filtered reference). **WER disfluency gain** is simply: WER *on disfluency filtered original transcript*—WER *on original transcript*. For preservation of disfluent material, the higher this gain the better. However for accuracy of filtering out disfluency, the lower the better.

4.2 Timing

Following [25] we use the First Occurrence (FO) and Final Decision (FD) measures to investigate timeliness, where:

FO is the time between the (true) beginning of a word and the *first* time it *occurs* in the output (regardless if it is afterwards changed). In Fig. 1c, d would perform poorly using this metric, in particular for 'take' which is reported only long after it has been spoken.

FD is the time between the (true) end of a word and the time when the recognizer *decides* on the word, without later revising it anymore. If an ASR correctly guesses a word before it is over, the value will be negative. Often, FD occurs simultaneously with FO. If not, a word is revised and later returned to, which can be a frequent occurrence at word boundaries.

Timeliness can only be measured for words that are correctly recognized or at least appear in the final output of the recognizer and timing distributions are reported below. FO and FD measure *when words are recognized*, but not how well-aligned these are to the actual timing of the word in the audio. However, our impression is that recognizers which report such timing information are very accurate (on the order of centiseconds). Thus, the availability of timing is mostly a binary decision and depends on the recognizer's interface.

4.3 Diachronic Evolution

The diachronic evolution of hypotheses is relevant to capture *how often* consuming processors have to re-consider their output and for *how long* hypotheses are likely to

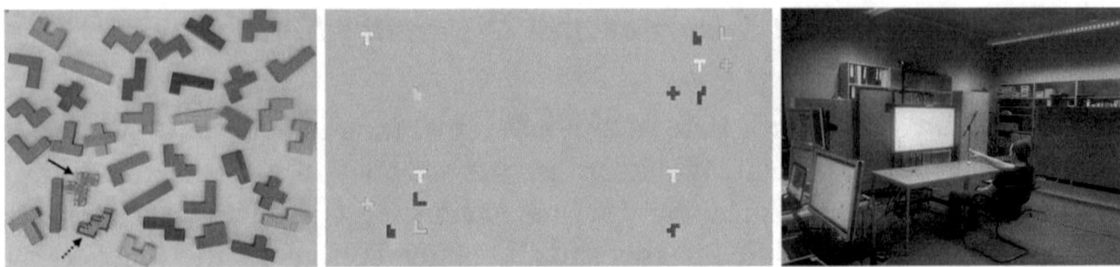

Fig. 2 Example game scenes and collection setup used in collecting Pentomino interaction data

still change. We have previously used *Edit Overhead* the proportion of unnecessary edits during hypothesis building, to account for the former. However, we disregard this aspect in the present work, as EO is mostly measuring computational overhead and there are effective measures to reduce EO [8].

We instead focus on the stability of hypotheses [9], which measures the 'temporal extent' of edits. For words that are added and later revoked or substituted we measure the "survival time" and report aggregated plots of *word survival rate* (*WSR*) after a certain age. These statistics can be used to estimate the likelihood of the recognizer being committed to a word during recognition.

5 Evaluation Domain: Pentomino Puzzle Playing Dialogue

The evaluations below make use of recorded human-human dialogue, and also interactions between humans and (wizard controlled) SDSs, where participants were instructed to play simple games with the "systems". In all cases, the games made use of geometric *Pentomino* puzzle tiles where participants referred to and instructed the systems or human interlocutors to manipulate the orientation and placement of those tiles. The interactions were all collected and utterances were segmented and transcribed. The corpora were originally described, respectively, in [26–28]. We make use of two sets of data in German and English. The German data yields 13,063 utterances (average length of 5 words; std 6.27) with a vocabulary size of 1,988. Example game scenes are shown in Fig. 2 and example utterances (with English glosses) are given in Examples (4), and (5) below. We use the German data for training and evaluating ASR models explained in Sect. 6.2. We also use English data (both UK and US) from this domain yielding 686 telephone-mediated utterances (6,157 words) for evaluating existing English models, as explained in Sect. 6.1.

(4) a. *drehe die Schlange nach rechts*
b. rotate the snake to the right

(5) a. *dann nehmen wir noch das zw- also das zweite t das oben rechts ist ... aus dieser gruppe da da mchteichögern das gelbe t haben ... ja*
b. then we take now the se- so the second t that is on the top right ... out of this group there I would like to have the yellow t ... yes

6 Evaluation of Three ASR Systems: Google, Sphinx-4, and Kaldi

6.1 Experiment 1: Off-the-shelf Models for a Dialogue Domain

In our first experiment we do not train or adapt any of our ASR systems but evaluate their off-the-shelf performance as in [15] but including incremental performance. We evaluate on 686 utterances from the English data explained above.

6.1.1 Systems

We evaluate Sphinx-4 [29] with most recent general AM and LM (version 5.2 PTM) for (US-)English, Google's web-based ASR API [30] (in the US-English setting) and Kaldi [31], for which we use the English Voxforge recipe (57,474 training utterances, avg 9.35 words per utterance, presumably dominated by US-English). We choose *Google* as the state-of-the-art ASR available via a Web-interface. We use *Sphinx-4* because it has previously been adapted for incremental output processing [8] and Kaldi as an open-source speech recognition system that is growing in popularity and has incremental capabilities [32].

Google partial results can consist of multiple segments, each of which is given a stability estimate [10]. In practice, Google only returns stabilities of 1 or 90 % (for both German and English). While incremental results are 1-best, the final (non-incremental) result contains multiple alternatives, with a confidence measure for the first (presumably most likely) alternative. This final hypothesis appears to make use of post-incremental re-scoring or re-ranking. While this is obviously intended to optimize the result quality (SER or WER), it means that incremental results are just a 'good guess' as to what the final result will be, with implications for the timing metrics as reported and discussed below.

We implemented multiple options for interpreting the Google output:

- **stable** use only those segments which have a high stability (we use a threshold of >50 %, but estimates as reported by Google are essentially binary),
- **quick** use all segments, including the material with low stability,
- **sticky** ignore the re-ranking from Google and choose the final hypothesis that best matches the previous 1-best incremental result (as generated by the **quick** setting). This setting is expected to result in lower non-incremental performance.

6.1.2 Non-incremental Quality and Disfluency Suitability

WER results across the reference variants are shown in Table 1. *Google-API* clearly outperforms the other systems. However, its WER does not degrade on disfluency-

Table 1 Word Error Rate (WER) results on English Pentomino data for the off-the-shelf systems under different transcript conditions with the WER disfluency gain in brackets

System	US English speakers		All English speakers	
	WER (all)	Disfluency filtered	WER (all)	Disfluency filtered
Google-API-stable/quick	25.46	28.16 (+2.70)	40.62	41.60 (+0.98)
Google-API-sticky	26.08	29.29 (+3.21)	41.23	42.82 (+1.59)
Sphinx-4	57.61	62.31 (+4.70)	72.08	75.34 (+3.26)
Kaldi	71.31	73.38 (+2.07)	77.57	79.05 (+1.48)

filtered transcripts as much as *Sphinx-4*, which has the largest WER disfluency gain of 4.70, showing it is preserving the disfluent material the most. Manual inspection shows Google filtering out many speech repairs and performing badly around them—see (6-a) versus (6-b). An improved model for filled pauses would also prevent errors like (7-b).

(6) a. **Reference**: and the and his front uh his le- the the the back
b. **Google-API-fast**: and the and the front of theater

(7) a. **Reference**: uh another L shape except it's um symmetrically
b. **Google-API-fast**: another L shape septic sam symmetrically

Also, we notice that performance varies substantially between UK and US speakers, which is a problem for a corpus that contains mixed speakers. Finally, the post-hoc re-scoring that is performed by Google-API in the stable and quick conditions only marginally improves WER over sticking with the strategy used for incremental processing (presumably SER-optimizing Viterbi decoding).

Finally, we note that the Google-API only provides a transcript of words, both Sphinx and Kaldi generate detailed word timings that can be used for analysis by downstream modules.

6.1.3 Incremental Quality

Figure 3 plots timing and stability for three recognizers (and Google's three settings). Timing metrics are shown for all hypothesized words (rather than just for words that match the transcript). As can be seen in Fig. 3a, b, both Kaldi and Sphinx often have a first impression (FO, Subfigure a) of the word right after it is being spoken, while Google is lagging a little. Google and Sphinx are a little quicker in deciding for a word (FD, Subfigure b) than Kaldi, but Google in particular is hurt by words being revised long after they have been hypothesized. This is clearly observable in Fig. 3c, which shows that a word still has a 5 %-chance of revision even after it has been hypothesized for 1 second (and Google is already slower in hypothesizing words in the first place). This ratio is even worse when limiting hypotheses to just the 'stable' part, but can be radically improved when ignoring the final, non-incremental

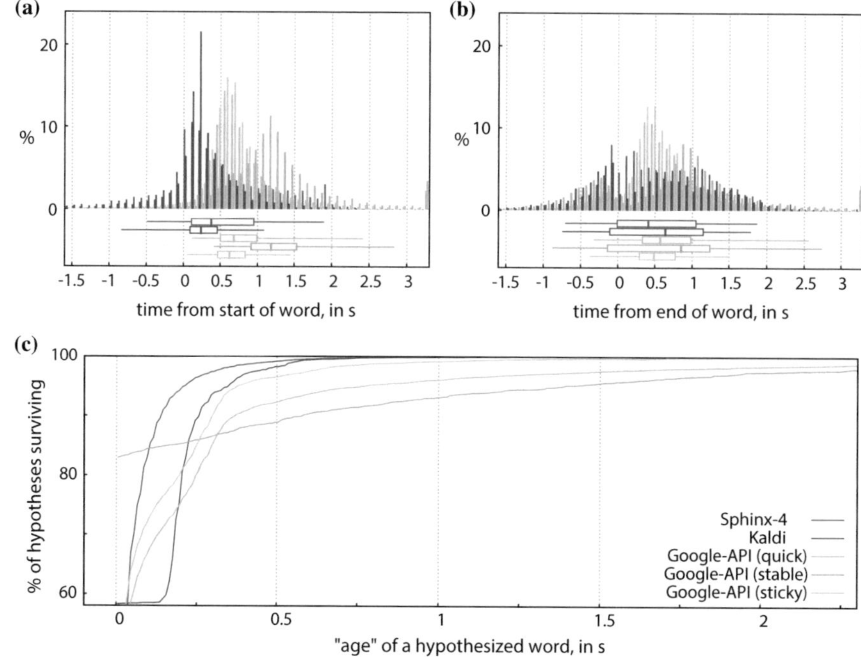

Fig. 3 **a** and **b** Histograms showing the distribution of **a** *first occurrence* of words and **b** *final decision* for words for the three recognizers (and Google's three settings). Box plots show the median, quartiles (*box*) and 5/95 % quantiles (*whiskers*). Some extreme (negative) values may be caused by alignment errors. **c** Stability of hypotheses expressed as *word survival rate* over time. A higher curve implies a higher stability

changes of Google ASR (the 'sticky' setting), albeit at the cost of about 2 % points WER relative. As Fig. 3c also shows, Kaldi most likely performs some variation of hypothesis smoothing [8] for 150 ms.

6.2 Experiment 2: Training Models on In-Domain Data

We found rather poor performance (in terms of WER) for the off-the-shelf open-source systems in our interaction-driven domain, presumably because this speaking style does not conform to the material used when training models for open-source systems. In this experiment, we trained models with in-domain data, under the hypothesis that these result in better performance.

Table 2 Word Error Rate (WER) results on German Pentomino data for the trained systems under different transcript conditions with the WER disfluency gain in brackets

System	German	
	WER (all)	Disfluency filtered
Google-API-stable/quick	22.00	21.86 (−0.14)
Google-API-sticky	20.51	20.44 (−0.07)
Sphinx-4	30.28	30.25 (−0.03)
Kaldi	38.95	38.91 (−0.04)

6.2.1 Systems and Data

We train acoustic and language models for German using 10.7 h of transcribed interactions (partly human-human, human-system, and human-wizard) from the Pentomino domain described above.[3] Our Kaldi model is based on an adaptation of the Voxforge recipe, while our Sphinx-4 model uses the standard settings of SphinxTrain. Both used the same data for training.

We evaluate our trained systems (and the Google systems) on 465 utterances (3,818 words) from randomly chosen speakers from the German data explained above (the rest was used for training). Given the human-Wizard interaction domain, compared to the English corpus above, it contains slower, more dictation-like speech with few disfluencies, so we would expect the accuracy results to be better, all things being equal in this domain. However, we find how the large gap to big data driven ASRs such as Google can be closed somewhat with in-domain trained models.

6.3 Results

WER results across the reference variants are shown in Table 2. Google-API's systems have comparable performance to the English data above, however the post-hoc rescoring actually hurts on this data, with a relative performance hit of 7 %. Sphinx-4 and Kaldi greatly improve through the in-domain training.

The disfluency results in this setting are not as interesting, given the lack of disfluency in the training files, and we take the analysis on the English data above to be indicative of the relative performance of the ASRs.

Incremental metrics are largely unchanged, with a tendency for Sphinx and Kaldi to perform even better which may be related to their better non-incremental performance (in terms of WER).

[3]In our effort, we tried reasonably hard to build well-performing models, but we did not strive for best performance, using as much material (whether in-domain or not) as we could get; e.g., blending our LMs with Wikipedia, or the like.

7 Conclusions

We claim that for suitability for incremental, interactive dialogue systems, ASR, in addition to having good utterance-final accuracy, must also exhibit good incremental properties, and offer a broad interface that either keeps or marks up disfluencies, and provides timing information for downstream processing.

In our evaluation, we find that Google-API offers the best non-incremental performance and almost as good incremental performance as Sphinx and Kaldi. However, Google tends to filter out disfluencies, does not provide word timing information, and limits access to 500 calls per API key a day. We also find that Google's post-hoc rescoring does not improve WER while considerably hurting incremental performance. Finally, Sphinx and Kaldi seem to be on par performance-wise, and at least when trained on in-domain data, these perform similarly well to the Google-API.

We have not, in the present article, factored out the difference between in-domain acoustic models and language models. LMs may already be enough to boost performance for open-source recognizers and are much easier to train. Finally, we want to look into how to incrementally combine recognizers (e.g. Google-API for lowest-possible WERs with Sphinx or Kaldi for timely and time-stamped responses).

Acknowledgements This work is supported by a Daimler and Benz Foundation PostDoc Grant to the first author, by the BMBF KogniHome project, DFG DUEL project (grant SCHL 845/5-1) and the Cluster of Excellence Cognitive Interaction Technology 'CITEC' (EXC 277) at Bielefeld University.

References

1. Schlangen, D., Skantze, G.: A general, abstract model of incremental dialogue processing. Dial. Discourse **2**(1), 83–111 (2011)
2. Aist, G., Allen, J., Campana, E., Galescu, L., Gallo, C.A.G., Stoness, S., Swift, M., Tanenhaus, M.: Software architectures for incremental understanding of human speech. In: Proceedings of Interspeech, pp. 1922–1925 (2006)
3. Skantze, G., Schlangen, D.: Incremental dialogue processing in a micro-domain. In: Proceedings of EACL, pp. 745–753 (2009)
4. Skantze, G., Hjalmarsson, A.: Towards incremental speech generation in dialogue systems. In: Proceedings of SIGdial (2010)
5. Asri, L.E., Laroche, R., Pietquin, O., Khouzaimi, H.: NASTIA: negotiating appointment setting interface. In: Proceedings of LREC, pp. 266–271 (2014)
6. Edlund, J., Gustafson, J., Heldner, M., Hjalmarsson, A.: Towards human-like spoken dialogue systems. Speech Commun. **50**(8–9), 630–645 (2008)
7. Aist, G., Allen, J., Campana, E., Gallo, C.G., Stoness, S., Swift, M.: Incremental understanding in human-computer dialogue and experimental evidence for advantages over nonincremental methods. In: Proceedings of SemDial, pp. 149–154 (2007)
8. Baumann, T., Atterer, M., Schlangen, D.: Assessing and improving the performance of speech recognition for incremental systems. In: Proceedings of NAACL-HTL 2009, pp. 380–388. ACL (2009)
9. Selfridge, E.O., Arizmendi, I., Heeman, P.A., Williams, J.D.: Stability and accuracy in incremental speech recognition. In: Proceedings of SigDial, pp. 110–119. ACL (2011)

10. McGraw, I., Gruenstein, A.: Estimating word-stability during incremental speech recognition. In: Proceedings of Interspeech (2012)
11. Baumann, T., Schlangen, D.: The inproTK 2012 release. In: Proceedings of SDCTD. ACL (2012)
12. Shriberg, E.: Disfluencies in switchboard. In: Proceedings of ICSLP (1996)
13. Fischer, K.: What computer talk is and isn't: Human-computer conversation as intercultural communication. In: Linguistics—Computational Linguistics, vol. 17. AQ-Verlag (2006)
14. Paul, D.B., Baker, J.M.: The design for the wall street journal-based CSR corpus. In: Proceedings of the Workshop on Speech and Natural Language, pp. 357–362. ACL (1992)
15. Morbini, F., Audhkhasi, K., Sagae, K., Artstein, R., Can, D., Georgiou, P., Narayanan, S., Leuski, A., Traum, D.: Which ASR should i choose for my dialogue system? In: Proceedings of SigDial, pp. 394–403 (2013)
16. Ginzburg, J., Tian, Y., Amsili, P., Beyssade, C., Hemforth, B., Mathieu, Y., Saillard, C., Hough, J., Kousidis, S., Schlangen, D.: The disfluency, exclamation and laughter in dialogue (DUEL) project. In: Proceedings of SemDial, pp. 176–178 (2014)
17. Meteer, M., Taylor, A., MacIntyre, R., Iyer, R.: Disfluency annotation stylebook for the switchboard corpus. ms. Technical report, Department of Computer and Information Science, University of Pennsylvania (1995)
18. Brennan, S., Schober, M.: How listeners compensate for disfluencies in spontaneous speech. J. Memory Lang. **44**(2), 274–296 (2001)
19. Hough, J., Purver, M.: Strongly incremental repair detection. In: Proceedings of EMNLP, pp. 78–89. ACL (2014)
20. Clark, H.H.: Using Language. Cambridge University Press (1996)
21. Core, M.G., Schubert, L.K.: A syntactic framework for speech repairs and other disruptions. In: Proceedings of ACL, pp. 413–420 (1999)
22. Clark, H.H., Fox Tree, J.E.: Using uh and um in spontaneous speaking. Cognition **84**(1), 73–111 (2002)
23. Ginzburg, J., Fernández, R., Schlangen, D.: Disfluencies as intra-utterance dialogue moves. Seman. Pragmat. **7**(9), 1–64 (2014)
24. von der Malsburg, T., Baumann, T., Schlangen, D.: TELIDA: a package for manipulation and visualisation of timed linguistic data. In: Proceedings of SigDial (2009)
25. Baumann, T., Buß, O., Schlangen, D.: Evaluation and optimisation of incremental processors. Dial. Discourse **2**(1), 113–141 (2011)
26. Fernández, R., Lucht, T., Schlangen, D.: Referring under restricted interactivity conditions. In: Proceedings of SIGdial, pp. 136–139. ACL (2007)
27. Kousidis, S., Kennington, C., Schlangen, D.: Investigating speaker gaze and pointing behaviour in human-computer interaction with the mint.tools collection. In: Proceedings of SIGdial (2013)
28. Kennington, C., Schlangen, D.: Simple learning and compositional application of perceptually grounded word meanings for incremental reference resolution. In: Proceedings of ACL (2015)
29. Walker, W., Lamere, P., Kwok, P., Raj, B., Singh, R., Gouvea, E., Wolf, P., Woelfel, J.: Sphinx-4: a flexible open source framework for speech recognition. Technical Report SMLI TR2004-0811, Sun Microsystems Inc. (2004)
30. Schalkwyk, J., Beeferman, D., Beaufays, F., Byrne, B., Chelba, C., Cohen, M., Kamvar, M., Strope, B.: Your word is my command: Google search by voice: a case study. In: Advances in Speech Recognition, pp. 61–90. Springer (2010)
31. Povey, D., Ghoshal, A., Boulianne, G., Burget, L., Glembek, O., Goel, N., Hannemann, M., Motlicek, P., Qian, Y., Schwarz, P., Silovsky, J., Stemmer, G., Vesely, K.: The Kaldi speech recognition toolkit. In: Proceedings of ASRU (2011)
32. Plátek, O., Jurčíček, F.: Free on-line speech recogniser based on Kaldi ASR toolkit producing word posterior lattices. In: Proceedings of SIGdial, pp. 108–112. ACL (2014)

Part VII
Dialogue State Tracking Challenge 4

The Fourth Dialog State Tracking Challenge

Seokhwan Kim, Luis Fernando D'Haro, Rafael E. Banchs, Jason D. Williams and Matthew Henderson

Abstract Dialog state tracking is one of the key sub-tasks of dialog management, which defines the representation of dialog states and updates them at each moment on a given on-going conversation. To provide a common test bed for this task, three dialog state tracking challenges have been completed. In this fourth challenge, we focused on dialog state tracking on human-human dialogs. The challenge received a total of 24 entries from 7 research groups. Most of the submitted entries outperformed the baseline tracker based on string matching with ontology contents. Moreover, further significant improvements in tracking performances were achieved by combining the results from multiple trackers. In addition to the main task, we also conducted pilot track evaluations for other core components in developing modular dialog systems using the same dataset.

Keywords Dialog state tracking · Dialog management · Challenge overview

S. Kim (✉) · L.F. D'Haro · R.E. Banchs
Institute for Infocomm Research, Fusionopolis, Singapore
e-mail: kims@i2r.a-star.edu.sg

L.F. D'Haro
e-mail: luisdhe@i2r.a-star.edu.sg

R.E. Banchs
e-mail: rembanchs@i2r.a-star.edu.sg

J.D. Williams
Microsoft Research, Redmond, WA, USA
e-mail: jason.williams@microsoft.com

M. Henderson
Google, Mountain View, CA, USA
e-mail: matthen@google.com

K. Jokinen and G. Wilcock (eds.), *Dialogues with Social Robots*,
Lecture Notes in Electrical Engineering 427,
DOI 10.1007/978-981-10-2585-3_36

1 Introduction

Dialog systems interact with users using natural language to help them achieve a goal, and are increasingly becoming a part of daily life, with examples including Apple's Siri, Google Now, Xbox and Cortana from Microsoft, Facebook M, among others. As the dialog progresses, the dialog system maintains a representation of the state of the dialog in a process called dialog state tracking (DST). For example, in a travel planning system, the dialog state might indicate the search parameters for the type of hotel the user has said they're searching for, such as their desired star rating, location, and price range. Dialog state tracking is difficult because automatic speech recognition (ASR) and spoken language understanding (SLU) errors are common, and can cause the system to misunderstand the user. Moreover, it can be difficult to determine when to retain information and at the same time, state tracking is crucial because the system relies on the estimated dialog state to choose actions, for example, which hotels to suggest.

To provide a common test bed for this task, three Dialog State Tracking Challenges (DSTCs) have been organized [1–3]. Different from the previous challenges which had focused on human-machine dialogs, in this fourth edition, we have focused on dialog state tracking in human-human dialogs. The goal of the main task in this challenge was to track dialog states for sub-dialog segments. For each turn in a given sub-dialog, the tracker was required to fill out a frame of slot-value pairs considering all dialog history prior to the turn. We expect these shared efforts on human dialog state tracking will contribute to progress in developing much more human-like dialog systems.

In addition to the main task, this fourth edition of the challenge also proposed a series of pilot tasks for evaluating each of the core components needed for developing end-to-end dialog systems. More specifically, four pilot tasks were offered: Spoken Language Understanding (SLU), Speech Act Prediction (SAP), Spoken Language Generation (SLG), and End-to-end system (EES). This effort constitutes a first step towards the construction of distributed modular systems and the development of a computational framework for collaborative end-to-end system evaluation. In the evaluation, one team participated in the SLU pilot task, but all the data and tasks remain available for research use.

The rest of the article is organized as follows. Section 2 provides a general overview of the challenge tasks and the used dataset. Sections 3, 4 and 5 describes evaluation results of the main task, while Sect. 6 describes evaluation results of the SLU pilot tasks. Finally, Sect. 7 presents our main conclusions and recommendations.

2 Challenge Overview

2.1 Problem Statement

2.1.1 Main Task

The goal of the main task is to evaluate state tracking for human-human dialogs between tourists and tour guides. Since each subject in these conversations tends to be expressed not just in a single turn, but through a series of multiple turns, a dialog state is defined for each sub-dialog segment level as a frame structure filled with slot-value pairs representing the main subject of a given segment. Figure 1 shows examples of segment-level dialog state frame structures.

Each frame could have two different kinds of slots: regular slots and *INFO* slot. While regular slots should be filled with particular values explicitly discussed in the segment, *INFO* slots indicate the subjects that are discussed but not directly related to any particular values of other slots. The possible slot types and the list of their candidate values vary by topic category, which are described in an ontology.

Speaker	Utterance	Dialog State
Tourist	Can you give me some uh- tell me some cheap rate hotels, because I'm planning just to leave my bags there and go somewhere take some pictures.	
Guide	Okay. I'm going to recommend firstly you want to have a backpack type of hotel, right?	
Tourist	Yes. I'm just gonna bring my backpack and my buddy with me. So I'm kinda looking for a hotel that is not that expensive. Just gonna leave our things there and, you know, stay out the whole day.	Type=*Hostel* Pricerange=*Cheap*
Guide	Okay. Let me get you hm hm. So you don't mind if it's a bit uh not so roomy like hotel because you just back to sleep.	
Tourist	Yes. Yes. As we just gonna put our things there and then go out to take some pictures.	
Guide	Okay, um-	
Tourist	Hm.	
Guide	Let's try this one, okay?	
Tourist	Okay.	
Guide	It's InnCrowd Backpackers Hostel in Singapore. If you take a dorm bed per person only twenty dollars. If you take a room, it's two single beds at fifty nine dollars.	Name=*InnCrowd Backpackers Hostel* Info=*Pricerange*
Tourist	Um. Wow, that's good.	
Guide	Yah, the prices are based on per person per bed or dorm. But this one is room. So it should be fifty nine for the two room. So you're actually paying about ten dollars more per person only.	
Tourist	Oh okay. That's- the price is reasonable actually. It's good.	

Fig. 1 Example human-human dialog and dialog state labels for the main task of DSTC4

In this challenge, a dialog session segmented into a series of sub-dialogs labeled with topic categories is given as an input to a tracker. For each turn in a given sub-dialog, the frame should be filled out considering all dialog history up to the current turn. The performance of a tracker is evaluated by comparing its outputs with reference annotations.

2.1.2 Pilot Tasks

In addition to the main task, the challenge included a series of optional pilot tracks for the core components in developing end-to-end dialog systems using the same dataset and considering either the information from the tourist or the tour guide. The four proposed tasks were:

- Spoken language understanding (SLU): The objective is to tag a given utterance with speech acts (slot values) and semantic slots.
- Speech act prediction (SAP): The objective is to predict the speech act of the next turn imitating the policy of one speaker. Here, the input to the systems will be the utterances and annotations (semantic tags and speech acts) from a given user (i.e. tourist or guide) along with the resulting semantic tags for the next opposite user (i.e. guide or tourist) utterances, and the system must produce the speech acts for the given user utterances.
- Spoken language generation (SLG): The objective is to generate a response utterance for one of the participants by using the corresponding speech act and semantic slot information.
- End-to-end system (EES) The objective is to develop an end-to-end system by pipelining and/or combining different SLU, SAP and SLG systems. Here, the input to the systems will be the one user utterances and the system must produce the other user utterances.

Different from the main task, in which dialog states are defined at the sub-dialog level and each of the sub-dialogs has a frame structure with slot value pairs to represent the subject discussed within it; in the pilot tasks, annotations are provided at the utterance level and, accordingly, systems must deal with slot value pairs at the utterance level.

2.2 Challenge Design

Similar to the previous challenges, both the main and pilot problems are studied as corpus-based tasks with static dialogs. In the development phase, a set of labelled dialogs are released to participants so that they train and optimize their models. And then, the developed components produce the outputs on the unlabelled test set in the evaluation phase. Since every participant uses the same shared datasets for both development and evaluation, the results on the test set can be directly compared to each other.

2.3 *Data*

The data used in the challenge is TourSG corpus which consists of 35 dialog sessions on touristic information for Singapore collected from Skype calls between three tour guides and 35 tourists. These 35 dialogs sum up to 31,034 utterances and 273,580 words. All the recorded dialogs with the total length of 21 hours were manually transcribed and annotated with speech act and semantic labels for each turn level.

For the main task, each full dialog session was divided into sub-dialog segments considering their topical coherence and then they were categorized by topics. Each sub-dialog assigned to one of the five major topic categories has an additional frame structure with slot value pairs to represent some more details about the subject discussed within the sub-dialog.

For the challenge, TourSG corpus were divided into four parts (Table 1). Training and development sets consist of manual transcriptions and annotations at both utterance and sub-dialog levels for training and optimizing the trackers, respectively. For the test sets, only manual transcriptions without annotations are provided during the evaluation period.

Along with the dialog corpus, an ontology was also created to provide the tagset definitions as well as the domain knowledge regarding tourism in Singapore. While subjects of human-machine conversations are inevitably restricted within the knowledge-base contents used in developing the system, human-human dialogs are much more flexible and broad in terms of the coverage of subjects. To make the resource as general as possible, the entries in the ontology were collected not only from the corpus itself, but also from external knowledge sources. First, the structured information were automatically extracted from the Wikipedia articles related to Singapore and the official website of Singapore Tourism Board. Then, the collected instances were validated by matching with the annotations in the corpus. Finally, all the missing parts in the ontology were completed manually to cover all the subjects discussed in the dialogs.

More detailed information about the data can be found from [4].

3 Main Task: Evaluation

3.1 *Evaluation Metrics*

A system for the main task should generate the tracking output for every turn in a given log file. While all the transcriptions and segment details provided in the log object from the beginning of the session to the current turn can be used, any information from the future turns are not allowed to be considered to analyze the state at a given turn.

Table 1 Overview of DSTC4 data. SG1, SG2, and SG3 are the three tour guides that participated in the data collection and acco (accommodation), attr (attraction), food, shop (shopping), trsp (transportation), and other are topic categories of dialog segments

	#Dialogs				#Segments							#Utterances
Set	SG1	SG2	SG3	Total	Acco	Attr	Food	Shop	Trsp	Other	Total	
Training	7	7	0	14	187	762	275	149	374	357	2104	12,759
Development	3	3	0	6	94	282	102	67	87	68	700	4,812
Test (main)	3	3	3	9	174	616	134	49	174	186	1333	7,848
Test (pilot)	2	2	2	6	126	352	124	49	119	107	877	5,615

Although the fundamental goal of this tracking is to analyze the state for each sub-dialog level, the execution should be done at each utterance level regardless of the speaker from the beginning to the end of a given session in sequence. It aims at evaluating the capabilities of trackers not only for understanding the contents mentioned in a given segment, but also for predicting its dialog states even at an earlier turn of the segment.

To examine these both aspects of a given tracker, two different schedules are considered to select the utterances for the target of evaluation:

- Schedule 1: all turns are included
- Schedule 2: only the turns at the end of segments are included

If some information is correctly predicted or recognized at an earlier turn in a given segment and well kept until the end of the segment, it will have higher accumulated scores than the other cases where the same information is filled at a later turn under schedule 1. On the other hand, the results under schedule 2 indicate the correctness of the outputs after providing all the turns of the target segment.

In this task, the following two sets of metrics are used for evaluation:

- Accuracy: Fraction of segments in which the tracker's output is equivalent to the gold standard frame structure
- Precision/Recall/F-measure
  - Precision: Fraction of slot-value pairs in the tracker's outputs correctly filled
  - Recall: Fraction of slot-value pairs in the gold standard labels correctly filled
  - F-measure: The harmonic mean of precision and recall

While the first metric is to check the equivalencies between the outputs and the references at the whole frame level, the others can show the partial correctness at each slot-value level.

3.2 Baseline Tracker

A simple baseline tracker is provided to participants. The baseline tracker determines the slot values by fuzzy string matching[1] between the entries in the ontology and the transcriptions of the utterances mentioned from the beginning of a given segment to the current turn. If a part of given utterances is matched with an entry for a slot in the ontology with over a certain level of similarity, the entry is simply assigned as a value for the particular slot in the tracker's output. Since this baseline does not consider any semantic or discourse aspects from given dialogs, its performance is very limited and there is much room for improvement.

[1] https://github.com/seatgeek/fuzzywuzzy.

4 Main Task: Results

Logistically, the training and development datasets, the ontology, and the scoring scripts were released to the participants on 15 April 2015. Then, the unlabelled test set for the main task was released on 17 August 2015. In this challenge, a web-based competition platform [2] was newly introduced for receiving submissions and evaluating them automatically. Once an entry was uploaded to the site, the evaluation results were immediately provided to the participant and compared to the others by posting them to the leaderboard page.

Teams were given two weeks to run their trackers on the test set and enter the outputs to the submission system. Following the tradition of the previous challenges, the number of entries submitted by each team was limited up to five. And also, all the results posted on the leaderboard were anonymized. After the evaluation phase, the test labels were released to the participants.

In total, 24 entries were submitted from 7 research teams participating in the main task. Teams were identified by anonymous team numbers team 1–7, and the baseline system was marked as team 0.

Table 2 shows the averaged results over the whole test set for each submitted entry. More specific scores by topic and slot type and all the submitted entries are available on the DSTC4 website [3] and the full details on the trackers themselves are published in individual papers at IWSDS 2016. Most submitted trackers outperformed the baseline in all the combinations of schedules and metrics. Especially, the best entries from team 3 achieved more than three times and almost twice as high performances as the baseline in accuracy and F-measure, respectively, under both schedules. Figure 2 reveals that the highly-ranked trackers in the overall comparison tend to produce evenly good results across all topic categories. The entry *team3.entry3* is ranked the best for all the topics except just one, and *team4.entry3* also yields competitive results in all the cases.

To investigate the reasons for the performance differences among the trackers, the slot-level errors under Schedule 2 from the best entry of each team were categorized into the three error types following [5]:

- Missing attributes: when the reference contains values for a slot, but the tracker does not output any value for the slot
- Extraneous attributes: when the reference does not contain any value for a slot, but the tracker outputs values for that slot
- False attributes: when the reference contains values for a slot, and the tracker outputs an incorrect value for that slot

The error distributions in Fig. 3 indicate that the missing slot errors act as a decisive factor in performance variations across teams.

The influences of these false negatives to the tracking performances are demonstrated also in the analysis of correct outputs. Figure 4 compares the distributions

[2] https://www.codalab.org/competitions/4971.

[3] http://www.colips.org/workshop/dstc4/results.html.

Table 2 Main task results on the test set. Team 0 is the rule-based baseline. Bold denotes the best result in each column

		Schedule 1				Schedule 2			
Team	Entry	Accuracy	Precision	Recall	F-measure	Accuracy	Precision	Recall	F-measure
0	0	0.0374	0.3589	0.1925	0.2506	0.0488	0.3750	0.2519	0.3014
1	0	0.0456	0.3876	0.3344	0.3591	0.0584	0.4384	0.3377	0.3815
	1	0.0374	0.4214	0.2762	0.3336	0.0584	0.4384	0.3377	0.3815
	2	0.0372	0.4173	0.2767	0.3328	0.0575	0.4362	0.3377	0.3807
	3	0.0371	0.4179	0.2804	0.3356	0.0584	0.4384	0.3426	0.3846
2	0	0.0487	0.4079	0.2626	0.3195	0.0671	0.4280	0.3257	0.3699
	1	0.0467	0.4481	0.2655	0.3335	0.0671	0.4674	0.3275	0.3851
	2	0.0478	0.4523	0.2623	0.3320	0.0706	0.4679	0.3226	0.3819
	3	0.0489	0.4440	0.2703	0.3361	0.0697	0.4634	0.3335	0.3878
3	0	**0.1212**	0.5393	0.4980	0.5178	**0.1500**	0.5569	0.5808	0.5686
	1	0.1210	0.5449	0.4964	0.5196	**0.1500**	0.5619	0.5787	0.5702
	2	0.1092	0.5304	**0.5031**	0.5164	0.1316	0.5437	**0.5875**	0.5648
	3	0.1183	**0.5780**	0.4904	**0.5306**	0.1473	0.5898	0.5678	**0.5786**
4	0	0.0887	0.5280	0.3595	0.4278	0.1072	0.5354	0.4273	0.4753
	1	0.0910	0.5314	0.3122	0.3933	0.1055	0.5325	0.3623	0.4312
	2	0.1009	0.5583	0.3698	0.4449	0.1264	0.5666	0.4455	0.4988
	3	0.1002	0.5545	0.3760	0.4481	0.1212	0.5642	0.4540	0.5031
5	0	0.0309	0.2980	0.2559	0.2754	0.0392	0.3344	0.2547	0.2892
	1	0.0268	0.3405	0.2014	0.2531	0.0401	0.3584	0.2632	0.3035
	2	0.0309	0.3039	0.2659	0.2836	0.0392	0.3398	0.2639	0.2971
6	0	0.0421	0.4175	0.2142	0.2831	0.0541	0.4380	0.2656	0.3307
	1	0.0478	0.5516	0.2180	0.3125	0.0654	0.5857	0.2702	0.3698
	2	0.0486	0.5623	0.2314	0.3279	0.0645	**0.5941**	0.2850	0.3852
7	0	0.0286	0.2768	0.1826	0.2200	0.0323	0.3054	0.2410	0.2694
	1	0.0044	0.0085	0.0629	0.0150	0.0061	0.0109	0.0840	0.0194

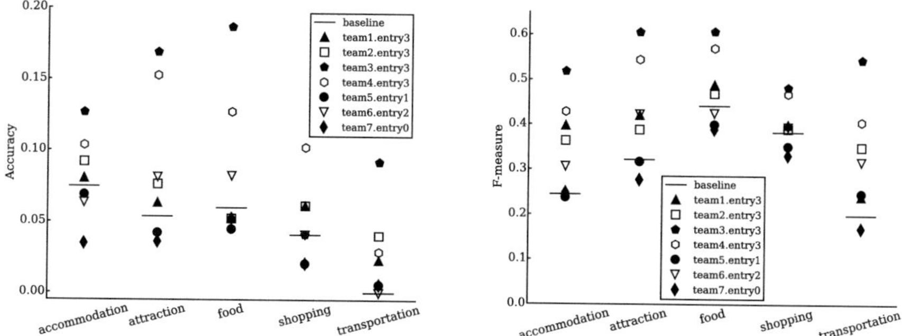

Fig. 2 Accuracy (*left panel*) and F-measure (*right panel*) on the test set per topic for the best tracker from each team in the main dialog state tracking task

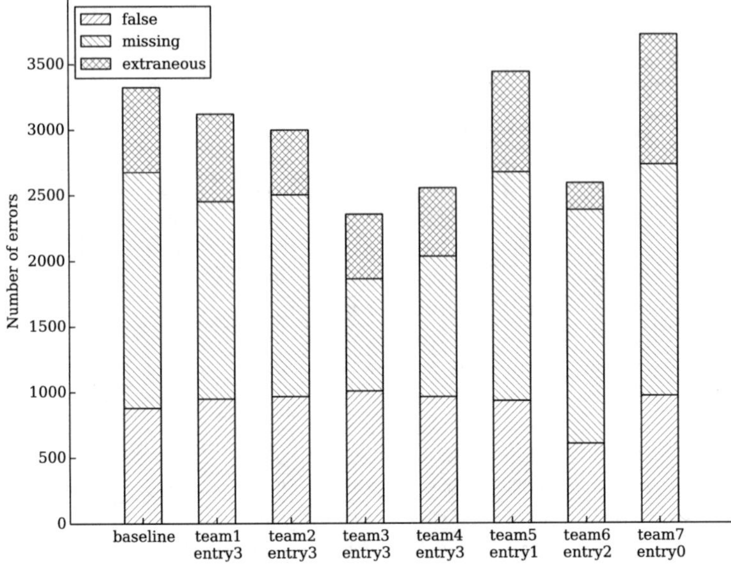

Fig. 3 Number of errors made by type for the best entry from each team in the main dialog state tracking task on the test set

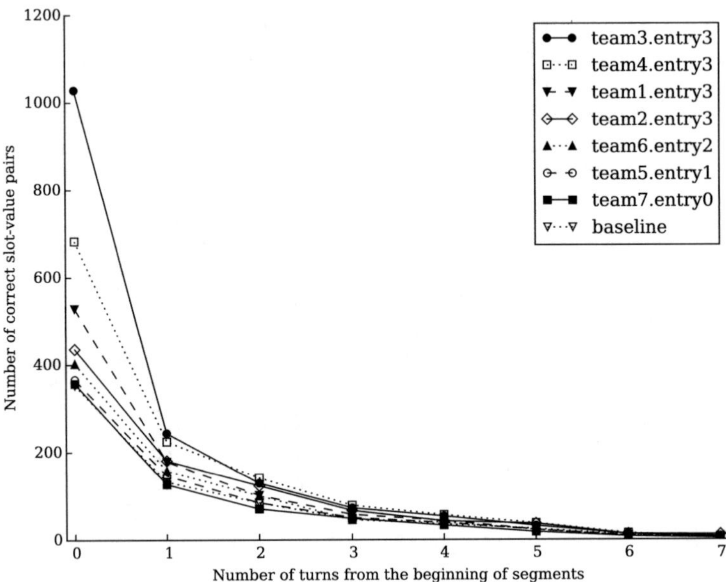

Fig. 4 Distributions of correct slot-value pairs in the best output from each team by turn offsets where each value is filled for the first time since the beginning of the sub-dialog

of the number of correct outputs by the turn offset where each value is filled from the beginning of the sub-dialog. Most of the differences in number of true positives among teams exist at earlier turns of dialog segments, which means that the highly-ranked trackers managed to rescue many slot-value pairs that were missed by others.

5 Main Task: Ensemble Learning

A merit of corpus-based tasks is that ensemble learning could be studied simply by synthesizing the multiple outputs on the same dataset to improve the performances compared to any single individual system. In the previous dialog state tracking challenges, ensemble learning techniques including score averaging [6] and stacking [2] contributed to improve the tracking performances.

Also for the main task of this challenge, we examined the effectiveness of ensemble learning based on the submitted entries. Since no score information was available in tracking outputs for DSTC4, we adopted the following three simple strategies for combining the outputs:

- Union: fill a slot with a value if the slot-value pair occurs in at least one of the tracking outputs to be combined
- Intersection: fill a slot with a value if the slot-value pair occurs in all the tracking outputs to be combined
- Majority: fill a slot with a value if the slot-value pair occurs in more than half the tracking outputs to be combined.

Table 3 compares the performances of combined outputs with the single best entry. The tracking outputs to be combined were selected based on single entry performances in F-measure under Schedule 2 without distinction of team. For example, entry 3, 1, and 0 from team 3 were considered as top 3 entries. The results show that most of the combinations failed to achieve performance improvement from the single best output. Only statistically significant improvement across all metrics was observed when top 3 entries were combined by intersection. This suggests that system combination without considering any correlations among the trackers does not guarantee better results.

To see how much the performances could be improved in case the optimal combination is somehow given considering their correlations, we run the evaluation on every possible combination of 25 entries including the baseline. Table 4 shows the performances of the best combination in each metric. These results are significantly better than the single best entry in most metrics. All the statistical significances in these analyses were computed using approximate randomization [7].

Table 3 Accuracy and F-measure for various combinations of trackers in the main task on the test set. Bold denotes the best result in each column

	Schedule 1		Schedule 2	
Tracker	Accuracy	F-measure	Accuracy	F-measure
Single best entry	0.1212	0.5306	0.1500	0.5786
Top 3 entries: union	0.1111^{-}	0.5147^{-}	0.1325^{-}	0.5619^{-}
Top 3 entries: intersection	0.1241^{+}	**0.5344**$^{+}$	**0.1561**$^{+}$	**0.5861**$^{+}$
Top 3 entries: majority voting	0.1172^{-}	0.5194^{-}	0.1421^{-}	0.5703
Top 5 entries: union	0.0980^{-}	0.5133^{-}	0.1107^{-}	0.5543^{-}
Top 5 entries: intersection	0.1157	0.4370^{-}	0.1369	0.5008^{-}
Top 5 entries: majority voting	0.1183^{-}	0.5210^{-}	0.1439	0.5711
Top 10 entries: union	0.0623^{-}	0.4719^{-}	0.0680^{-}	0.5014^{-}
Top 10 entries: intersection	0.0300^{-}	0.1816^{-}	0.0453^{-}	0.2275^{-}
Top 10 entries: majority voting	**0.1268**$^{+}$	0.4741^{-}	0.1456	0.5380^{-}
All entries: union	0.0077^{-}	0.1320^{-}	0.0078^{-}	0.1366^{-}
All entries: intersection	0.0132^{-}	0.0229^{-}	0.0192^{-}	0.0331^{-}
All entries: majority voting	0.0646^{-}	0.3535^{-}	0.0898^{-}	0.4135^{-}

+/− indicates statistically significantly better/worse than the single best entry ($p < 0.01$), computed with approximate randomization

Table 4 The best possible (oracle) combination of trackers in the main task on the test set. All the listed performances were achieved by the majority voting strategy. Bold denotes the best result in each metric. +/- indicates statistically significantly better/worse than the single best entry in Table 3 ($p < 0.01$), computed with approximate randomization

	Schedule 1		Schedule 2	
Combination	Accuracy	F-measure	Accuracy	F-measure
T3E0 + T3E2 + T3E3 + T4E1 + T4E3 + T6E0 + T6E2	**0.1310**$^{+}$	0.4870^{-}	0.1517	0.5534^{-}
T3E1 + T3E3 + T4E2	0.1241^{+}	**0.5359**$^{+}$	0.1569^{+}	0.5885^{+}
T2E3 + T3E0 + T3E2 + T3E3	0.1230^{+}	0.5351^{+}	**0.1587**$^{+}$	0.5878^{+}
T2E3 + T3E0 + T3E1 + T3E2 + T3E3 + T4E2	0.1242^{+}	0.5354^{+}	0.1587^{+}	**0.5893**$^{+}$

6 Pilot Tasks

6.1 Evaluation Metrics

Two different families of metrics were used for evaluating the pilot tasks: classification accuracy metrics used for SLU and SAP tasks, and semantic similarity metrics used for SLG and EES tasks. For all subtasks in the pilot tasks, evaluation schedule 1 was used (i.e. system outputs are evaluated at all turns). In more detail, the following evaluation metrics were used:

- SLU and SAP tasks:
  - Precision: Fraction of semantic tags and/or speech acts that are correct.
  - Recall: Fraction of semantic tags and/or speech acts in the gold standard that are generated.
  - F-measure: The harmonic mean of precision and recall.
- SLG and EES tasks:
  - BLEU: Geometric average of n-gram precision (for n = 1, 2, 3, 4) of the system generated utterance with respect to the reference utterance [8].
  - AM-FM: Weighted mean of (1) the cosine similarity between the system generated utterance and the reference utterance and (2) the normalized n-gram probability of the system generated utterance [9].

6.2 Web-Based Evaluation

Regarding operational aspects of pilot task evaluation, participants were required to implement a web-service (WS) to run their systems. During the evaluation, a master evaluation script was used to call the corresponding web-services at specified time slots during the evaluation dates. In order to facilitate these implementations, a server and client python scripts were provided with default configuration to check that the systems were working and reachable from outside local network.

During the evaluation, the participant's server received a JSON object containing the input parameters required for the given task and role and the server used the input parameters to generate a corresponding answer that was send back to the organizer's client using a JSON message. Then, based on the retrieved result, the client calculated the actual values for the proposed metrics. For debugging purposes, both the server and client included a logging module to keep record of all the requests and answers interchanged between both modules. For additional information about the pilot task, messages, and provided scripts please refer to [10].

6.3 Results

Given that the pilot tasks were optional, we only received answers from a single team that submitted up to 5 different systems only for the NLU task considering the tourist and guide users. Table 5 shows the results extracted for this team. A baseline was not available for this task.

In past DSTCs, the evaluation was done by having teams submit a file with tracker output. In DSTC4, evaluations were conducted by having teams provide trackers as a web service. However, occasionally the web connection would timeout. For future evaluations, we suggest incorporating automatic reconnections when timeouts occur, and to add better handling of asynchronous communication data and packet-loss.

Table 5 Results from 5 entries submitted by one team to the NLU task, on the test set

		Speech Act			Semantic Tag		
Speaker	Entry	Precision	Recall	F-measure	Precision	Recall	F-measure
Guide	1	0.629	0.519	0.569	0.565	0.489	0.524
	2	0.633	0.523	0.573	0.565	0.489	0.524
	3	**0.745**	**0.615**	**0.674**	0.565	0.489	0.524
	4	0.631	0.521	0.571	0.565	0.489	0.524
	5	0.676	0.558	0.612	0.565	0.489	0.524
Tourist	1	0.358	0.298	0.325	0.574	0.476	0.521
	2	0.293	0.244	0.266	0.574	0.476	0.521
	3	0.563	0.468	0.511	0.574	0.476	0.521
	4	0.294	0.244	0.267	0.574	0.476	0.521
	5	**0.574**	**0.477**	**0.521**	0.574	0.476	0.521

7 Conclusions

We have presented the official evaluation results of the Fourth Dialog State Tracking Challenge (DSTC4). This edition of the challenge has continued the tradition of its previous editions by providing a common testbed for the evaluation of Dialog State Tracking, one of the key tasks in Dialog Management. However, different from previous editions, which focused on human-machine dialogs, this edition has focused on dialog state tracking in human-human dialogs. The goal of the main task was to track dialog states for sub-dialog segments, which means that for each turn in a given sub-dialog, the tracker was required to fill out a frame of slot-value pairs considering all dialog history prior to the turn.

A total of seven teams participated in the main task with an overall number of twenty four entries submitted. Most of the submitted entries outperformed the provided baseline tracker system, which was based on a string matching strategy for identifying mentions of contents using the provided ontology as a reference. In a post-evaluation exercise of ensemble learning, results from multiple tracker submissions were combined. As a result, further significant improvements on dialog state tracking performance were observed.

In addition to the main task, this fourth edition of the challenge also proposed four pilot tasks with the objective of evaluating each of the core components needed for developing end-to-end dialog systems. More specifically, the proposed pilot tasks included: Spoken Language Understanding (SLU), Speech Act Prediction (SAP), Spoken Language Generation (SLG), and End-to-End System (EES). Only one team participated in the SLU pilot tasks with five different submissions for each of the two speaker roles involved in the provided datasets. This evaluation interestly showed that guide speech acts are significantly more predictable than tourist speech acts, while semantic tags are similarly predictable for both roles.

As final remarks, we would like to highlight that this challenge results have confirmed the feasibility of the state tracking task in human-human dialogs, which are much more unstructured and noisy than human-machine dialogs. We expect these shared efforts on human dialog state tracking will contribute to progress in developing much more human-like dialog systems. Regarding the pilot task, on the other hand, we were able to test a new evaluation modality for dialog technology, which in our opinion constitutes a first step towards the effective development of distributed modular systems and a computational framework for collaborative end-to-end system evaluation.

As final recommendation, we suggest to continue pursuing the study of human-human dialogs as a means for better modeling and understanding the complexity of the pragmatic phenomena, as well as to include new languages to explore the feasibility of cross-language and/or multilingual approaches to dialog management. Similarly, we recommend to continue the efforts on pilot tasks for the next editions of the challenges, to continuing moving in the direction of distributed and modular end-to-end system construction and evaluation.

References

1. Williams, J., Raux, A., Ramachandran, D., Black, A.: The dialog state tracking challenge. In: Proceedings of the SIGDIAL 2013 Conference, pp. 404–413 (2013)
2. Henderson, M., Thomson, B., Williams, J.: The second dialog state tracking challenge. In: Proceedings of the 15th Annual Meeting of the Special Interest Group on Discourse and Dialogue, p. 263 (2014)
3. Henderson, M., Thomson, B., Williams, J.D.: The third dialog state tracking challenge. In: Proceedings of the 2014 IEEE Spoken Language Technology Workshop (SLT), pp. 324–329. IEEE (2014)
4. Kim, S., D'Haro, L.F., Banchs, R.E., Williams, J., Henderson, M.: Dialog state tracking challenge 4 handbook. http://www.colips.org/workshop/dstc4/Handbook_DSTC4.pdf (2015)
5. Smith, R.W.: Comparative error analysis of dialog state tracking. In: Proceedings of the 15th Annual Meeting of the Special Interest Group on Discourse and Dialogue, p. 300 (2014)
6. Lee, S., Eskenazi, M.: Recipe for building robust spoken dialog state trackers: dialog state tracking challenge system description. In: Proceedings of the SIGDIAL 2013 Conference, pp. 414–422 (2013)
7. Yeh, A.: More accurate tests for the statistical significance of result differences. In: Proceedings of the 18th Conference on Computational linguistics, vol. 2. pp. 947–953. Association for Computational Linguistics (2000)
8. Papineni, K., Roukos, S., Ward, T., Zhu, W.J.: Bleu: a method for automatic evaluation of machine translation. In: Proceedings of the 40th Annual Meeting on Association for Computational linguistics, pp. 311–318. Association for Computational Linguistics (2002)
9. Banchs, R.E., D'Haro, L.F., Li, H.: Adequacy-fluency metrics: evaluating mt in the continuous space model framework. IEEE/ACM Trans. Audio Speech Lang. Process. **23**(3), 472–482 (2015)
10. Kim, S., D'Haro, L.F., Banchs, R.E., Williams, J., Henderson, M.: Dialog state tracking challenge 4 pilot task guidelines. http://www.colips.org/workshop/dstc4/DSTC4_pilot_tasks.pdf (2015)

Convolutional Neural Networks for Multi-topic Dialog State Tracking

Hongjie Shi, Takashi Ushio, Mitsuru Endo, Katsuyoshi Yamagami and Noriaki Horii

Abstract The main task of the fourth Dialog State Tracking Challenge (DSTC4) is to track the dialog state by filling in various slots, each of which represents a major subject discussed in the dialog. In this article we focus on the 'INFO' slot that tracks the general information provided in a sub-dialog segment, and propose an approach for this slot-filling using convolutional neural networks (CNNs). Our CNN model is adapted to multi-topic dialog by including a convolutional layer with general and topic-specific filters. The evaluation on DSTC4 common test data shows that our approach outperforms all other submitted entries in terms of overall accuracy of the 'INFO' slot.

Keywords Dialog state tracking · Convolutional neural networks · Multi-topic dialog state tracking · Domain adaptation · Semi-supervised learning

1 Introduction

The selection of an appropriate action (i.e. "what to say next" in a conversation) is the core problem of dialog management [1]. To address this problem, statistical machine learning approaches, such as reinforcement learning, are often employed. Such machine learning approaches offer several potential advantages over tradi-

H. Shi (✉) · T. Ushio · M. Endo · K. Yamagami · N. Horii
Intelligence Research Laboratory, Panasonic Corporation, Osaka, Japan
e-mail: shi.hongjie@jp.panasonic.com

T. Ushio
e-mail: ushio.takashi@jp.panasonic.com

M. Endo
e-mail: endo.mitsuru@jp.panasonic.com

K. Yamagami
e-mail: yamagami.katsuyoshi@jp.panasonic.com

N. Horii
e-mail: horii.noriaki@jp.panasonic.com

K. Jokinen and G. Wilcock (eds.), *Dialogues with Social Robots*,
Lecture Notes in Electrical Engineering 427,
DOI 10.1007/978-981-10-2585-3_37

tional rule-based hand-crafted approaches, however, they rely on the availability of quantities of appropriately annotated dialog corpus for training. It is well known that hand annotations are expensive, time-consuming and require human experts. On the other hand, unannotated dialog corpus is relatively inexpensive and abundant. For this reason, a system capable of automatic annotation for large dialog corpus will be very useful for learning human-like dialog strategies.

The fourth Dialog State Tracking Challenge (DSTC4) based on human-human dialog provides a common test bed for developing such automatic annotation systems. The main task of this challenge is to track the dialog states by filling out a frame of slot-value pairs for sub-dialog segments with regard to various topics (e.g. Shopping, Accommodation, Transportation). There are two types of slot to be filled in: regular slots and 'INFO' slot. The regular slots indicate detailed information discussed in that subject, such as [PLACE: *Chinatown*] or [DISH: *Dim sum*]. The 'INFO' slot, on the other hand, indicates the subject that are generally discussed in the segment if *no* specific information is mentioned in that subject, such as [INFO: *Place*] or [INFO: *Dish*]. The baseline system provided by DSTC4, which uses fuzzy string matching to identify the value name, performs particularly poorly in this 'INFO' slot. This is because the name of subject itself rarely shows up in a dialog. For instance, we do not always use the exact term 'price range' when we talk about the 'price range' of a hotel. Instead, a word or phrase like 'dollars', 'expensive' or 'price is reasonable' is likely to be appeared in the context. An improved system should be capable of learning these word-subject correlations and making use of them to predict a value.

In this article, we focus on this 'INFO' slot filling task and propose an approach using convolutional neural networks (CNNs). In our CNN model, we use a convolutional layer which consists of general and topic-specific filters to achieve an improvement of performance for the multi-topic dialog state tracking. During the training process we also apply semi-supervised learning to make use of unlabelled data on the internet. The evaluation on unseen test data shows that our approach outperforms the baseline method by 24 % prediction accuracy, which is a competitive result among all 7 participants of DSTC4.

2 Data Characteristics and Problem Description

The forth Dialoge State Tracking Challenge (DSTC4) is based on a TourSG corpus, which consists of 35 dialog sessions on touristic information for Singapore collected from Skype calls between three tour guides and 35 tourists [2]. This corpus is divided into **train**, **dev** and **test** sets. Every participant is asked to develop their own system based on labelled **train** and **dev** dataset, and all submitted systems are evaluated by the common unlabelled **test** dataset.

A full dialog session is divided into sub-dialog segments considering their topical coherence. Each sub-dialog segment is assigned to one of the following five major topic categories: 'Accommodation', 'Attraction', 'Food', 'Shopping' and 'Transportation'. The set of candidate values for 'INFO' slot is defined by ontology for

Table 1 Complete lists of candidate values for the 'INFO' slot in each topic

Topic	Candidate values for the 'INFO' slot
Accommodation	Amenity, Architecture, Booking, Check-in, Check-out, Cleanness, Facility, History, Hotel rating, Image, Itinerary, Location, Map, Meal included, Name, Preference, Pricerange, Promotion, Restriction, Room size, Room type, Safety, Type
Attraction	Activity, Architecture, Atmosphere, Audio guide, Booking, Dresscode, Duration, Exhibit, Facility, Fee, History, Image, Itinerary, Location, Map, Name, Opening hour, Package, Place, Preference, Pricerange, Promotion, Restriction, Safety, Schedule, Seat, Ticketing, Tour guide, Type, Video, Website
Food	Cuisine, Delivery, Dish, History, Image, Ingredient, Itinerary, Location, Opening hour, Place, Preference, Pricerange, Promotion, Restriction, Spiciness, Type of place
Shopping	Cuisine, Delivery, Dish, History, Image, Ingredient, Itinerary, Location, Opening hour, Place, Preference, Pricerange, Promotion, Restriction, Spiciness, Type of place
Transportation	Deposit, Distance, Duration, Fare, Itinerary, Location, Map, Name, Preference, Pricerange, Schedule, Service, Ticketing, Transfer, Type

each topic. A complete list of 'INFO' slot values for each topic is shown in Table 1 and a more detailed descriptions of each value can be found in [2]. In total, there are 54 distinct 'INFO' slot values in all five topics: 23 in 'Accommodation', 31 in 'Attraction', 16 in 'Food', 16 in 'Shopping', 15 in 'Transportation'. Some of these values appear in more than one topic. For example, the value 'Pricerange', which indicates the subject of price ranges, appears in all five topics. In this paper we consider such values to indicate exactly the same subject, but in different topics.

The 'INFO' slot filling task is similar to the multi-domain text classification problem, as we can regard each 'INFO' slot value as a class and each topic as a distinct domain. Furthermore, since multiple 'INFO' slot values can be assigned to a single sub-dialog segment, the 'INFO' slot-filling task is essentially a multi-label classification problem. In this paper, we apply a recently proposed approach of text classification to this 'INFO' slot-filling task.

3 Related Work

Our approach is based on recent work by Kim which proposed to use convolutional neural networks to classify sentences [3]. It has been shown that this CNN model improves state of the art performance on several major text classification tasks including sentiment analysis and question classification, despite of its simple architecture which requires little tuning. Furthermore, this model is robust to variable length of input text, which is a desirable property for dealing with sub-dialog segment consisting of an arbitrary number of utterances.

To our knowledge, there is no previous work directly addressed the *multi-domain* text classification problem using convolutional neural networks. A well-known easy domain adaptation method proposed by Daumé III [4], which is generally applicable to any machine learning algorithm, was widely used in multi-domain problems including the dialog state tracking [5, 6]. However, this easy domain adaptation method requires dimensional augmentation of feature space, which will dramatically increase the complexity of the CNN model.

A more recent related work on multi-domain dialog state tracking using recurrent neural networks (RNNs) was proposed by Mrkšić et al. [7]. Their idea for domain adaptation is to pre-train the RNN models using out-of-domain data. However, in the CNN case,[1] we did not observe a significant improvement in performance with this method (the results will be shown in Sect. 6.1). A possible reason for this may be that a shallow CNN model does not take advantage of pre-training as much as a RNN model does.

4 Convolutional Neural Network Model

4.1 Model Definition

Our model architecture, as shown in Fig. 1, is a slight variant of the CNN architecture of Kim [3]. In the model, a convolutional layer is applied to the dialog segment matrix $\mathbf{s}$, where each row corresponds to the k-dimensional feature vector $\mathbf{w}_{ij} \in \mathbb{R}^k$ of the j-th word in the i-th utterance. Moreover, a zero vector is inserted between every two adjacent utterances:

$$\mathbf{s} = \begin{bmatrix} | & & | & | & | & & | \\ \mathbf{w}_{11}, & \cdots, & \mathbf{w}_{1l_1}, & \mathbf{0}, & \mathbf{w}_{21}, & \cdots, & \mathbf{w}_{nl_n} \\ | & & | & | & | & & | \end{bmatrix}^T. \tag{1}$$

Here l_i is the length of i-th utterance of the dialog segment and we define $N \stackrel{\text{def}}{=\!=} \sum_i l_i + i - 1$ as the total length of this dialog segment matrix. A feature map $\mathbf{h} \in \mathbb{R}^{N-d+1}$ in the convolutional layer is obtained by convolving a filter $\mathbf{m} \in \mathbb{R}^{d\times k}$ with this dialog segment matrix, followed by a rectified linear unit (ReLU). That is, the i-th component of this feature map is calculated by:

$$h_i = \max(0, (\mathbf{m} * \mathbf{s})_i + b), \tag{2}$$

[1]The original RNN models used in [7] are not designed for the text classification problem, so we did not apply those models to the 'INFO' slot filling task.

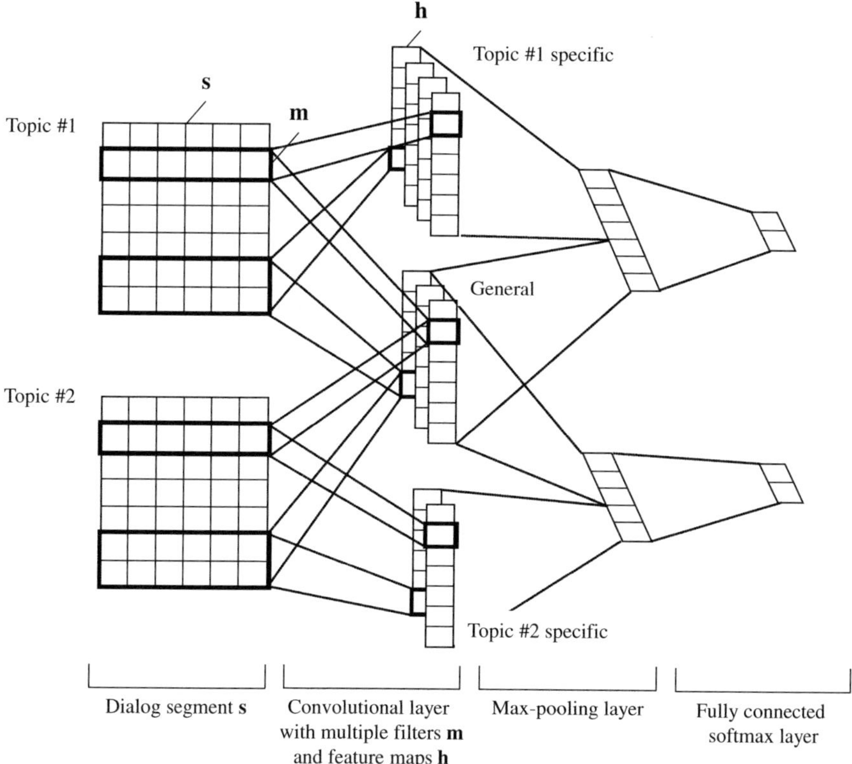

Fig. 1 Model architecture for 2 topics with general and topic-specific convolutional layers

where $b \in \mathbb{R}$ is a bias term for this particular feature map. To address the problem of varying dialog segment length, a max-pooling layer takes the maximum value over this feature map:

$$\hat{h} = \max\{h_i\}. \tag{3}$$

The resulting $\hat{h}$ is the most relevant feature corresponding to this filter **m**. The model uses multiple filters (with varying window size) to generate multiple features. These features form the penultimate layer and are passed to a fully connected softmax layer for classification.

4.2 Multi-topic Model

For the dialog data which is categorized into different topics, it is intuitive to independently train one topic-specific model for each topic. However, topic-specific models do not share information with each other. This is a disadvantage because there may be considerable amount of sharable common features across topics. For example, a common word like 'expensive' may be related to 'Pricerange' in all topics. A general model which is trained regardless of topics, on the other hand, can not be specialised for each topic.

To address these problems, we propose a *multi-topic* CNN model. In this model, we join two separate topic-specific models by sharing some of the filters in each convolutional layer, as shown in Fig. 1. For more than two topics, the filters are divided into one general set and multiple topic-specific sets, where the general filters are *trained and used* by all topics, and topic-specific filters are *trained and used* only by one particular topic. By doing so, the model can learn and use general features and topic-specific features at once. In addition, the number of filters in each set can be easily adjusted to best fit each topic and achieve an optimal balance between the general and topic-specific feature learning.

A simple modification can be done to this model by also sharing the learned weights in the fully connected softmax layer. This ensures that the learned general features contribute to the output equally for different topics, which may be desirable for the case that the amount of in-topic data is very limited compared to out-topic data.

5 Experiment Setup

Since more than one value can be assigned to a single dialog segment, the 'INFO' slot-filling task is a multi-label classification problem. The above CNN model is not designed for multiple outputs, therefore we transform the multi-label problem into a set of binary classification problems. That is, we independently train one *value-specialised* CNN model for each value. These value-specialised models can also ensure that for different values, each model learns different features. For those values that occur across different topics such as 'Pricerange' mentioned in Sect. 2, we apply the mulit-topic CNN model discussed in Sect. 4.2 with also sharing the weights in the fully connected layer which corresponds to the general filters.

5.1 Feature Representation

The feature vector representation of each word in the dialog segment is obtained by combining multiple features:

$$\mathbf{w} = \mathbf{f}_{w1} \oplus \mathbf{f}_{w2} \oplus \mathbf{f}_{\text{slot}} \oplus \mathbf{f}_{\text{speaker}}, \tag{4}$$

where $\oplus$ denotes the concatenation of two vectors. The details of these features are:

- $\mathbf{f}_{w1}$: a 50-dimensional word vector representation trained on text8 corpus (the first 100MB of the cleaned English Wiki corpus), using `word2vec` [8].
- $\mathbf{f}_{w2}$: similar to $\mathbf{f}_{w1}$ but trained on DSTC4 training data corpus along with 50,000 Singapore travel-related reviews on Tripadvisor.
- $\mathbf{f}_{\text{slot}}$: a 16-dimensional slot vector. We first tag the substring in a dialog segment with the slot it belongs to, if the substring matches with some entries in the ontology. The tagged slot is then indicated by this slot vector in the form of $[0, 0, \ldots, 1, \ldots, 0]$. The purpose of adding this feature is to allow the model to use the information in the ontology.
- $\mathbf{f}_{\text{speaker}}$: a 2-dimensional the speaker vector which indicates speaker of this word.

5.2 Hyperparameters and Training

In our experiment we use filters with windows of 1 and 2 dimensions so that the model can learn both unigram and bigram features. There are 100 filters (50 with each size) in the general filter set, and 20 filters (10 with each size) in each topic-specific filter set. This configuration is determined by performing a rough grid search. In the experiment, we find that as long as the number of filters is not set to be extremely small (less than 20) or extremely large (more than 500), the performance is stable. This result is consistent with a recent report of the same one-layer CNN model, which shows that the change in prediction accuracy is less than 1 % over a wide range of different numbers of filters [9].

To avoid over-fitting during training, we employ *dropout* on both convolutional layer and penultimate layer, and include a regularization term which penalise the $l2$ norm of all the weights [10]. The mini-batch training of size 20 and the gradient optimization method *RMSprop* are also used during the training process [11]. A comprehensive practitioners' guide on training this CNN model can be found in [9].

5.3 Semi-supervised Learning

The amount of available training data in DSTC4 dataset is far from sufficient compared to other typical text classification task.[2] To address this problem, we apply self-training semi-supervised learning process to make use of the huge amount of unlabeled data available on the internet [12]. We randomly chose 50,000 Singapore

[2]There are total 54 distinct values in the 'INFO' slot, and the average number of dialog segments related to each value is around 15, which we considered as insufficient compared to other typical text classification task such as '20 newsgroups'.

travel-related reviews from Tripadvisor[3] as the unlabeled dataset for the self-training. The self-training algorithm we used is summarized in Algorithm 1.

Algorithm 1 Self-training

1. Let L to be the **train** dataset, D the **dev** dataset, U the unlabeled dataset.
2. Train f from L using supervised learning.
3. Test f on D and calculate the prediction accuracy r.
4. Repeat until r stops increasing:
5. Apply f to the unlabeled instances in U.
6. Remove a subset S from U; add $\{(\mathbf{x}, f(\mathbf{x}))|\mathbf{x} \in S\}$ to L.
7. Train f from L using supervised learning.
8. Test f on D and calculate r.

6 Results and Discussion

The evaluation results of our model and other submitted entries are shown in Fig. 2 and Table 2. Figure 2 is a comparison of F-measure for each topic between the top 4 teams with the highest scores. Our result (under the team ID 'team1') was top for the 'Accommodation' and 'Attraction' topics and the second best for the remaining three topics. The overall results, which are equivalent to the weighted average scores across all topics, are listed in the Table 2. Our model was top for all 4 evaluation metrics with 2 evaluation schedules.[4]

Our model performed relatively poorly in the 'Shopping' topic. One possible reason for this is the lack of test data in this topic. Another likely reason is that the most frequently occurring 'INFO' slot values in the 'Shopping' topic, such as 'Item' or 'Tax refund', did not benefit from our multi-topic model, because they appeared exclusively in this 'Shopping' topic. The baseline system, on the other hand, achieved a relatively high F-score for this topic. This suggests that a hybrid approach which combines unsupervised string matching based method and the supervised machine learning based method may be preferred for this topic.

We did not use any other dialog information besides the value of the 'INFO' slot for training our model. However it is highly likely that other dialog information such as 'dialog acts', 'semantic tag' or the values of other regular slots will also be useful for predicting the 'INFO' slot value. Furthermore, no dialog data outside the current dialog segment (e.g. previous utterances) was included in the input of our model. A more sophisticated model capable of handling these information may achieve a higher prediction accuracy.

[3]http://www.tripadvisor.com/Tourism-g294265-Singapore-Vacations.html.

[4]For full details of the evaluation metrics see [2].

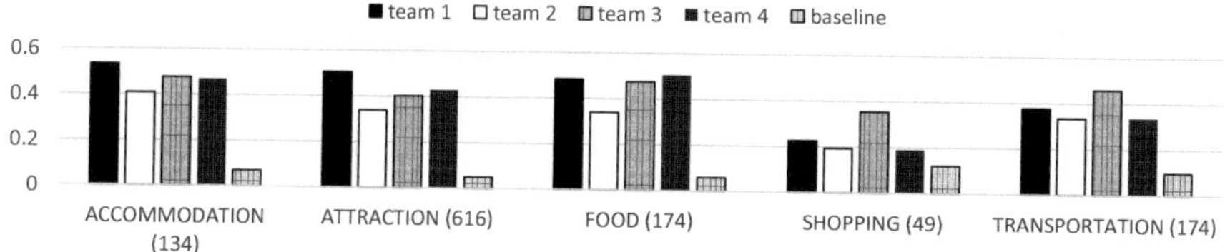

Fig. 2 The F-measure of the 'INFO' slot by topic. Our results are identified as team 1. The number in brackets is the count of dialog segments in that topic within the **test** data set

Table 2 Overall results for the 'INFO' slot of 4 teams with the highest scores

	Schedule1				Schedule2			
Team	Accuracy	Precision	Recall	F-score	Accuracy	Precision	Recall	F-score
1-entry3	**0.27**	0.58	**0.35**	**0.44**	**0.31**	0.68	**0.37**	**0.48**
1-entry1	0.27	**0.61**	0.34	0.43	0.30	**0.69**	0.36	0.47
2-entry3	0.15	0.38	0.27	0.32	0.16	0.39	0.31	0.35
3-entry3	0.22	0.53	0.31	0.39	0.26	0.53	0.36	0.43
4-entry3	0.23	0.55	0.28	0.38	0.26	0.56	0.34	0.42
Baseline	0.03	0.30	0.04	0.07	0.03	0.26	0.04	0.07

'1-entry3' is our results with semi-supervised learning, and 1-entry1 is the results without semi-supervised learning

6.1 Multi-topic Model

To investigate the effectiveness of the proposed multi-topic CNN model, we trained three sets of models: one general model, five topic-specific models and one multi-topic model, and compare their performance on certain 'INFO' slot values. We chose the 'Pricerange' value for this comparison, because it appears in all five topics and is one of the most frequently occurring value of the 'INFO' slot. Table 3 shows the comparison results, which is the average score of 10 runs. The multi-topic model outperformed the other two models, despite that the same number of filters were used in all three models.

In this experiment, we found that compared to the topic specific model, a general model tended to improve the accuracy of the topic with relatively small amount of data ('Shopping'), however by compromising the performance in the topic with relatively large amount of data ('Accommodation', 'Food'). In contrast, the multi-topic model was able to improve the performance consistently regardless of the amount of data available in a topic.

For more details, we also observed that the topic-specific model was able to capture the topic-related terms such as 'bargain' in the 'Shopping' topic and 'room rates' in the 'Accommodation' topic, while failed to learn the general terms such as 'cheaper' for the 'Shopping' topic due to inadequate data available for training in this topic. In contrast, the multi-topic model captured both general and topic-specific features,

Table 3 A comparison of the F-measure of 'Pricerange' value between different models which are trained on **train** dataset and evaluated by the **dev** dataset

Topic	General model	Topic-specific model	Out-of-topic pre-training	Multi-topic model
Accommodation (42/30)	0.87	0.92	0.92	**0.95**
Attraction (2/4)	0.00	0.00	0.00	0.00
Food (42/30)	0.82	0.90	0.86	**0.91**
Shopping (8/8)	0.49	0.37	0.43	**0.51**
Transportation (1/0)	N/A	N/A	N/A	N/A
Overall	0.76	0.80	0.79	**0.83**

The numbers in brackets refer to (the number of dialog segments assigned to 'Pricerange' in **train** dataset/**dev** dataset). Bold indicates statistical significance over all non-shaded results in the same row using t-test ($p = 0.05$)

which leads to an improvement of performance. A more detailed analysis of errors between all three models are attached in the Appendix (Tables 4 and 5).

The domain adaptation procedure of training RNN models proposed in [7] was also examined in our experiment. In our case, we applied their method by pre-training each topic-specific model with the data from all topics. The results can be found in Table 3 under the 'Out-of-topic pre-training' column. The out-of-topic pre-training did not lead to a significant improvement of performance against the Topic-specific model. This may be because the one-layer CNN model does not take advantage of pre-training as much as a RNN model used in [7] does.

6.2 *Semi-supervised Learning*

During the training process of 'entry3', we applied semi-supervised learning for the following 7 'INFO' slot values: 'Exhibit', 'Itinerary', 'Map', 'Preference', 'Pricerange', 'Restriction' and 'Ticketing'. The 'INFO' slot values and the amount of unlabeled data used were determined by the self-training Algorithm 1. The result in which we did not apply the semi-supervised learning process, was submitted as 'entry1' listed in Table 2. We achieved a slight improvement in overall accuracy and F-score by conducting the semi-supervised learning. We believe that the effect of semi-supervised learning can be further improved if the unlabeled data is more carefully chosen.

7 Conclusion

We have described a multi-topic convolutional neural network model for the 'INFO' slot filling of DSTC4. Our model is a combination of the general and topic-specific models, in which we use topic-shared (general) and topic-specific filters to capture general and topic-specific features at once. Our model is shown to outperform both the general and topic-specific models and is one the most competitive approach for the 'INFO' slot-filling submitted to DSTC4.

In our future work, we intended to extend our model to the regular slots of DSTC4, and also extend our work to more general multi-domain text classification problems.

Appendix

Table 4 The most 'Pricerange'-correlated words (uni-gram features) learned by different models, in the topic of 'Shopping'

General model	Topic-specific model	Multi-topic model
Price(s)	Dollar(s)	Price(s)
Expensive	Cost	**Expensive**
Dollar(s)	Price(s)	Dollar(s)
Cost	Cent	Cost
Low	Per	Per
Cheap	Bargain	Cent
Money	Rebate	Bargain
	Money	Low
	Hundred	Hundred
	Thousand	Money
	Low	**Cheap**
	Refund	
	Purchase	
	More	

Only the proposed multi-topic model learns both the topic-independent words (e.g. expensive) and topic-dependent words (e.g. bargain)

Table 5 Example **test** dialog segments (in 'Shopping' topic) which are hand-annotated with the 'Pricerange' label

Speaker	Transcription	G	T	M
Tourist	Cause I've heard that's really **expensive** when we go to shopping malls in Singapore	✓	×	✓
Guide	Shopping malls not really **expensive** in the sense that the shopping malls have a mix of **expensive** as well as not so **expensive** items.			
Tourist	And also Little India	×	✓	✓
Guide	And you can do a bit of bargaining, but usually the prices are very good			
Tourist	And they are not too **expensive**, are they?	×	×	✓
Guide	No. No they are not **expensive** at all			
Guide	In fact some of them when they hang big signs outside their stalls uh usually I don't bargain because they are already telling you that this is a bargain			
Tourist	I think I better get one of those			
Tourist	That's so nice			
Guide	Yes you can get one for yourself and for your daughter as well			
Tourist	I just might uh do that			
Guide	A lot of tourists do that because they can wear their souvenir			
Tourist	Right			
Tourist	Food street market			
Guide	Okay			

Topic-specific model and General model failed to give the correct answers in No.1&3 and No.2&3 dialog segments respectively, because they were unable to detect the general features and topic-specific features (bold and underlined words as shown in Table 4) at the same time. Only the proposed Multi-topic model correctly predicted the 'Pricerange' label in all three dialog segments (G = General model, T = Topic-specific model, M = Multi-topic model, ✓ = True positive, × = False negative)

References

1. Lemon, O., Pietquin, O.: Data-Driven Methods for Adaptive Spoken Dialogue Systems. Springer (2012)
2. Kim, S., D'Haro, L.F., Banchs, R.E., Williams, J.D., Henderson, M.: The fourth dialog state tracking challenge. In: Proceedings of the 7th International Workshop on Spoken Dialogue Systems (IWSDS) (2016)
3. Kim, Y.: Convolutional neural networks for sentence classification. arXiv preprint arXiv:1408.5882 (2014)
4. Daumé III, H.: Frustratingly easy domain adaptation. arXiv preprint arXiv:0907.1815 (2009)
5. Williams, J.: Multi-domain learning and generalization in dialog state tracking. In: Proceedings of SIGDIAL (2013)
6. Pan, S.J., Yang, Q.: IEEE Trans. Surv. Transfer Learn. **22**(10), 1345–1359 (2010)

7. Mrkšić, N., Séaghdha, D.O., Thomson, B., Gašić, M., Su, P.H., Vandyke, D.: Multi-domain dialog state tracking using recurrent neural networks. In: Proceedings of SIGDIAL (2013)
8. Mikolov, T., Sutskever, I., Chen, K., Corrado, G.S., Dean, J.: Distributed representations of words and phrases and their compositionality. Adv. Neural Inf. Process. Syst. 3111–3119 (2013)
9. Zhang, Y., Wallace, B.: A sensitivity analysis of (and practitioners' guide to) convolutional neural networks for sentence classification. arXiv preprint arXiv:1510.03820 (2015)
10. Hinton, G.E., Srivastava, N., Krizhevsky, A., Sutskever, I., Salakhutdinov, R.R.: Improving neural networks by preventing co-adaptation of feature detectors. arXiv preprint arXiv:1207.0580 (2012)
11. Tieleman, T., Hinton, G.E.: Lecture 6.5-rmsprop: Divide the gradient by a running average of its recent magnitude. COURSERA Neural Netw. Mach. Learn. **4** (2012)
12. Yarowsky, D.: Unsupervised word sense disambiguation rivaling supervised methods. In: Proceedings of the 33rd annual meeting on Association for Computational Linguistics, pp. 189–196 (1995)

The MSIIP System for Dialog State Tracking Challenge 4

Miao Li and Ji Wu

Abstract This article presents our approach for the Dialog State Tracking Challenge 4, which focuses on a dialog state tracking task on human-human dialogs. The system works in an turn-taking manner. A probabilistic enhanced frame structure is maintained to represent the dialog state during the conversation. The utterance of each turn is processed by discriminative classification models to generate a similar semantic structure to the dialog state. Then a rule-based strategy is used to update the dialog state based on the understanding results of current utterance. We also introduce a slot-based score averaging method to build an ensemble of four trackers. The DSTC4 results indicate that despite the simple feature set, the proposed method is competitive and outperforms the baseline on all evaluation metrics.

Keywords Spoken dialog system · Dialog state tracking · Spoken language understanding · Iterative alignment

1 Introduction

Spoken dialog systems enable users to achieve their goals by interacting with a computer agent via natural language. To build a robust dialog system, it needs to maintain a distribution over multiple hypotheses of the true dialog state, which is called dialog state tracking [1].

Dialog state tracking has attracted many research these days. Conventional commercial systems simply choose the most probable hypothesis and discard other hypotheses to build a hand-crafted method, which can't handle the uncertainty introduced by the Automatic Speech Recognition (ASR) and Spoken Language

M. Li (✉) · J. Wu
Multimedia Signal and Intelligent Information Processing Laboratory,
Department of Electronic Engineering, Tsinghua University, 100084 Beijing, China
e-mail: miao-li10@mails.tsinghua.edu.cn

J. Wu
e-mail: wuji_ee@mail.tsinghua.edu.cn

K. Jokinen and G. Wilcock (eds.), *Dialogues with Social Robots*,
Lecture Notes in Electrical Engineering 427,
DOI 10.1007/978-981-10-2585-3_38

Understanding (SLU). Following the Partially Observable Markov Decision Process (POMDP) framework, many generative methods are investigated [2–5]. Recently, discriminative models have been found that yield better performance, such as Maximum Entropy [6], Conditional Random Field [7, 8], Deep Neural Network (DNN) [9], and Recurrent Neural Network [10]. Some domain-independent rule-based approaches [11, 12] are also attractive due to their efficiency and portability.

The DSTC provides a common test bed for the dialog state tracking task. Different with the previous DSTC tasks, the DSTC4 task focus on a state tracking task on human-human dialogs. The domain of this task is much larger than the previous tasks, which involves food, traffic, accommodation, etc. The value space is also quite big and makes the task more difficult.

In this article, we present a turn-taking framework to solve this task. In our turn-taking algorithm, a probabilistic enhanced frame structure is initialized to represent the sub-dialog state in the beginning of a sub-dialog. A probabilistic semantic structure is extracted from each turn of the sub-dialog by a SLU model and then used to update the sub-dialog state.

The rest of this article is organized as follows. Since the DSTC4 is quite different with previous DSTC tasks, we briefly introduce the DSTC4 task in Sect. 2 for the sake of readability. Then, the details of our turn-taking algorithm are described in Sect. 3. A SLU is presented in Sect. 4. Our slot-based score averaging ensemble method is introduced in Sect. 5, and the evaluation results on DSTC4 are illustrated in Sect. 6. Finally, we conclude the paper in Sect. 7.

2 Task Description

Different with previous DSTC tasks, DSTC4 [13] focuses on a dialog state tracking task on human-human dialogs. The goal of the main task is to track dialog states for sub-dialog segments. Each sub-dialog is related to one specific topic about traveling. The boundary and topic of each sub-dialog in conversation are given and the tracker should produce a frame of slot-value pairs for each turn in the given sub-dialog. The slots and their potential values for a topic are defined by an ontology of this task. In conventional human-machine spoken dialog system literature, a “turn” represents one interaction between a user and a dialog system, which includes a user action and a system action. While in the rest of this paper, a “turn” includes only one utterance from a speaker since this is a human-human dialog state tracking task. The value space is quite large compared with previous DSTC tasks and the slots of this task are multi-value slots. Figure 1 illustrates an example of sub-dialog and its frame.

The dataset [14] consists of 35 dialog sessions on touristic information for Singapore. It is collected from Skype calls between three tour guides and 35 tourists. 14 dialogs are used as training set, 6 dialogs are used as development set and 15 dialogs are used as test set. These 35 dialogs sum up to 31,034 utterances and 273,580 words.

A baseline tracker is provided for all participants. Given a topic and current utterance, the baseline tracker uses a fuzzy string matching method to match current

Fig. 1 An example of a sub-dialog and its frame. The topic of this sub-dialog is given as "ATTRACTION". Two slots are involved and slot "TYPE_OF_PLACE" has two values in this sub-dialog

Sub-Dialog	Topic: ATTRACTION
Turn 1	Guide: Okay and ha- %uh okay, what do you and your friends like to do?
Turn 2	Tourist: %Um for me, I like to know about the culture, history and of course historical places in Singapore.
Turn 3	Guide: Okay.

Frame/Slot	Values
INFO	Preference
TYPE_OF_PLACE	Historic site Cultural site

utterance with all possible values defined in the ontology. If the matching score of a value exceeds a predefined threshold, the value and its slot will be added in the output frame.

3 Probabilistic Framework of the Turn-Taking Algorithm

In this section, we will describe the probabilistic framework of our turn-taking algorithm. The Fig. 2 illustrates our framework. In each turn t, the tracker maintains a probabilistic dialog state s_t, the tracker's output o_t is based on current state s_t. When the tracker receives an utterance u_t, a SLU module is used to extract semantic structures h_t from u_t. Then the tracker updates a new dialog state s_{t+1} based on the understanding result h_t and current state s_t.

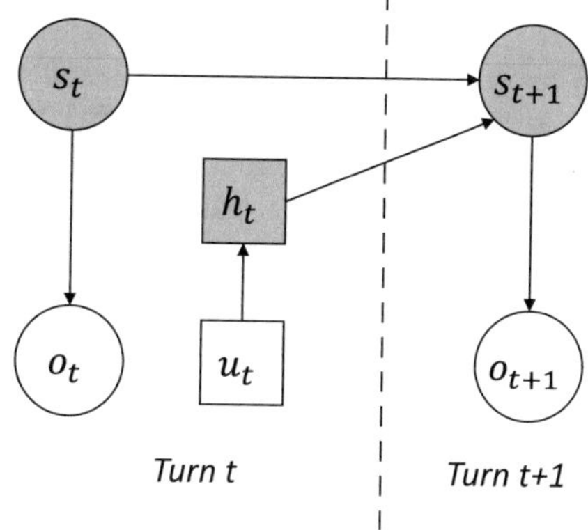

Fig. 2 The flow diagram of our turn-taking algorithm. *Shaded circles* represent dialog states of a certain turn and *clear circles* represent the frame structure output by the tracker. The *clear squares* represent utterances and the *shaded squares* are the understanding results of utterances

Fig. 3 An example of a probabilistic enhanced frame structure

Frame/Slot	Values
INFO (0.6)	Preference (0.8)
TYPE_OF_PLACE (0.7)	Historic site (0.85)
	Cultural site (0.95)

Two structures and three models are contained in this framework, one each for the state structure s_t, the semantic structure h_t, the SLU model $p(h_t|u_t)$, the update model $p(s_{t+1}|s_t, h_t)$ and the output model $p(o_t|s_t)$. We start from the structures.

3.1 Probabilistic Enhanced Frame Structure

Dialog states for previous DSTC tasks always maintain a distribution for each slot which is called a marginal distribution. A joint-distribution for the combination of all the slots is maintained at the same time. Since a slot can assign more than one possible values at one time in this task (such as slot "TYPE_OF_PLACE" in Fig. 1), maintains a distribution for a slot is intractable and not necessary. Therefore, following the idea of DS-State in [12], we introduce a probabilistic enhanced frame structure which is shown in Fig. 3 to represent a sub-dialog state. The structure has two types of probabilities, a probability attached to a slot indicates how likely the slot is involved in the sub-dialog and a probability attached to a slot-value pair indicates how likely the slot-value pair is involved in the sub-dialog up to current turn. Instead of using an N-best list of SLU hypotheses, we use the same probabilistic frame structure to represent the semantic structure h_t. This structure can be easily converted from an N-best list.

3.2 State Update Model and Output Model

The update method is shown in (1), which is similar to [12]. Because we use the same structure to represent dialog state and SLU result, the state update model $p(s_{t+1}|s_t, h_t)$ is more likely to combine the two probabilistic structures. And the update process will not suffer from the rule applying ordering problem in [11].

$$\begin{aligned} p^t_{state}(s) &= 1 - (1 - p^{t-1}_{state}(s))(1 - p^t_{turn}(s)) \\ q^t_{state}(s, v) &= 1 - (1 - q^{t-1}_{state}(s, v))(1 - q^t_{turn}(s, v)) \end{aligned} \tag{1}$$

where $p^t_{state}(s)$ is the probability of slot s in sub-dialog state in time t, $p^t_{turn}(s)$ is the probability of slot s in turn t, $q^t_{state}(s, v)$ is the probability of slot-value pairs s, v in sub-dialog in time t and $q^t_{turn}(s, v)$ is the probability of slot-value pairs s, v in turn t. If the slot or slot-value pair is not in the structure, we think its probability is 0. When we observe that some slots or slot-value pairs are involved in the understanding results of current turn, their probabilities in the sub-dialog state will increase. Because the values or slots are not mutually exclusive, the probabilities of other slots or slot-value pairs will not decrease.

The output model $p(o_t|s_t)$ we used is quite simple. We output the slot-value pairs whose probabilities of slots exceed a threshold T_s and probabilities of slot-value pairs exceed a threshold T_v.

Given the framework we introduced, any SLU models that can produce such probabilistic structures can be used in this framework. Systems that use different SLU models can be easily combined to build a more powerful system. In the DSTC4 task, four SLU models are used to build four trackers. In the next section, we will introduce one of our SLU model implemented in this task.

4 Slot Value Classification Parser

In this section, we will introduce a SLU model which is called a Slot Value Classification (SVC) parser. Similar to the STC model in [15], we use a set of discriminative classification models to predict whether a slot or a slot-value pairs is involved in an utterance. The set of classification models is determined by the ontology provided by the task. There are two types of classification models corresponding to the frame structure, namely slot models and slot-value pair models. There exists a binary slot model for each slot in the ontology. Only values of enumerable slots have their relative binary slot-value pair models. A slot is enumerable when the number of values of the slot is bounded, such as slot "INFO", "TYPE_OF_PLACE" and "LINE". While the slot "CUISINE", "FROM" and "NEIGHBOURHOOD" are non-enumerable.

4.1 *Training and Parsing Algorithms of the SVC Model*

In the training stage, for each pair of sub-dialog and its annotated frame, features are first extracted from the sub-dialog, then the frame is decomposed into a set of slots and slot-value pairs. For example, the frame illustrated in Fig. 1 will be decomposed into a slot set: {*INFO*, *TYPE_OF_PLACE*} and a slot-value pair set: {(*INFO*, *Preference*), (*TYPE_OF_PLACE*, *Historic site*), (*TYPE_OF_PLACE*, *Cultural site*)}. If a slot or slot-value pair is in the decomposed set of the frame, the sub-dialog will be considered as a positive sample for the classification model, otherwise the sub-dialog will be considered as a negative sample.

While in the parsing stage, we first run all the classification models on the utterance's feature. Then a probabilistic enhanced frame structure is built based on the posterior probability of each model. Similar to [15], there are two ways to construct the frame structure given a set of slots and slot-value pairs with their probabilities:

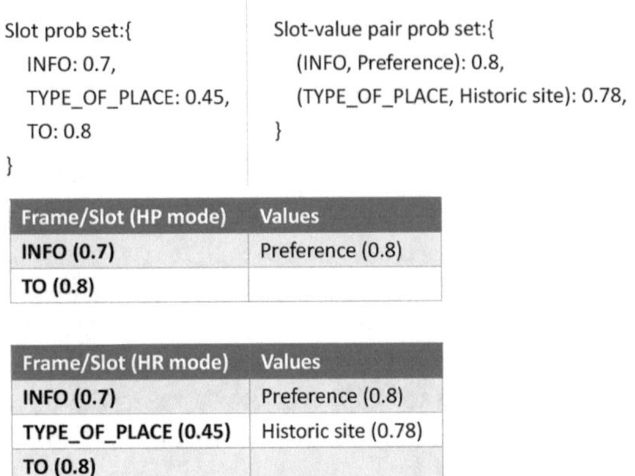

Fig. 4 An example of the two construction mode, the slot threshold T_s is 0.5 in this example. The slot set and the slot-value pair set are produced by all the classification models. Because the probability of slot "TYPE_OF_PLACE" is lower than the threshold, the "high precision mode" will discard all slot-value pairs with slot "TYPE_OF_PLACE". While the "high recall mode" will not discard these slot-value pairs

1. `High precision mode`: first, add all slots with their probabilities into the frame. Then add the slot-value pairs with their probabilities into the frame when the slot is already in the frame and its probability is larger than a threshold T_s.
2. `High recall mode`: first, add all slots with their probabilities into the frame. Then add all the slot-value pairs with their probabilities into the frame.

Figure 4 shows the difference between these two modes.

Up to now, we can construct a probabilistic frame structure with slots and values of enumerable slots. For non-enumerable slots (for example slot "TO" in Fig. 4), if the SVC parser believes a non-enumerable slot is involved in current utterance (its probability from the slot classification model is larger than a threshold T_s), a value extractor which is based on fuzzy string matching method will be used to extract possible values for the slot. The fuzzy string matching method is similar to the baseline method and the matching scores are used as the probabilities for possible values. After that, we can get a complete probabilistic frame structure for an utterance of a turn in a sub-dialog.

4.2 An Iterative Alignment Method for SVC

A problem of the SVC parser is there exists a mismatch between training and parsing. The classification models are trained by sub-dialogs but are used for turns. The problem is because annotation frames are annotated for sub-dialogs, not for turns in the sub-dialogs. To solve this problem, we try to use an Iterative Alignment Method

Sub-Dialog	Topic: ATTRACTION	Frame
Turn 1	Guide: Okay and ha- %uh okay, what do you and your friends like to do?	**INFO**: Preference
Turn 2	Tourist: %Um for me, I like to know about the culture, history and of course historical places in Singapore.	**INFO**: Preference **TYPE_OF_PLACE:** [Historic site, Cultural site]
Turn 3	Guide: Okay.	**NULL**

Fig. 5 An example of aligned sub-dialog. The first turn only involve the slot-value pair (*INFO*, *Preference*), the third turn actually involves no slot-value pairs

(IAM) to extract an aligned annotation for each sub-dialog. Figure 5 shows an example of aligned annotation of a sub-dialog.

The Iterative Alignment Method is shown as below. It works in a semi-supervised way, we only use the annotation in the sub-dialogs and don't have any handcraft annotations.

1. Base models are trained based on sub-dialogs.
2. For each sub-dialog and frame pairs. We decompose the frame into slot-value pairs.
   (a) For each turn in the sub-dialog, classification models which correspond to the frame are used to parse the utterance of the turn.
   (b) If the probability of a slot-value pairs is larger than a threshold T_h, this turn will be considered as a positive sample of the slot-value pairs.
   (c) If the probability of a slot-value pairs is less than a threshold T_l, this turn will be considered as a negative sample of the slot-value pairs.
   (d) Otherwise, this turn will be discarded to train the model.
3. New models are trained based on the original sub-dialogs together with the new aligned data.
4. Go to step 2.

Using the Iterative Alignment Method, we can get a fine-grained annotation for each turn, which leads to better classification models. The experiments in Sect. 6 will show the benefits from IAM.

In the DSTC4 task, we use Support Vector Machine (SVM) as the discriminative classification model. LibSVM package [16] with linear kernel is used to train the models. There are only two features used, one is the unigram feature, the other is the slot-value pairs extracted by the baseline tracker. All the thresholds mentioned above are chosen empirically based on the development set.

5 Slot-Based Score Averaging Ensemble

It has been proven that ensemble methods are very useful to improve the performance in dialog state tracking tasks [17]. Score averaging [10, 18] is one of the most popular and powerful ensemble methods, which uses a weighted vector corresponding to the score of each system to build an ensemble. Following the representation and update methods we mentioned in Sect. 3, different trackers can be built when using different SLU models. It is straight forward to build an ensemble tracker based on these trackers. Different with the conventional score averaging method, we don't use a single weighted vector to combine those trackers. For different slots, we construct slot-specific weighted vectors based on trackers' f-scores on these slots to combine the trackers.

In DSTC4 task, four trackers are used to build the ensemble tracker. These four trackers all follow the framework we mentioned before, only the SLU models are different. The first tracker uses the same fuzzy string matching method as the baseline to produce a probabilistic frame structure, the probabilities are generated by normalising the matching scores. The second tracker uses probability association rules between slot-value pairs and semantic tags of utterances. The rules are mined by association rule mining algorithms. A semantic tagger is trained based on the semantic annotations in the training set to extract semantic tags. In parsing step, the semantic tagger is first used to extract semantic tags for each utterance, and then slot-value pairs and probabilities are generated based on association rules and their confidence scores. The third tracker uses SVC models with only unigram features and the fourth tracker uses SVC models with unigram and baseline features. Both of the third tracker and the fourth tracker use the IAM process. To perform the score averaging ensemble, all the probabilities produced by different models are normalized to scale different thresholds (T_s, T_v) to a global one, the global threshold is set to 0.5 in our system.

6 Experiment

The evaluation results are shown in Table 1. We are team2 in the DSTC4 results. "SVC_uB" represent the system using SVC models with unigram and baseline features but without IAM which is mentioned in Sect. 4.2. We didn't submit "SVC_uB" to the evaluation, the performance of "SVC_uB" in Table 1 is calculated based on the annotation of test set. "SVC_uB" represent the system using SVC models with with unigram and baseline features and IAM process. It is entry0 of our submission. "Ensemble" represents the slot-based score averaging ensemble tracker mentioned in Sect. 5 and it is entry3 of our submission.

In the evaluation results, all of our submissions outperformed the baseline tracker, which shows the effectiveness of our algorithms. Without the IAM process, the system "SVC_uB", which is trained based on annotation on sub-dialogs, can't have

Table 1 Experimental results

Model	Schedule 1				Schedule 2			
	Acc.	Pre.	Rec.	F1	Acc.	Pre.	Rec.	F1
Baseline	0.0374	0.3589	0.1925	0.2506	0.0488	0.3750	0.2519	0.3014
SVC_uB[a]	0.0412	0.3281	**0.2758**	0.2997	0.0523	0.3540	**0.3415**	0.3476
SVC_uB_a(e0)	0.0487	0.4079	0.2626	0.3195	0.0671	0.4280	0.3257	0.3699
Ensemble(e3)	**0.0489**	**0.4440**	0.2703	**0.3361**	**0.0697**	**0.4634**	0.3335	**0.3878**

[a]The performance of SVC_uB is calculated based on the annotation of test set

precise predictions about what slots or slot-value pairs are involved in an utterance. So the precision of system "SVC_uB" is the lowest of the four systems presented in Table 1. Comparing "SVC_uB" with "SVC_uB_a", we can find that the iterative alignment method can significantly improve the precision without losing much recall. Using our slot-based score averaging method, the ensemble tracker leads to a 4.8 % relative improvement compared with the best single tracker.

7 Conclusion

This article has presented our algorithms for DSTC4. Our algorithms work in a turn-taking manner. A probabilistic enhanced frame structure is used to represent and update the dialog states. A SVC parser together with an iterative alignment method are introduced for spoken language understanding. Following the framework we presented above, it is very easy to construct different trackers with different SLU modules. Finally, a slot-based score averaging method is used to build an ensemble tracker, which can help improving the performances of all single trackers.

References

1. Williams, J., Raux, A., Ramachandran, D., Black, A.: The dialog state tracking challenge. In: Proceedings of the SIGDIAL 2013 Conference, pp. 404–413 (2013)
2. Henderson, J., Lemon, O.: Mixture model POMDPs for efficient handling of uncertainty in dialogue management. In: Proceedings of the 46th Annual Meeting of the Association for Computational Linguistics on Human Language Technologies: Short Papers, pp. 73–76. Association for Computational Linguistics (2008)
3. Williams, J.D., Young, S.: Partially observable Markov decision processes for spoken dialog systems. Comput. Speech Lang. **21**(2), 393–422 (2007)
4. Thomson, B., Young, S.: Bayesian update of dialogue state: a POMDP framework for spoken dialogue systems. Comput. Speech Lang. **24**(4), 562–588 (2010)
5. Young, S., Gašić, M., Keizer, S., Mairesse, F., Schatzmann, J., Thomson, B., Kai, Y.: The hidden information state model: a practical framework for POMDP-based spoken dialogue management. Comput. Speech Lang. **24**(2), 150–174 (2010)

6. Metallinou, A., Bohus, D., Williams, J.: Discriminative state tracking for spoken dialog systems. Proceedings of ACL **1**, 466–475 (2013)
7. Lee, S.: Structured discriminative model for dialog state tracking. In: Proceedings of the SIGDIAL 2013 Conference, pp. 442–451 (2013)
8. Ren, H., Xu, W., Zhang, Y., Yan, Y.: Dialog state tracking using conditional random fields. In: Proceedings of the SIGDIAL 2013 Conference. Metz, France, August (2013)
9. Sun, K., Chen, L., Zhu, S., Yu, K.: The SJTU system for dialog state tracking challenge 2. In: Proceedings of the 15th Annual Meeting of the Special Interest Group on Discourse and Dialogue, pp. 318 (2014)
10. Henderson, M., Thomson, B., Young, S.: Word-based dialog state tracking with recurrent neural networks. In: Proceedings of the 15th Annual Meeting of the Special Interest Group on Discourse and Dialogue, p. 292 (2014)
11. Wang, Z., Lemon, O.: A simple and generic belief tracking mechanism for the dialog state tracking challenge: on the believability of observed information. In: Proceedings of the SIGDIAL 2013 Conference (2013)
12. Ji, W., Li, M., Lee, C.-H.: A probabilistic framework for representing dialog systems and entropy-based dialog management through dynamic stochastic state evolution. IEEE/ACM Trans. Audio Speech Lang. Process. **23**(11), 2026–2035 (2015)
13. Kim, S., D'Haro, L.F., Banchs, R.E., Williams, J., Henderson, M.: Handbook Dialog State Tracking Challenge 4, vs 3.0 (2015)
14. Kim, S., D'Haro, L.F., Banchs, R.E., Williams, J., Henderson, M.: The fourth dialog state tracking challenge. In: Proceedings of the 7th International Workshop on Spoken Dialogue Systems (IWSDS) (2016)
15. Mairesse, F., Gašić, M., Jurčíček, F., Keizer, S., Thomson, B., Yu, K., Young, S.: Spoken language understanding from unaligned data using discriminative classification models. In: Proceedings of the IEEE International Conference on Acoustics, Speech and Signal Processing, 2009. ICASSP 2009, pp. 4749–4752. IEEE (2009)
16. Chang, C.-C., Lin, C.-J.: Libsvm: a library for support vector machines. ACM Trans. Intell. Syst. Technol. **2**(3), 27 (2011)
17. Henderson, M., Thomson, B., Williams, J.: The second dialog state tracking challenge. In: Proceedings of the 15th Annual Meeting of the Special Interest Group on Discourse and Dialogue, p. 263 (2014)
18. Williams, J.D.: Web-style ranking and slu combination for dialog state tracking. In: Proceedings of the 15th Annual Meeting of the Special Interest Group on Discourse and Dialogue, p. 282 (2014)

Robust Dialog State Tracking for Large Ontologies

Franck Dernoncourt, Ji Young Lee, Trung H. Bui and Hung H. Bui

Abstract The Dialog State Tracking Challenge 4 (DSTC 4) differentiates itself from the previous three editions as follows: the number of slot-value pairs present in the ontology is much larger, no spoken language understanding output is given, and utterances are labeled at the subdialog level. This article describes a novel dialog state tracking method designed to work robustly under these conditions, using elaborate string matching, coreference resolution tailored for dialogs and a few other improvements. The method can correctly identify many values that are not explicitly present in the utterance. On the final evaluation, our method came in first among 7 competing teams and 24 entries. The F1-score achieved by our method was 9 and 7 percentage points higher than that of the runner-up for the utterance-level evaluation and for the subdialog-level evaluation, respectively.

Keywords Dialog state tracking · Dialog management

1 Introduction

Spoken dialog systems are computer-based systems that interact with users through natural language to help them achieve a goal. Since they provide a convenient and natural way for the users to interact with computers, they have become increasingly popular recently. Examples of such systems are Apple Siri, Google Now, Microsoft Cortana, and Amazon Echo.

F. Dernoncourt (✉) · J.Y. Lee · T.H. Bui · H.H. Bui
Adobe Research and MIT, San Jose, CA, USA
e-mail: francky@mit.edu

J.Y. Lee
e-mail: jjylee@mit.edu

T.H. Bui
e-mail: bui@adobe.com

H.H. Bui
e-mail: hubui@adobe.com

K. Jokinen and G. Wilcock (eds.), *Dialogues with Social Robots*,
Lecture Notes in Electrical Engineering 427,
DOI 10.1007/978-981-10-2585-3_39

A dialog state tracker is a key component of a spoken dialog system and its goal is to maintain the dialog states throughout a dialog. For example, in a tourist information system, the dialog state may indicate a tourist attraction that the user is interested in, as well as the types of information related to the attraction, such as entrance fees and location. Although it might be easy for humans, dialog state tracking is difficult for computers because they do not "understand" natural human language as humans do. Yet, dialog state tracking is crucial for reliable operations of a spoken dialog system because the latter relies on the estimated dialog state to choose an appropriate response, for example, which tourist attractions to suggest.

The Dialog State Tracking Challenge (DSTC) is a series of community challenges that allow researchers to study the state tracking problem on a common corpus of dialogs using the same evaluation methods. DSTC 4, which is the focus of this paper, differentiates itself from previous editions in several ways. First, the ontology contains many more slot-value pair: for example, the ontology for the DSTC 3 corpus contains only 9 slots and 271 values, while DSTC 4's ontology has 30 slots and 1667 values, most of which are shared across several slots. Second, the output of the Spoken Language Understanding (SLU) component, is not available. Third, the states are labeled at the subdialog level only, which disadvantages machine-learning-based approaches.

Best approaches in the previous DSTCs include neural networks [1–3], web-style ranking and SLU combination [4], maximum entropy models [5] and conditional random fields [6]. However, these prior solutions are limited to domain ontologies with a small number of slots and values. Furthermore, most of the above methods and other rule-based methods [7] require the output of the SLU.

This article proposes a novel dialog state tracking method designed to work robustly under the DSTC 4 conditions. The method is composed of an elaborate string matching system, a coreference resolution mechanism tailored for dialogs, and a few other improvements. The paper is organized as follows. Section 2 describes briefly the DSTC 4 dataset. Section 3 presents in detail several trackers we used for the challenge. Section 4 compares the performances on the test set of our trackers with those of the trackers submitted by other teams participated in the challenge. Section 5 summarizes our work and proposes further improvements.

2 The DSTC 4 Dataset

The corpus used in this challenge consists of 35 dialog sessions on touristic information for Singapore, collected from Skype calls between three tour guides and 35 tourists. Each dialog session is a dialog between a guide and a tourist, where the guide helps the tourist plan for a trip to Singapore by providing recommendations based on the tourist's preferences. These 35 dialogs sum up to 31,034 utterances and 273,580 words. All the recorded dialogs with the total length of 21 h have been manually transcribed and annotated with speech act and semantic labels for each utterance as well as dialog states for each subdialog.

Each dialog is divided into subdialogs, each of which has one topic and contains one or several utterances. Dialog states are annotated for each subdialog. A dialog state is represented as a list of slot-value pairs. The slot is a general category, while the value indicates more specifically what the dialog participants have in mind. For example, one possible slot is "TYPE OF PLACE" and a possible value of this slot is "Beach". The DSTC 4 corpus is provided with an ontology that specifies the list of slot-value pairs that a subdialog of a given topic may contain.

Following the official split, the train, development and test sets contain 14, 6 and 9 dialogs respectively. The remaining 6 dialogs are used as a test set for another task. The test set labels were released only after the final evaluation.

The goal of the main task of DSTC4 is to track dialog states, considering all dialog history up to and including the utterance. Trackers are evaluated based on the predicted state for either each utterance (utterance-level evaluation) or for each subdialog (subdialog-level evaluation). Since the gold labels are available only at the subdialog level, in the utterance-level evaluation the predicted state for each utterance is compared against the gold labels of the subdialog that contains the utterance. Four performance metrics are used: subset accuracy, precision, recall and F1-score. For the subset accuracy, for a given utterance, the list of all slot-value pairs in the dialog state must exactly match the subdialog gold label to be counted as a true positive. Kim et al. [8, 9] contain further information pertaining to the dataset.

3 Method

This section presents the dialog state trackers we used for the challenge. We describe two rule-based trackers, two machine-learning-based trackers and a hybrid tracker.

3.1 Fuzzy Matching Baseline

A simple rule-based tracker was provided by the organizers of the challenge. It performs string fuzzy matching between each value in the ontology and the utterance. If the matching score is above a certain threshold for a value, then any slot-value pair with that value is considered as present.

3.2 Machine-Learning-Based Trackers

3.2.1 Cascade Tracker

The cascade tracker aims to refine the fuzzy matching tracker. For each slot, a classifier is trained to detect whether the slot is present or absent given an utterance. If a slot is predicted as present for a given utterance, then the fuzzy matching score is computed between each value of the detected slot and the utterance. For the classifier, we tried logistic regression (LR), support vector machines (SVM), and random forests (RF): we kept RF as it yields the best results. The features used are unigrams, bigrams, word vectors and named-entities. The word vector features are obtained by first mapping each word of the utterance to a word vector, then summing them all. We use pre-trained word vectors provided on the word2vec website.[1]

3.2.2 Joint Tracker

The main weakness of the cascade tracker is that in order to detect the value, it relies on fuzzy matching instead of utilizing more meaningful features. To address this issue, the joint tracker predicts the slot and the value jointly. For each slot, an RF classifier is trained to detect whether a given value is present or absent. The features used are the same as in the cascade tracker. Since the vast majority of values are absent in a given utterance, the negative examples are downsampled in the training phase.

3.3 *Elaborate Rule-Based Tracker*

Since the machine-learning-based approaches using traditional features were performing poorly, an elaborate rule-based tracker was constructed in order to overcome the shortcomings of the machine-learning-based approaches. The main pipeline of the elaborate rule-based tracker is described in Fig. 1. The tracker makes use of the knowledge present in the ontology as well as the synonym list that is defined for each slot-value pair. The inputs of the dialog state tracker are the current utterance (i.e. the utterance for which the tracker should predict the slot-value pairs), and the dialog history. The dialog history contains the list of previous utterances, as well as the list of slot-value pairs that the tracker predicted for the previous utterances. Lastly, based on the input and the knowledge, the tracker outputs a list of slot-value pairs for the current utterances.

This tracker tries to model how a human would track the dialog states, and therefore is very intuitive and interpretable. Figure 2 presents the four main steps the tracker follows to predict an output based on the input and the knowledge. The first step detects the presence of each slot-value pair in the utterance, by finding a match between any substring of the utterance and any of the synonyms of each slot-value pair. The second step resolve coreferences of certain type and detects additional slot-value pairs associated with them. Among the slot-value pairs detected from synonym

[1] https://code.google.com/p/word2vec/: GoogleNews-vectors-negative300.bin

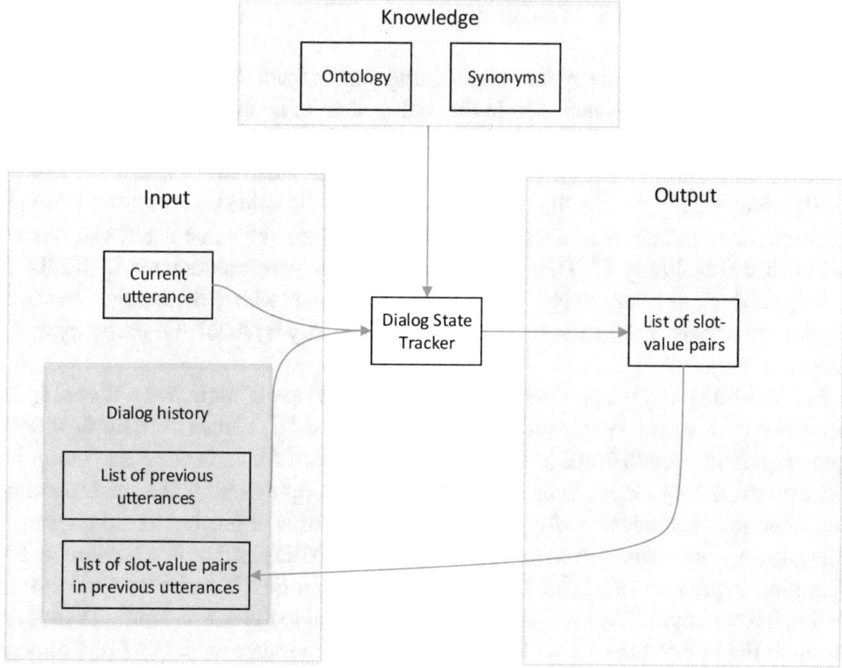

Fig. 1 Overview of the dialog state tracking system

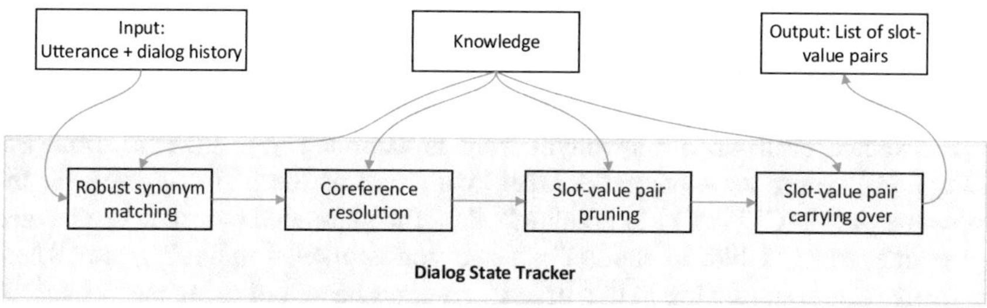

Fig. 2 The four main steps of the elaborate rule-based tracker

matching and coreference resolution, there often exist a group of slot-value pairs that are closely related, but only one of them is present in the dialog state. In the third step, the tracker selects the most likely slot-value pair among the closely-related slot-value pairs and eliminates all others. In the last step, slot-value pairs are carried over from the previous dialog state, whenever the topic continues and no new slot-value pair is detected for certain slots. The following four subsections present each step in more details.

3.3.1 Robust Synonym Matching

The motivation behind the synonym matching comes from the observation that even when a slot-value pair is included in the dialog state of an utterance, the value itself does not appear in the utterance. Instead, a synonym of the value often appears as a substring of the utterance. For example, for the slot-value pair "PLACE: Amoy by Far East Hospitality 4", it is likely that a synonym of the value such as "Amoy Hotel" is present in an utterance of a spoken dialog, rather than the value itself, viz. "Amoy by Far East Hospitality 4". Therefore, each slot-value pair is associated with a list of synonyms that are likely to be present in the utterances whose dialog state contains the slot-value pair. The synonym list was created partly by hand and partly by using a set of rules.

For flexibility and better detection, each synonym may contain two optional specifications: first, a synonym could be expressed as an AND clause of multiple words; second, part-of-speech (verb or noun) may be specified for each word that comprise a synonym. If a synonym is an AND clause of multiple words, the corresponding slot-value pair is detected only when each of the words is present in the utterance. This allows a slot-value pair to be detected even when the word order is switched. For example, for the slot-value pair "PLACE: Amoy by Far East Hospitality 4" discussed earlier, if a synonym ("Amoy" AND "Hotel") is added to the corresponding synonym list, then the slot-value pair will be detected in an utterance such as "I recommend the hotel called Amoy."

If a word that comprises a synonym is specified with a part-of-speech, then the corresponding slot-value pair is detected only when the word appears in the utterance and is tagged as having the specified part-of-speech. To take into account for the part-of-speech, the tracker performs part-of-speech tagging on each utterance prior to the synonym matching. The part-of-speech specification allows a slot-value pair to be detected even when a synonym word is exhibited in a different form. For example, if we use a synonym "Snorkel" with verb as the part-of-speech for the slot-value pair "ACTIVITY: Snorkeling", then the slot-value pair will be detected in the utterances "I like to snorkel", "Have you snorkeled before?", and "There are many people snorkeling in this beach". Another benefit of using part-of-speech specification is that it helps reduce many false positives by distinguishing between noun and verb instances of a word. For example, specifying noun as the part of speech in a synonym "Show" for the slot-value pair "ACTIVITY: Show" will prevent the incorrect detection of the slot-value pair in an utterance such as "I would like to show you this picture."

Moreover, we adopted two simple strategies to enhance the detection of slot-value pairs. First, we lemmatized each word in both the synonyms and the utterances before matching, increasing the chance of detecting the plural as well as singular form of a synonym. Second, in order to account for misspellings while preserving the precision of the tracker, we permitted one spelling mistake on long synonyms only. Specifically, we allowed a synonym to be detected if a substring of the utterance had Levenshtein distance of 1 from the synonym, only when a synonym has more than 5 characters and each word in a synonym has more than 3 characters.

3.3.2 Coreference Resolution for Dialogs

Coreferences are numerous in dialogs but are harder to detect than in formal written text, as the existing coreference resolution systems typically perform well on the latter, but not on the former.

The tracker contains a coreference resolution system for place-related anaphoras. This system is customized for the slot-filling tasks and works as follows. For each utterance, from its syntactic parsing tree the tracker detects the presence of the following three templates:

- Template 1: possessive adjective (my/your/our) + a type of place
- Template 2: demonstrative pronoun (the/this/that/these/those) + a type of place
- Template 3: here/there

For example, "our hotel" and "your museums" belong to Template 1, and "this garden" and "these parks" belong to Template 2. If Template 1 or Template 2 is present, then the tracker considers as present the last detected slot-value pair of the same type in the dialog history. The type of each place-related slot-value pair is specified in the provided ontology. If Template 3 is present, then the tracker considers as present the last detected value of any place-related slots (e.g. "PLACE" or "NEIGHBOURHOOD") in the dialog history.

3.3.3 Ontology-Based Slot-Value Pair Pruning

Among the slot-value pairs detected from synonym matching and coreference resolution, there often exist a group of slot-value pairs that are closely related, e.g. different branches of the same hotel chain. In most situations, however, only one of these slot-value pairs is present in the dialog state. In order to select the most likely slot-value pair among the closely related slot-value pairs, the tracker utilizes the domain knowledge present in the ontology as well as the observations from the training data.

For example, for each hotel listed as a possible value, the ontology also contains additional information about the hotel such as its neighborhood and price range. When multiple hotel branches are detected from the synonym matching step, then the tracker checks whether other related information about the branch is found in the context and selects the most likely branch based on the observation. If no relevant information is found, the tracker selects the most likely branch based on prior observations from the training data.

Another kind of closely-related slot-value pairs are those with the values that overlap with each other, such as "Park Hotel" and "Grand Park Hotel". If the utterance is "I will stay at the Grand Park Hotel", then the synonym matching step will detect both "Park Hotel" and "Grand Park Hotel" values. To avoid this issue, the tracker deletes any slot-value pair whose value is a (strict) substring of the value of another slot-value pair, among the detected slot-value pairs.

For slot-value pairs of special slots such as "TO" and "FROM", syntactic parsing trees are used in order to determine whether each value follows a preposition such as "to", "into", "towards", and "from". Based on this and the order in which the values appear in the utterances of a subdialog, the most likely slot-value pair(s) are determined.

3.3.4 Slot-Value Pair Carrying over

As a dialog progresses, many slot-value pairs remain present for several subsequent utterances as the subject of the dialog continues. As a result, the tracker implements the following rule: for certain slots, whenever a slot-value is detected as present in the previous utterance, the slot-value pair remains present until another value appears for the same slot or the topic changes. The tracker learns for which slots it is optimal to do so by using the training data and comparing the slot-value pair prediction results with and without the rule for a given slot.

3.4 Hybrid Tracker

In order to take advantage of the strength of both the rule-based and the machine-learning-based approaches, the hybrid tracker uses the rule-based tracker's outputs as features for the joint tracker. The output of each of the four main steps of the elaborate rule-based tracker is used as features, as Fig. 3 illustrates.

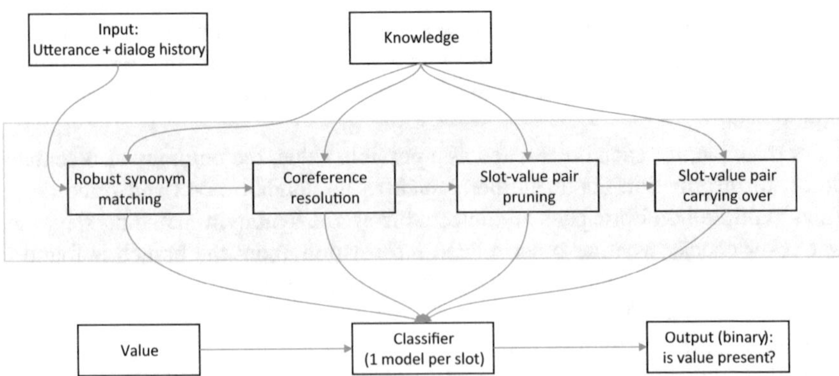

Fig. 3 The hybrid tracker uses the output of each of the four main steps of the elaborate rule-based tracker as features

4 Results

Table 1 compares the performances on the test set of our trackers as well as the best tracker of each other team that took part in the challenge. For the utterance-level evaluation, all teams but two including us obtained an F1-score below 0.35, which reflects the difficulty of the task. Team 4 reached 0.4481, while our best entry scored 0.5306. By the same token, for the subdialog-level evaluation, all teams but two including us obtained an F1-score below 0.40. Team 4 reached 0.5031, while our best entry scored 0.5786.

Looking at the results for various trackers described in Sect. 3, we observe that the cascade tracker and the joint tracker both perform poorly. The joint tracker has a much lower recall than the cascade tracker, which may be due to the fact that the same classifier is used for all values of a slot. However, the elaborate rule-based tracker yields a much higher performance, far above the fuzzy matching baseline. The hybrid tracker, which uses the output of the rule-based tracker as features, further increases the F1-score, but has a lower subset accuracy. Unlike the joint tracker, the hybrid tracker improves the F1-score, which may result from the higher quality and density of the features used.

The results for the utterance-level evaluation are lower than for the subdialog-level evaluation, which is expected since for the utterance-level evaluation the predicted state for each utterance is compared to the gold state of the subdialog that contains the utterance. It is often hard or impossible in the first utterances of a subdialog to guess what state the subdialog (i.e. the state of the last utterance of the subdialog) will have, since the tracker is allowed to access only the current and previous utterances, but not the upcoming utterances.

Table 1 Comparison of results for various dialog state trackers on the test set

Tracker	Utterance-level				Subdialog-level			
	Accuracy	Precision	Recall	F1-score	Accuracy	Precision	Recall	F1-score
Baseline	0.0374	0.3589	0.1925	0.2506	0.0488	0.3750	0.2519	0.3014
Cascade	0.0227	0.2962	0.2145	0.2488	0.0314	0.3138	0.2734	0.2922
Joint	0.0260	0.4682	0.1170	0.1872	0.0357	0.4648	0.1602	0.2383
Elaborate	**0.1210**	0.5449	**0.4964**	0.5196	**0.1500**	0.5619	**0.5787**	0.5702
Hybrid	0.1183	**0.5780**	0.4904	**0.5306**	0.1473	**0.5898**	0.5678	**0.5786**
Team 4	0.1002	0.5545	0.3760	0.4481	0.1212	0.5642	0.4540	0.5031
Team 2	0.0489	0.4440	0.2703	0.3361	0.0697	0.4634	0.3335	0.3878
Team 6	0.0486	0.5623	0.2314	0.3279	0.0645	**0.5941**	0.2850	0.3852
Team 1	0.0371	0.4179	0.2804	0.3356	0.0584	0.4384	0.3426	0.3846
Team 5	0.0268	0.3405	0.2014	0.2531	0.0401	0.3584	0.2632	0.3035
Team 7	0.0286	0.2768	0.1826	0.2200	0.0323	0.3054	0.2410	0.2694

The poor performances of trackers that solely rely on machine-learning can be partly explained by the lack of dialog state labels at the utterance level. If a tracker is trained using the features extracted for each utterance and the subdialog label as the utterance label, then the tracker will learn many incorrect associations between features and labels. In other words, using subdialog labels as utterance labels introduce much noise.

For example, if a subdialog comprises the two utterances "Good morning" and "Hi! I am on my way to Paris!", and the subdialog label is "TO: Paris", the first training sample will be the features extracted from "Good morning" and the label "TO: Paris". This will tend to create false positives, since it is likely that "Good morning" in other subdialogs will not be labeled with "TO: Paris".

However, if a tracker is trained only for subdialogs to avoid this issue, then it results in having much fewer samples to train on. This brings us to the issue of data scarcity: even though the train set contains 14 labeled dialogs, the number of training samples is still quite small. This certainly gives a significant advantage to hybrid trackers over machine-learning-based trackers.

In addition to the dialog states, the labels also contain semantic tags for each utterance. We tried to take advantage of the finer granularity of the semantic tagging: as an experiment, we used the gold semantic tags as features, but our results on the development set did not show any improvement.

5 Conclusion and Future Work

This article describes and compares several dialog state trackers on the DSTC 4 corpus. Due to the size of the ontology and the utterances being labeled at the subdialogue-level only, the rule-based approach yields better results than the pure machine learning approaches. However, using the rule-based tracker as features for the machine-learning-based tracker allows to further improve the results. On the final evaluation of the main task, our method came in first among 7 competing teams and 24 entries. Our method achieved an F1-score 9 and 7 percentage points higher than the runner-up for the utterance-level evaluation and for the subdialog-level evaluation, respectively.

Modeled after how humans would track dialog states, our elaborate rule-based tracker is not only intuitive and interpretable, but also has potential to be further improved by combining machine-learning-based approaches. One such example is our hybrid tracker, but there are many other ways that machine learning techniques could be used to improve our system.

First, the synonyms list was mostly manually curated, as using existing synonym lists such as Wordnet was causing the precision to decrease significantly. Some general rules were used to automatically generate the synonyms, but one could further automate the generation of synonym list. Moreover, extending the coreference resolution system to general slot-value pairs can improve the performance of the tracker. Furthermore, instead of blindly carrying over slot-value pairs when no new value is detected for certain slots, it would be interesting to implement algorithms that can detect when the subject of the dialog has changed and only carry over slot-value pairs when the subject has not changed.

Another weakness of our system is that it detects all slot-value pairs that are mentioned in the utterances, rather than selectively detecting those that are not only mentioned, but also the main subject of discussion. One example is when a value is mentioned but negated, e.g. "I recommend Keong Saik Hotel, not The Fullerton Hotel". Then according to our system both slot-value pairs "PLACE: Keong Saik Hotel" and "PLACE: The Fullerton Hotel" will be detected as present, but the gold dialog state will only include the former. Such mistakes result in many false positives. Implementing algorithms to detect the main subject may greatly improve the precision.

Acknowledgements The authors would like to warmly thank the DSTC 4 team for organizing the challenge and being so prompt to respond to emails. The authors are also grateful to the anonymous reviewers as well as to Walter Chang for their valuable feedback.

References

1. Henderson, M., Thomson, B., Young, S.: Deep neural network approach for the dialog state tracking challenge. In: Proceedings of the SIGDIAL 2013 Conference, pp. 467–471 (2013)
2. Henderson, M., Thomson, B., Young, S.: Robust dialog state tracking using delexicalised recurrent neural networks and unsupervised adaptation. In: Proceedings of the 2014 IEEE Spoken Language Technology Workshop (SLT), pp. 360–365. IEEE (2014)
3. Henderson, M., Thomson, B., Young, S.: Word-based dialog state tracking with recurrent neural networks. In: Proceedings of the 15th Annual Meeting of the Special Interest Group on Discourse and Dialogue, p. 292 (2014)
4. Williams, J.D.: Web-style ranking and SLU combination for dialog state tracking. In: Proceedings of the 15th Annual Meeting of the Special Interest Group on Discourse and Dialogue, p. 282 (2014)
5. Lee, S., Eskenazi, M.: Recipe for building robust spoken dialog state trackers: dialog state tracking challenge system description. In: Proceedings of the SIGDIAL 2013 Conference, Metz, France, August (2013)
6. Ren, H., Xu, W., Zhang, Y., Yan, Y.: Dialog state tracking using conditional random fields. In: Proceedings of the SIGDIAL 2013 Conference, Metz, France, August (2013)
7. Sun, K., Chen, L., Zhu, S., Yu, K.: A generalized rule based tracker for dialogue state tracking. In: Proceedings of the 2014 IEEE Spoken Language Technology Workshop (SLT), pp. 330–335. IEEE (2014)
8. Kim, S., D'Haro, L.F., Banchs, R.E., Williams, J., Henderson, M.: Dialog State Tracking Challenge 4. Handbook (2015)
9. Kim, S., D'Haro, L.F., Banchs, R.E., Williams, J., Henderson, M.: The fourth dialog state tracking challenge. In: Proceedings of the 7th International Workshop on Spoken Dialogue Systems (IWSDS) (2016)

Erratum to: Dialogues with Social Robots

Kristiina Jokinen and Graham Wilcock

Erratum to:
K. Jokinen and G. Wilcock (eds.), *Dialogues with Social Robots*, Lecture Notes in Electrical Engineering 427, DOI 10.1007/978-981-10-2585-3

The original version of the book was published with book volume number "999" and now has been changed to "427". The erratum book has been updated with the change.

The updated original online version for this book can be found at 10.1007/978-981-10-2585-3

K. Jokinen (✉) · G. Wilcock
University of Helsinki, Helsinki, Finland
e-mail: kristiina.jokinen@helsinki.fi

G. Wilcock
e-mail: graham.wilcock@helsinki.fi

K. Jokinen and G. Wilcock (eds.), *Dialogues with Social Robots*,
Lecture Notes in Electrical Engineering 427,
DOI 10.1007/978-981-10-2585-3_40

Author Index

K. Jokinen and G. Wilcock (eds.), *Dialogues with Social Robots*,
Lecture Notes in Electrical Engineering 427,
DOI 10.1007/978-981-10-2585-3

Printed in the United States
By Bookmasters